Financial Econometrics Models and Methods

This is a thorough exploration of the models and methods of financial econometrics by one of the world's leading financial econometricians and is for students in economics, finance, statistics, mathematics, and engineering who are interested in financial applications.

Based on courses taught around the world, the up-to-date content covers developments in econometrics and finance over the past twenty years while ensuring a solid grounding in the fundamental principles of the field.

Care has been taken to link theory and application to provide real-world context for students, worked exercises and empirical examples have also been included to make sure complicated concepts are solidly explained and understood.

OLIVER LINTON is a fellow of Trinity College and is Professor of Political Economy at Cambridge University. Formerly, Professor of Econometrics at the London School of Economics and Professor of Economics at Yale University. He obtained his PhD in Economics from the University of California at Berkeley in 1991. He has written more than a hundred articles on econometrics, statistics, and empirical finance. In 2015 he was a recipient of the Humboldt Research Award of the Alexander von Humboldt Foundation. He has been a Co-editor at the Journal of Econometrics since 2014. He is a Fellow of: the Econometric Society, the Institute of Mathematical Statistics, the Society for Financial Econometrics, the British Academy, and the International Foundation of Applied Econometrics. He was a lead expert in the U.K. Government Office for Science Foresight project: "The future of Computer Trading in Financial Markets", which published in 2012. He has appeared as an expert witness for the FSA and the FCA in several cases involving market manipulation.

Financial Econometrics
Models and Methods

OLIVER LINTON

University of Cambridge

CAMBRIDGE
UNIVERSITY PRESS

University Printing House, Cambridge CB2 8BS, United Kingdom

One Liberty Plaza, 20th Floor, New York, NY 10006, USA

477 Williamstown Road, Port Melbourne, VIC 3207, Australia

314–321, 3rd Floor, Plot 3, Splendor Forum, Jasola District Centre, New Delhi – 110025, India

79 Anson Road, #06-04/06, Singapore 079906

Cambridge University Press is part of the University of Cambridge.
It furthers the University's mission by disseminating knowledge in the pursuit of
education, learning, and research at the highest international levels of excellence.

www.cambridge.org
Information on this title: www.cambridge.org/9781107177154
DOI: 10.1017/9781316819302

© Oliver Linton 2019

This publication is in copyright. Subject to statutory exception
and to the provisions of relevant collective licensing agreements,
no reproduction of any part may take place without the written
permission of Cambridge University Press.

First published 2019

Printed in the United Kingdom by TJ International Ltd, Padstow Cornwall

A catalogue record for this publication is available from the British Library.

ISBN 978-1-107-17715-4 Hardback
ISBN 978-1-316-63033-4 Paperback

Additional resources for this publication at www.cambridge.org/linton

Cambridge University Press has no responsibility for the persistence or accuracy
of URLs for external or third-party internet websites referred to in this publication
and does not guarantee that any content on such websites is, or will remain,
accurate or appropriate.

To my wife, Jianghong Song.

Short Contents

List of Figures	page xv
List of Tables	xix
Preface	xxi
Acknowledgments	xxv
Notation and Conventions	xxvii

1	Introduction and Background	1
2	Econometric Background	55
3	Return Predictability and the Efficient Markets Hypothesis	75
4	Robust Tests and Tests of Nonlinear Predictability of Returns	134
5	Empirical Market Microstructure	152
6	Event Study Analysis	201
7	Portfolio Choice and Testing the Capital Asset Pricing Model	238
8	Multifactor Pricing Models	279
9	Present Value Relations	314
10	Intertemporal Equilibrium Pricing	337
11	Volatility	358
12	Continuous Time Processes	422
13	Yield Curve	463
14	Risk Management and Tail Estimation	476
15	Exercises and Complements	497
16	Appendix	524
	Bibliography	533
	Index	553

Contents

List of Figures	*page* xv
List of Tables	xix
Preface	xxi
Acknowledgments	xxv
Notation and Conventions	xxvii

1	**Introduction and Background**	**1**
	1.1 Why Do We Have Financial Markets?	1
	1.2 Classification of Financial Markets	3
	1.3 Types of Markets and Trading	8
	1.4 Financial Returns	12
	1.5 Risk Aversion	27
	1.6 Mean Variance Portfolio Analysis	39
	1.7 Capital Asset Pricing Model	45
	1.8 Arbitrage Pricing Theory	49
	1.9 Appendix	53
2	**Econometric Background**	**55**
	2.1 Linear Regression	55
	2.2 Time Series	61
3	**Return Predictability and the Efficient Markets Hypothesis**	**75**
	3.1 Efficient Markets Hypothesis	75
	3.2 The Random Walk Model for Prices	81
	3.3 Testing of Linear Weak Form Predictability	85
	3.4 Testing under More General Conditions than rw1	106
	3.5 Some Alternative Hypotheses	118
	3.6 Empirical Evidence regarding Linear Predictability based on Variance Ratios	121
	3.7 Trading Strategy Based Evidence	124
	3.8 Regression Based Tests of Semi-Strong Form Predictability	129
	3.9 Summary of Chapter	132

Contents

4 Robust Tests and Tests of Nonlinear Predictability of Returns — 134

- 4.1 Robust Tests — 134
- 4.2 Nonlinear Predictability and Nonparametric Autoregression — 146
- 4.3 Further Empirical Evidence on Semistrong and Strong Form EMH — 148
- 4.4 Explanations for Predictability — 150
- 4.5 Summary of Chapter — 151

5 Empirical Market Microstructure — 152

- 5.1 Stale Prices — 152
- 5.2 Discrete Prices and Quantities — 164
- 5.3 Bid, Ask, and Transaction Prices — 168
- 5.4 What Determines the Bid–Ask Spread? — 173
- 5.5 Strategic Trade Models — 183
- 5.6 Electronic Markets — 187
- 5.7 Summary of Chapter — 196
- 5.8 Appendix — 196

6 Event Study Analysis — 201

- 6.1 Some Applications — 201
- 6.2 Basic Structure of an Event Study — 202
- 6.3 Regression Framework — 218
- 6.4 Nonparametric and Robust Tests — 221
- 6.5 Cross-sectional Regressions — 222
- 6.6 Time Series Heteroskedasticity — 223
- 6.7 Panel Regression for Estimating Treatment Effects — 224
- 6.8 Matching Approach — 228
- 6.9 Stock Splits — 229
- 6.10 Summary of Chapter — 237
- 6.11 Appendix — 237

7 Portfolio Choice and Testing the Capital Asset Pricing Model — 238

- 7.1 Portfolio Choice — 238
- 7.2 Testing the Capital Asset Pricing Model — 241
- 7.3 Maximum Likelihood Estimation and Testing — 246
- 7.4 Cross-sectional Regression Tests — 259
- 7.5 Portfolio Grouping — 264
- 7.6 Time Varying Model — 268
- 7.7 Empirical Evidence on the CAPM — 270
- 7.8 Summary of Chapter — 273
- 7.9 Appendix — 274

Contents

8 Multifactor Pricing Models — 279

8.1	Linear Factor Model	279
8.2	Diversification	280
8.3	Pervasive Factors	286
8.4	The Econometric Model	288
8.5	Multivariate Tests of the Multibeta Pricing Model with Observable Factors	290
8.6	Which Factors to Use?	293
8.7	Observable Characteristic Based Models	299
8.8	Statistical Factor Models	300
8.9	Testing the APT with Estimated Factors	312
8.10	The MacKinlay Critique	312
8.11	Summary of Chapter	312
8.12	Appendix	312

9 Present Value Relations — 314

9.1	Fundamentals versus Bubbles	314
9.2	Present Value Relations	316
9.3	Rational Bubbles	319
9.4	Econometric Bubble Detection	321
9.5	Shiller Excess Volatility Tests	323
9.6	An Approximate Model of Log Returns	326
9.7	Predictive Regressions	329
9.8	Summary of Chapter	335
9.9	Appendix	336

10 Intertemporal Equilibrium Pricing — 337

10.1	Dynamic Representative Agent Models	337
10.2	The Stochastic Discount Factor	338
10.3	The Consumption Capital Asset Pricing Model	339
10.4	The Equity Premium Puzzle and the Risk Free Rate Puzzle	345
10.5	Explanations for the Puzzles	346
10.6	Other Asset Pricing Approaches	353
10.7	Summary of Chapter	357

11 Volatility — 358

11.1	Why is Volatility Important?	358
11.2	Implied Volatility from Option Prices	359
11.3	Intra Period Volatility	362
11.4	Cross-sectional Volatility	370
11.5	Empirical Studies	371

11.6	Discrete Time Series Models	374
11.7	Engle's ARCH Model	377
11.8	The GARCH Model	380
11.9	Asymmetric Volatility Models and Other Specifications	389
11.10	Mean and Variance Dynamics	392
11.11	Estimation of Parameters	395
11.12	Stochastic Volatility Models	402
11.13	Long Memory	404
11.14	Multivariate Models	407
11.15	Nonparametric and Semiparametric Models	412
11.16	Summary of Chapter	419
11.17	Appendix	419

12 Continuous Time Processes — 422

12.1	Brownian Motion	422
12.2	Stochastic Integrals	427
12.3	Diffusion Processes	428
12.4	Estimation of Diffusion Models	436
12.5	Estimation of Quadratic Variation Volatility from High Frequency Data	450
12.6	Levy Processes	459
12.7	Summary of Chapter	462

13 Yield Curve — 463

13.1	Discount Function, Yield Curve, and Forward Rates	463
13.2	Estimation of the Yield Curve from Coupon Bonds	464
13.3	Discrete Time Models of Bond Pricing	469
13.4	Arbitrage and Pricing Kernels	471
13.5	Summary of Chapter	475

14 Risk Management and Tail Estimation — 476

14.1	Types of Risks	476
14.2	Value at Risk	477
14.3	Extreme Value Theory	479
14.4	A Semiparametric Model of Tail Thickness	482
14.5	Dynamic Models and VAR	487
14.6	The Multivariate Case	489
14.7	Coherent Risk Measures	492
14.8	Expected Shortfall	493
14.9	Black Swan Theory	494
14.10	Summary of Chapter	496

Contents

15	**Exercises and Complements**		**497**
16	**Appendix**		**524**
	16.1	Common Abbreviations	524
	16.2	Two Inequalities	526
	16.3	Signal Extraction	527
	16.4	Lognormal Random Variables	528
	16.5	Data Sources	529
	16.6	A Short Introduction to Eviews	530
	Bibliography		533
	Index		553

List of Figures

1.1	Level of the S&P500 index	19
1.2	Daily return on the S&P500 daily index	23
1.3	Euro/dollar daily exchange rate	25
1.4	Return on the euro/dollar daily exchange rate	26
1.5	Daily Tbill Rates	26
1.6	Daily Federal Funds rate annualized	27
1.7	Price of West Texas Oil, monthly frequency	27
1.8	Daily return on West Texas Oil	28
3.1	Shows "Head and Shoulders" pattern in artificial dataset	79
3.2	Correlogram of S&P500 daily returns	92
3.3	Correlogram of S&P500 daily returns by decade	93
3.4	Correlogram of FTSE100 daily returns from 1984–2017	93
3.5	Correlogram of FTSE100 daily returns long horizon	94
3.6	ACF(1) of daily Dow stock returns against market capitalization	95
3.7	Average ACF of Dow stocks daily returns	97
3.8	Daily return on the Malaysian Ringgit	104
3.9	Variance ratio of S&P500 daily returns, 1950–2017	123
3.10	Variance ratio of FTSE100 daily returns, 1984–2017	123
3.11	ACF of the winner and loser stocks in sample	127
3.12	ACF of winner and loser stocks in and out of sample	128
3.13	ACF of the max versus max of the ACF	129
4.1	Cowles–Jones statistic with standard errors	139
4.2	Quantilogram of daily S&P500 returns	142
4.3	Fraction of positive returns on the stocks of the S&P500 index	143
4.4	The AD line on the daily S&P500 return	144
4.5	Length of daily runs on the S&P500	145
5.1	Prices and discretization	166
5.2	Histogram of integer remainders	166
5.3	Price changes on S&P500 Emini contract	170
5.4	Autocovariance of daily Dow	172
5.5	Efficient price volatility of the daily Dow	172
5.6	Price trajectory 1	180
5.7	Price trajectory 2	181
5.8	Daily Amihud illiquidity on S&P500	186
5.9	Daily Amihud illiquidity of S&P500 index since 2008	186
5.10	FTSE100 one minute data during nonfarm payroll announcement	191
5.11	Flash crash	192

List of Figures

5.12	Feedback loop	192
6.1	Shows examples of price trajectories around event at time 0	205
6.2	Dow Jones stock splits	232
6.3	Exxon stock splits CAR	233
6.4	Exxon splits average CAR	233
6.5	Exxon splits, longer window	234
6.6	Exxon splits average CAR over splits	234
6.7	Exxon splits, shorter window	235
6.8	Exxon splits, shorter window average CAR	235
7.1	Efficient frontier of Dow stocks	241
7.2	Skewness of Dow Jones returns by sampling frequency	245
7.3	Kurtosis of Dow Jones returns by frequency	245
7.4	Risk return relation	264
7.5	Time varying betas of IBM	269
7.6	Time varying alphas of IBM	269
7.7	SMB portfolio beta with market return	270
8.1	Quantiles of ordered (absolute) cross-correlations between S&P500 stocks	283
8.2	Quantiles of ordered (absolute) cross-correlations between S&P500 stocks idiosyncratic errors	284
8.3	Solnik curve of Dow stocks	285
8.4	GMV of Dow stocks	286
8.5	Rolling window correlation between the FTSE100 and FTSE250	287
8.6	Fama–French factors returns and implied prices	295
8.7	Variance ratios of Fama–French factors	297
8.8	US monthly CPI percentage change	298
8.9	Eigenvalues of $\hat{\Sigma}$ for daily S&P500 stocks, $N=441$ and $T=2732$	309
8.10	Eigenvalues of $\hat{\Psi}$ for monthly S&P500 returns	309
8.11	Dominant principal component for monthly data	310
9.1	NASDAQ price level	315
9.2	Bitcoin price level from 2010–2017	315
9.3	Bubble time series	320
9.4	The logarithm of S&P500 index with trend line fitted from 1950 to 2017	325
9.5	Gross return on S&P500	333
9.6	Dividend yield on the S&P500	333
9.7	Rolling window (± 20 years) R^2 of predictive regression	334
9.8	Rolling window (± 20 years) slope coefficent of predictive regression	335
9.9	Overlapping daily 5 year returns on the S&P500	335
10.1	Growth rate of annual real PCE in 2009 dollars	346
10.2	Rolling window trailing 10 year gross nominal returns on the CRSP value weighted index	347
10.3	Distribution of the annual risk premium on the FF market factor from ten years of daily data	349
10.4	Quarterly US differenced log of PCE (seasonally adjusted)	349
10.5	Autocorrelation of quarterly US differenced log of PCE	350

List of Figures

10.6	Quarterly time series CAY_t	355
11.1	US commercial airline fatalities per mile	359
11.2	Daily VIX index	361
11.3	Histogram of the VIX index	361
11.4	ACF of the VIX	362
11.5	Annual volatility of the S&P500	364
11.6	Parkinson estimator of daily volatility	368
11.7	ACF of intraday volatility	368
11.8	Histogram of intraday volatility	369
11.9	Cross-section volatility	371
11.10	Correlogram of returns (first panel) and absolute returns (second panel)	375
11.11	Conditional standard deviation of daily returns	382
11.12	Standardized residuals	383
11.13	S&P500 daily return cross autocovariance $\text{cov}(Y_t^2, Y_{t-j}), j = -10, \ldots, 10$	389
11.14	Comparison of the estimated news impact curves from GARCH(1,1) and GJR(1,1) for daily S&P500 returns	392
11.15	Density of S&P500 returns	413
11.16	Shows the nonparametrically estimated conditional comulants $cum_j(y_t|y_{t-k})$, for $j = 1, 2, 3, 4$ and $k = 1, \ldots, 50$	415
11.17	Time varying cumulants of S&P500 returns	418
12.1	Crossing time density	425
12.2	ACF of daily federal funds rate 1954–2017	434
12.3	Conditional mean of exponential	435
12.4	Volatility signature plot	454
13.1	Time series of one month and ten year yields since 2000	470
13.2	Conditional cumulants of T-bill	474
13.3	Conditional cumulants of 10-year bonds	475
14.1	Estimated tail index of daily S&P500 returns 1950–2017	487
14.2	Expected shortfall	495
14.3	Daily S&P500 stock return events exceeding 6σ	496

List of Tables

1.1	Stock exchanges ranked by value traded	11
1.2	Descriptive statistics 1950–2017	23
1.3	Daily CRSP market returns descriptive statistics	24
1.4	Daily returns on individual stocks descriptive statistics, 1995–2017	25
3.1	Dow Jones stocks with market capitalization in January 2013	94
3.2	Parameter estimates of AR(22) model	104
3.3	Autocorrelation of squared returns and kurtosis of returns	113
3.4	The off-diagonal terms	113
3.5	Variance ratios for weekly small-size portfolio returns	114
3.6	Variance ratios for weekly medium-size portfolio returns	115
3.7	Variance ratios for weekly large-size portfolio returns	116
3.8	Day of the week effect	131
5.1	Tick size on the LSE	165
5.2a	Limit order book example	188
5.2b	Limit order book example	188
5.3	Round trip speed of the LSE	193
6.1	Recent stock splits on NASDAQ	230
6.2	Dow stocks split size distribution	232
7.1	Tangency and GMV portfolio weights for Dow Stocks	240
7.2	Market model estimates of Dow Stocks, daily data	247
7.3	Market model estimates of Dow Stocks, monthly data	248
8.1	Fama–French factors, mean and std	296
8.2	Fama–French factors correlation matrix	296
9.1	Dividend yield on the Dow stocks	336
10.1	Market risk premium	348
11.1	The FTSE100 top 20 most volatile days since 1984	369
11.2	The S&P500 top 20 most volatile days since 1960	370
11.3	Idiosyncratic volatility of Dow stocks	374
11.4	GARCH(1,1) parameter estimates	382
11.5	Estimated EGARCH model	391
11.6	Estimation of asymmetric GJR GARCH model	391
11.7	Estimates of GARCH in mean model	394
11.8	Correlation Matrix of estimated parameters from GARCH model	399
11.9	Daily GARCH in mean t-error	400
11.10	Estimated d by frequency	406
11.11	Estimated GARCH model by decade	417
13.1	Summary statistics of daily yields	471
13.2	Autocorrelation of daily yields	471

Preface

This work grew out of my teaching and research. It started from my considerable admiration of the seminal financial econometrics book Campbell, Lo, and MacKinlay (1997), henceforth CLM, and my teaching of that material to Master's students. I have kept along a similar line to that work in terms of selection of material and development, and have updated the material in several places. I have tried to adapt the material to a Master's audience by reducing the peerless literature review that is in that book and by amplifying some of the econometric discussions. I have included some theorems of varying degrees of rigor, meaning that I have not in every case specified all the required regularity conditions. I apologize for any upset this may cause, but the interests of time and space kept me from doing this. I hope the use of theorems can help to focus the material and make it easier to teach.

Financial econometrics has grown enormously as a discipline in the 20 years since CLM was published, and the range of authors engaged in its development has also increased. Computer scientists and so-called econo-physicists have taken an interest in the field and made major contributions to our understanding of financial markets. Mathematicians and statisticians have established rigorous proofs of many important results for our field and developed new tools for analyzing and understanding big financial data. The academic landscape has also become more international with a big expansion in the study of finance and financial statistics in China.

Data is the plural of anecdote, and happily there has recently been a massive increase in the amount of data, which has in itself stimulated research. Simultaneously computer power, both hardware and software, has increased substantially, allowing the analysis of much larger and more complex datasets than was previously possible. Econometric methodology has also expanded in many relevant areas, notably: volatility measurement and modelling; bounteous variate statistics where the size of the cross-section and time series is large; tools for extreme risk management; and quantifying causal effects. Despite the improvement in tools, the Global Financial Crisis following 2008, led to some skepticism about the value of economic theory and econometrics in predicting the armageddon that then ensued and in managing its aftermath, but this has in turn led to development of more relevant methodology, and hopefully some humility. Has all this attention and development improved our understanding of how financial markets work and how to change them for the better? I think so, but the subject is far from complete or satisfactory. The quantity and quality of empirical work and its presentation has improved substantially over time, but this has to some extent just made it harder for the reader to tell where the "bodies are buried" and what is the permanent value added of a particular study. Statistical methods are vital in this endeavor and in many cases their contribution is to provide a framework for acknowledging the limitations of our knowledge. Some of the empirical regularities that were cited in CLM have not stood the test of time, or at least they had to be qualified and their limitations acknowledged. This makes it hard to give a clear and simple picture that one could explain to a teenager.

This book is intended to be used as a text for advanced undergraduates and graduate students in economics, finance, statistics, mathematics, and engineering who are interested in financial applications and the methodology needed to understand and interpret those applications. I have taught part of this material at Yale University, the London School of Economics, the University of Cambridge, Humboldt University, Shandong University, SHUFE, and Renmin University. I provide two introductory chapters on financial institutions, financial economics, and econometrics, which provide some essential background that will be drawn on in later chapters. The main material begins with the efficient markets hypothesis and the predictability of asset returns, which is a central question of financial economics. I provide some updates on the empirical regularities found in CLM. I then provide a separate chapter on robust methods, which are important because large observations such as the October 1987 stock market crash can have an undue influence on estimation and hypothesis tests. I then cover some topics in market microstructure, which is a very active area struggling with new developments in market structure, technology, and regulation. I cover the classical topics of stale and discrete prices as well as the models for adverse selection and market impact that form the language of this area. I use some matrix algebra in Chapters 6 to 8; it is hard if not impossible to present this material cogently without these tools, and to understand big data without the basics of linear algebra is like trying to assemble an IKEA cupboard in the dark without an Allen key. Luckily there are many excellent books that provide suitable coverage of the necessary material, for example Linton (2016). The material on event studies has been expanded to include recent work coming from microeconometrics that can be used to evaluate the effects of policy changes on outcomes other than stock returns. I also include the standard methodology based on the market model but provide a more detailed discussion of the effects of overlapping estimation and event windows. I cover the CAPM next with some discussion of portfolio grouping methods and the two main testing methodologies. The chapter on multifactor models covers the main approaches including the Fama–French approach, which was still in its early days when CLM was published but is now one of the dominant methodologies. I also cover statistical factor models and characteristic based models. The next two chapters consider some intertemporal asset pricing material and the associated econometrics such as predictive regressions, volatility tests, and generalized method of moments (GMM). The chapter on volatility describes the three main approaches to measuring and defining volatility based on option prices, high frequency data, and dynamic time series modelling. The chapter on continuous time models develops some of this material further, but also introduces the models and methods widely used in this area. I cover some material on yield curve estimation and its application to pricing. The final chapter considers risk management including extreme value theory and dynamic modelling approaches. I use a number of datasets of different frequencies in the book to illustrate various points and to report on the main features of financial data. As usual, results can vary.

Preface

I have included short biographies of authors who have influenced me regarding financial econometrics particularly. My prediction is that at least one of them will win the Nobel Prize in economics.

The book contains many terms in bold face, which can then be investigated further by internet search. I have provided some computer code in different languages, such as MATLAB, GAUSS, and R, pertinent to various parts of the book. I am told that STATA can accomplish many things, but I have yet to see the light. A lot of analysis can be done by EVIEWS, and I provide a short introduction to its use in handling daily stock return data. I also provide a link to a number of data sources that can help with student projects and the like.

Acknowledgments

No man is an island, and I would like to thank the people who have had an influence on my research career. Even though they may not have been directly involved in the development of this book, I have drawn heavily on my interactions with them throughout this project. In roughly chronological order they are: Haya Friedman, Jan Magnus, Tom Rothenberg, Jens Perch Nielsen, Peter Bickel, Greg Connor, Peter Robinson, Wolfgang Härdle, Enno Mammen, Per Mykland, Miguel Delgado, Neil Shephard, Don Andrews, Peter Phillips, Xiaohong Chen, Arthur Lewbel, Yoon-Jae Whang, Zhijie Xiao, Christian Hafner, Frank Diebold, Eric Ghysels, Jean-Marie Dufour, Haim Levy, Andrew Patton, Jiti Gao, Jon Danielsson, Jean-Pierre Zigrand, Alexei Onatskiy, Andrew Harvey, Andrew Chesher, Hashem Pesaran, Richard Smith, and Mark Salmon. I would like also to thank my current and former PhD and Master's students who have contributed to the development of this book. I would like to thank Greg Connor, Katja Smetanina, and anonymous referees for comments.

Notation and Conventions

In this book I use the dating convention yyyymmddhhmmss. A visiting time traveller would surely prefer to know the year before the month or day, although he might ask why we have we have chosen 24 hours in a day and 60 minutes in an hour, etc. I use $\xrightarrow{P}$ to denote convergence in probability and $\implies$ to denote convergence in distribution. $\log(x)$ is the natural logarithm unless otherwise stated. $\mathbb{R}$ is the set of real numbers, $'$ denotes differentiation, and $^{\top}$ denotes matrix transpose. We use $X_n = O(n)$ to mean that X_n/n is bounded for a deterministic sequence X_n. I use $\simeq$ to generically denote an approximation and $\sim$ to mean to have the same distribution as. I do not have a bracketing convention like some journals, but I do have a preference for round curved brackets over square ones. Dollars or $ are US unless otherwise specified.

1 Introduction and Background

1.1 Why Do We Have Financial Markets?

The purpose of financial markets can be broadly defined to be to channel funds from savers to borrowers. Individuals have a preference for **consumption smoothing**, both intertemporal (saving and dissaving over time so that consumption is less affected by time-varying income) and intratemporal, i.e., across states of nature (to avoid or mitigate the effects of disaster or large price swings). Diversification of **idiosyncratic** (individual specific) liquidity risks enables the mobilization of society's savings for long-term projects (which may often be the most productive). There is economic efficiency and growth if entrepreneurs with productive ideas/projects borrow from savers with inferior investment opportunities of their own (households are typically net savers; firms are net borrowers; government can be either, it depends on revenues versus expenditures). To provide the **insurance** or **hedging** demanded by some requires that someone, perhaps called **speculators**, take the other side thereby enabling the **risk transfer**. Ultimately these transactions can increase welfare, so financial markets are not necessarily in total a **zero sum game**.

The productive capacity of the economy is linked to real assets, to land, buildings, machinery, technology and knowledge, and goods and services production. Financial assets are pieces of paper and/or electronic entries that do not contribute directly to the productive capacity of the economy; they are titles to income generated by real assets. Real assets constitute net wealth and generate the net income of the economy. Financial assets define the allocation of wealth and income among households. Financial assets are liabilities of the issuers and so aggregation of household and firm balance sheets leaves only real assets as the net wealth. Domestic net worth can be obtained by summing the value of residential and nonresidential real estate, equipment and software, inventories, and consumer durables. **Human capital** (education and health) is also part of a nation's wealth but its stock is difficult to value. Households own real assets such as houses and consumer durables but also financial assets such as bank deposits, life insurance and pension reserves, corporate and noncorporate equity, debt instruments, mutual fund shares, etc. Savings are the part of current income that is not consumed. Investing is about which assets to acquire with the saved income.

The efficiency of the capital allocation process matters. A country can save and invest a large proportion of its output, but if the financial system allocates it inefficiently, a shortfall in growth and welfare ensues. Some examples of inefficiencies might include: preferential flow of credit to Party members' companies, poor funding options for small and medium sized enterprises (SMEs), and poor legal systems such as those that allow **insider trading** with impunity. There is some evidence that enforcement of insider trading laws reduces the cost of equity/capital (Bhattacharya and Daouk (2002)).

Some authors have distinguished between two types of financial systems: **bank-based** systems (China, Germany, Japan, etc.) and **market-based** (US, UK, etc.). In the bank-based system, credit is primarily obtained through the banking system, whereas in the market-based system, borrowers derive much credit from issuing securities to raise funds. Levine (2002) describes the relative merits of both systems and their relative prevalence in different places over time. He argues that it is unclear which system is superior and that the distinction may not be very useful in determining sources of growth; instead the legal and institutional environment matters more.

Securities markets are physical or virtual venues where demand meets supply. Ideally, they guide scarce resources towards their most productive use. Investors analyze companies and bid prices up or down, which influences the cost of capital (both debt and equity). Ultimately, this determines which companies will live and which will die. Ideally, this enables efficient **risk management** – the slicing and dicing of risks and their transfer to parties most willing to bear them – which improves household welfare and reduces firms' cost of capital. A key feature of security markets is that they allow the separation of ownership and management. A small business can be owner-managed. A large corporation has capital requirements that exceed the possibilities of single individuals, and possibly has hundreds of thousands of shareholders. They elect the board of directors, which hires managers who, ideally, run the business efficiently and maximize its value for the shareholders. In principle, this separation ultimately leads to improvement of economic efficiency and welfare. However, conflicts of interests arise often in financial markets, and this can lead to negative outcomes.

The **principal-agent problem** is a classic example. Shareholders want the most productive projects, i.e. those with the highest positive "net present value," to be undertaken. Instead, managers may pursue their own interests: empire building, taking excessive risks to generate short-term profits and ignoring long-term consequences, keeping inefficient suppliers in exchanging for kickbacks, engaging in corporate book cooking, etc. Some well known recent examples include: WorldCom, Enron, and Parmalat. Arthur Andersen, the demised accounting company, received more income from consulting for Enron during the 1990s than from auditing it. Enron used special purpose entities and vehicles to get debt off its books, and the auditor might have been lenient to protect its consulting profits. In principle, boards of directors may force underperforming managers out. They may design compensation contracts that align incentives, or they may not. Security analysts, large shareholders, and creditors monitor the management, and they can sometimes affect the direction it takes. Some other examples of conflicts of interest include: optimistic security research in exchange for investment banking business (hence information barriers between corporate finance operations and retail or trading business); trading against or ahead of clients (**front running**); and insider trading where insiders or their proxies trade ahead of product news, earnings announcements, and merger and acquisition deals.

1.2 Classification of Financial Markets

The financial crisis of the late 2000s demonstrated a range of weaknesses in the financial markets architecture including **moral hazard** where executives had incentives towards taking excessive risks: big profits yield big bonuses; big losses do not coincide with negative bonuses. Government bailouts were needed to stabilize the system with large future fiscal consequences. Additional regulation and supervision were introduced. Many fines were issued to investment banks for misconduct. Turner (2009) provides some analysis of what happened.

1.2 Classification of Financial Markets

We can classify financial markets into the following: **money markets** – debt instruments with maturity < 1 year; **bond markets** – debt instruments with maturity ≥ 1 years; **stock/equity markets** – shares of listed companies; **foreign exchange markets** – currency pairs; **derivatives markets** – futures contracts, options; **commodity markets** – pork bellies, copper, etc. We will discuss these different markets and the instruments that they trade below.

1.2.1 Money Market

The money market is a subsector of the fixed-income market. Small investors can invest via money market mutual funds. Instruments include:

Treasury bills (T-bills). Highly liquid short-term government debt. Sold at a discount from the stated maturity value (investors' earnings are the face value minus purchase value).

Certificates of deposit (CDs). Time deposit with a bank which cannot be withdrawn prior to maturity but can be traded.

Eurodollar CDs. Dollar-denominated but issued by a foreign bank or a foreign branch of a US institution, hence not Fed-regulated (likewise Euroyens). Nowadays, there's nothing European about the prefix although the origin of the term goes back to US$ deposits in European banks after WWII.

Commercial paper. Short-term unsecured debt of large companies (instead of direct bank borrowing).

Bankers' acceptances. Payment order (time draft, postdated cheque) endorsed as accepted by a bank and hence bearing its credit rating. Frequently used in international trade to substitute creditworthiness of the bank rather than that of an unknown trading partner (e.g. an exporter selling products to an unknown importer receives acceptance from the importer's bank to pay in 90 days). Traded at a discount (depends on interest rates and the bank's credit quality); the holder may keep it until maturity or sell it in the open market or to the bank.

Repurchase agreements (repos, RPs). Agreement to sell a security now and buy it back later (at a higher price). Effectively collateralized borrowing (although legally distinct since there is automatic ownership transfer). Reverse repo: effectively collateralized lending (repo for the counterparty).

There are some well known and often-quoted money market rates:

The **Federal funds rate**. Commercial banks required to hold a certain minimum amount of Fed Funds on reserve accounts with the Federal Reserve (depending on customer deposits). Banks with excess funds lend to banks with a shortage, usually overnight. Nowadays not just to meet reserve requirements but a general source of funding.

LIBOR (London Interbank Offered Rate). Unsecured funds rate in the wholesale money market. Important reference rate for global money markets. Calculated for several currencies and maturities. This "calculation" was the subject of a recent investigation that uncovered some abuse of the system by the participants whose quotes defined the rate.

Money market securities tend to have low risk but they are not riskfree, and they may carry different risk premia. Also, differences in the ease of trading can result in additional premia. During the credit crunch in the late 2000s, a lack of trust in counterparties led to a severe upward jump in risk premia, and some segments of money markets completely froze, i.e. became illiquid. Large-scale intervention by central banks was needed to slowly restore credit conditions and the ability of corporations to obtain funding.

1.2.2 Bond Market

Bonds are contracts that specify fixed payments of cash at specific future dates. They are a big market: US Treasury Securities 21×10^{12} outstanding in 2017; Japanese securities 10.46×10^{12} outstanding in 2013. There are several types of bonds in the US market: **zero coupon bonds** (single payment at a specified future time), also known as zeros; **T-bills** with original maturities of less than a year; **coupon bonds** (these have coupon payments c expressed as a percentage of redemption value, which is set equal to $100; coupon is paid every six months (or year) until maturity at which date the holder receives $100); **notes** with original maturities from one to ten years; **bonds** with original maturities more than ten years. There are also options on bonds, futures, swaps, swaptions, cap, floor, collar, etc. In the UK, government debt securities are called gilts (or gilt-edged securities) and may have maturity up to 50 years. Treasury Inflation-Protected Securities (TIPS) in the USA and Index-Linked Gilts in the UK are both designed to deliver returns protected against inflation. Types of bonds also include:

Federal agency debt. Ginnie Mae, Fannie Mae, Freddie Mac – US mortgage-related agencies that issue securities and channel the raised funds to savings and loans institutions that lend it to individual mortgage borrowers; they back the majority of US mortgages. They are supposed to improve the availability of credit for housing and make it less dependent on local conditions. Government guarantees (explicit for Ginnie Mae, implicit for the other two which are now under conservatorship) improve credit rating and thus reduce the cost of borrowing. The financial crisis of the late 2000s revealed that there were enormous taxpayer liabilities from this insurance.

1.2 Classification of Financial Markets

Eurobonds. Bonds denominated in a currency other than that of the country in which they are issued. In contrast, some bonds are issued abroad in local currencies:

Yankee bonds: dollar-denominated bonds sold in the USA by non-US issuers;
Samurai bonds: yen-denominated bonds sold in Japan by non-Japanese issuers.

Municipal bonds. Issued by states, cities, counties, or other local governments. They have a tax-exempt status and so high-tax-bracket investors often hold them.

Corporate bonds. Wide spectrum of credit quality; default risk different from government bonds. Secured bonds are those with specific collateral. Unsecured bonds are also known as debentures in the USA (in the UK debentures are often secured). Junior (subordinated) bonds/debentures have lower-priority claim to the firm's assets than existing senior debt (bonds/debentures or loans) and are hence more risky. **Callable bonds** give the firm the option to repurchase the bonds at a stipulated call price. **Convertible bonds** give the holder the option to convert each bond into a stipulated number of shares.

Asset-backed securities. Bonds or notes collateralized by a pool of underlying assets, assets such as home equity loans (hence mortgage-backed securities), credit cards, auto loans, and student loans. Cash flows can be directly passed through to investors after administrative fees have been subtracted (pass-through securities). Alternatively, cash flows can be carved up according to specified rules (structured securities). **Securitization** converts a pool of loans to asset-backed securities (ABS). Banks originate loans backed by assets, service them for a fee but pass them through to ABS holders. The innovation is that the availability of funds is not dependent on local credit conditions, which means better terms for borrowers. Investment banks create special purpose vehicles (SPVs), which are "bankruptcy remote," to isolate a loans pool, segment it by credit risk, and issue structured securities against it. Infamous CDOs (collateralized debt obligations) are structured asset-backed securities with cash flows dependent on the underlying fixed-income assets. There is a separation between the originator and the distributor – incentives to earn fees through volume regardless of credit quality (credit risk was not borne by the originator, the consequence of which was exacerbated by rating errors and conflicts of interest).

1.2.3 Equity Market

Common stocks and equities. Each share of common stock (i.e. a piece of the firm) entitles the owner to one vote at the AGM (annual general meeting) and to a share in the financial benefits of ownership. Shareholders elect the board of directors which selects and oversees managers who run the corporation on a day-to-day basis. Shareholders not attending the annual meeting can vote by proxy, empowering another party to vote in their name. A **proxy fight or contest** is when some shareholders attempt to replace the management which fights back via other shareholders and various defensive tactics (generally hard and the real threat to management comes from takeovers).

Common stock is a residual claim: if liquidation happens, the residual claimants are the last in the line for the proceeds from the assets. Common stock is sometimes called junior equity (subordinated to preferred stock). **Limited liability** is where the maximum loss is the original investment, unlike owners of unincorporated businesses whose creditors can (unless the business owners have taken out directors' liability insurance) lay claim to their personal assets (their house, car, furniture, favorite mug). If the stock is **listed** at and hence traded on a public stock exchange or traded **over the counter** (OTC), then it is publicly traded equity. If it is not **publicly traded**, then it is called **private equity**. Common stocks usually pay dividends to shareholders several times a year. Alternatively, companies can repurchase their stock, which should raise its price and lead to a capital gain for the shareholders.

Preferred stocks. Features of both equity and debt (hybrid securities like convertible bonds; preferred equity and subordinated debt are also known as mezzanine finance/capital). Promises to pay a fixed amount each year and carries no voting power, hence similar to perpetuities (perpetual bonds, infinite-maturity bonds). The difference is that the firm retains discretion over dividend payments; there is no contractual obligation. Dividend payments are cumulative, i.e. unpaid dividends cumulate and must be paid in full before any dividends are paid to common stockholders. May be redeemable (like callable bonds), convertible (into common stock at a pre-specified conversion ratio), or adjustable-rate (dividend rate tied to market interest rates). Unlike coupon payments on bonds (and interest on bank loans), dividends from preferred stocks are not tax-deductible expenses for the corporation.

Depository receipts. Represent ownership of a predefined number of foreign shares (depositary shares; preferred shares or bonds also possible) held by a domestic depositary bank (through its foreign branch or local custodian bank), which issues the receipts. Listed on a domestic exchange or traded OTC, they are an easier and cheaper way for investors to get exposure to foreign securities. Typically denominated in US$, but also euros, so that they carry a foreign exchange risk for most of the interested parties. They make it easier for foreign firms to satisfy otherwise stringent security registration requirements for listing their securities e.g. in New York or London. Global depository receipts (GDRs) in one or more markets outside of the USA; American depository receipts (ADRs) in the USA.

Market indices reflect the broader valuation of the stock market as a whole, some narrower (a few dozen securities), some broader (thousands of securities), such as the Dow 30, S&P500, and Russell 5000. There are different weighting schemes, which we will discuss more below. **Exchange-traded funds** (ETF's) allow investors to trade an asset whose return mimics the return on broad indices, for example the SPDR ETF tracks the S&P500.

Why are securities publicly held and traded? Firms raise equity and debt capital for investment and growth; governments issue debt to finance deficits. There can be a public offering – primary offering sold to the general public – or private placement (nonpublic) – primary offering, sold to one or a few institutional investors (banks, insurance companies, mutual funds, pension funds, etc. A **seasoned equity offering** (SEO) is the selling of additional equity of publicly traded

1.2 Classification of Financial Markets

corporations. An **initial public offering** (IPO) is the selling of shares to the public for the first time. Public offerings of stocks and bonds are typically managed by more than one investment bank. There are various ways of doing it. Typically, there is an underwriting syndicate headed by a lead underwriter(s) or manager(s). They give advice on the terms of selling securities. There is a preliminary prospectus (also known as a **red herring** prospectus since it is a statement in red that has not been approved yet), then a road show to publicize the offering and gauge demand (which is entered in a book; hence this is called bookbuilding); once it is approved by the Securities and Exchange Commission (SEC), there is a final prospectus that states the offering price. There is a **tombstone announcement** (upright rectangle like a cemetery tombstone) in the financial press. Commercial banks accept deposits and provide loans to firms. Investment banks are engaged in the securities business. Universal banks do both. Underwriters typically buy the securities from the issuer for less than the offering price and resell them to the public (their compensation is through the underwriting spread). This procedure is called **firm commitment**. Some alternative contracts that are used include: **best-efforts deal** – cheaper, not really underwriting since banks do not bear the risk of not selling the entire issue; **all-or-none deal** – the stock must be bought in its entirety, or not at all (raising funds for only half a factory is no good); and **auction-based offering** – e.g. modified Dutch auction, **OpenIPO**, used by Google – attempt to avoid the usual initial underpricing of IPOs (issuer may feel money has been left on the table in addition to explicit IPO costs) and provide fairer access (online) to investors across the board (rather than favour large institutional investors who can "flip" their allocations for profit).

Trading of already-issued securities among investors takes place in the **secondary market**, in which the number of outstanding securities is not affected, as there is just a transfer of ownership. The performance of the secondary market is important for the primary market and affects the cost of raising capital. Liquidity of public issues implies a greater willingness to commit funds because there is a well functioning market for you to exit your position. This ultimately implies a lower cost of long-term capital for the issuer.

Equity futures market. The E-mini S&P, often abbreviated to **E-mini** (despite the existence of many other E-mini contracts), and designated by the commodity ticker symbol ES, is a stock market index futures contract traded on the Chicago Mercantile Exchange's (CME) Globex electronic trading platform. The notional value of one contract is 50 times the value of the S&P500 stock index. It was introduced by the CME on September 9, 1997, after the value of the existing S&P contract (then valued at 500 times the index, or over $500,000 at the time) became too large for many small traders. The E-mini quickly became the most popular equity index futures contract in the world. Hedge funds often prefer trading the E-mini over the big S&P since the latter still uses the open outcry pit trading method, with its inherent delays, versus the all-electronic Globex system. The current average daily implied volume for the E-mini is over $200 billion, far exceeding the combined traded dollar volume of the underlying 500 stocks. Following the success of this product, the exchange introduced the E-mini NASDAQ-100 contract, at one fifth of the original NASDAQ-100 index based contract, and many other "mini" products geared primarily towards small speculators, as opposed to large hedgers. In June 2005 the exchange introduced a yet smaller product based on the S&P, with the underlying asset being 100 shares of the highly-popular SPDR exchange-traded fund. However, due to the different regulatory requirements, the performance bond (or margin) required for one such contract is almost as high as that for the five times larger E-mini contract. The product never became popular, with volumes rarely exceeding ten contracts a day. The E-mini contract trades 23 hours a day from 5:00pm – 4:15pm the next day (excluding the 3:15pm – 3:30pm maintenance shutdown), five days a week, on the March quarterly expiration cycle.

1.3 Types of Markets and Trading

Here, we discuss some of the details of trading.

Definition 1 *Brokered markets. Brokers with special knowledge offer search services for a fee, matching supply with demand. Some examples include: the real estate market; the primary market for securities (investment bankers act as brokers between issuers and investors); and the 'upstairs' market for large blocks of listed shares where brokers locate counterparties to conduct trade off exchange.*

Definition 2 *Dealer markets. Dealers (market makers) trade for their own account, build an inventory of assets, and make markets by quoting bid prices at which they are willing to buy and ask or offer prices at which they are willing to sell. Investors just look up and compare prices quoted by dealers. Dealers maintain liquidity by providing immediacy for those who want to trade (hence quote-driven market). For example OTC market, i.e. a decentralized network of brokers and dealers who negotiate sales of securities (not a formal exchange).*

Definition 3 *Electronic auction markets. All traders converge to a single (physical or "electronic") venue to buy and sell. They submit orders that are executed if and when matching orders arrive, nowadays by a computer system called the matching engine. This is called an order-driven market. Investors need not search across dealers to find the best price (and do not need to pay such intermediaries).*

The leading trading venues include the following:

NASDAQ – OTC market – dealers quote prices and brokers execute trades on behalf of their clients by contacting dealers with an attractive quote. Before 1971, all OTC quotations were recorded manually and published daily on so-called **pink sheets**. In 1971, the National Association of Securities Dealers Automated Quotation (NASDAQ) developed to link brokers and dealers in a computer network where prices could be displayed and revised. Originally it was just a quotation system, not a trading system (direct negotiation between brokers and dealers is still required). Nowadays, it has an electronic trading platform that allows electronic execution of trades without the need for direct negotiation, and the bulk of trades is executed that way.

New York Stock Exchange (NYSE, the Big Board). Trading in each security used to be managed by a specialist who maintains a limit order book for that security but steps in as a dealer when liquidity is insufficient (he should maintain a fair and orderly market) and earns commissions and the bid–ask spread. Floor trading – brokers representing their clients come to the relevant specialist on the floor of the exchange (each security has a unique specialist but specialists make a market in multiple securities). Electronic trading (the vast majority) – electronic trading platform enabling direct submission of orders to specialists over the computer network and their automatic execution. Trades of very large blocks that cannot be handled by specialists "downstairs" on the floor are negotiated and matched by specialized brokers "upstairs" (such brokerages known as "block houses").

The NYSE also operates a corporate bond exchange, the electronic trading platform Automated Bond System. However, the vast majority of bond trading (even for NYSE-listed bonds) occurs in the OTC market. There is a network of bond dealers linked by an electronic quotation system. Dealers may not carry an extensive inventory of the wide range of bonds available. They provide a brokerage service by locating counterparties. The market in many issues is quite "thin" and trading is infrequent.

London Stock Exchange (LSE). Until 1997 it operated similarly to NASDAQ with an automated quotations system for an OTC market with security firms acting as broker-dealers. Now it is mostly electronic limit order book trading although some transactions (large blocks, less liquid securities) continue to be carried out through dealers.

There has been recent development of electronic trading replacing human to human trading: the development of **algorithmic trading** strategies and **high frequency trading (HFT)**. This has coincided with reduced costs of trading and communication, information availability, online brokerage (lower commissions), etc. **Reg NMS** (National Market System) in the USA and **MiFID** (Market in Financial Instruments Directive) in Europe permitted and encouraged the introduction of new electronic venues for trading equities. In the USA, new exchanges were created such as Direct Edge, BATS, etc. In the UK, likewise with Chi-X.

Dark pools and **broker crossing networks** are electronic venues with no **pre trade transparency**, i.e. the order book is not visible to participants. Price is usually determined by the midpoint of the prevailing bid and ask prices on some reference **lit exchange**. The purpose of these venues is to facilitate trading of large blocks of securities in relative secrecy, thereby avoiding price impact.

There are a number of different types of orders that traders can use in electronic markets:

A **market order** is a buy (bid) or sell (ask) order that is to be executed immediately at the current market price. There are also price-contingent orders such as **limit orders** – these are executed only above or below a stipulated price limit. Bid – buy at or below a stated price. Ask – sell at or above a stated price. Limit orders may not execute if they are not competitive and/or the market moves away.

Stop orders – not executed until the market price reaches a stipulated limit. Stop-buy order – buy at or above a stated price. Stop-loss order – sell at or below a stated price.

Iceberg orders are limit orders with only a fraction of the order visible to other participants. If part of the order is executed, some of the remaining quantity may become visible.

Pegged orders – price driven by a reference price such as the midpoint of the bid–ask spread on some other trading venue.

Orders may also be limited by time. Day orders expire at the end of the trading day. Open orders (good-till-cancelled, fill-or-kill) – remain in force for up to six months unless cancelled.

There have been **institutional changes**, including the demutualization of stock exchanges at the turn of millennium (they became listed companies in their own right, often listed on their own markets). Competition was introduced between exchanges due to MiFID in Europe in 2007 and earlier in the US by reg NMS in 2005. Smart order routing technology facilitated linking market places together and enabling competition between and within market places. Competition between liquidity providers replaced essentially monopoly provision of market making. The decimalization of tick size in the US come in at the end of the 1990s. Transnational mergers between exchanges have also come about – NYSE Euronext, LSEG – NYSE Euronext bought by ICE of Atlanta.

There has also been **globalization of investment**. Emerging market economies and their stock markets have seen substantial growth. The ten biggest stock markets in the world by market capitalization in (US$ millions), according to the World Federation of Exchanges (FESE), at the end of 1999 and 2010 are shown in Table (1.1). The top four positions have not changed (ignoring the name branding changes), although both NASDAQ and Tokyo have seen declines in market capitalization over the decade, and NYSE and LSE have both seen only relatively modest increases in the market value. The striking feature of the 2010 picture is that positions 6–10 have been taken by emerging economy stock markets, like China, India, and Brazil, and the smaller European ones have been replaced by these larger capitalized exchanges, which have evidently grown enormously throughout the decade. The growth of these emerging market exchanges has been due to the increase in market capitalization of their domestic firms, and this is likely to continue for the foreseeable future.

1.3.1 Margin Trading

Investors can borrow part of the purchase price of a security from a broker (broker's call loan) and the security serves as a collateral, provided they have a margin account

1.3 Types of Markets and Trading

Table 1.1 Stock exchanges ranked by value traded

Rank	Exchange 1999	Value (US$ millions)	Exchange 2010	Value (US$ millions)
1	NYSE	11,437,597.3	NYSE Euronext (US)	13,394,081.8
2	NASDAQ	5,204,620.4	NASDAQ OMX	3,889,369.9
3	Tokyo	4,463,297.8	Tokyo SE Group	3,827,774.2
4	London	2,855,351.2	London SE Group	3,613,064.0
5	Paris	1,496,938.0	NYSE Euronext (Europe)	2,930,072.4
6	Deutsche Börse	1,432,167.0	Shanghai SE	2,716,470.2
7	Toronto	789,179.5	Hong Kong Exchanges	2,711,316.2
8	Italy	728,240.4	TSX Group	2,170,432.7
9	Amsterdam	695,196.0	Bombay SE	1,631,829.5
10	Switzerland	693,133.0	National Stock Exchange India	1,596,625.3
10=			BM&FBOVESPA	1,545,565.7

with the broker. The **initial margin** is the ratio of initial cash on the margin account to the purchase value of the securities. For example, in the US, the Federal Reserve imposes a minimum of 50% (can be set higher by brokers). The **maintenance margin** is the minimum required ratio of cash on the margin account and current market value of securities. Margin serves as a buffer protecting the broker against losses. Once below the maintenance level, the investor receives a **margin call** asking her to deposit additional cash or liquidate some securities. If not, the broker is authorized to restore satisfaction of margin requirements by the latter method. The leverage implied by margining (also known as gearing, or debt-financing of trades) increases volatility and risk, i.e. both the upside and the downside return potential increases.

Example 1 *Suppose initial margin requirement 50%, maintenance margin 25%. Investor buys 100 shares at the price of $100, i.e. pays $10,000 in total, borrowing $4,000. Own capital/equity is $6,000 and the initial margin is 60%. If the share price fell to $70, the investor's equity would become $3,000 and the percentage margin $3,000/$7,000 = 43%. Investor's equity would become negative if the price P fell below $40 (implies a loss for the broker). Margin call if the price fell $53.33:*

$$\frac{100P - 4,000}{100P} = 0.25.$$

If the borrowing rate is 5% and the price goes up by 20%, the investor's return is

$$\frac{12,000 - 4,200 - 6,000}{6,000} = 30\% > 20\%.$$

Note we repay the principal ($4,000) and the interest ($200) on the margin loan. If the price goes down by 20%, the investor's return is

$$\frac{8,000 - 4,200 - 6,000}{6,000} = -36.67\% < -20\%.$$

1.3.2 Short-Selling

Using long strategies, one wishes to buy low and sell high, thus profiting from a price increase. With short-selling, one borrows a security (implies debt in security units) hoping to sell high and later buy back low, thus profiting from a price fall. This can be viewed as borrowing money with the rate of interest being the return on the security in question, where the short-sellers wish the return to be as low as possible. Securities are borrowed from a broker. Short-sellers can cover the short position at any time by purchasing the security back (they pay fees as well as any dividends to the lender). Proceeds from short sales kept on the margin account plus additional margin (cash or securities) must be posted.

Example 2 *Investor borrows 100 shares from a broker who locates the securities, i.e. borrows from other clients or outside institutional investors. She sells the shares short at the current price $100 so that $10,000 is credited to her account. To satisfy the initial margin requirement of 50%, she must additionally deposit cash or securities (e.g. T-bills) worth at least $5,000 (5,000/10,000 = 50%) so that in total $15,000 are on the account. If the price P declines to $75, she covers the position and takes a profit of $2,500. If the maintenance margin is 30% there would be a margin call if the share price rises above $115.38:*

$$\frac{15,000 - 100P}{100P} = 0.3.$$

Long versus short positions. Long positions have limited losses (limited liability means that prices can't be negative) and unlimited profits (price can theoretically rise towards infinity). Short positions have limited profits, but unlimited losses. One can limit the potential losses associated with short selling by using **stop-loss** orders. For example, in the above example, a stop-buy order at $105 would aim to limit the loss to $500.

1.4 Financial Returns

1.4.1 Definition and Measurement

The concept of return is central to finance. This is the benefit of holding the asset that accrues to its owner. The return depends on the price of the asset and the holding period considered. In classical economics one is taught about the centrality of the price system and the law of one price. In practice, prices are formed in markets of different types (auction markets, dealer markets, etc.). They could be transaction prices

1.4 Financial Returns

or merely quoted prices. They may be the weighted averages of prices of different transactions. It is important to know which prices are used and hence how returns are constructed, and we will discuss this in more detail in Chapter 5. For now we suppose that prices P are observed at some times t. Some assets also pay the holders a dividend, which should be included in the definition of the total return.

Definition 4 *The simple gross return between times s and t is calculated as*

$$\mathcal{R}_{s,t} = \frac{P_t + D_t}{P_s},$$

where D_t are any dividends paid out during the holding period $[s, t]$. The simple net return is defined

$$R_{s,t} = \mathcal{R}_{s,t} - 1.$$

The gross return is non-negative because prices are non-negative; the net return can be negative, but not less than -1. Total return can be divided into the **dividend return** D_t/P_s and the **capital gain** P_t/P_s. The dividend component has some special methodology of its own. If the interval $[s, t]$ is long, then one should take account of precisely when the dividend was paid, with an earlier paid dividend more valuable than a later paid dividend because of the time value of money. This will be addressed by looking at multiperiod returns, which we explain below. In practice, it is the **dividend yield** of a stock that is reported.

Definition 5 *The dividend yield on a stock at the end of month t is the total cash dividend paid out over the preceding 12 months divided by the current price level.*

Handling dividends in practice is quite tricky, since for individual stocks these are only paid infrequently, and we shall often rely on data providers to have included dividends in the return calculation. The price series is easy to work with and so we mostly focus on capital gains, and mostly drop dividends from the remaining discussion and highlight it at particular points.

Taxes, inflation, and exchange rates may also be relevant to investors when calculating their net return. In many countries both dividend income and realized capital gains are taxed, but at different rates. Furthermore, the effective tax rate varies across individuals and corporations, making the computation of after tax returns complex. Many investors invest abroad in foreign currency denominated assets, and so in calculating their rate of return they should adjust for changes in the currency during the investment period.

We shall often consider the case where $s = t - 1$ and $t = 1, 2, \ldots$, i.e. the one period case. In practice, however, we observe prices at some discrete set of times $\{t_1, \ldots, t_n\}$, which may not be equally spaced. For example, at the daily frequency, we usually take the closing price of each day, which is observed at Monday 4.30pm (say), Tuesday 4.30pm, Wednesday 4.30pm, Thursday 4.30pm, and Friday 4.30pm, but not on

Saturday or Sunday and not during holiday periods. In this case the price observations are not taken at exactly equal intervals. This may not pose an obvious problem as yet, but we will see later that it is convenient to have observations of roughly equal contribution. There is a marvelous fix to this, which is called **trading time**.

Definition 6 *Trading time hypothesis. Valuations of assets only change during trading so that returns are only generated when trading is taking place.*

The consequence of this assumption is that we can relabel the observation times as $1, \ldots, n$, say, and let $\mathcal{R}_t = \mathcal{R}_{t-1,t}$ and $R_t = R_{t-1,t}$ for simplicity for the one period return. The Friday to Monday return is considered to be generated in the same way as the Monday to Tuesday return etc. This assumption simplifies the analysis of daily data. In some cases, this assumption is extended in the context of intraday data to mean that where we have a sequence of prices $P_{t_1}, P_{t_2}, \ldots, P_{t_n}$ made at time points $t_1, \ldots, t_n$ (during the trading day) that are not equally spaced, we may just relabel the times as $1, 2, \ldots, n$ and treat the resulting price series as equally spaced. The logic here is that during trading, **time is deformed** so that consecutive transactions are considered to be equally spaced in time.

The alternative assumption to trading time is called **calendar time**.

Definition 7 *Calendar time hypothesis. Valuations of assets change continuously and returns are generated in calendar time.*

In this case at the daily frequency we may observe a sequence of prices $P_1, P_2, P_3, P_4, P_5, P_8, P_9, \ldots$. We do not observe P_6 or P_7 and so we do not have a full record of the daily returns. Alternatively, we may say we have a series that contains some daily returns and some three day returns etc. In that case, we may need to make an adjustment to the Friday to Monday return to make it comparable with the Monday to Tuesday return. This requires some modelling assumptions, which we discuss below.

The k-period (gross) return relates to the one-period returns as follows

$$1 + R_{t,t+K} = \mathcal{R}_{t,t+K} = \frac{P_{t+K}}{P_t} = \mathcal{R}_{t+K} \times \cdots \times \mathcal{R}_t = (1 + R_{t+K-1,1}) \times \cdots \times (1 + R_{t,1}). \tag{1.1}$$

In the presence of dividends in intervening time periods, we use $\mathcal{R}_s = (P_s + D_s)/P_{s-1}$ in (1.1) for $s = t+1, \ldots, t+K$.

Multiyear returns are often annualized for easy comparability (likewise daily returns are compounded to give annual figures). This is best understood without the subscripts, so suppose $\mathcal{R}_K$ is the K-period gross return, then the geometric average $(\mathcal{R}_K)^{1/K}$ is the implied one period constant gross return. If $\mathcal{R}$ is the daily gross return, then $\mathcal{R}^K$ is the K-period gross return. We may also ask what is the implied per period return that if compounded over K periods gives the one period return. This is the

1.4 Financial Returns

value $r_{K,1}$ that satisfies

$$1 + R_1 = \left(1 + \frac{r_{K,1}}{K}\right)^K. \tag{1.2}$$

If one takes $K \to \infty$ one obtains the continuously compounded rate of return, which we denote by r, because

$$\lim_{K \to \infty} \left(1 + \frac{r}{K}\right)^K = \exp(r),$$

which you know from High School.

Definition 8 *The continuously compounded return or log return over the interval $[t-1, t]$ is*

$$r_t \equiv \log \mathcal{R}_t = p_t - p_{t-1},$$

where $p_t \equiv \log P_t$. The multiperiod log return is

$$r_{t,t+k} = \log \mathcal{R}_{t,t+k} = \log(\mathcal{R}_{t-k+1} \times \mathcal{R}_{t-k+2} \times \cdots \times \mathcal{R}_t) = r_{t-k+1} + r_{t-k+2} + \cdots + r_t.$$

Using logarithmic returns it is easy to annualize daily returns. Common practice is to multiply average daily returns by 252 to get an average annual return, since there are roughly 252 trading days in the NYSE trading year. (This is based on the calculation $(365.25 \times (5/7)) - 9$, where there are currently 9 public holidays per year. The current primary opening hours of the NYSE are 9.30am–4pm, although there are three days close to holiday periods with shorter hours of 9.30am–1pm.).

We are often concerned with portfolios of returns or stock indexes. Most indexes are weighted averages of prices (although the Value Line index was a weighted average of logarithmic prices; see Harris (2003)).

Definition 9 *Index values are defined on a set of stocks $j = 1, \ldots, n$ based on some weighting scheme $\{w_{jt}\}$ that is generally time varying*

$$I_t = \sum_{j=1}^{n} w_{jt} P_{jt}.$$

For example, $w_{jt} = 1/n$ corresponds to an equal weighted index.

The Dow Jones Industrial Average (DJIA), which consists of 30 "bluest of the blue chips," is a **price-weighted index**, that is, (sum of security prices)/divisor. The amount invested in each security included in the index is proportional to its price. This measures the return on a portfolio consisting of one unit of each security. In a **value-weighted index** (also known as cap-weighted) the amount invested in each security is proportional to its market capitalization (the price × quantity of tradeable shares). The quantity may not be the number of securities outstanding but the so-called free float (excluding securities that are not available for public trading if, e.g., held by the government or founding families). The S&P500 index is constructed using value weighting, and we give some more details on this important index. The formula to calculate the S&P500 is

$$I_t = \frac{\sum_{i=1}^{n} P_{it} Q_{it}}{Divisor}, \qquad (1.3)$$

where Q_{it} is the number of shares of stock i that are used in the calculation and P_{it} is the price of each unit. The formula is created by a modification of the **Laspeyres index**, which uses base period quantities (share counts) to calculate the price change. In the modification to the Laspeyres index, the quantity measure in the numerator is replaced by the current quantity, so the numerator becomes a measure of the current market value, and the product in the denominator is replaced by the divisor that both represents the initial market value and sets the base value for the index. An index is not exactly the same as a portfolio. For instance, when a stock is added to or deleted from an index, which happens quite regularly for the S&P500, the index level should not jump up or drop down. By comparison, a portfolio's value would usually change as stocks are swapped in and out. To assure that the index's value, or level, does not change when stocks are added or deleted, the divisor is adjusted to offset the change in market value of the index. Thus, the divisor plays a critical role in the index's ability to provide a continuous measure of market valuation when faced with changes to the stocks included in the index. In a similar manner, some corporate actions that cause changes in the market value of the stocks in an index should not be reflected in the index level. Adjustments are made to the divisor to eliminate the impact of these corporate actions. The S&P500 index reflects capital gain. There is also a total return S&P500 index that also reflects the income flow generated by dividends.

Definition 10 *The dividend yield on an index with current weights w_i is the weighted average of the cash dividends divided by the weighted average of the current prices (the current level of the index).*

The Shiller website listed in the data appendix gives several versions of the yield on the S&P500. The Wilshire 5000 (tracks over 6,000 NYSE, NASDAQ and Amex stocks) is equally-weighted. The amount invested in each security is the same. Examples of non-US indices are: FTSE 100 (UK, cap-weighted, about 80% of the LSE), Nikkei 225 (Japan, price-weighted), DAX (Germany), TSX (Canada), and MSCI (Morgan Stanley Capital International) international indices. Note that the return we calculate on an index does not correspond to the return of a buy and hold strategy, because of the constant rebalancing that is needed as individual prices change.

Prices can only take certain values, not all real numbers. In fact, prices of individual stocks lie in a discrete grid determined by the **tick size** (for example, this is 1 cent for all US stocks over $1 in value). Specifically, $P_t \in \{0, 0.01, 0.02, \ldots\}$. This means that prices can only change between consecutive transactions or quote updates by some multiple of 1 cent, if at all, so that

$$P_t - P_{t-1} \in \{0, \pm 0.01, \pm 0.02, \ldots\}.$$

1.4 Financial Returns

The question is: how much should this feature affect how we model prices and returns? The answer partly depends on the frequency of the data, because for low frequency there may be many intermediate transactions that have taken place between consecutive prices in the series so that the price grid observed in the data contains many distinct points (and could therefore be well approximated by a continuous random variable). On the other hand, consecutive transactions typically only take a small number of increment values, e.g., $0, \pm 1$, that is, for **high frequency data**, discreteness may be a dominant feature. (It is common practice (which we will uphold) to call any intraday data high frequency. Data frequency of a month or more is usually called low frequency. Perhaps there is room for a further terminology, **medium frequency**, to refer to daily frequency.) Nevertheless, most analysis is made assuming continuously distributed random variables and the discreteness issue is ignored. There are several justifications for this. Returns $(P_t - P_{t-1})/P_{t-1}$ or continuously compounded returns $p_t - p_{t-1}$ can appear to be almost continuous, even when $P_t - P_{t-1}$ is discrete. Furthermore, for indexes, discreteness is also less of an issue, because if there are many stocks the weighted increments to the stock index can take a large number of values.

1.4.2 Returns versus Log Returns

A mathematically convenient feature of continuously compounded (log) returns is that they can take any value on the whole real line, $r_t \in \mathbb{R}$, whereas actual returns are limited from below by limited liability, i.e., the fact that you can't lose more than your stake means that $R_t \geq -1$. Therefore, r_t is logically consistent with a normal distribution, whereas R_t is not consistent with a normal distribution. Since $r = \log(1 + R)$, dropping subscripts, we have $r \leq R$ for all values (of R). This follows because $\log(1 + R) - R$ is a strictly concave function of R with the global maximum achieved at $R = 0$. We can think of r as the tangent curve to the reverse relationship $R = \exp(r) - 1$ at the point $r = R = 0$, and so for values of r, R close to zero, the two return measures give similar values. On the other hand, this linear approximation works terribly for large values of returns such as accrue over long horizons or when bankruptcy or near bankruptcy is an issue (returns are close to minus one).

Theorem 1 *Logarithmic returns have the property of time additivity*

$$r_{t,t+k} = r_t + r_{t+1} + \cdots + r_{t-k+1}.$$

This is very convenient when it comes to statistical testing. Weekly returns are the sum of the five daily returns. This is not true for actual returns, i.e., $R_{t,t+k} \neq R_t + R_{t+1} + \cdots + R_{t-k+1}$.

Example 3 *Suppose that you gain 10% and lose 10%, then you are down 1%*

$$(1 + 0.1) \times (1 - 0.1) = 0.99,$$

but log returns says you are evens.

Theorem 2 *Actual returns have the property of portfolio additivity, i.e., for portfolio weights $w_{1t}, w_{2t}, \ldots, w_{nt}$ fixed at time t,*

$$R_t(w) = w_{1t} R_{1t} + w_{2t} R_{2t} + \cdots + w_{nt} R_{nt}. \tag{1.4}$$

Proof. Let n_i be the number of shares purchased in asset $i = 1, 2$, and let $V_t = n_1 P_{1t} + n_2 P_{2t}$ be the amount invested at time t, and let $w_{it} = n_i P_{it}/V_t$ be the weight of the portfolio invested in asset i at time t. The value at time $t+1$ is given by

$$\begin{aligned}
V_{t+1} &= n_1 P_{1,t+1} + n_2 P_{2,t+1} \\
&= \frac{w_{1t} V_t}{P_{1t}} P_{1,t+1} + \frac{w_{2t} V_t}{P_{2t}} P_{2,t+1} \\
&= \frac{w_{1t} V_t}{P_{1t}} P_{1t}(1 + R_{1,t+1}) + \frac{w_{2t} V_t}{P_{2t}} P_{2t}(1 + R_{2,t+1}) \\
&= w_{1t} V_t (1 + R_{1,t+1}) + w_{2t} V_t (1 + R_{2,t+1}) \\
&= V_t (1 + w_{1t} R_{1,t+1} + w_{2t} R_{2,t+1}),
\end{aligned}$$

from which we obtain that

$$\frac{V_{t+1} - V_t}{V_t} = w_{1t} R_{1,t+1} + w_{2t} R_{2,t+1}.$$

Logarithmic returns are not portfolio additive because

$$r_t(w) = \log\left(\frac{w_{1t} P_{1t} + \cdots + w_{nt} P_{nt}}{w_{1t} P_{1,t-1} + \cdots + w_{nt} P_{n,t-1}}\right) \neq w_{1t} r_{1t} + \cdots + w_{nt} r_{nt},$$

although in practice they may be approximately so over short horizons.

When we come to calculate multiperiod returns, neither logarithmic returns nor actual returns are multiperiod portfolio additive. There are several commonly used ways (which are approximately equivalent when returns are small) to compute multiperiod portfolio returns.

Definition 11 *For simplicity just consider the equally weighted case, $w_{it} = 1/n$ for $i = 1, \ldots, n$. The arithmetic, buy and hold, and rebalanced returns, are respectively:*

$$\overline{\mathcal{R}}_{AR} = \frac{1}{n\tau} \sum_{i=1}^{n} \sum_{t=1}^{\tau} \mathcal{R}_{it} \quad ; \quad \overline{\mathcal{R}}_{BH} = \frac{1}{n} \sum_{i=1}^{n} \left(\prod_{t=1}^{\tau} \mathcal{R}_{it}\right) \quad ; \quad \overline{\mathcal{R}}_{RB} = \prod_{t=1}^{\tau} \left(\frac{1}{n} \sum_{i=1}^{n} \mathcal{R}_{it}\right). \tag{1.5}$$

Roll (1983) showed how the choice of a multiperiod return can matter in practice for estimation of the small firm premium, which we will discuss in Chapters 7 and 8. He finds that the annual small firm premium was 7.5% using the buy and hold method but nearly 14% using the other two methods.

1.4 Financial Returns

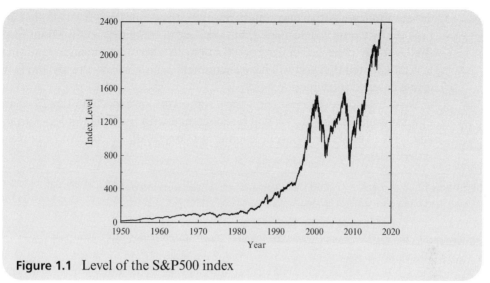

Figure 1.1 Level of the S&P500 index

We next consider the returns on the daily S&P500 index over the past 60 years.

Example 4 *We consider the daily S&P500 index from 19500103 to 20170321, a total of $n = 16,913$ daily prices. Figure 1.1 shows the time series. The dominant feature of the price sequence shown below is the substantial increase over time. We have $P(1) = 16.66$ and $P(n) = 2344.02$, and so the full return over the period is*

$$R_{1,n} = \frac{P(n) - P(1)}{P(1)} = 139.0.$$

On the other hand, using the logarithmic definition of a multiperiod return

$$r_{1,n} = \log(P(n)) - \log(P(1)) = 4.947.$$

This shows the substantial difference between the actual return and the logarithmic return (in this case just the capital gain) over the long horizon. Which would you prefer? Exactly. The annualized returns are $r_{year} = 255 \times 4.947/16913 = 0.0746$, whereas the annualized $R_{year} = \log(139)/66 = 0.075$, which are almost identical.

By comparison, annual US Gross Domestic Product rose from $0.300 trillion in 1950 to $18.569 trillions in 2016, a growth of around 61 fold. The consumer price index (CPI) rose from 24.1 in 1950 to 240.01 in 2016 (annual averages), around a 10 fold increase. So the growth of the level of the stock market (and this indeed only includes capital gain) is substantial indeed over the long horizon.

For historical reference, Malthus (1798) argued that population growth was exponential (R), whereas food production technology grew at a linear rate (r).

1.4.3 Returns as Random Variables and their Properties

We treat returns as random variables, that is, their realization is not known beforehand but the rules of what values they may take are set and understandable. There is some debate about **Knightian uncertainty**, whereby these rules may not be fully known at the time the outcome is determined, but we have nothing new to say about this here. We shall work with the well understood framework where outcomes are random variables. It is anticipated that you will have encountered this concept already and so we do not give a detailed introduction here.

We will suppose that you know the notation for the expected value $E(X)$ and $\mathrm{var}(X)$ of a random variable X. We will be concerned with multiple random variables and a key issue is how they relate to each other. We recall the concept of correlation and independence.

Definition 12 *We say that random variables X and Y are uncorrelated if $\mathrm{cov}(X,Y) = 0$.*

Definition 13 *We say that random variables X and Y are mutually independent if their joint c.d.f. satisfies*

$$\Pr(X \leq x, Y \leq y) = F_{X,Y}(x,y) = F_X(x) \times F_Y(y) = \Pr(X \leq x) \times \Pr(Y \leq y)$$

for all $x, y \in \mathbb{R}$. We say that random variables X and Y are identically distributed if $F_X(x) = F_Y(x)$ for all $x \in \mathbb{R}$. We say that random variables X and Y are independent and identically distributed if both these conditions hold.

If X, Y are jointly normally distributed, then they are uncorrelated if and only if they are independent. But this is not true more generally. Independence implies uncorrelatedness, but the converse is not true.

We first compare returns with logarithmic returns, considering them as random variables. Note that by **Jensen's inequality**

$$E(r) = E\left(\log(1+R)\right) \leq \log(1 + E(R)),$$

which gives a tighter bound on $E(r)$ that is than the bound $E(R)$ implied by the ordering $r \leq R$, because $\log(1 + E(R)) \leq E(R)$ with strict whenever $E(R) \neq 0$.

We will think of returns as being random variables that may take a wide range of values. Even in cases where returns can be quite large, expected returns may be quite small and in particular $-1 < E(R) \leq 1$ so that we may approximate

$$\log(1 + E(R)) = E(R) - \frac{1}{2}E^2(R) + \frac{1}{3}E^3(R) \cdots$$

in terms that are getting smaller and smaller, and in that case $\log(1 + E(R))$ is close to $E(R)$. We may further apply a **Taylor approximation** to the function $\log(1+R)$ around the point $\log(1 + E(R))$ to obtain

$$E(r) = E\left(\log(1+R)\right) = \log(1 + E(R)) + E\left(R - E(R)\right) \frac{1}{1 + E(R)}$$

$$- \frac{1}{2} E\left(R - E(R)\right)^2 \frac{1}{(1 + E(R))^2} + \cdots$$

$$\simeq E(R) - \frac{1}{2}\mathrm{var}(R). \tag{1.6}$$

1.4 Financial Returns

The approximation is good provided $\Pr(-1 < R \leq 1)$ is very close to one and the third moments exist and are even smaller than $E(R)$ and $\mathrm{var}(R)$, something that is questionable in practice; see Chapter 14 for more discussion.

The expected value of returns is sometimes called just return as this is a central quantity of interest to investors. Also of interest is the variance of returns, which is often called the risk. Risk averse investors would prefer less risk and more return. We suppose that there is also a riskless asset with return R_f. This is non-random over the period specified, i.e. the payoff is known at the time of purchase. The **Sharpe ratio** gives a simple risk adjusted rate of return that is easy to understand and analyze.

Definition 14 *The Sharpe ratio of asset with return R when the riskless asset return is R_f is*

$$S = \frac{E(R) - R_f}{\sqrt{\mathrm{var}(R)}}.$$

The Sharpe ratio is invariant to scaling of excess returns, meaning that if $R - R_f \to c(R - R_f)$ for some $c > 0$, then the Sharpe ratio S is the same. It is widely used to measure the performance of an investment relative to the benchmark provided by the risk free rate. One may prefer to compare performance relative to an alternative benchmark. The **information ratio** of a security with payoff R relative to a benchmark security with return R_b is

$$IR = \frac{E(R - R_b)}{\sqrt{\mathrm{var}(R - R_b)}}.$$

This quantity is widely used in portfolio management (Grinold and Kahn (1999)).

Example 5 *Consider a casino roulette wheel with 37 slots, 18 red, 18 black, and 1 void, and suppose that you bet on black. We can describe the gain per play of the Casino through the random variable X, where*

$$X = \begin{cases} 1 & 19/37 \\ -1 & 18/37. \end{cases}$$

The expected rate of return for the Casino is $1/37$. The variance per play is

$$E(X^2) - E^2(X) = 1 - \frac{1}{37^2} \simeq 1,$$

which gives a Sharpe ratio per play of around 0.027 (we may assume $R_f = 0$ here). Compare this with stock returns. For the S&P500 from 1950–2014, the daily Sharpe ratio is around 0.035, while the two day ratio is 0.049, and the three day ratio is 0.061.

We next discuss some specific distributions that are used to describe random returns. The gross return satisfies $\mathcal{R}_t \geq 0$ due to limited liability, so that normality of gross returns is not logically possible. Also, even if $\mathcal{R}_t$ were normal, then $\mathcal{R}_{t,t+k}$ would not be normal, since a product of normal variables is not normal. On the other hand, if log returns r_t are normal, then the multiperiod continuously compounded return is normal as well (a sum of normals is normal). The simple return is lognormal if and only if the continuously compounded return is normal, i.e.

$$\mathcal{R} = 1 + R \sim \log N(\mu, \sigma^2) \text{ if and only if } r = \log(1 + R) \sim N(\mu, \sigma^2).$$

Note that in this case $E(\mathcal{R}) = \exp(\mu + \sigma^2/2)$ and $E(r) = E(R) - \text{var}(R)/2$, and so the approximation in (1.6) is exact in this case. The lognormal model is used, for example, in standard continuous-time finance (see Chapter 12). Empirically it is not satisfactory, so we need to consider distributions other than normal and log normal such as the Pareto, which are treated in Chapter 14.

We may measure departures from normality through the **cumulants**. **Skewness** is a measure of the asymmetry of the distribution and **Kurtosis** is a measure of the peakedness of the distribution or its heavy tailedness. These quantities are defined below:

$$\kappa_3 \equiv E\left[\frac{(r-\mu)^3}{\sigma^3}\right] \quad ; \quad \kappa_4 \equiv E\left[\frac{(r-\mu)^4}{\sigma^4}\right]. \tag{1.7}$$

For a normal random variable, $\kappa_3 = 0$ and $\kappa_4 = 3$. Some authors work with **excess kurtosis**, which is $\kappa_4 - 3$. If a distribution has positive excess kurtosis (leptokurtosis), the tails of the distribution are fatter than those of a normal distribution. If a distribution has negative excess kurtosis (platykurtosis), then the tails of the distribution are thinner than those of a normal distribution. Platypuses are quite rare animals, and platykurtotic distributions are also rather rare in practice.

We will generally use frequentist statistical methodology to infer answers to questions about hypothetical populations from samples drawn from that population. Suppose we have a sample of returns $\{r_1, \ldots, r_T\}$ drawn from the population. The sample cumulants (mean, variance, skewness and kurtosis) are:

$$\bar{r} = \frac{1}{T}\sum_{t=1}^{T} r_t \quad ; \quad s^2 = \frac{1}{T}\sum_{t=1}^{T}(r_t - \bar{r})^2.$$

$$\widehat{\kappa}_3 = \frac{\frac{1}{T}\sum_{t=1}^{T}(r_t - \bar{r})^3}{\left(\frac{1}{T}\sum_{t=1}^{T}(r_t - \bar{r})^2\right)^{3/2}} \quad ; \quad \widehat{\kappa}_4 = \frac{\frac{1}{T}\sum_{t=1}^{T}(r_t - \bar{r})^4}{\left(\frac{1}{T}\sum_{t=1}^{T}(r_t - \bar{r})^2\right)^{2}}.$$

Empirically, for daily logarithmic returns, there is weak evidence of non zero skewness (negative for indices, positive or close to zero for individual stocks) and strong evidence of large positive excess kurtosis (both indices and individual stocks). It is time to examine some data. In Table 1.2 we give the sample moments and cumulants of the S&P500 daily stock index return series from 1950–2017 in percentage terms.

In Figure 1.2 we show the time series plot of daily returns. The main features are the extreme negative value in October 1987, which was around 21 standard deviation units. According to the normal distribution the likelihood of such an event would be **machine zero**, that is so small as to be indistinguishable from zero by most computer

1.4 Financial Returns

Table 1.2 Descriptive statistics 1950–2017

	Daily	Weekly	Monthly	Annual
Mean	0.0339	0.1619	0.6985	8.5967
St. Deviation	0.9653	2.0585	4.1278	15.341
Diameter of c.i.	0.0145	0.0681	0.2850	3.701
Skewness	−0.6410	−0.3531	−0.4185	−0.4956
Excess Kurtosis	20.8574	4.9182	1.7371	0.4864
Minimum	−20.467	−18.195	−21.763	−40.091
Maximum	11.580	14.116	16.305	40.452
Number Zeros	125	6	1	0

Descriptive statistics for the returns on the S&P500 index for the period 1950–2017 for four different data frequencies. Total return over the full period is 139.70. Values in first three rows multiplied by 100. Minimum and maximum are measured in standard deviations and from the mean. The third row contains the diameter of the 95% confidence interval (c.i.) for the percentage mean of returns, $1.96 \times \sqrt{\frac{s^2}{n}}$, where s^2 is the standard deviation of returns. We are assuming here that returns are i.i.d.; more discussion of whether this assumption is appropriate will be given in Chapters 3 and 11.

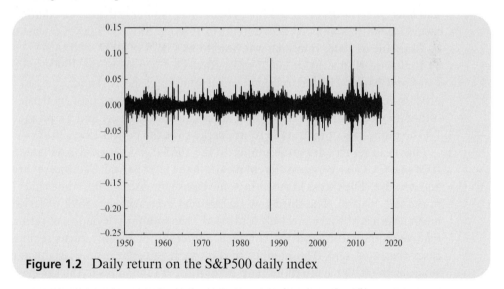

Figure 1.2 Daily return on the S&P500 daily index

calculations. This is why a lot of research has gone into developing models for returns in which such events are not so unlikely and can be given some rationale.

In Table 1.3 we show the figures for the CRSP (Center for Research in Security Prices) total value weighted return. This is a broader index than the S&P500 and includes dividend income by reinvestment. This is in percentages and is return minus the risk free rate. The total return over the period 1950–2017 was 1,213, if you had

Table 1.3 Daily CRSP market returns descriptive statistics

	Daily	Weekly	Monthly	Annual
Mean	0.04618	0.2247	0.9767	12.608
St. Deviation	0.9302	2.0583	4.2427	17.560
Diameter of c.i.	0.0135	0.0682	0.2930	4.1924
Skewness	−0.5669	−0.493	−0.5085	−0.4249
Excess Kurtosis	16.5957	5.6315	1.9043	2.9068–3
Minimum	−17.4130	−17.980	−22.640	−36.7491
Maximum	11.3540	13.586	16.610	50.193
Number Zeros	23	1	1	0

Total ruturn over the full period is 1213.06.

invested a thousand dollars in 1950, you would now be more than a millionaire, before taxes and fees anyway.

We also investigated the individual stocks whose properties are described in Table 1.1 in CLM. Nearly all firms have a relatively short lifetime and ultimately end up bankrupt or absorbed into some other entity, which raises the issue of **survivorship bias**, which can affect the interpretation of some statistical calculations. In this case, General Signal Corp was bought in 1998 by SPX, Wrigley Co was taken over by Mars Inc in 2008, Interlake was bought by GKN in 1999, Garan Inc was bought by Berkshire Hathaway in 2002, while Raytech Corp spent the 1990s under administration **Chapter 11** (due to litigation surrounding asbestos) and emerged from that process in 2001 under the condition that 90% of its stock be held by trustees against future litigation. The remaining four stocks are still trading, and we provide in Table 1.4 the properties of their daily return series over the period 1995–2017.

The mean returns are quite similar to the 1962–1994 period for AP and EGN, but IBM and CUO have quite different means in the later period. Skewnesses are all small and positive. The excess kurtosis in some cases increases and in some cases decreases in the later period. The third row reveals that even with $n = 5592$ observations, the mean return is not very precisely estimated – the standard deviation of returns is large relative to the mean, and the confidence interval for mean returns includes in two cases the value zero. The number of days with zero returns varies substantially by stock, with CUO having more than 50% of the trading days in the sample returning the same closing price as the previous day. The total return over the period also varies substantially and tells a slightly different story – the buy and hold investor would have much preferred CUO to AP even though its mean daily return is a little lower, because this is partly caused by the many zeros in the daily return series.

We also consider two other financial time series, the daily EURUSD (euro/US dollar) exchange rate and the daily return on the three month T-bill rate. In Figure 1.3 the exchange rate series is notably different from the stock price series, even accounting

1.4 Financial Returns

Table 1.4 Daily returns on individual stocks descriptive statistics, 1995–2017

	IBM	AP	EGN	CUO
Mean ($\times 100$)	0.0615	0.0556	0.0723	0.0529
St. Deviation ($\times 100$)	1.770	2.713	2.216	2.269
Diameter of c.i.	0.0464	0.0711	0.0582	0.0595
Skewness	0.189	0.319	0.422	0.497
Excess Kurtosis	7.321	6.569	8.000	16.10
Minimum ($\times 100$)	−15.54	−22.83	−16.25	−18.68
Maximum ($\times 100$)	13.16	18.28	22.26	22.26
Number Zeros	39	496	236	3009
Total Return	11.977	1.887	13.445	3.594

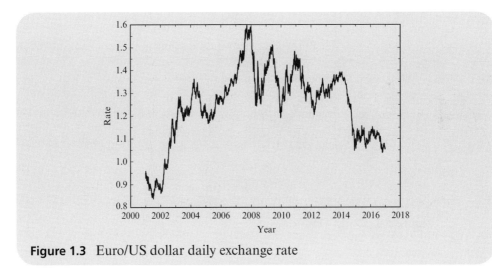

Figure 1.3 Euro/US dollar daily exchange rate

for the much shorter time frame. Specifically, there is very little growth in the series over the twenty year period, around 10% appreciation. We show the return in Figure 1.4. The rate goes up and it comes down. The mean daily return is very small, 0.00494, in percentage terms, and the standard deviation is large, 0.6636. There is some positive skewness and excess kurtosis but overall it is closer to the normal distribution than daily stock returns. This is not true for all currency pairs, so that some rates display substantial appreciation or depreciation over time. Note also that in comparing exchange rates as investments one also has to look at the relative interest rates in the two countries, as the **carry trade** involves borrowing at low interest rates in one currency and then lending at high interest rates in another currency. That is, the total

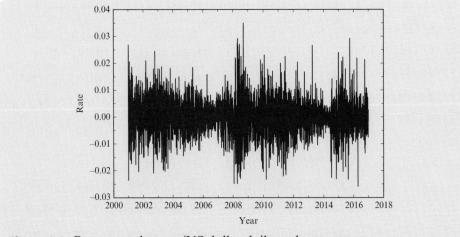

Figure 1.4 Return on the euro/US dollar daily exchange rate

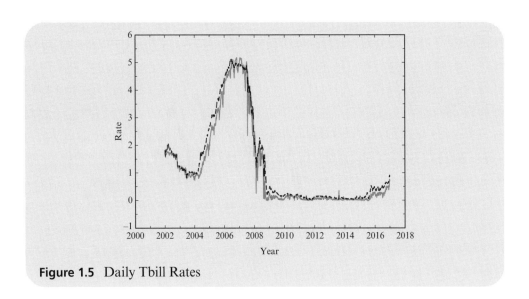

Figure 1.5 Daily Tbill Rates

return is $r_H - r_L + R$, where R is the appreciation of the currency, while r_H, r_L are the two interest rates.

In Figure 1.5 we examine the daily one month and one year T-bill interest rate series over the period 2002–2017. The series looks quite unusual, especially since 2009 as both have been close to zero.

The price of oil has been an important bellwether of industrialized economies. The price of oil was pretty low and varied very little from year to year until the OPEC price hike of 1974 as shown in Figure 1.7. Thereafter the price has been determined partly by financial market activity and partly by the production decisions of OPEC and other nations. The daily return series (from 1986) shown in Figure 1.8 look very similar to the stock return series.

1.5 Risk Aversion

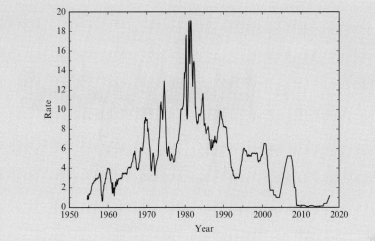

Figure 1.6 Daily Federal Funds rate annualized

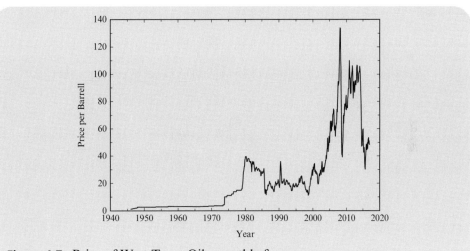

Figure 1.7 Price of West Texas Oil, monthly frequency

1.5 Risk Aversion

As your grandma probably told you, stock market investing is gambling, so we turn to the standard economic theory concerning gambling and insurance, or rather choice under uncertainty. We represent the random outcome of a gamble by the random variable X.

Definition 15 *A fair gamble is a gamble that is (actuarially) fair, i.e., is zero in expectation, $E(X) = 0$.*

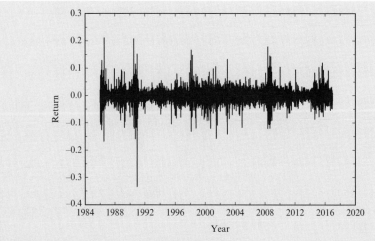

Figure 1.8 Daily return on West Texas Oil

Example 6 *Roulette again. If you bet $k on red, and red comes up, you receive $2k in return (and no taxes), if red does not come up, then you lose your stake. Let X be your gain, then*

$$X = \begin{cases} k & \text{with prob } 18/37 \\ -k & \text{with prob } 19/37. \end{cases}$$

Therefore,

$$E(X) = k\frac{18}{37} + (-k)\frac{19}{37} = \frac{-k}{37}.$$

This gamble is not fair. The expected rate of return for the Casino is 1/37 per play. For n (independent) plays, the variance is approximately 1/n, which gets very small very rapidly as n increases.

We wish to study rational choice among alternative risky prospects. Utility is not a primitive in economic theory; what is assumed is that each consumer can "value" goods in terms of subjective preferences. It is assumed that preferences satisfy certain axioms, and this implies that they can be represented by a utility function. The theory was developed by John von Neumann and Oskar Morgenstern in the 1940s. It is a formalization and reinterpretation of what Daniel Bernoulli did in the eighteenth century to resolve the **St Petersburg Paradox**.

1.5 Risk Aversion

Example 7 *St Petersburg Paradox. Consider the following game of chance: you pay a fixed fee to enter and then a fair coin is tossed repeatedly until a tail appears, ending the game. The pot starts at 1 dollar and is doubled every time a head appears. You win whatever is in the pot after the game ends. Thus you win 1 dollar if a tail appears on the first toss, 2 dollars if a head appears on the first toss and a tail on the second, 4 dollars if a head appears on the first two tosses and a tail on the third, etc. What would be a fair price P to pay for entering the game? The expected value is*

$$P = \frac{1}{2} \times 1 + \frac{1}{4} \times 2 + \frac{1}{8} \times 4 + \cdots = \infty.$$

This seems counterintuitive. Certainly, no-one is going to pay this. Bernoulli solved this problem by computing the logarithm of earnings (sort of utility) instead and taking its expectation

$$\frac{1}{2} \times \log 1 + \frac{1}{4} \times \log 2 + \frac{1}{8} \times \log 4 + \cdots = \sum_{k=1}^{\infty} \frac{k-1}{2^{k-1}} = \log 2 < \infty.$$

An alternative way of looking at the St Petersburg game, is to look at the problem of bankruptcy for the other side of the bet given that they have finite capital. Could they really pay out $\$2^{100}$ if there were 99 heads in a row and then a tail? The following example illustrates the bankruptcy question.

Example 8 *Suppose that you have $1 and bet $1 on red each time or until you become devoid of funds. What is the probability you become bankrupt? Let p be the probability of a red. Let $P_{j,k}$ be the probability of going from $j to $k. Now note that*

$$P_{1,0} = 1 - p + pP_{2,0}.$$

Furthermore, by a symmetry argument $P_{2,1} = P_{1,0}$ and by independence $P_{2,0} = P_{2,1}P_{1,0}$, so that $P_{2,0} = P_{1,0}^2$. Therefore

$$P_{1,0} = 1 - p + pP_{1,0}^2,$$

and the quadratic equation has solutions

$$P_{1,0} = 1; \quad P_{1,0} = \frac{1-p}{p}.$$

The first solution is relevant when $p \leq 1/2$, while the second is relevant otherwise. When you go to a casino, $p < 1/2$, which says that you are eventually devoid of funds with probability one.

We now consider expected utility theory (EUT) in more detail. Suppose that $v(x) = a + b \times u(x)$, where a and $b > 0$ are constant. Then for any gambles X and Y, we clearly have

$$E(v(X)) \geq E(v(Y)) \text{ if and only if } E(u(X)) \geq E(u(Y)).$$

The ordinal utility function representing preferences under certainty is invariant to positive increasing transformations. In contrast, the von Neumann–Morgenstern (VNM) utility function in the EU framework is invariant only to positive affine transformations. If v is a positive increasing but nonaffine (e.g. quadratic or cubic) transformation of u, the two may give rise to different choices (i.e. they do not represent the same preferences).

Definition 16 *An individual is risk-averse if she dislikes a fair gamble (she will not accept it for free). Similarly, she is risk-seeking/loving if she likes it and risk-neutral if she is indifferent about it.*

Theorem 3 *A decision maker with utility function u is locally (globally) risk-averse if and only if u is strictly concave at a given wealth level (all wealth levels).*

Proof. Let X be a fair gamble. We have strict concavity if and only if

$$E(u(W+X)) < u(W+E(X)) = u(W)$$

by Jensen's inequality, where W is a given wealth level. The local case is for given $W > 0$, the global case is for all $W > 0$.

Definition 17 *The compensatory risk premium Π_c is defined by*

$$E(u(W + \Pi_c + X)) = u(W).$$

It is the amount of money that compensates the individual for accepting the gamble X.

Definition 18 *The insurance risk premium Π_i is defined by*

$$E(u(W+X)) = u(W - \Pi_i).$$

It is the amount of money that the individual is willing to pay to avoid the gamble X.

Definition 19 *The Certainty equivalent (CE) of gamble X is defined by*

$$E(u(W+X)) = u(W + CE).$$

It is the amount of money that gives the individual the same welfare as the gamble. Clearly, $CE = -\Pi_i$. If the individual is risk-averse, then $CE < E(X)$. Also, $W^{CE} \equiv W + CE$ is the certainty equivalent wealth of the wealth prospect $\tilde{W} \equiv W + X$. This says what certain wealth level the individual considers just as satisfying as the risky wealth prospect. The risk premium is then $\Pi = E(\tilde{W}) - W^{CE}$.

1.5 Risk Aversion

Theorem 4 *Suppose that X is a given gamble. Then, we can relate the compensatory and insurance premium as follows:*

$$\Pi_c\left(W - \Pi_i(W)\right) = \Pi_i(W);$$

$$\Pi_c(W) = \Pi_i\left(W + \Pi_c(W)\right).$$

Proof. First, we have $E(u(W+X)) = E(u(W - \Pi_i + \Pi_i + X)) = u(W - \Pi_i)$. Setting $W' \equiv W - \Pi_i$, we get $E(u(W' + \Pi_i + X)) = u(W')$. Second, $E(u(W + \Pi_c + X)) = u(W + \Pi_c - \Pi_c)$. Setting $W'' \equiv W + \Pi_c$, we get $E(u(W'' + X)) = u(W'' - \Pi_c)$. □

In this proposition X is assumed to be fair. If not, we could write $X = E(X) + \varepsilon$, i.e. we can view X as a package consisting of a certain (as in non-random) amount $E(X)$ and a fair gamble ε. Then we would just redefine the wealth level as $W + E(X)$ so that the risk premia captures the effect of risk exclusively (as is the case of Π reflecting the displeasure of volatility around expected wealth). Here, X is considered a generic gamble, so that $\Pi_i = \Pi - E(X)$.

Arrow (1965) and Pratt (1964) developed a local measure of risk aversion, i.e. applicable for small gambles. Let X be a small fair gamble, i.e. $E(X) = 0$. Then $\Pi = \Pi_i$. Take the first-order Taylor approximation of the right hand side $u(W - \Pi)$ in terms of Π around 0 to obtain

$$u(W - \Pi) \simeq u(W) - u'(W) \times \Pi.$$

We take the second-order Taylor approximation of $u(W + X)$ in X to obtain

$$E\left(u(W+X)\right) \simeq u(W) + u'(W) \times E(X) + \frac{1}{2} \times u''(W) E(X^2).$$

It follows that

$$\Pi \simeq \frac{1}{2} \times \frac{-u''(W)}{u'(W)} E(X^2) \equiv \frac{\sigma^2}{2} \times A(W),$$

where $\sigma^2 = E(X^2) = \text{var}(X)$.

Definition 20 *The Arrow–Pratt function of absolute and relative risk aversion is*

$$A(W) = -\frac{u''(W)}{u'(W)}, \quad \mathcal{R}(W) = W \times A(W). \tag{1.8}$$

Risk aversion is driven by the fact that marginal utility decreases with wealth. The function $A(W) = -(u'(W))'/u'(W)$ is the rate of decay of marginal utility; it says by what percentage marginal utility decreases if wealth is increased by one dollar. $\mathcal{R}(W)$ (also known as proportional risk aversion) is the elasticity of marginal utility with respect to wealth. It says by what percentage marginal utility decreases if wealth is increased by one percent. Note that $A(W)$ and hence $\mathcal{R}(W)$ are invariant to positive affine transformations of the utility function.

Example 9 *Quadratic utility*

$$u(W) = W - \frac{b}{2} \times W^2, \quad \text{for } W < \frac{1}{b}.$$

In this case, $A(W) = b/(1 - bW)$.

Example 10 *CARA (constant absolute risk aversion) utility:*

$$u(W) = \frac{e^{-\alpha W}}{-\alpha},$$

where α is the coefficient of absolute risk aversion, $A(W) = \gamma$.

Example 11 *CRRA (constant relative risk aversion) utility:*

$$u(W) = \begin{cases} \frac{W^{1-\gamma}-1}{1-\gamma} & \text{for } \gamma \neq 1 \\ \log W & \text{for } \gamma = 1, \end{cases}$$

where γ is the coefficient of relative risk aversion, $\mathcal{R}(W) = \gamma$.

The above examples are special cases of HARA (hyperbolic absolute risk aversion) utility

$$u(W) = \begin{cases} \frac{1-\gamma}{\gamma}\left(\frac{aW}{1-\gamma} + b\right)^{\gamma} & \text{for } \gamma \neq 1 \\ \log W & \text{for } \gamma = 1. \end{cases} \quad (1.9)$$

In this case, $u'(W) = a(\frac{aW}{1-\gamma} + b)^{\gamma-1}$ and $u''(W) = -a^2(\frac{aW}{1-\gamma} + b)^{\gamma-2}$ so that the risk tolerance $T(W) = -u'(W)/u''(W) = \alpha W + \beta$ for constants α, β. We have $\mathcal{R}'(W) > 0$ if $b > 0$ and $\mathcal{R}'(W) < 0$ if $b < 0$, which caters to a wide variety of attitudes to risk.

Now we focus on any risk, i.e. not just small ones. The following result characterizes risk aversion.

Theorem 5 *The following three conditions are equivalent:*

(A) Individual 1 is more risk-averse than individual 2, i.e. at any common wealth level, the former requires a larger premium for any risk than the latter;
(B) For all W, $A_1(W) \geq A_2(W)$;
(C) The utility function u_1 is a concave transformation of u_2: there exists ϕ with $\phi' > 0$ and $\phi'' \leq 0$ such that $u_1(W) = \phi(u_2(W))$ for all W.

Proof. We first show that (B) holds if and only if (C) holds. We have:

$$u_1'(W) = \phi'(u_2(W)) \times u_2'(W)$$

$$u_1''(W) = \phi''(u_2(W)) \times (u_2'(W))^2 + \phi'(u_2(W)) \times u''(W),$$

which implies that

$$A_1(W) = A_2(W) - \frac{\phi''(u_2(W))}{\phi'(u_2(W))} \times u_2'(W).$$

1.5 Risk Aversion

Thus, A_1 is uniformly greater than A_2 if and only if ϕ is increasing and concave.

Because $\Pi \simeq \frac{\sigma^2}{2} \times A(W)$, it follows that (A) implies (B). Now it is enough to show that (C) implies (A). Any gamble X gives rise to risk premia Π_1 and Π_2 for individuals 1 and 2, respectively. We have

$$u_1(W - \Pi_1) = E(u_1(W + X)) = E(\phi(u_2(W + X))),$$

and denoting $Y = u_2(W + X)$, Jensen's inequality implies that

$$u_1(W - \Pi_1) = E(\phi(Y)) \leq \phi(E(Y)) = \phi(u_2(W - \Pi_2)) = u_1(W - \Pi_2).$$

Because $u_1' > 0$, we must have $\Pi_1 \geq \Pi_2$. □

1.5.1 The Canonical Portfolio Allocation Problem

We now consider a situation where an individual has to invest their wealth W into different assets, i.e. to choose the allocation of her wealth between them. Suppose there are two assets in the economy – a risky one with (net) return R, where $0 < \mathrm{var}(R) < \infty$, and a riskfree one with return R_f. An investor with wealth W decides what amount to invest in a risky asset. Her final wealth is

$$\tilde{W} = (1 + R_f) \times (W - \omega) + (1 + R) \times \omega = W \times (1 + R_f) + \omega \times (R - R_f),$$

where ω is the amount of initial wealth invested in risky assets. She must solve the portfolio problem

$$\max_{\omega \in \mathbb{R}} V(\omega), \quad \text{where } V(\omega) = E\left(u(\tilde{W})\right).$$

Note that if $\omega \notin [0, W]$, then borrowing or short-selling takes place. By selecting ω, given her preferences, she structures the most desirable distribution of $\tilde{W}$ subject to the investment opportunities.

The first order condition for an interior solution to the problem is

$$V'(\omega^*) = E\left[u'(\tilde{W}) \times (R - R_f)\right] = 0. \tag{1.10}$$

We assume that such a solution exists. Note that since $u' > 0$, we must have $\Pr(R > R_f) \in (0, 1)$. Neither asset can dominate the other. The second order condition is satisfied as the function V is concave (it is a weighted average of concave functions) so that $V''(\omega^*) < 0$.

Theorem 6 *Suppose that $u' > 0$ and $u'' < 0$, and let ω^* be the solution satisfying (1.10). Then*

$$\omega^* > 0 \text{ if and only if } E(R) > R_f.$$

This says: invest a positive fraction of wealth in the risky asset if and only if its expected return is greater than the risk free rate.

Proof. Since $u'' < 0$, we have $V''(\omega) = E(u''(\tilde{W}) \times (R - R_f)^2) < 0$, and so the function $V'(\omega)$ is strictly decreasing. Therefore,

$$V'(0) = E\left(u'(W \times (1 + R_f))\right) \times (E(R) - R_f) \begin{cases} > V'(\omega^*) & \text{if } \omega^* > 0 \\ < V'(\omega^*) & \text{if } \omega^* < 0. \end{cases}$$

So $\omega = 0$ can't be a solution. Likewise $w = W$. Since $u' > 0$, the proof is complete.

We also give an alternative argument because it provides a useful interpretation. Note that

$$E\left(\tilde{W}(\omega)\right) = W \times (1 + R_f) + \omega \times (E(R) - R_f)$$

is greater than $W \times (1 + R_f)$ if and only if ω and $E(R) - R_f$ have the same sign. If $E(\tilde{W}(\omega)) \leq W \times (1 + R_f)$, then $\omega^* = 0$. Write the first order condition as

$$0 = E\left(u'(\tilde{W}(\omega^*))\right) \times (E(R) - R_f) + \text{cov}(u'(\tilde{W}(\omega^*)), R),$$

so that when $E(R) - R_f > 0$, we have $\text{cov}(u'(\tilde{W}(\omega^*)), R) < 0$, which means that you value returns when they are negatively correlated with marginal utility.

Changes of Risk Aversion with Wealth

We next consider how risk aversion changes with wealth. There are several concepts according to whether we are talking about absolute or relative risk aversion and whether the result increases, decreases, or stays the same.

Definition 21 *We abbreviate increasing/constant/decreasing absolute risk aversion ($A(W)$) by IARA, CARA, DARA, respectively. Increasing/constant/decreasing relative risk aversion ($\mathcal{R}(W)$) is abbreviated as IRRA, CRRA, DRRA, respectively.*

Regarding the absolute measure, Arrow argues for DARA, which sounds intuitively reasonable. In such a case, the risky asset is a normal good (as the individual's wealth increases, her demand for it rises; in contrast, it would be an inferior good under IARA). Regarding the relative measure, it seems less clear: IRRA/CRRA/DRRA imply that as wealth increases, the fraction of wealth allocated to the risky asset decreases/remains constant/increases.

Theorem 7 *(Arrow). Let $\omega^*(W)$ solve the portfolio problem (1.10), and let*

$$\eta(W) = \frac{W}{\omega^*} \times \frac{d\omega^*}{dW}(W) = \frac{d\log \omega^*}{d\log W}.$$

1.5 Risk Aversion

Then for all W

$$A'(W) \begin{cases} < 0 \ (i.e.\ DARA) & \text{implies } \frac{d\omega^*}{dW}(W) > 0 \\ = 0 \ (i.e.\ CARA) & \text{implies } \frac{d\omega^*}{dW}(W) = 0 \\ > 0 \ (i.e.\ IARA) & \text{implies } \frac{d\omega^*}{dW}(W) < 0 \end{cases}$$

$$\mathcal{R}'(W) \begin{cases} < 0 (i.e.\ DRRA) & \text{implies } \eta(W) > 1 \\ = 0 (i.e.\ CRRA) & \text{implies } \eta(W) = 1 \\ > 0 (i.e.\ IRRA) & \text{implies } \eta(W) < 1. \end{cases}$$

Proof. We prove the DARA case (the proofs for CARA and IARA are similar). We assume that $E(R) > R_f$ and that $\omega^* > 0$. Define

$$F(W, \omega) \equiv E\left(u'\left(W(1 + R_f) + \omega(R - R_f)\right)(R - R_f)\right),$$

and note that the first order condition implies that $F(W, \omega^*(W)) = 0$.
By the **implicit function theorem** we have

$$\frac{d\omega^*}{dW} = -\frac{\frac{\partial F}{\partial W}}{\frac{\partial F}{\partial \omega^*}} = \frac{(1 + R_f)E\left(u''(\tilde{W})(R - R_f)\right)}{-E\left(u''(\tilde{W})(R - R_f)^2\right)}. \tag{1.11}$$

Since $u'' < 0$, we have

$$\frac{d\omega^*}{dW} > 0 \text{ if and only if } E\left(u''(\tilde{W})(R - R_f)\right) > 0.$$

When $R \geq R_f$, $\tilde{W} \geq W(1 + R_f)$, and so DARA implies that $A(\tilde{W}) \leq A(W(1 + R_f))$. Similarly, when $R < R_f$, $\tilde{W} < W(1 + R_f)$ and hence $A(\tilde{W}) > A(W(1 + R_f))$. In each case, multiplying both sides of $A(\tilde{W})$ by $-u'(\tilde{W})(R - R_f)$ gives:

$$u''(\tilde{W})(R - R_f) \begin{cases} \leq -A(W(1 + R_f))\left(u'(\tilde{W})(R - R_f)\right), & \text{when } R \geq R_f, \\ > -A(W(1 + R_f))(u'(\tilde{W})(R - R_f)), & \text{when } R < R_f. \end{cases} \tag{1.12}$$

Remember that for the first order condition to hold, we had $\Pr(R < R_f) \in (0, 1)$. So the two previous expressions imply that

$$E\left(u''(\tilde{W})(R - R_f)\right) > -A(W(1 + R_f))E(u'(\tilde{W})(R - R_f)).$$

By the first order condition, the expectation on the right hand side is zero. Using the result to sign the expression proves the DARA case.

Now we prove the IRRA case (CRRA and DRRA similar). Let us rewrite

$$\eta = \frac{d\omega^*}{dW}\frac{W}{\omega^*} = 1 + \frac{\frac{d\omega^*}{dW}W - \omega^*}{\omega^*}.$$

Substituting for $\frac{d\omega^*}{dW}$ from (1.11) and rearranging, we obtain

$$\eta - 1 = \frac{(1+R_f)W \times E\left(u''(\tilde{W})(R-R_f)\right) + \omega^* E\left(u''(\tilde{W})(R-R_f)^2\right)}{-\omega^* E\left(u''(\tilde{W})(R-R_f)^2\right)}$$

$$= \frac{E\left(u''(\tilde{W})\tilde{W}(R-R_f)\right)}{-\omega^* E\left(u''(\tilde{W})(R-R_f)^2\right)}.$$

Since $u'' < 0$, we have $\eta > 1$ if and only if $E(u''(\tilde{W})\tilde{W}(R-R_f)) > 0$. Under IRRA, $\mathcal{R}(\tilde{W}) \geq \mathcal{R}(W(1+R_f))$, when $R \geq R_f$ and $\mathcal{R}(\tilde{W}) < \mathcal{R}(W(1+R_f))$ when $R < R_f$. Multiplying both sides by $-u'(\tilde{W})(R-R_f)$ gives

$$u''(\tilde{W})\tilde{W}(R-R_f) \begin{array}{l} \leq -\mathcal{R}(W(1+R_f))u'(\tilde{W})(R-R_f), \text{ when } R \geq R_f, \\ > -\mathcal{R}(W(1+R_f))u'(\tilde{W})(R-R_f), \text{ when } R < R_f. \end{array}$$

Again, this and the first order condition implies that $E(u''(\tilde{W})\tilde{W}(R-R_f)) < 0$, which gives $\eta < 1$. □

Example 12 *Consider the special case of CRRA utility with $W=1$, $R_f=0$, and*

$$R = \begin{cases} 1 & \text{with prob } p \\ -1 & \text{with prob } 1-p. \end{cases}$$

Then wealth is

$$\tilde{W}(\omega) = 1 + \omega R = \begin{cases} 1 + \omega & \text{with prob } p \\ 1 - \omega & \text{with prob } 1-p. \end{cases}$$

The first order condition is

$$\frac{p}{(1+\omega)^\gamma} = \frac{(1-p)}{(1-\omega)^\gamma}.$$

It follows that

$$\omega^* = \frac{x-1}{1+x} \quad ; \quad x = \left(\frac{p}{1-p}\right)^{1/\gamma}.$$

Therefore, provided $p > 1/2$, we have $x > 1$ and $\omega^* > 0$. Since $\omega^* < 1$ always, we have $\omega^* \in (0,1)$ in this case.

1.5.2 Stochastic Dominance

So far we have discussed ranking of investment opportunities on the premise that the utility function is given. We are interested in the circumstances under which one asset can be said to be "better" than another asset for a whole class of investors and

1.5 Risk Aversion

consumers, i.e. for a class of utility functions. This is called **preference-free ranking**; if we cannot rank using such general criteria, we have to make specific assumptions about individual preferences.

Definition 22 *State-by-state dominance. In every state of the world, the payoff of the dominant asset is greater than the payoff of the dominated asset.*

Example 13 *Consider two assets with payoffs X_1 and X_2 and let states be denoted as ω. The following example shows state-by-state dominance of X_1 over X_2*

	ω_1	ω_2
X_1	2	4
X_2	1	3

Definition 23 *Mean-variance (MV) dominance. The dominant asset has a greater mean and a smaller variance than the dominated asset (it is possible for either the means or the variances to be equal if the other moments differ).*

This reflects the fact that individuals like more wealth but dislike its variability; the measure of the latter here is variance. We show that the MV framework has some drawbacks.

Example 14 *An example of MV dominance drawbacks:*

	ω_1	ω_2
X_1	2	6
X_2	1	3

Let each state be equally likely. Then $E(X_1) = 4 > 2 = E(X_2)$ while $var(X_1) = 4 > 1 = var(X_2)$. Thus, although X_1 state-by-state dominates X_2 and clearly nonsatiated individuals would prefer it, there is no MV dominance.

Thus, the mean variance dominance concepts provide only an incomplete ordering or ranking. Rothschild and Stiglitz (1970) note that variance is not an exact measure of risk in the expected utility framework and propose a more appropriate definition of risk. We discuss this further in Chapter 14.

We turn to the concept of **stochastic dominance** as an alternative way of ranking risk prospects (Levy (2006)). Let X_1 and X_2 be two random variables, and let $F_1(x)$ and $F_2(x)$ denote the cumulative distribution functions of X_1, X_2, respectively.

Definition 24 *Let $\mathcal{U}_1$ denote the class of all VNM type utility functions, u, such that u is increasing. Also, let $\mathcal{U}_2$ denote the class of all utility functions in $\mathcal{U}_1$ for which u is concave.*

Theorem 8 X_1 *First order stochastic dominates (FSD) X_2, denoted $X_1 \stackrel{FSD}{\succeq} X_2$, if and only if:*

(1) $E(u(X_1)) \geq E(u(X_2))$ for all $u \in \mathcal{U}_1$; or
(2) $1 - F_1(x) \geq 1 - F_2(x)$ for all x.

Proof. We show that these are equivalent definitions. Suppose that (2) does not hold. Then there exists an x^* such that $F_1(x^*) > F_2(x^*)$. Let $u(x) = 1$ $(x > x^*)$, which is one if and only if $x > x^*$. This is a weakly increasing utility function. Furthermore,

$$E(u(X_1)) = 1 - F_1(x^*) < 1 - F_2(x^*) = E(u(X_2)),$$

which implies that (1) does not hold for this u. To prove the other direction, suppose that (2) holds. Then for all x

$$\Pr(u(X_1) > x) = \Pr\left(X_1 > u^{-1}(x)\right)$$
$$\geq \Pr\left(X_2 > u^{-1}(x)\right)$$
$$= \Pr(u(X_2) > x),$$

which implies that (1) holds. Here, u^{-1} is the inverse (or generalized inverse of a continuous weakly increasing function). □

This means that there is equivalence between saying that all nonsatiated expected utility maximizers prefer asset 1 to asset 2 and saying that for any level of payoff, the probability of a higher payoff is greater with asset 1 than with asset 2. FSD does not mean that the realized payoff is always greater (FSD is not state-by-state dominance). The utility based interpretation is most natural in economics terms, but the characterization in terms of the c.d.f is useful in practical applications, since we can estimate this quantity from data.

We next turn to the weaker concept of second order stochastic dominance.

Theorem 9 X_1 *Second order stochastic dominates (SSD) X_2, denoted $X_1 \stackrel{SSD}{\succeq} X_2$, if and only if one of the following equivalent conditions holds:*

(1) $E(u(X_1)) \geq E(u(X_2))$ for all $u \in \mathcal{U}_2$; or:
(2) $\int_{-\infty}^{x} F_1(t)dt \leq \int_{-\infty}^{x} F_2(t)dt$ for all x.

This means that there is equivalence between saying that all nonsatiated risk averse expected utility maximizers prefer asset 1 to asset 2 and saying that for any level of

1.6 Mean Variance Portfolio Analysis

payoff, the integral of the probability of a higher payoff is greater with asset 1 than with asset 2. We may use stochastic dominance criteria to compare risky payoffs. If X state-by-state dominates Y, then X first order dominates Y, so this criterion does not suffer from the particular defect that afflicts mean variance. However, stochastic dominance theory becomes more complicated in the context of portfolio choice, and for this we will rely on the mean variance framework.

1.6 Mean Variance Portfolio Analysis

We next consider the general mean-variance portfolio choice problem with n risky assets. In this setting the mean variance framework is very simple to apply compared to alternative approaches. The implicit assumption is that investors have a (derived) preference function over just the mean and the variance of the portfolio, which is increasing in mean and decreasing in variance. They have to balance these two contributions to their utility.

1.6.1 Without a Risk Free Asset

Suppose we have n assets with random returns $R_1, \ldots, R_n$ with means $E(R_j) = \mu_j$, variances $var(R_j) = \sigma_{jj}$, and covariances $cov(R_j, R_k) = \sigma_{jk}$. Let $R = (R_1, \ldots, R_n)^\top$ be the $n \times 1$ vector of returns with mean vector and covariance matrix

$$E(R) = \mu = \begin{pmatrix} \mu_1 \\ \vdots \\ \mu_n \end{pmatrix}, \tag{1.13}$$

$$var(R) = E\left((R-\mu)(R-\mu)^\top\right) = \Sigma = \begin{pmatrix} \sigma_{11} & & \\ & \ddots & \sigma_{jk} \\ & & \ddots \\ & & & \sigma_{nn} \end{pmatrix}. \tag{1.14}$$

Let $R(w)$ denote the random portfolio return

$$R(w) = \sum_{j=1}^{n} w_j R_j = w^\top R,$$

where $w = (w_1, \ldots, w_n)^\top$ are weights with $\sum_{j=1}^{n} w_j = 1$. The mean and variance of the portfolio are

$$\mu_w = E(R(w)) = w^\top \mu = \sum_{j=1}^{n} w_j \mu_j; \quad \sigma_w^2 = var(R(w)) = w^\top \Sigma w = \sum_{j=1}^{n} \sum_{k=1}^{n} w_j w_k \sigma_{jk}.$$

Example 15 *Consider the equal weighting portfolio $w_i = 1/n$, $i = 1, \ldots, n$. This has always positive weights and*

$$\mu_{ew} = \frac{i^T \mu}{i^T i} = \frac{1}{n} \sum_{i=1}^{n} \mu_i \quad ; \quad \sigma_{ew}^2 = \frac{i^T \Sigma i}{(i^T i)^2} = \frac{1}{n^2} \sum_{i=1}^{n} \sum_{j=1}^{n} \sigma_{ij}.$$

With regard to the choice of portfolio weights there is a trade-off between mean and variance, meaning that as we increase the portfolio mean, which is good, we end up increasing its variance, which is bad. To balance these two effects we choose a portfolio that minimizes the variance of the portfolio subject to the mean being a certain level. Assume that the matrix Σ is nonsingular, so that Σ^{-1} exists with $\Sigma^{-1}\Sigma = \Sigma\Sigma^{-1} = I$ (otherwise, there exists w such that $\Sigma w = 0$). We first consider the **global minimum variance** (GMV) portfolio.

Definition 25 *The GMV portfolio w is the solution to the following minimization problem*

$$\min_{w \in \mathbb{R}^n} w^T \Sigma w \quad \text{subject to } w^T i = 1,$$

where $i = (1, \ldots, 1)^T$.

This is a classic problem of constrained optimization, and to solve this problem we form the **Lagrangian**, which is the objective function plus the constraint multiplied by the Lagrange multiplier λ

$$\mathcal{L}(w, \lambda) = \frac{1}{2} w^T \Sigma w + \lambda(1 - w^T i). \tag{1.15}$$

This has first order condition

$$\frac{\partial \mathcal{L}}{\partial w}(w, \lambda) = \Sigma w - \lambda i = 0 \text{ implies } w = \lambda \Sigma^{-1} i.$$

Then premultiplying by the vector i and using the constraint, we have $1 = i^T w = \lambda i^T \Sigma^{-1} i$, so that $\lambda = 1/i^T \Sigma^{-1} i$, and the optimal weights are

$$w_{GMV} = \frac{\Sigma^{-1} i}{i^T \Sigma^{-1} i}. \tag{1.16}$$

This portfolio has mean and variance

$$\mu_{GMV} = \frac{i^T \Sigma^{-1} \mu}{i^T \Sigma^{-1} i} \quad ; \quad \sigma_{GMV}^2 = \frac{1}{i^T \Sigma^{-1} i}. \tag{1.17}$$

The GMV portfolio may sacrifice more mean return than you would like, so we consider the more general problem where we ask for a minimum level m of the mean return.

1.6 Mean Variance Portfolio Analysis

Definition 26 *The portfolio that minimizes variance for a given level of mean return solves*

$$\min_{w \in \mathbb{R}^n} w^\mathsf{T} \Sigma w$$

subject to the constraints $w^\mathsf{T} i = 1$ and $w^\mathsf{T} \mu \geq m$.

The Lagrangian is

$$\mathcal{L}(w, \lambda, \gamma) = \frac{1}{2} w^\mathsf{T} \Sigma w + \lambda(1 - w^\mathsf{T} i) + \gamma(m - w^\mathsf{T} \mu), \quad (1.18)$$

where $\lambda, \gamma \in \mathbb{R}$ are the two Lagrange multipliers. The first order condition with respect to w is

$$\frac{\partial \mathcal{L}}{\partial w} = \Sigma w - \lambda i - \gamma \mu = 0,$$

which yields

$$w_{opt} = \lambda(m) \Sigma^{-1} i + \gamma(m) \Sigma^{-1} \mu \quad (1.19)$$

for the optimal λ, γ. Then imposing the two restrictions, $1 = i^\mathsf{T} w_{opt} = \lambda i^\mathsf{T} \Sigma^{-1} i + \gamma i^\mathsf{T} \Sigma^{-1} \mu$ and $m = \mu^\mathsf{T} w_{opt} = \lambda \mu^\mathsf{T} \Sigma^{-1} i + \gamma \mu^\mathsf{T} \Sigma^{-1} \mu$, we obtain a system of two equations in λ, γ, which can be solved exactly to yield

$$\lambda = \frac{C - Bm}{\Delta} \quad ; \quad \gamma = \frac{Am - B}{\Delta}$$

$$A = i^\mathsf{T} \Sigma^{-1} i, \; B = i^\mathsf{T} \Sigma^{-1} \mu, \; C = \mu^\mathsf{T} \Sigma^{-1} \mu, \; \Delta = AC - B^2, \quad (1.20)$$

provided $\Delta > 0$. This portfolio has mean m and variance

$$\sigma_{opt}^2(m) = \frac{Am^2 - 2Bm + C}{\Delta},$$

which is a quadratic function of m. The set of portfolios $w_{opt}(m)$, $m \geq 0$ is called the **mean variance efficient set**. This can be represented as a hyperbola in mean and standard deviation space

$$\left\{ (m, \sigma_{opt}(m)), \text{ where } \sigma_{opt}(m) = \sqrt{\frac{Am^2 - 2Bm + C}{\Delta}}, \; m \geq 0 \right\}. \quad (1.21)$$

We note that there is a dual problem, in which one maximizes the portfolio mean given that it has no more risk than a specified level. The two problems yield exactly the same efficient frontier. The equally weighted portfolio is not generally on the efficient frontier, and is not designed to optimize any criterion. However, this method typically improves on full weighting on the least risky asset or the highest return asset, say, and is very simple to compute.

Example 16 *We may give the mean variance approach an interpretation using expected utility theory. Suppose that u is quadratic utility, i.e. $u(x) = x - \frac{\gamma}{2}x^2$, or CARA utility $u(x) = -\exp(-\gamma x)$ plus the assumption that stock returns are jointly normal. The utility maximizing solution in both cases depends only on the unconditional mean vector μ and the unconditional covariance matrix Σ of the return series, and we obtain the optimal portfolio weights*

$$w = \frac{1}{\gamma}\Sigma^{-1}(\mu - \theta i), \quad \text{where } \theta = \frac{\mu^{\mathsf{T}}\Sigma^{-1}i - \gamma}{i^{\mathsf{T}}\Sigma^{-1}i}.$$

In the n asset case, many assets lie inside the frontier. The optimal portfolio weights can contain negative values and in practice often do. There can be many stocks that have negative mean (over the sample period where this is estimated), and it is desirable to short these assets. One can impose the additional restrictions that $w_i \geq 0$, which makes the portfolio choice problem more complicated to solve – there is no closed form solution in general. There are many practical issues associated with this line of argument. For example, the discreteness issue – you generally have to buy integer numbers of shares – would appear to rule out the use of calculus altogether.

The next result is useful because it allows one to represent the efficient frontier parsimoniously.

Theorem 10 *(Two fund separation.) Any portfolio in the mean variance efficient set can be expressed as a linear combination of just two (any two) efficient portfolios.*

Proof. From (1.19) we see that minimum variance portfolios form a straight line in $\mathbb{R}^n$. Let m be the mean of a given optimal portfolio, and let m_1 and m_2 be any distinct real numbers. There is a unique α such that $m = \alpha m_1 + (1-\alpha)m_2$. It follows from the equations for λ, γ that:

$$\lambda(m) = \alpha\lambda(m_1) + (1-\alpha)\lambda(m_2)$$
$$\gamma(m) = \alpha\gamma(m_1) + (1-\alpha)\gamma(m_2)$$
$$w(m) = \alpha w(m_1) + (1-\alpha)w(m_2).$$

Therefore, we can express the weights of the optimal portfolio in terms of the weights of the two other efficient portfolios.

We consider the covariance properties of portfolios. For any portfolio w we have

$$\text{cov}(w_{GMV}^{\mathsf{T}}R, w^{\mathsf{T}}R) = \sigma_{GMV,w} = w_{GMV}^{\mathsf{T}}\Sigma w = \frac{i^{\mathsf{T}}\Sigma^{-1}\Sigma w}{i^{\mathsf{T}}\Sigma^{-1}i} = \frac{i^{\mathsf{T}}w}{i^{\mathsf{T}}\Sigma^{-1}i} = \frac{1}{i^{\mathsf{T}}\Sigma^{-1}i} = \sigma_{GMV}^2.$$

Any asset has the same covariance with the minimum variance portfolio. Suppose there were an asset w with lower (higher) covariance with the GMV portfolio, then we would buy (sell) a portfolio with this asset and the GMV portfolio, and this would have variance $\sigma_p^2 = \alpha^2\sigma_{ww} + (1-\alpha)^2\sigma_{GMV} + \alpha(1-\alpha)\sigma_{GMV,w}$, and we would find some $\alpha > 0$ that would result in less variance than σ_{GMV}^2 (which corresponds to $\alpha = 0$).

1.6 Mean Variance Portfolio Analysis

Now consider a portfolio in the efficient mean variance set with weights w and mean μ_w. Its weights are of the form $w = \lambda(\mu_w)\Sigma^{-1}i + \gamma(\mu_w)\Sigma^{-1}\mu$. Suppose that w is not the GMV portfolio and let $p \neq w$ be any other portfolio. We have

$$\text{var}(w^\mathsf{T} R) = \sigma_w^2 = w^\mathsf{T}\Sigma w = \lambda(\mu_w)w^\mathsf{T} i + \gamma(\mu_w)w^\mathsf{T}\mu = \lambda(\mu_w) + \gamma(\mu_w)\mu_w,$$
$$\text{cov}(p^\mathsf{T} R, w^\mathsf{T} R) = \sigma_{pw} = p^\mathsf{T}\Sigma w = \lambda(\mu_w)p^\mathsf{T} i + \gamma(\mu_w)p^\mathsf{T}\mu = \lambda(\mu_w) + \gamma(\mu_w)\mu_p.$$

It follows that

$$\sigma_w^2 = \sigma_{pw} + \gamma(\mu_w)(\mu_w - \mu_p).$$

We have $\sigma_{pw} = 0$ if and only if $\sigma_w^2 = \gamma(\mu_w)(\mu_w - \mu_p)$. There exists a unique minimum variance portfolio $b(w)$ such that $\sigma_{w,b(w)} = 0$ for which $\sigma_w^2 = \gamma(\mu_w)(\mu_w - \mu_{w,b(w)})$. Theorem 11 follows.

Theorem 11 *We have for any portfolio p and any minimum variance portfolio w except the GMV one, that*

$$\mu_p = \mu_{w,b(w)} + \beta_{pw}(\mu_w - \mu_{w,b(w)}), \qquad \beta_{pw} = \frac{\sigma_{pw}}{\sigma_w^2}. \tag{1.22}$$

The **beta**, β_{pw}, is a risk measure. Take, for example, the portfolio that weights 100% on asset i. Then the above equation holds with β_{iw}, which measures the contribution of the security i to the risk of the portfolio w.

1.6.2 Portfolio Choice with Risk Free Asset

Now suppose there is a risk free security, with payoff R_f, in addition to the risky assets. The investor puts w_0 into the risk free asset and w_j invested into risky asset j with $j = 1, \ldots, n$ as before. We have

$$w_0 = 1 - \sum_{j=1}^{n} w_j = 1 - i^\mathsf{T} w.$$

Therefore the weights $w \in \mathbb{R}^n$ do not need to sum to one, but to $1 - w_0$. Then the return on the portfolio is $w_0 R_f + w^\mathsf{T} R$. This has expected return

$$\mu_w = w_0 R_f + w^\mathsf{T}\mu = w_0 R_f + w^\mathsf{T}\mu = R_f + w^\mathsf{T}(\mu - R_f i).$$

The variance of this portfolio is $\sigma_w^2 = w^\mathsf{T}\Sigma w$, because the risk free asset does not contribute to that.

We consider the problem how to choose w to minimize variance subject to achieving mean m.

Definition 27 *The portfolio that minimizes variance for a given level m of mean return solves*

$$\min_{w \in \mathbb{R}^n} w^\mathsf{T} \Sigma w$$

subject to the constraint $R_f + w^\mathsf{T}(\mu - R_f i) = m$.

The Lagrangian is

$$\mathcal{L}(w, \kappa) = \frac{1}{2} w^\mathsf{T} \Sigma w + \kappa(m - R_f - w^\mathsf{T}(\mu - R_f i)), \quad (1.23)$$

where κ is the Lagrangian multiplier. The first order condition is $0 = \Sigma w - \kappa(\mu - R_f i)$, which yields that the optimal weights satisfy

$$w^* = \kappa \Sigma^{-1}(\mu - R_f i). \quad (1.24)$$

To solve for κ, we use $m - R_f = (\mu - R_f i)^\mathsf{T} w^* = \kappa(\mu - R_f i)^\mathsf{T} \Sigma^{-1}(\mu - R_f i) \equiv \kappa D$. Therefore,

$$\kappa = \frac{m - R_f}{(\mu - R_f i)^\mathsf{T} \Sigma^{-1}(\mu - R_f i)}.$$

The fraction of the portfolio held in risky assets is

$$i^\mathsf{T} w^* = \frac{m - R_f}{(\mu - R_f i)^\mathsf{T} \Sigma^{-1}(\mu - R_f i)} \times i^\mathsf{T} \Sigma^{-1}(\mu - R_f i)$$

$$= \frac{(m - R_f) i^\mathsf{T} \Sigma^{-1} i}{(\mu - R_f i)^\mathsf{T} \Sigma^{-1}(\mu - R_f i)} \times \frac{1}{i^\mathsf{T} \Sigma^{-1} i} i^\mathsf{T} \Sigma^{-1}(\mu - R_f i)$$

$$= \psi(\mu_{GMV} - R_f)$$

for some constant ψ with $w_0 = 1 - i^\mathsf{T} w^*$. This portfolio has mean m and variance

$$\sigma^2_{w^*}(m) = \frac{(m - R_f)^2}{(\mu - R_f i)^\mathsf{T} \Sigma^{-1}(\mu - R_f i)}.$$

It follows that the relationship between the portfolio mean and standard deviation is as follows

$$\left\{ (m, \sigma_{w^*}(m)), \text{ where } \sigma_{w^*}(m) = \frac{|m - R_f|}{\sqrt{(\mu - R_f i)^\mathsf{T} \Sigma^{-1}(\mu - R_f i)}} \right\}. \quad (1.25)$$

The minimum variance set can be generated from any two distinct minimum variance portfolios. Now there is a natural choice of funds, the riskless asset and a particular portfolio with risky assets only. There are two cases: $\mu_{GMV} \neq R_f$ and $\mu_{GMV} = R_f$.

1.7 Capital Asset Pricing Model

Definition 28 *Suppose that $\mu_{GMV} \neq R_f$. The tangency portfolio has weights proportional to w^* but rescaled to satisfy the portfolio constraint*

$$w_{\tan} = \frac{1}{i^T \Sigma^{-1}(\mu - R_f i)} \Sigma^{-1}(\mu - R_f i) = \frac{1}{B - AR_f} \Sigma^{-1}(\mu - R_f i),$$

which has

$$\mu_{\tan} = \frac{C - BR_f}{B - AR_f} \quad ; \quad \sigma_{\tan}^2 = \frac{D}{(B - AR_f)^2}.$$

The constants A, B and C are defined in (1.20). In the case where $R_f < \mu_{GMV}$, then $\mu_{\tan} > \mu_{GMV}$. In the case where $R_f > \mu_{GMV}$, then $\mu_{\tan} < \mu_{GMV}$.

If $R_f = \mu_{GMV}$, then the minimum variance portfolio is self financing or zero wealth. There is no tangency point.

1.7 Capital Asset Pricing Model

Sharpe (1964) proposed the Capital Asset Pricing Model (CAPM) as an equilibrium theory built on the Modern Portfolio Theory. Investors have mean-variance preferences and care about their end-of-period wealth (one-period model). They have homogenous expectations. A riskfree asset exists and investors can borrow and lend at the riskfree rate (later relaxed). All assets are tradable and perfectly divisible. There are no market frictions, information is costless and available to everyone. There are no taxes or short-selling restrictions. The assumptions are not realistic but the theory provides an important insight: in equilibrium, not all risks carry rewards; there is no compensation for diversifiable risks, only the nondiversifiable market-wide risk is priced. The CAPM is a theory of financial equilibrium with no direct link to the real economy. It assumes that equilibrium prices clear the market for all assets so that demand for each asset equals its supply. Demand and supply functions are not calculated explicitly; only the equilibrium prices and returns are characterized. Given a perfect capital market, risk averse investors, and a two-parameter return distribution (or quadratic utility), Markowitz (1952ab) found that the optimal portfolio for any investor must be efficient in the sense that no other portfolio with the same expected return (or higher) has lower variance of return. Given homogeneous expectations and short selling, Black (1972) has shown that in a market equilibrium the value weighted market index is efficient. The economic model is static, meaning there is only one decision time. Since investors have the same beliefs, they perceive the same investment opportunity set. By the Two-Fund Separation Theorem, they will invest in a combination of cash and the tangency portfolio. Based on this, it is possible to characterize the equilibrium without detailing the structure in terms of supply and demand. Since all investors hold the same tangency portfolio of risky assets, it must consist of all risky assets. If some were missing, it would be in positive supply but there would be zero demand for it, which cannot be an equilibrium. Similarly, each asset's relative value in the tangency portfolio must be the same as its market capitalization relative to the total market capitalization. Thus, the tangency portfolio is the market portfolio M. Where M lies on the efficient frontier which is called the **capital market line** (CML); this is where optimal portfolios of individuals lie.

Suppose that an individual's preferences are described solely by mean and variance, i.e. expected utility of final wealth is solely a function of the mean and variance of final wealth, i.e. there is a two argument function $V(\mu, \sigma^2)$ that equals the utility function over final wealth. In the case where a risk free asset exists, she solves the following problem

$$\max_{w \in \mathbb{R}^n} V(R_f + w^\mathsf{T}(\mu - R_f i), w^\mathsf{T} \Sigma w). \tag{1.26}$$

This problem has a first order condition

$$V_\mu(\mu_w, \sigma_w^2)(\mu - R_f i) + 2V_{\sigma^2}(\mu_w, \sigma_w^2)\Sigma w = 0,$$

so that the optimal weights satisfy the nonlinear equation

$$w^* = \frac{-V_\mu(\mu_{w^*}, \sigma_{w^*}^2)}{2V_{\sigma^2}(\mu_{w^*}, \sigma_{w^*}^2)} \Sigma^{-1}(\mu - R_f i). \tag{1.27}$$

Therefore, the optimal investment in risky assets is proportional to the tangency portfolio. Since the tangency portfolio is the market portfolio M in equilibrium, it follows that

$$\mu_i - R_f = \beta_i(\mu_m - R_f), \tag{1.28}$$

where $\beta_i = \sigma_{iM}/\sigma_M^2$, which is the CAPM equation or security market line (SML).

The CAPM beta is a scaled covariance of asset (or portfolio) i with the market (the scaling factor is the constant $1/\sigma_M^2$) and this is the relevant measure of risk according to the CAPM. The graphical depiction of the CAPM is the SML; its slope is $\mu_M - R_f$, which is also called the market price of risk. The CML is a set of portfolios p such that

$$\mu_p = R_f + \frac{\mu_p - R_f}{\sigma_p} \times \sigma_p, \tag{1.29}$$

and these have the maximum Sharpe ratio $\frac{\mu_p - R_f}{\sigma_p}$.

How can we justify restricting attention to mean-variance preferences in (1.26)? There are two ways: either preferences over outcomes are quadratic or exponential, or we restrict the distribution of returns. A sufficient condition is that the vector of returns is jointly normal, in which case the distribution of returns on any portfolio is also normal and determined only by its mean and variance. It can be shown that for risk averse nonsatiated investors, the mean outcome is beneficial, while the variance is bad. However, it has been shown that although in the absence of strong restrictions on investor preferences the assumption of normality is sufficient to generate (1.28), it is not a necessary condition. In particular, Chamberlain (1983), Owen and Rabinovitch (1983), and Berk (1997) show that the CAPM can be obtained under the assumption of elliptically symmetric return distributions without strongly restricting preferences.

1.7 Capital Asset Pricing Model

Definition 29 *A random variable X is elliptically symmetrically distributed if its density f can be written in the following fashion*

$$f(x) = (\det(\Sigma))^{-1/2} g((x-\mu)^\top \Sigma^{-1}(x-\mu)),$$

for some function $g(\cdot)$ and positive definite, symmetric matrix Σ. Here, $\det(A)$ is the determinant of the square matrix A.

Berk (1997) shows that elliptical symmetry is the most general distributional assumption that will imply the CAPM when agents maximize expected utility, that is elliptical symmetry is both necessary and sufficient for the CAPM. The elliptically symmetric family contains the Gaussian distribution as a special case, but many well-known thick-tailed distributions also belong to this class – the Student t, logistic, and scale mixed-normal being examples (see Chapter 14 for more discussion).

1.7.1 CAPM Properties

In equilibrium, all assets lie on the SML although they lie inside the set of feasible portfolios formed from risky assets and not on the CML. This is because not all variance of an asset matters to risk-averse investors; they can diversify away some of the risk (idiosyncratic risk) of an asset by holding other assets in their portfolio so that some of their movements offset (because they are not perfectly positively correlated). There is a limit to such diversification, however; the economy-wide risk (systematic risk) cannot be diversified away and so investors must be rewarded for bearing it. In sum, the risk of an asset can be decomposed as follows: total risk = systematic (non-diversifiable) risk + idiosyncratic (diversifiable) risk. For an efficient portfolio, total risk = systematic risk.

We can write

$$R_i = a_i + \beta_i \times R_M + \varepsilon_i$$

such that $\text{cov}(\varepsilon_i, R_M) = 0$, which implies $\sigma_i^2 = \beta_i^2 \times \sigma_M^2 + \sigma_\varepsilon^2$. The first term reflects systematic risk, the second term idiosyncratic risk.

We can rewrite the CAPM equation as

$$\mu_i - R_f = \frac{\mu_M - R_f}{\sigma_M} \times \rho_{iM} \times \sigma_i,$$

which shows that the portion of systematic risk that is priced is measured by $\rho_{iM} \times \sigma_i \leq \sigma_i$. For an inefficient asset or portfolio, the inequality is strict, i.e. only part of its variance is priced. In sum, covariance (or its scaled version beta) and not variance measures the risk relevant for asset pricing. The reward (risk premium) that a holder of an asset receives depends on how much the asset contributes to the overall market risk. This can be shown as follows:

$$\sigma_M^2 = \text{cov}(R_M, R_M) = \text{cov}\left(\sum_{t=1}^n w_i \times R_i, R_M\right) = \sum_{i=1}^n w_i \times \sigma_{iM}.$$

Thus, the (weight-adjusted) contribution of asset i to the variance of the market portfolio is σ_{iM}. Dividing by σ_M^2, we obtain $\beta_M = 1 = \sum_{i=1}^n w_i \times \beta_i$.

The CML shows the risk return relationship of the set of efficient portfolios p that combine the tangency portfolio (on the efficient frontier of the risky assets) with the risk free asset such that

$$E(R_p) = R_f + \frac{E(R_m) - R_f}{\sigma_m} \times \sigma_p,$$

where the slope here is the Sharpe ratio of the market portfolio. For well diversified portfolios, the risk return relationship is positive with a slope given by the market Sharpe ratio and the corresponding risk measure of the portfolio is its standard deviation.

In the same way, we have for any general portfolio p consisting of K assets,

$$\beta_p = \sum_{i=1}^K w_i \times \beta_i.$$

This follows because $\sigma_{pM} = \text{cov}(R_p, R_M)$, and the covariance is additive, meaning

$$\text{cov}\left(\sum_{i=1}^K w_i R_i, R_M\right) = \sum_{i=1}^K w_i \text{cov}(R_i, R_M).$$

1.7.2 CAPM Without Riskfree Asset

Black (1972) presents a version of the CAPM with no riskfree asset. We use the following results without a proof.

A convex combination of frontier portfolios is also a frontier portfolio. The set of efficient portfolios is a convex set. Note this does not mean that the efficient frontier is a convex-shaped set in the (μ, σ) space; it means that a convex combination of efficient portfolios always belongs to the set of efficient portfolios. If every investor holds an efficient portfolio, the market portfolio, which is a weighted average of individual portfolios, is also efficient. For any frontier portfolio p except the minimum variance portfolio, there exists a unique frontier portfolio Z_p which has zero covariance with it. The latter is called the zero covariance portfolio relative to p. If p is efficient (inefficient), Z_p is inefficient (efficient).

Theorem 12 *For any asset or portfolio i and any frontier portfolio p, the following relationship holds:*

$$E(R_i) = E(R_{Z_p}) + \beta_{ip} \times \left(E(R_p) - E(R_{Z_p})\right),$$

$$\beta_{ip} = \frac{\text{cov}(R_i, R_p)}{\text{var}(R_p)}.$$

This is a result derived without any equilibrium considerations; it simply reflects the necessary relationships between assets and portfolios in the feasible set. Note that i

does not have to be on the minimum variance frontier. With the same assumptions as in the standard CAPM, investors will only hold efficient portfolios and so the market portfolio is efficient in equilibrium. Thus, we can select $p = M$ and, denoting $Z_p = 0$, we get

$$E(R_i) = E(R_0) + \beta_i \times (E(R_M) - E(R_0)), \qquad (1.30)$$

where R_0 is the return on the **zero beta portfolio**.

It is only now that we have employed some economic reasoning; the general relationship with an arbitrary p is just a technical statement, whereas here we have a relationship that holds in an equilibrium with optimizing investors (who, depending on preferences, hold different mixes of risky assets). Note that the name of the model comes from the fact that $\beta_0 = 0$. The zero beta CAPM is also called the two-factor model since it can be written as

$$E(R_i) = \beta_i \times E(R_M) + (1 - \beta_i) \times E(R_0), \qquad (1.31)$$

which shows that the expected return on an asset is a linear function of two risk factors: the market portfolio and the zero beta portfolio.

1.7.3 Criticisms of Mean-Variance Analysis and the CAPM

There are some high level criticisms of the underlying framework. It has been argued that expected utility as a decision criterion is flawed, some of the assumptions which underline this framework are invalid, and the mean-variance criterion may lead to paradoxical choices. Allais (1953) shows that using EUT in making choices between pairs of alternatives, particularly when small probabilities are involved, may lead to some paradoxes. Some other fundamental papers question the validity of the risk aversion assumption within the EUT framework. Friedman and Savage (1948), Markowitz (1952a), and Kahneman and Tversky (1979) claim that the typical preference must include risk-averse as well as risk-seeking segments. One implication of this is that the variance is not a good measure of risk, which casts doubt on the validity of the CAPM.

1.8 Arbitrage Pricing Theory

The arbitrage pricing theory was introduced by Ross (1976). Connor (1984), and Chamberlain and Rothschild (1983).

1.8.1 The Linear Factor Model

The main assumption of the APT is that there are a large number of assets and they all obey the **linear factor model**, which says that the return realization R_i for each asset $i \in \{1, \ldots, n\}$ is generated as follows

$$R_i = \alpha_i + \sum_{k=1}^{K} b_{ik} f_k + \varepsilon_i, \qquad (1.32)$$

where f_k are random K **common factors** that may or may not be observed, b_{ik} are called factor loadings (sensitivity of the return on asset i to factor k), while ε_i is a random variable denoting idiosyncratic risk (as opposed to systematic risk of the economy-wide factors) with $E(\varepsilon_i) = 0$ and $\text{var}(\varepsilon_i) < \infty$ and for each $i = 1, \ldots, n$

$$\text{cov}(f_k, \varepsilon_i) = 0, \quad k = 1, \ldots, K.$$

This is a population model.

1.8.2 Portfolios

Now we define some useful portfolio notions.

Definition 30 *A (unit cost) portfolio is $w_1, \ldots, w_n$ such that*

$$\sum_{i=1}^{n} w_i = 1.$$

Definition 31 *An (zero cost) arbitrage portfolio is $w_1, \ldots, w_n$ such that*

$$\sum_{i=1}^{n} w_i = 0.$$

Definition 32 *A well-diversified portfolio is such that*

$$\lim_{n \to \infty} \sum_{i=1}^{n} w_i^2 = 0.$$

Note that $\sum_{i=1}^{n} w_i^2 = w^\top w$ is the **Herfindahl index** of the weight vector.

Definition 33 *A portfolio that is hedged against factor k risk is one whose weighting sequence satisfies*

$$\sum_{i=1}^{n} w_i b_{ik} = 0.$$

A portfolio that is hedged against all factor risk is one where this holds for all $k = 1, \ldots, K$.

1.8 Arbitrage Pricing Theory

1.8.3 Restrictions Implied by the Absence of Arbitrage

The factor model structure implies that for $i = 1, \ldots, n$

$$\mu_i = E(R_i) = \alpha_i + \sum_{k=1}^{K} b_{ik} E(f_k)$$

$$R_i = E(R_i) + \sum_{k=1}^{K} b_{ik} (f_k - E(f_k)) + \varepsilon_i.$$

Consider the well diversified portfolio p that is hedged against factor risk,

$$\begin{aligned} R_p &= \sum_{i=1}^{n} w_i R_i \\ &= \sum_{i=1}^{n} w_i \alpha_i + \sum_{k=1}^{K} \sum_{i=1}^{n} w_i b_{ik} \times f_k + \sum_{i=1}^{n} w_i \varepsilon_i \\ &= \sum_{i=1}^{n} w_i \alpha_i + \sum_{i=1}^{n} w_i \varepsilon_i \\ &\simeq \sum_{i=1}^{n} w_i \alpha_i. \end{aligned} \quad (1.33)$$

If the portfolio is an arbitrage portfolio then $R_p \simeq 0$, otherwise you make money for nothing (if the portfolio is a unit cost portfolio and there exists a risk free asset, then $R_p \simeq R_f$). The two approximations denoted by $\simeq$ hold as $n \to \infty$, i.e. for **large economies**. We will discuss the approximation $\sum_{i=1}^{n} w_i \varepsilon_i \simeq 0$ in detail in Chapter 8.

We next apply some standard arguments of linear algebra to obtain restrictions on the vector α. First note that the matrix

$$(i_n, B) = \begin{pmatrix} 1 & b_{11} & \cdots & b_{1K} \\ \vdots & \vdots & & \vdots \\ 1 & b_{n1} & \cdots & b_{nK} \end{pmatrix} \quad (1.34)$$

generates a (proper) subspace of the vector space $\mathbb{R}^n$ of dimension $K + 1$ (assuming that $n > K + 1$ and that there is no redundancy in B); let's call it $\mathbb{B}$. This subspace has an orthogonal complement, denoted $\mathbb{B}^\perp$, of dimensions $n - K$, which contains all the vectors orthogonal to $\mathbb{B}$ and hence to (i_n, B) in particular. A diversified arbitrage portfolio p that is hedged against factor risk satisfies $w^\top (i_n, B) = 0$, i.e. its weighting sequence w is in the **null space** of (i_n, B). A diversified unit cost portfolio that is hedged against factor risk satisfies $w^\top B = 0$, i.e. its weighting sequence w is in the null space of B.

Since the vector α is orthogonal to the arbitrage portfolio weights w, it must lie in the space spanned by (i_n, B). Therefore for some constants $\theta_0, \theta_1, \ldots, \theta_K$ we have

$$\alpha_i = \theta_0 + \sum_{k=1}^{K} b_{ik}\theta_k \qquad (1.35)$$

for each $i = 1, \ldots, n$. In matrix notation this could be written as

$$\alpha = \theta_0 i_n + B\theta,$$

where $\theta = (\theta_1, \ldots, \theta_K)^\mathsf{T} \in \mathbb{R}^K$.

It follows that $(\mu_1, \ldots, \mu_n)^\mathsf{T} \in \mathrm{span}(i_n, B) \subset \mathbb{R}^n$, i.e. there exist constants $\lambda_0, \lambda_1, \ldots, \lambda_K$ such that for any asset i

$$\mu_i = \alpha_i + \sum_{k=1}^{K} b_{ik} E(f_k) = \theta_0 + \sum_{k=1}^{K} b_{ik}(E(f_k) + \theta_k) = \lambda_0 + \sum_{k=1}^{K} b_{ik}\lambda_k,$$

where $\lambda_0 = \theta_0$, and $\lambda_k = E(f_k) + \theta_k$, $k = 1, \ldots, K$ are risk premia associated with the factor k, i.e.

$$\mu = \lambda_0 i_n + B\lambda. \qquad (1.36)$$

There are $n - K - 1$ restrictions on α and μ. These are the main restrictions of the APT, and we give a more rigorous explanation for this in the appendix to this chapter, where it is acknowledged that these arrive from an approximation argument that holds when n is large.

Note that any portfolio $R_p = \sum_{i=1}^{n} w_{pi} R_i$ of the assets (with $\sum_{i=1}^{n} w_{pi} = 1$) would satisfy

$$\mu_p = E(R_p) = \sum_{i=1}^{n} w_{pi} E(R_i) = \theta_0 \sum_{i=1}^{n} w_{pi} + \sum_{i=1}^{n} w_{pi} \sum_{k=1}^{K} b_{ik}(E(f_k) + \theta_k),$$

i.e. it is a known function of the risk premia $\lambda = (\lambda_0, \lambda_1, \ldots, \lambda_K)^\mathsf{T}$, its own weights w_p, and the betas. Suppose that the factors are portfolios of traded assets and let W be the $K \times n$ matrix containing their weights, and let $\mu_f = E(f)$ be the $K \times 1$ vector containing the means of the factors. Then

$$\mu_f = \theta_0 W i_n + W B (\mu_f + \theta).$$

We must have $WB = I_K$ because if we regress a factor portfolio on all the factor portfolios we get a loading of one on the corresponding factor and zeros on the others. We also must have $W\alpha = 0$ so that $\theta_0 W i_n + \theta = 0$. If the factors are unit cost portfolios, then we may assume that $W i_n = i_K$, in which case we have $\theta = -\theta_0 i_K$. In that case, we have

$$\alpha = \theta_0 (i_n - B). \qquad (1.37)$$

1.9 Appendix

If the factors are zero cost portfolios (like the Fama–French factors), then $Wi_n = 0$ and $\theta = 0$. In that case, we have

$$\alpha = \theta_0 i_n. \tag{1.38}$$

In the case where the first factor is the market portfolio, so unit cost and the others are zero cost, we have $Wi_n = e_1 = (1, 0, \ldots, 0)^\mathsf{T} \in \mathbb{R}^K$ so that $\theta_0 = -\theta_1$ and $\theta_k = 0$, $k = 2, \ldots, K$. In that case we write

$$\alpha = \theta_0 (i_n - Be_1).$$

Suppose there is a risk free rate R_f. Then the factor model (1.32) can be rewritten as

$$R - R_f i_n = \alpha_f + B(f - R_f i_K) + \varepsilon,$$

where $\alpha = \alpha_f + R_f(i_n - Bi_K)$. If we take a unit cost portfolio w, then $w^\mathsf{T} \alpha_f = 0$, which implies that $\alpha_f = B\theta$ or $\alpha = R_f(i_n - Bi_K) + B\theta$ in the original model, since now we do not have $w^\mathsf{T} i_n = 0$.

The null restriction is that $\alpha_f = 0$ or $\alpha = R_f(i_n - Bi_K)$ in the case with traded assets or $\alpha_f = B\theta$ in the case without traded assets as factors. If the factors are also traded portfolios we obtain $\lambda_k = E(f_k) - R_f$, and it follows that

$$\mu_i - R_f = \sum_{k=1}^{K} b_{ik}(E(f_k) - R_f).$$

	Risk Free	No Risk Free
Traded Factors	$\alpha = (i_n - Bi_n)R_f$, $\alpha_f = 0$	$\alpha = (i_n - Bi_n)\theta_0$
Non-Traded Factors	$\alpha = (i_n - Bi_K)R_f + B\theta$, $\alpha_f = B\theta$	$\alpha = i_n \gamma_0 + B\theta$

1.9 Appendix

Consider the residuals from a least-squares projection of asset expected returns on i_n and B:

$$\mu = \theta_0 i_n + B\theta + \eta,$$

where the n superscripts (since this regression is for a given n in the sequence) is left implicit for notational simplicity. Consider the scaled portfolio from the expected return pricing errors:

$$\omega = \left(\frac{1}{\eta^\mathsf{T} \eta}\right) \eta.$$

Regression residuals are orthogonal to the explanatory variables, so this portfolio has zero cost and zero exposures to all the factors. It has an expected return of

Chapter 1 Introduction and Background

$\omega^\mathsf{T}\mu = \omega^\mathsf{T}\eta = 1$ and a total return variance of

$$E\left((\omega^\mathsf{T}\varepsilon)^2\right) \leq \left(\frac{1}{\eta^\mathsf{T}\eta}\right)\sigma_\varepsilon^2,$$

where σ_ε^2 denotes the upper bound imposed on the eigenvalues of the idiosyncratic return covariance matrix. Unless there is a finite c such that $\lim_{n\to\infty} \sup \eta^\mathsf{T}\eta \leq c$ then there exists a sequence of zero cost portfolios with expected return of one and variance of approximately zero. Assume that there does not exist a sequence of zero cost portfolios with expected return of one and variance going to zero, by the absence of statistical arbitrage. This then implies that the average squared pricing error goes to zero with n:

$$\lim_{n\to\infty} \frac{1}{n}\eta^\mathsf{T}\eta = 0.$$

And that is all that can be shown. For any chosen precision level $\delta > 0$ the proportion of assets with model-predicted expected returns within that precision level goes to 1.

2 Econometric Background

Here, we give some of the main econometric methods and results we rely on in what follows.

2.1 Linear Regression

Linear regression is an essential tool for describing how a variable x affects a variable y. Suppose that we observe data that is collected into matrices as follows

$$y = \begin{pmatrix} y_1 \\ \vdots \\ y_n \end{pmatrix} \quad ; \quad X = \begin{pmatrix} x_{11} & \cdots & x_{1K} \\ \vdots & & \vdots \\ x_{n1} & & x_{nK} \end{pmatrix} = \begin{pmatrix} x_1^\mathsf{T} \\ \vdots \\ x_n^\mathsf{T} \end{pmatrix}.$$

We shall suppose that $\text{rank}(X) = K < n$ (this is an assumption, but it is immediately verifiable from the data in contrast to some other assumptions we will make), which implies that the $K \times K$ matrix $X^\mathsf{T} X$ is nonsingular. There is a lot of recent work in modern statistics concerned with the case where $K > n$ (in which case this condition is violated), but we shall not consider this here.

The general linear regression model can be written in several ways

$$\begin{aligned} y &= X\beta + \varepsilon, \\ y_i &= x_i^\mathsf{T} \beta + \varepsilon_i, \quad i = 1, \ldots, n, \\ y_i &= \sum_{j=1}^{K} x_{ij} \beta_j + \varepsilon_i, \quad i = 1, \ldots, n, \end{aligned} \quad (2.1)$$

where ε_i is a random error term. The simplest case is when $K = 2$ and $x_{i1} = 1$ for $i = 1, \ldots, n$, which can be rewritten in a simpler notation

$$y_i = \alpha + \beta x_i + \varepsilon_i, \quad (2.2)$$

where α is called the intercept and β is called the slope. We will consider both bivariate regressions and multivariate ($K > 2$) regressions and will use the vector and matrix notation for convenience.

There are several type of assumptions that are made here about the data and error terms. We first consider the case where y_i, x_i are independent and identically distributed (i.i.d.) and satisfy the model (2.1). The weakest assumption we can make with regard to the error term is that $E(\varepsilon_i x_i) = 0$. In this case we may interpret $x_i^T \beta$ as the **best linear predictor** of y_i in the sense of minimizing mean squared error $E((y_i - m(x_i))^2)$ with respect to all linear functions $m(x) = x^T b$. This is sometimes denoted as $E_L(y_i|x_i) = x_i^T \beta$. The β have a population definition

$$\beta = E(x_i x_i^T)^{-1} E(x_i y_i).$$

The assumption that $E(\varepsilon_i x_i) = 0$ is almost no assumption at all since any y_i, x_i for which the moments exist will have a linear relation with uncorrelated errors. A stronger assumption is the conditional moment restriction that $E(\varepsilon_i|x_i) = 0$, in which case

$$E(y_i|x_i) = x_i^T \beta.$$

In this case $x_i^T \beta$ is the regression function and is the **best predictor** of y_i in the sense of minimizing mean squared error with respect to all (measurable) functions of x_i. In general, the best predictor of y_i by x_i may be nonlinear, so it is a special model restriction to assume that it is linear. Best linear predictors are by definition linear, whereas regression functions may or may not be linear – they have to be affirmed so as a matter of higher belief.

In the simple case (2.2), we may write the regression model as

$$y_i - E(y_i) = \frac{\text{cov}(y_i, x_i)}{\text{var}(x_i)} (x_i - E(x_i)) + \varepsilon_i,$$

and also

$$\frac{y_i - E(y_i)}{\sqrt{\text{var}(y_i)}} = \text{corr}(y_i, x_i) \times \frac{x_i - E(x_i)}{\sqrt{\text{var}(x_i)}} + \varepsilon_i^*,$$

where $\varepsilon_i^* = \varepsilon_i / \sqrt{\text{var}(y_i)}$. This representation shows the **regression towards the mean** property. That is, if x_i exceeds $E(x_i)$, then there is a tendency for the amount (in standardized units) by which y_i exceeds $E(y_i)$ to be reduced according to the magnitude of $\text{corr}(y_i, x_i)$, which we know to be between -1 and 1. If a father is taller than average, his son tends to be taller than average but by a smaller margin. If a mutual fund does better than average this period, then next period the performance is likely to be less good than average.

In many cases, the first element of x_i is a constant, i.e. $x_{1i} = 1$. Let $\widetilde{x}_i = (x_{2i}, \ldots, x_{Ki})^T$ be the remaining elements of x_i. One way towards regression is by treating $(y_i, \widetilde{x}_i^T)$ as random variables and considering their joint distribution and deriving the conditional expectation from that. Suppose that $(y_i, \widetilde{x}_i^T)$ are multivariate normal with mean vector $\mu_y, \mu_{\widetilde{x}}^T$ and covariance matrix

$$E\left[\begin{pmatrix} y_i - \mu_y \\ \widetilde{x}_i - \mu_{\widetilde{x}} \end{pmatrix} \begin{pmatrix} y_i - \mu_y \\ \widetilde{x}_i - \mu_{\widetilde{x}} \end{pmatrix}^T\right] = \Sigma = \begin{pmatrix} \sigma_{yy} & \sigma_{y\widetilde{x}}^T \\ \sigma_{\widetilde{x}y} & \Sigma_{\widetilde{x}\widetilde{x}} \end{pmatrix},$$

2.1 Linear Regression

where $\Sigma_{\tilde{x}\tilde{x}}$ is nonsingular. It follows that the regression of y_i on $\tilde{x}_i$ is linear

$$y_i = \mu_y + \beta^{\mathsf{T}}(\tilde{x}_i - \mu_y) + \tilde{\varepsilon}_i,$$

$$\beta = \Sigma_{\tilde{x}\tilde{x}}^{-1}\sigma_{\tilde{x}y}.$$

For the normal distribution, absence of correlation is equivalent to independence, and indeed the $\tilde{\varepsilon}_i$ here is not just uncorrelated with $\tilde{x}_i$ but is independent of it and mean zero. This shows the relationship between the covariance matrix Σ and the regression problem.

Finally, in some cases the i.i.d. assumption may not be appropriate. In this case $E(\varepsilon_i|x_i) = 0$ need not imply that $E(\varepsilon_i|x_1,\ldots,x_n) = 0$. We therefore consider explicitly the stronger assumption

$$E(\varepsilon|x_1,\ldots,x_n) = E(\varepsilon|X) = 0. \tag{2.3}$$

This may not be appropriate in a time series setting, where it would be possibly appropriate to condition only on the past. Let

$$E(\varepsilon\varepsilon^{\mathsf{T}}|X) = \Omega, \tag{2.4}$$

where Ω is an $n \times n$ covariance matrix. A commonly adopted special case is where

$$E(\varepsilon\varepsilon^{\mathsf{T}}|X) = \sigma^2 I. \tag{2.5}$$

The weighted least squares estimator of β is

$$\widehat{\beta}_W = (X^{\mathsf{T}}WX)^{-1}X^{\mathsf{T}}Wy,$$

where W is some $n \times n$ weighting matrix. The ordinary least squares (OLS) estimator $\widehat{\beta}$ is $\widehat{\beta}_W$ with $W = I_n$. It is linear in y and unbiased ($E(\widehat{\beta}) = \beta$ for all β) under the conditional moment restriction assumption (2.3). It is best linear unbiased under (2.5) and best unbiased when ε is also normally distributed. We have

$$\text{var}(\widehat{\beta}|X) = (X^{\mathsf{T}}X)^{-1}X^{\mathsf{T}}\Omega X(X^{\mathsf{T}}X)^{-1} = V \tag{2.6}$$

in general. If (2.5) holds, then $\text{var}(\widehat{\beta}|X) = \sigma^2(X^{\mathsf{T}}X)^{-1}$.

If on the other hand we only assume $E(\varepsilon_i|x_i) = 0$ or $E(\varepsilon_i|x_1,\ldots,x_i) = 0$, then the OLS estimator may be biased, although consistent so that the bias disappears with large sample size.

We next discuss briefly the conditions under which the OLS estimator is consistent as the sample size n increases. For a square symmetric matrix A, let $\lambda_{\min}(A)$, and $\lambda_{\max}(A)$ denote the smallest and largest eigenvalues respectively of A. For consistency of the OLS estimator (as the sample size increases) it suffices that $\mathrm{var}(c^{\mathsf{T}}\widehat{\beta}) \to 0$ for any linear combination c. We have, under (2.3) and (2.4)

$$\begin{aligned}
\mathrm{var}(c^{\mathsf{T}}\widehat{\beta}) &= c^{\mathsf{T}}(X^{\mathsf{T}}X)^{-1}X^{\mathsf{T}}\Omega X(X^{\mathsf{T}}X)^{-1}c \\
&= \sigma^2 \times c^{\mathsf{T}}c \times \frac{c^{\mathsf{T}}(X^{\mathsf{T}}X)^{-1}X^{\mathsf{T}}\Omega X(X^{\mathsf{T}}X)^{-1}c}{c^{\mathsf{T}}c} \\
&\leq c^{\mathsf{T}}c \times \frac{\lambda_{\max}(X^{\mathsf{T}}X)\lambda_{\max}(\Omega)}{\lambda_{\min}^2(X^{\mathsf{T}}X)},
\end{aligned}$$

by the properties of eigenvalues. It follows that consistency of $\widehat{\beta}$ holds provided that the right hand side goes to zero. A sufficient condition is that $\lambda_{\min}(X^{\mathsf{T}}X) \to \infty$, while $\lambda_{\max}(X^{\mathsf{T}}X)/\lambda_{\min}(X^{\mathsf{T}}X) \leq c < \infty$ and $\lambda_{\max}(\Omega) \leq c < \infty$. In the special case $\Omega = \sigma^2 I$, it suffices that $\lambda_{\min}(X^{\mathsf{T}}X) \to \infty$. Intuitively, this is saying that the variation of the signal (the regressors) should increase with the sample size. This condition is satisfied when x_i are i.i.d. random variables with the matrix $E(x_i x_i^{\mathsf{T}})$ being finite and nonsingular. Under general conditions $\widehat{\beta}$ is consistent and asymptotically normal, i.e.

$$\sqrt{n}(\widehat{\beta} - \beta) \Longrightarrow N(0, V_\infty),$$

where $V_\infty = \lim_{n \to \infty} nV$. The OLS estimator is consistent but not efficient; the generalized least squares (GLS) estimator $\widehat{\beta}_W$ with $W = \Omega^{-1}$ is efficient but infeasible without knowledge of Ω. We focus mostly just on the OLS estimator for its simplicity and its widespread applicability.

An intermediate case between $\Omega = \sigma^2 I$ and the general Ω is the case of diagonal Ω, denoted by $D = \mathrm{diag}(\sigma_1^2, \ldots, \sigma_n^2)$. This specification is widely adopted in applications; it corresponds to allowing heteroskedasticity across error terms but not correlation between them. In that case, there is a simple and well established method of conducting inference. White's standard errors (se_W) are the square root of the diagonal elements of

$$\widehat{V}_W = (X^{\mathsf{T}}X)^{-1}X^{\mathsf{T}}SX(X^{\mathsf{T}}X)^{-1}, \quad S = \mathrm{diag}(\widehat{\varepsilon}_1^2, \ldots, \widehat{\varepsilon}_n^2). \tag{2.7}$$

They are consistent under general conditions for the case (2.4), meaning that $n\widehat{V}_W - V_\infty \xrightarrow{P} 0$.

The least squares standard errors (se_{LS}) are the square root of the diagonal elements of

$$\widehat{V}_{OLS} = s^2(X^{\mathsf{T}}X)^{-1}, \quad s^2 = \frac{1}{n-K}\widehat{\varepsilon}^{\mathsf{T}}\widehat{\varepsilon} = \frac{1}{n-K}\sum_{t=1}^{n}\widehat{\varepsilon}_t^2.$$

2.1 Linear Regression

They are consistent in the special case $\Omega = \sigma^2 I$. In fact,

$$\frac{1}{n-K}E(\hat{\varepsilon}^\mathsf{T}\hat{\varepsilon}) = \frac{1}{n-K}E(\varepsilon^\mathsf{T}\varepsilon) - \frac{1}{n-K}E(\varepsilon^\mathsf{T}X(X^\mathsf{T}X)^{-1}X^\mathsf{T}\varepsilon)$$

$$= \frac{n\sigma^2}{n-K} - \frac{\sigma^2 K}{n-K} = \sigma^2.$$

Typically, $se_W > se_{LS}$, but it is not necessarily the case. Likewise, t-statistics computed with se_W are usually but not always smaller in magnitude than t-statistics computed with se_{LS}.

The **bootstrap** is a very popular method for obtaining confidence intervals or performing hypothesis tests. It is related to the **jackknife** and to **permutation** tests, which historically precede it. There can be computational reasons why this method is preferred to the usual approach based on estimating the unknown quantities of the asymptotic distribution. There can also be statistical reasons why the bootstrap is better than the asymptotic plug-in approach. The bootstrap has been shown to work in many different situations. We just give a procedure for linear regression. Let $H_n(x) = \Pr(c^\mathsf{T}\hat{\beta} \leq x | X)$ be the distribution of interest.

Multiplier Bootstrap Algorithm

1. Generate a sample of i.i.d. random variables $z_1^{(s)}, \ldots, z_n^{(s)}$ with mean zero and variance one (for example standard normals) for each s.
2. Let

$$T_s^* = c^\mathsf{T}(X^\mathsf{T}X)^{-1}\sum_{i=1}^n x_i \hat{\varepsilon}_i z_i^{(s)}. \tag{2.8}$$

3. Repeat S times and calculate the (empirical) distribution of $\{T_1^*, \ldots, T_S^*\}$ and use this distribution to approximate the unknown distribution $H_n(x)$, i.e.

$$\widehat{H}_B(x) = \frac{1}{S}\sum_{s=1}^S \mathbf{1}(T_s^* \leq x). \tag{2.9}$$

Critical values can be calculated as quantiles of this distribution. Specifically, we let $\widehat{H}_B^{-1}(\alpha) = \inf\{x : \widehat{H}_B(x) \geq \alpha\}$ and the $1-\alpha$-coverage confidence interval for $c^\mathsf{T}\hat{\beta}$ given by the closed interval

$$\left[c^\mathsf{T}\hat{\beta} - \widehat{H}_B^{-1}(\alpha/2), c^\mathsf{T}\hat{\beta} + \widehat{H}_B^{-1}(1-\alpha/2)\right]. \tag{2.10}$$

This works because conditional on the data T_s^* is mean zero and has variance $c^\mathsf{T}(X^\mathsf{T}X)^{-1}X^\mathsf{T}SX(X^\mathsf{T}X)^{-1}c$.

Classical Tests

The classical theory of testing in the likelihood framework suggests three generic approaches: the Wald test, the score test or LM test, and the likelihood ratio (LR) test.

Chapter 2 Econometric Background

We let $\theta \in \mathbb{R}^p$ denote all the parameters and let L denote the likelihood function of the data. Consider the linear restrictions

$$H_0 : Q\theta = c,$$

where Q is a $q \times p$ matrix and $c \in \mathbb{R}^p$. The maximum likelihood estimator (MLE) of θ is denoted by $\widehat{\theta}$, while the restricted MLE is denoted by $\widetilde{\theta}$ (this maximizes L subject to the restrictions $Q\theta = c$). Now define the following test statistics:

$$\text{LR} : 2\left[\log \frac{L(\widehat{\theta})}{L(\theta^*)}\right] = 2\{\ell(\widehat{\theta}) - \ell(\widetilde{\theta})\},$$

$$\text{Wald} : (Q\widehat{\theta} - c)^{\mathsf{T}} \left\{ Q\left[-\frac{\partial^2 \ell}{\partial\theta\partial\theta^{\mathsf{T}}}\bigg|_{\widehat{\theta}}\right]^{-1} Q^{\mathsf{T}} \right\}^{-1} (Q\widehat{\theta} - c),$$

$$\text{LM} : \frac{\partial \ell}{\partial \theta}\bigg|_{\widetilde{\theta}}^{\mathsf{T}} \left[-\frac{\partial^2 \ell}{\partial\theta\partial\theta^{\mathsf{T}}}\bigg|_{\widetilde{\theta}}\right]^{-1} \frac{\partial \ell}{\partial \theta}\bigg|_{\widetilde{\theta}}.$$

The Wald test only requires computation of the unrestricted estimator, while the Lagrange multiplier only requires computation of the restricted estimator. The likelihood ratio requires computation of both. There are circumstances where the restricted estimator is easier to compute, and there are situations where the unrestricted estimator is easier to compute. These computational differences are what has motivated the use of either the Wald or the LM test. The LR test has certain advantages; but computationally it is the most demanding.

Testing is usually carried out by setting the null rejection frequency (Type I error) to some level α, for example $\alpha \in (0, 1)$. A more modern approach is to use the *p*-value.

Definition 34 *For a test statistic the p-value is the probability of just rejecting the hypothesis. A p-value close to one means no evidence against the null; a p-value close to zero means strong evidence against the null.*

Goodness of Fit

It is often of interest to measure the **Goodness of Fit** of the regression model. Define the residual sum of squares

$$RSS = \sum_{i=1}^{n}(y_i - \widehat{y}_i)^2.$$

The idea is to compare the residual sum of squares from a regression on given X with the residual sum of squares of the regression with an X that contains only ones. Let $\overline{\varepsilon} = y - \overline{y}i$, where $i = (1, \ldots, 1)^{\mathsf{T}}$.

Definition 35 *For a general $X = (x_1, \ldots, x_K)$, we define*

$$R^2 = 1 - \frac{\widehat{\varepsilon}^\top \widehat{\varepsilon}}{\overline{\varepsilon}^\top \overline{\varepsilon}}.$$

Remarks
1. When X contains a column vector of ones, $0 \leq R^2 \leq 1$. If X does not contain a column vector of ones, R^2 could be less than zero.
2. In the bivariate case, R^2 is the squared sample correlation between y and x.
3. R^2 is invariant to some changes of units. If $y \mapsto ay + b$ for any constants a, b, then $\widehat{y}_i \mapsto a\widehat{y}_i + b$ and $\overline{y} \mapsto a\overline{y} + b$, so R^2 is the same in this case. Clearly, if $X \mapsto XA$ for a nonsingular matrix A, then $\widehat{y}$ is unchanged, as is y and $\overline{y}$.
4. R^2 always increases with the addition of variables. With $K = n$ we can make $R^2 = 1$.
5. R^2 can't be used to compare across models with different y.

2.2 Time Series

We now consider time series data, i.e. data that are produced in sequence. There are some special features of these models. We start with univariate time series $\{y_t\}_{t=1}^T$.

2.2.1 Some Fundamental Properties

There are two main features of time series that we shall investigate: stationarity/nonstationarity and dependence. We first define stationarity, which comes in two flavours.

Definition 36 *Strong Stationarity. The stochastic process y is said to be strongly stationary if the vectors $(y_t, \ldots, y_{t+r})$ and $(y_{t+s}, \ldots, y_{t+s+r})$ have the same distribution for all t, s, r.*

Definition 37 *Weak Stationarity. The stochastic process y is said to be weakly stationary if the vectors $(y_t, \ldots, y_{t+r})$ and $(y_{t+s}, \ldots, y_{t+s+r})$ have the same mean and variance for all t, s, r.*

Most of what we say is restricted to (weakly) stationary series, but in the past 40 years there have been major advances in the theory of nonstationary time series. There has also been some exploration of the difference between weak and strong stationarity (see Chapter 11).

Definition 38 *Dependence. One measure of dependence is given by the autocovariance (also called the covariogram) of a stationary process or the autocorrelation function (also called the correlogram or ACF)*

$$\operatorname{cov}(y_t, y_{t-j}) = \gamma(j) \quad ; \quad \rho(j) = \frac{\gamma(j)}{\gamma(0)}.$$

Note that stationarity was used here in order to assert that these moments only depend on the gap j and not on calendar time t as well. For i.i.d. series, $\gamma(j) = 0$ for all $j \neq 0$, while for positively (negative) dependent series $\gamma(j) > (<)0$. Economics data often appear to come from positively dependent series, but some series have negative dependence for some horizons. Mixingness is the property that dependence dies out with horizon. It can be measured in different ways.

Definition 39 *Covariance mixing. A stationary sequence $\{y_t, t = 0, \pm 1, \ldots\}$ is said to be covariance mixing if*

$$\gamma(j) = \operatorname{cov}(y_t, y_{t+j}) \to 0 \quad as \ j \to \infty.$$

This just says that the dependence (as measured by the covariance) on the past shrinks with horizon. This is an important property that is possessed by many models. Covariance mixing property is only well-defined for weakly stationary processes, i.e. those for which the second moments exist. In Chapter 11 we find that some commonly used processes (GARCH) do not necessarily satisfy this restriction and so we may need some more general definitions.

Definition 40 *Strong mixing. A stationary sequence $\{y_t, t = 0, \pm 1, \ldots\}$ is said to be strong mixing (or α–mixing) if*

$$\alpha(k) = \sup_{A \in \mathcal{F}_{-\infty}^n, B \in \mathcal{F}_{n+k}^\infty} |\Pr(A \cap B) - \Pr(A)\Pr(B)| \to 0$$

as $k \to \infty$, where $\mathcal{F}_{-\infty}^n$ and $\mathcal{F}_{n+k}^\infty$ are two σ–fields generated by $\{y_t, t \leq n\}$ and $\{y_t, t \geq n + k\}$, respectively. We call $\alpha(\cdot)$ the mixing coefficient.

Definition 41 *Beta mixing. A stationary sequence $\{y_t, t = 0, \pm 1, \ldots\}$ is said to be β–mixing if*

$$\beta(k) = \sup_{A \in \mathcal{F}_{-\infty}^n, B \in \mathcal{F}_{n+k}^\infty} |\Pr(A|B) - \Pr(A)| \to 0$$

as $k \to \infty$. We call $\beta(\cdot)$ the mixing coefficient.

2.2 Time Series

We have $2\alpha(k) \leq \beta(k)$; see Doukhan (1994). For independent series we have $\Pr(A \cap B) = \Pr(A)\Pr(B)$ and $\Pr(A|B) = \Pr(A)$ for all events A, B, so that $\alpha(k) = \beta(k) = \gamma(k) = 0$.

2.2.2 Some Models

Definition 42 *ARMA models. The following is a very general class of models called* ARMA(p,q):

$$y_t = \mu + \phi_1 y_{t-1} + \cdots + \phi_p y_{t-p} + \varepsilon_t - \theta_1 \varepsilon_{t-1} - \cdots - \theta_q \varepsilon_{t-q},$$

where ε_t is i.i.d., mean zero, and variance σ^2. We shall for convenience usually assume that $\mu = 0$. We also assume for convenience that this model holds for $t = 0, \pm 1, \ldots$. It is convenient to write this model using lag polynomial notation

$$A(L)y_t = B(L)\varepsilon_t, \tag{2.11}$$

where the lag polynomials $A(L) = 1 - \phi_1 L - \cdots - \phi_p L^p$ and $B(L) = 1 - \theta_1 L - \cdots - \theta_q L^q$. Here, $Ly_t = y_{t-1}$. The process $\{y_t\}$ is stationary provided the roots of the polynomial $A(z) = 1 - \phi_1 z - \cdots - \phi_p z^p$ (that is, the values of z for which $A(z) = 0$) lie outside of the unit circle on the complex plane.

Special Case AR(1)

Suppose that

$$y_t = \phi y_{t-1} + \varepsilon_t,$$

where ε_t is i.i.d. with mean zero and variance σ^2 for $t = 0, \pm 1, \ldots$. Here, $A(L) = 1 - \phi L$ and $B(L) = 1$ in (2.11).

The condition $|\phi| < 1$ is necessary and sufficient for y_t to be a stationary process. Write $y_{t-1} = \phi y_{t-2} + \varepsilon_{t-1}$, and substituting back we obtain

$$\begin{aligned} y_t &= \varepsilon_t + \phi \varepsilon_{t-1} + \phi^2 y_{t-2} \\ &= \varepsilon_t + \phi \varepsilon_{t-1} + \phi^2 \varepsilon_{t-2} + \cdots \\ &= \sum_{j=0}^{\infty} \phi^j \varepsilon_{t-j}. \end{aligned}$$

This shows that y_t depends on all the past shocks. Note that we need $|\phi| < 1$ for the above sum to exist, i.e. this condition is necessary and sufficient for the limit $\lim_{k \to \infty} \sum_{j=0}^{k} |\phi|^j$ to be finite.

Now we calculate the moments of y_t using the stationarity property. Let $\mu = E(y_t) = E(y_{t-1})$. Then we must have $\mu = E(y_t) = \phi E(y_{t-1}) = \phi \mu$, which can be true if and only if $\mu = 0$. Furthermore,

$$\text{var}(y_t) = \phi^2 \text{var}(y_{t-1}) + \sigma^2,$$

which implies by a similar logic that

$$\gamma(0) = \frac{\sigma^2}{1-\phi^2},$$

where $\gamma(0) = \mathrm{var}(y_t) = \mathrm{var}(y_{t-1})$. This last calculation of course requires that $|\phi| < 1$. Finally,

$$\mathrm{cov}(y_t, y_{t-1}) = E(y_t y_{t-1}) = \phi E(y_{t-1}^2) + 0,$$

which implies that

$$\gamma(1) = \phi \frac{\sigma^2}{1-\phi^2},$$

while

$$\mathrm{cov}(y_t, y_{t-2}) = E(y_t y_{t-2}) = \phi E(y_{t-1} y_{t-2}) = \phi^2 \frac{\sigma^2}{1-\phi^2}.$$

In general

$$\gamma(k) = \sigma^2 \frac{\phi^k}{1-\phi^2}; \quad \rho(k) = \frac{\phi^k}{1-\phi^2}. \tag{2.12}$$

The correlation function decays very fast (called exponential or geometric decay) towards zero as $s \to \infty$.

For the AR(2) case $y_t = \phi_1 y_{t-1} + \phi_2 y_{t-2} + \varepsilon_t$, we can likewise calculate the mean, variance, and autocovariance, which is left as an exercise. Regarding the stationarity condition, we must look at the solutions to the equation $1 - \phi_1 z - \phi_2 z^2 = 0$. These solutions are given explicitly by

$$z_1 = \frac{-\phi_1 + \sqrt{\phi_1^2 + 4\phi_2}}{2\phi_2} \quad ; \quad z_2 = \frac{-\phi_1 - \sqrt{\phi_1^2 + 4\phi_2}}{2\phi_2}.$$

It follows that the stationary region is

$$\{(\phi_1, \phi_2) : -1 < \phi_2 < 1, -2 < \phi_1 < 2, \phi_1 + \phi_2 < 1, \phi_2 - \phi_1 < 1\} \subset \mathbb{R}^2. \tag{2.13}$$

Note that it is not necessary that $|\phi_1| < 1$.

Moving Average MA(1)

Suppose that

$$y_t = \varepsilon_t - \theta \varepsilon_{t-1}, \tag{2.14}$$

where ε_t are i.i.d. with mean zero and variance σ^2. In this case, $E(y_t) = 0$ and $\mathrm{var}(y_t) = \sigma^2(1 + \theta^2)$. Furthermore, $\mathrm{cov}(y_t, y_{t-j}) = 0$ for $j \geq 2$, while

$$\mathrm{cov}(y_t, y_{t-1}) = E((\varepsilon_t - \theta \varepsilon_{t-1})(\varepsilon_{t-1} - \theta \varepsilon_{t-2})) = -\theta E(\varepsilon_{t-1}^2) = -\theta \sigma^2.$$

2.2 Time Series

Therefore,
$$\rho(1) = \frac{-\theta}{1+\theta^2}, \quad \rho(k) = 0, \ k = 2, \ldots.$$

This is a 1-dependent series. Note that the process is automatically stationary for any value of θ. If $|\theta| < 1$, we say that the process is **invertible** and we can write the process as

$$\sum_{j=0}^{\infty} \theta^j y_{t-j} = \varepsilon_t. \tag{2.15}$$

That is, $E(y_t | y_{t-1}, y_{t-2}, \ldots) = \sum_{j=1}^{\infty} \theta^j y_{t-j}$.

For general stationary and invertible ARMA(p,q) processes (2.11), we can write

$$\frac{A(L)}{B(L)} y_t = C(L) y_t = \sum_{j=0}^{\infty} \gamma_j y_{t-j} = \varepsilon_t,$$

$$y_t = \frac{B(L)}{A(L)} \varepsilon_t = D(L) \varepsilon_t = \sum_{j=0}^{\infty} \delta_j \varepsilon_{t-j}.$$

The first line is called the $AR(\infty)$ representation, and expresses y in terms of its own past. The second line is called the $MA(\infty)$ representation, and expresses y in terms of the past history of the random shocks.

2.2.3 Long Run Variance

We consider the properties of the sum of a time series, which is important in many contexts. Suppose that y_t satisfies $E(y_t) = 0$ and $E(y_t^2) \leq C < \infty$. Then, with $S_T = \sum_{t=1}^{T} y_t$, we have $E(S_T) = 0$ and

$$E(S_T^2) = \sum_{t=1}^{T} E(y_t^2) + \sum_{t=1}^{T} \sum_{\substack{t=1 \\ t \neq s}}^{T} E(y_t y_s),$$

which just follows directly. If the series is stationary (or at least the mean, variances, and covariances do not depend on time), then we may write $E(y_t y_s) = E(y_0 y_{t-s})$ for any t and by a change of variable argument we obtain exactly

$$E(S_T^2) = T E(y_0^2) + 2T \sum_{k=1}^{T} \left(1 - \frac{k}{T}\right) E(y_0 y_k). \tag{2.16}$$

Provided the series is weakly dependent, i.e. $\sum_{k=1}^{\infty} |E(y_0 y_k)| < \infty$, it follows that

$$\frac{1}{T} E(S_T^2) = E\left[\left(\frac{1}{\sqrt{T}} \sum_{t=1}^{T} y_t\right)^2\right] \to E(y_0^2) + 2 \sum_{k=1}^{\infty} E(y_0 y_k)$$

and by the law of large numbers (LLN) $S_T/T \xrightarrow{P} 0$. If the series, y_t is not mean zero, we just subtract the mean and obtain

$$\text{var}\left(\frac{1}{\sqrt{T}}S_T\right) \to \text{var}(y_0) + 2\sum_{k=1}^{\infty} \text{cov}(y_0, y_k) = \text{lrvar}, \qquad (2.17)$$

which is called the **long run variance** (as opposed to the short-run variance or just variance). In spectral analysis, this is 2π times the spectral density at frequency zero. Parzen (1957, equation 6.5), following earlier work of Bartlett (1950), proposed an estimator of this quantity

$$\widehat{\text{lrvar}} = \sum_{|k| \leq M_T} \left(1 - \frac{k}{M_T}\right) \widehat{\gamma}(k), \qquad (2.18)$$

where $\widehat{\gamma}(j)$ is the sample autocovariance (see below) and M_T is a sequence with $M_T \to \infty$ and $M_T/T \to 0$, and he established its consistency. This methodology was extended by Newey and West (1987) to allow y_t to be residuals from a general estimation problem and to allow some kinds of nonstationarity.

Herrndorf (1984, Corollary 1) established the following result for a general sequence of random variables not necessarily stationary (see also White and Domowitz (1984) for a similar result).

Theorem 13 *Suppose that $E(y_t) = 0$ and $E(|y_t|^{2+\delta}) \leq C < \infty$ for some $\delta > 0$ for $t = 1, 2, \ldots$. Suppose that for some $\sigma^2 = \text{lrvar} > 0$*

$$\frac{1}{T}E(S_T^2) \to \sigma^2,$$

and that

$$\sum_{k=1}^{\infty} \alpha(k)^{\frac{\delta}{2+\delta}} < \infty.$$

Then

$$\frac{1}{\sigma\sqrt{T}} S_T \Longrightarrow N(0, 1).$$

This central limit theorem (CLT) is commonly used in econometrics. It embodies a trade-off between the moment conditions and the rate of decay of the mixing coefficients. If $\delta = \infty$, then we only need $\sum_{k=1}^{\infty} \alpha(k) < \infty$, which allows $\alpha(k) = (k \log k)^{-1}$ for example, a slow rate of decay. On the other hand if δ is small, we require that $\alpha(k) \to 0$ fast as $k \to \infty$.

Hall and Heyde (1980) provide a comprehensive theory for martingale difference sequences including laws of large numbers and central limit theorems.

2.2 Time Series

2.2.4 Mean Reversion

The term mean reversion is widely used in financial econometrics. The common sense meaning of this term is that when the process is above its mean, the future value tends to be below the current value, i.e., it tends back towards its mean. Kim and Park (2014) give a formal definition in the context of a weakly stationary process. A special case of their definition 4.1 is given below.

Definition 43 *Strong mean reversion. A weakly stationary process is strongly mean reverting if for some $\delta < 0$*

$$\frac{1}{T} \sum_{t=1}^{T-1} (y_t - E(y_t))(y_{t+1} - y_t) \xrightarrow{P} \delta.$$

Example 17 *Consider a stationary AR(1) process*

$$y_t = \mu + \phi y_{t-1} + \varepsilon_t.$$

This may be written in two different ways:

$$y_t - E(y_t) = \phi (y_{t-1} - E(y_{t-1})) + \varepsilon_t,$$

$$y_{t+1} - y_t = \beta(y_t - E(y_t)) + \varepsilon_{t+1},$$

where $E(y_t) = \mu/(1-\phi)$ and $\beta = \phi - 1$. In this interpretation $E(y_t)$ is the long run value of the process and $\beta < 0$ measures the speed of convergence to the long run value: we see that values of $y_{t-1} > E(y_t)$ lead to in general $y_t - y_{t-1} < 0$ so that the process is drawn back to the mean.

This applies to short term mean reversion; any stationary process is mean reverting at the long horizon.

2.2.5 Nonstationary Processes

Suppose that

$$y_t = \mu + y_{t-1} + \varepsilon_t, \qquad (2.19)$$

where ε_t is i.i.d with mean zero and finite variance. This is called a unit root (plus drift) process. The special case with $\mu = 0$ is called the random walk; see Chapter 3 for more discussion. This process is nonstationary. In particular we have

$$y_t = y_0 + t\mu + \sum_{s=1}^{t} \varepsilon_s$$

for every t, where y_0 is the **initial condition**. Therefore,

$$E(y_t|y_0) = y_0 + t\mu$$

$$\text{var}(y_t|y_0) = t\sigma^2$$

and both of these moments tend to infinity linearly in t as $t \to \infty$ (except the mean when $\mu = 0$).

Letting Δ denote the **differencing operator**, we have $\Delta y_t = y_t - y_{t-1} = \mu + \varepsilon_t$, which is stationary, and we say that y_t is **difference stationary**.

Definition 44 *An explosive process is generated by*

$$y_t = \mu + (1+\delta)y_{t-1} + \varepsilon_t,$$

where $\delta > 0$. Then

$$y_t = y_0 + \mu \sum_{s=1}^{t}(1+\delta)^{t-s} + \sum_{s=1}^{t}(1+\delta)^{t-s}\varepsilon_s.$$

In this case both the mean and variance grow rapidly to infinity. Furthermore, y_t is not difference stationary.

2.2.6 Forecasting

We first consider forecasting inside some simple time series models. Consider the AR(1) process

$$y_t = \phi y_{t-1} + \varepsilon_t,$$

where ε_t is i.i.d. with mean zero and variance $\sigma^2 < \infty$. We want to forecast $y_{T+1}, y_{T+2}, \ldots, y_{T+r}$ given the sample information $\{y_1, \ldots, y_T\}$. We first assume that ϕ is known. We have $y_{T+1} = \phi y_T + \varepsilon_{T+1}$. Therefore, we forecast y_{T+1} by the conditional expectation

$$\widehat{y}_{T+1|T} = E(y_{T+1}|y_1, \ldots, y_T) = \phi y_T.$$

The reason for choosing the conditional expectation is that it minimizes the mean squared forecast error. The forecast error in this case is $y_{T+1} - \widehat{y}_{T+1|T} = \varepsilon_{T+1}$, which is mean zero and has variance σ^2. Any other forecast of y_{T+1} using the sample information $\{y_1, \ldots, y_T\}$ will have larger mean squared error.

What about forecasting r periods ahead? We have $y_{T+r} = \phi^r y_T + \phi^{r-1}\varepsilon_{T+1} + \cdots + \varepsilon_{T+r}$. Therefore, let

$$\widehat{y}_{T+r|T} = \phi^r y_T \tag{2.20}$$

be our forecast. The forecast error is $y_{T+r} - \widehat{y}_{T+r|T} = \phi^{r-1}\varepsilon_{T+1} + \cdots + \varepsilon_{T+r}$, which has mean zero and variance $\sigma^2(1 + \phi^2 + \cdots + \phi^{2r-2})$.

2.2 Time Series

In practice, we must use an estimate of ϕ, so that we compute instead

$$\widehat{y}_{T+r|T} = \widehat{\phi}^r y_T,$$

where $\widehat{\phi}$ is estimated from sample data. The forecast error is

$$y_{T+r} - \widehat{y}_{T+r|T} = y_{T+r} - \widehat{y}_{T+r|T}(\phi) + \widehat{y}_{T+r|T}(\widehat{\phi}) - \widehat{y}_{T+r|T}(\phi)$$
$$= \phi^{r-1}\varepsilon_{T+1} + \cdots + \varepsilon_{T+r} + \left(\widehat{\phi}^r - \phi^r\right) y_T.$$

Provided $\widehat{\phi}$ is a consistent estimator of ϕ (the sample size is large), the second term disappears as the estimation sample size increases. That is, the estimation error is relatively small compared to the intrinsic forecast error. The forecast error variance becomes $\sigma^2(1 + \phi^2 + \cdots + \phi^{2r-2})$.

We may provide a forecast interval around our point forecast under some conditions. Specifically, let

$$\widehat{y}_{T+r|T} \pm z_{\alpha/2} \times \sqrt{\widehat{\sigma}^2(1 + \widehat{\phi}^2 + \cdots + \widehat{\phi}^{2r-2})}, \qquad (2.21)$$

where $\widehat{\sigma}^2$ and $\widehat{\phi}$ are estimates of σ^2 and ϕ constructed from the sample. This is to be interpreted like a confidence interval. Notice that in large samples the interval becomes

$$\widehat{y}_{T+r|T} \pm z_{\alpha/2} \times \sqrt{\sigma^2(1 + \phi^2 + \cdots + \phi^{2r-2})},$$

which does not shrink to zero. The forecast interval is asymptotically valid provided $\varepsilon_t \sim N(0, \sigma^2)$. It says that with probability $1 - \alpha$ the realization of y_{T+r} should lie in this range.

We now suppose that the data are generated by a more general process

$$y_t = \mu + \sum_{j=1}^{\infty} \phi_j(y_{t-j} - \mu) + \varepsilon_t, \qquad (2.22)$$

where ϕ_j are coefficients. This is consistent with y_t being an ARMA(p,q) process. We would like to forecast y_{T+1} by

$$\widehat{y}_{T+1|T} = \mu + \sum_{j=1}^{\infty} \phi_j(y_{T+1-j} - \mu) = \mu + \sum_{j=1}^{T} \phi_j(y_{T+1-j} - \mu) + \sum_{j=T+1}^{\infty} \phi_j(y_{T+1-j} - \mu).$$

In practice, we do not observe $y_0, y_{-1}, \ldots$, and we do not know the parameters μ, ϕ_j, so we use only the first sum with estimated parameters

$$\widehat{y}_{T+1|T} = \widehat{\mu} + \sum_{j=1}^{T} \widehat{\phi}_j(y_{T+1-j} - \widehat{\mu}) \qquad (2.23)$$

to make the forecast. Of course we need to have good estimates of $\widehat{\phi}_j$, $j = 1, \ldots, T$, which generally require more structure such as provided by an ARMA(p,q) model in which the ϕ_j are functions of a small number of underlying parameters.

EWMA Forecasting

The exponentially weighted moving average (EWMA) forecast method is due to Holt (1957) and Winters (1960). The idea is to use a recursive structure, that is let for each $t > 1$

$$\widehat{y}_{t+1|t} = \alpha y_t + (1 - \alpha)\widehat{y}_{t+1|t}, \qquad (2.24)$$

and let $\widehat{y}_{1|0} = y_0$ some chosen initial value such as the sample average. The parameter $\alpha \in [0, 1]$ and determines the weight given to past observations. When $\alpha = 1$, $\widehat{y}_{T+1|T} = y_T$, while if $\alpha = 0$, $\widehat{y}_{T+1|T} = y_0$. In general we have

$$\widehat{y}_{T+1|T} = \sum_{t=1}^{T-1} \alpha(1-\alpha)^j y_{T-j} + (1-\alpha)^T y_0.$$

Many applications of this method chose the value α from past experience or by trial and error so that it is not an estimated quantity. This is what makes the method simple and attractive to use. This approach eschews modelling.

Forecast Evaluation

Forecast evaluation is a very important activity. This is usually done by splitting the sample into an estimation sample and an evaluation sample.

Definition 45 *Let the full sample be denoted* $\{y_1, \ldots, y_n\}$ *with* $n = T + K$, *and let the estimation sample be* $\{y_1, \ldots, y_T\}$, *and the evaluation sample* $\{y_{T+1}, \ldots, y_{T+K}\}$.

We then calculate forecasts $\widehat{y}_{T+1|T}, \ldots, \widehat{y}_{T+K|T}$ for all the evaluation samples and compute some measure of performance. Campbell and Thompson (2008) propose the out-of-sample R^2, which is defined in this case as

$$R^2_{OOS} = 1 - \frac{\sum_{j=1}^{K}(y_{T+j} - \widehat{y}_{T+j|T})^2}{\sum_{j=1}^{K}(y_{T+j} - \overline{y})^2}, \qquad (2.25)$$

where $\overline{y}$ is the sample mean of observations using the sample information.

If we are considering one-step ahead forecasts we may want to revise the estimation sample each time. That is, for forecasting $T + j$ we use the estimation sample $\{y_1, \ldots, y_{T+j-1}\}$ and compare with the mean $\overline{y}_{1:T+j-1}$ of the new estimation sample, that is,

$$R^2_{OOS,j} = 1 - \frac{\sum_{j=1}^{K}(y_{T+j} - \widehat{y}_{T+j|T+j-1})^2}{\sum_{j=1}^{K}(y_{T+j} - \overline{y}_{1:T+j-1})^2}.$$

2.2 Time Series

The choice between these two methods depends on what the purpose of the exercise is.

These measures can compare the forecasting performance of different methods. An alternative way of comparing two forecasts is to carry out a statistical test of the null hypothesis that the two forecasts have equal predictive power against the alternative that one is superior. Diebold and Mariano (1995) propose a test of **equal predictive accuracy**. Let there be two forecasts of a series y_{T+j} with errors $\widehat{\varepsilon}_j$ and $\widehat{\varepsilon}_j^*$, $j = 1, \ldots, K$. Under the null hypothesis the two errors have the same mean, but they may both be highly autocorrelated, which requires that one take account of this. Let

$$S = \frac{\frac{1}{\sqrt{K}} \sum_{j=1}^{K} d_j}{\sqrt{\widehat{\mathrm{lrvar}}(d_j)}}, \qquad (2.26)$$

where $d_j = \widehat{\varepsilon}_j - \widehat{\varepsilon}_j^*$ and $\widehat{\mathrm{lrvar}}(d_j)$ is an estimate of the long run variance $\mathrm{lrvar}(d_j)$ of the series d_j. Provided the evaluation sample size K is large and provided the forecast errors are stationary or approximately so, S can be approximated by a standard normal random variable.

Francis X. Diebold (born November 12, 1959) is an American economist known for his work in predictive econometric modelling, financial econometrics, and macroeconometrics. He earned both his BS and PhD degrees at the University of Pennsylvania ("Penn"), where his doctoral committee included Marc Nerlove, Lawrence Klein, and Peter Pauly. He has spent most of his career at Penn, where he has mentored approximately 75 PhD students. Presently he is Paul F. and Warren S. Miller Professor of Social Sciences and Professor of Economics at Penn's School of Arts and Sciences, and Professor of Finance and Professor of Statistics at Penn's Wharton School. He is also a Faculty Research Associate at the National Bureau of Economic Research in Cambridge, MA, and author of the No Hesitations blog. Diebold is an elected Fellow of the Econometric Society, the American Statistical Association, and the International Institute of Forecasters, and the recipient of Sloan, Guggenheim, and Humboldt Fellowships. He has served on the editorial boards of *Econometrica*, *Review of Economics and Statistics*, and *International Economic Review*. He has held visiting professorships at Princeton University, University of Chicago, Johns Hopkins University, and New York University. He was President of the Society for Financial Econometrics (2011–2013) and Chairman of the Federal Reserve System's Model Validation Council (2012–2013). In predictive econometric modelling Diebold is well known for the **Diebold–Mariano test** for assessing point forecast accuracy, methods for assessing density forecast conditional calibration, and his textbook, *Elements of Forecasting*. In financial econometrics Diebold is well known for his contributions to volatility modelling, including the Diebold–Nerlove latent-factor ARCH model and the Andersen–Bollerslev–Diebold–Labys extraction of realized volatility from high-frequency asset returns.

It is not necessary to have a fully specified model to forecast. In fact, a model that fits the data well in sample may do poorly at forecasting **out of sample**. At the extreme case one may perfectly fit the sample data (the in-sample R^2 is one), but such approaches generally fail to forecast the future well. Unfortunately it is equally true that a procedure that forecasts well in one period is not guaranteed to forecast well in another period.

2.2.7 Multivariate Time Series

Suppose that $y_t \in \mathbb{R}^n$ is a vector (weakly) stationary series, then the autocovariance matrix is defined as

$$\Gamma(j) = E\left((y_t - \mu)(y_{t-j} - \mu)^\mathsf{T}\right)$$

for $j = 0, \pm 1, \ldots$. In the bivariate case X_t, Y_t we have the cross-covariance functions

$$\gamma_{XY}(k) = \operatorname{cov}(X_t, Y_{t+k}) \quad ; \quad \gamma_{YX}(k) = \operatorname{cov}(Y_t, X_{t+k}).$$

In the univariate case we have $\gamma_{XX}(k) = \gamma_{XX}(-k)$, but the multivariate analogue of this is not necessarily true. In particular, we have $\gamma_{XY}(k) \neq \gamma_{XY}(-k)$, although $\gamma_{XY}(k) = \gamma_{YX}(-k)$. Suppose that $\gamma_{XY}(k) \neq 0$ and $\gamma_{YX}(k) = 0$. Then we interpret this as meaning that X is causing future Y but not vice versa. This is an important property that is possessed by many models.

Definition 46 *Vector autoregression (VAR) models. The following is a very general class of models called VAR(p):*

$$\Phi_0 X_t = \mu + \Phi_1 X_{t-1} + \cdots + \Phi_p X_{t-p} + \varepsilon_t,$$

where ε_t is i.i.d., mean zero, and variance matrix Ω_ε. We also assume for convenience that this model holds for $t = 0, \pm 1, \ldots$. It is convenient to write this model using lag polynomial notation

$$A(L)(X_t - \mu) = \varepsilon_t,$$

where the lag polynomial $A(L) = \Phi_0 - \Phi_1 L - \cdots - \Phi_p L^p$.

This model is widely used to describe mutual dependence in a time series setting. Note that we may write this model in **reduced form**

$$X_t = \widetilde{\mu} + \widetilde{\Phi}_1 X_{t-1} + \cdots + \widetilde{\Phi}_p X_{t-p} + \widetilde{\varepsilon}_t,$$

where: $\widetilde{\mu} = \Phi_0^{-1}\mu$, $\widetilde{\Phi}_1 = \Phi_0^{-1}\Phi_1, \ldots, \widetilde{\Phi}_p = \Phi_0^{-1}\Phi_p$, and $\widetilde{\varepsilon}_t = \Phi_0^{-1}\varepsilon_t$.

2.2 Time Series

In the VAR(1) case the process is stationary provided the matrix Φ has eigenvalues below one (note this doesn't mean that elements of A are less than one).

The univariate AR(p) process can be written as a VAR(1)

$$y_t = \phi_1 y_{t-1} + \cdots + \phi_p y_{t-p} + \varepsilon_t,$$

$$x_t = A x_{t-1} + u_t,$$

$$x_t = \begin{pmatrix} y_t \\ \vdots \\ y_{t+1-p} \end{pmatrix}, \quad u_t = \begin{pmatrix} \varepsilon_t \\ 0 \\ \vdots \end{pmatrix}, \quad A = \begin{pmatrix} \phi_1 & \phi_2 & \cdots & \phi_p \\ 1 & 0 & \cdots & \\ 0 & 1 & & \\ 0 & \cdots & & \ddots \end{pmatrix}.$$

On the other hand, the marginal process derived from a VAR(1) may be very complicated, i.e., $E(y_{1t}|y_{1,t-1}, y_{1,t-2}, \ldots)$ may depend on all lags. In some sense this shows the value of the VAR process in reducing the number of parameters needed to describe the process.

Definition 47 *Vector moving average (VMA) models. The following is a very general class of models called VMA(q):*

$$X_t = \mu + \varepsilon_t + \Theta_1 \varepsilon_{t-1} + \cdots + \Theta_q \varepsilon_{t-q},$$

where ε_t is i.i.d., mean zero, and variance matrix Ω_ε. We also assume for convenience that this model holds for $t = 0, \pm 1, \ldots$ It is convenient to write this model using lag polynomial notation

$$X_t - \mu = B(L) \varepsilon_t,$$

where $B(L) = I - \Theta_1 L - \cdots - \Theta_q L^q$. We have for $j = 1, \ldots, q$

$$\Gamma(j) = B(L) \Omega_\varepsilon B(L)^\intercal.$$

The **impulse response function** is a common way of measuring the dynamic effects of shocks (ε_t) on outcomes.

2.2.8 Granger Causality

Let Y and X be stationary time series. The concept of **Granger causality** is widely used to describe the temporal relationship between two series in terms of causality. We suppose that we may write

$$Y_t = a_0 + a_1 Y_{t-1} + \cdots + a_p Y_{t-p} + b_1 X_{t-1} + \cdots + b_p X_{t-p} + \varepsilon_{yt},$$

$$X_t = c_0 + c_1 X_{t-1} + \cdots + c_p X_{t-p} + d_1 Y_{t-1} + \cdots + d_p Y_{t-p} + \varepsilon_{xt},$$

where ε_{yt} and ε_{xt} are taken to be two uncorrelated white-noise series, $E(\varepsilon_{yt}\varepsilon_{ys}) = 0 = E(\varepsilon_{xt}\varepsilon_{xs})$ for all $t \neq s$. In theory p can equal infinity but in practice, of course, due to the finite length of the available data, p will be assumed finite and shorter than the given time series. We may say that X is causing Y provided some b_j is not zero. Similarly, we may say that Y is causing X if some d_j is not zero. If both of these events occur, there is said to be a feedback relationship between X and Y. One may test for the absence of causality by applying an F-test on the coefficients. This is automatically done in EVIEWS.

2.2.9 Rolling Window

We have a very long time series for daily stock returns, nearly a hundred years of daily data, and for other series likewise. This suggests that large sample approximations should be good, provided the model we apply is true. But there is the rub; it is hard to believe in many cases that means, variances, and covariances stay constant over such a long time period. To counter this issue, many authors use the rolling window method. Suppose that we have observations $X_1, \ldots, X_T$, where T is large.

Definition 48 *For each t with $t \geq k+1$ and $t \leq T-j$ we consider a window of time periods*

$$\mathcal{W}_{t;-k,j} = \{t-k, \ldots, t, \ldots, t+j\}, \tag{2.27}$$

which are centered at the time t and contain $k+j+1$ points, and the corresponding observations on the process X_t.

Common choices include backward windows with $j=0$ and $k>0$, forward windows with $k=0$ and $j>0$, and symmetric two sided windows with $k=j>0$. Instead of computing an estimator or test statistic from the whole sample we compute them for each window of data. In that way we can track the time variation of the statistic and compare with the value computed from the whole sample. Common choices of k, j are such that the window contains a year of data (252 observations), five years of observations, or ten years of observations. Note that rolling windows overlap a lot, e.g. $\mathcal{W}_{t;-k,k}$ and $\mathcal{W}_{t+1;-k,k}$ contain $2k-1$ common observations, and so the statistics computed on $\mathcal{W}_{t;-k,k}$ and $\mathcal{W}_{t+1;-k,k}$ are usually quite similar, and if they are not this calls attention to some potential issue. An alternative scheme is called **recursive windows**, where essentially $k = t-1$ so that the windows grow as time t increases.

3 Return Predictability and the Efficient Markets Hypothesis

In this chapter we discuss predictability of asset prices. The **efficient markets hypothesis** essentially predicts that there is no predictability, which to the untrained eye looks a bit like a non-theory. On closer scrutiny however this hypothesis is potentially quite powerful in terms of the predictions it makes. Of course there are many details that have to be filled in. There are several different statistical notions of predictability in widespread usage. We may categorize these into linear notions, which we will consider in this chapter, and nonlinear notions, which will be considered in Chapter 4. In either case one also has to consider a benchmark amount of predictability that is normal and then decide whether that has been exceeded or not in the data. A key issue is the frequency of the data at which predictability is measured, and that governs the type of analysis that is suitable for answering these questions.

We first discuss the efficient markets hypothesis and its implications, and then we talk about statistical models that are consistent with it, before considering how to test the hypothesis based on data.

3.1 Efficient Markets Hypothesis

The efficient markets hypothesis (EMH) is apparently very old, in some sense going back to Bachelier (1900) and before. Fama (1970) is often credited with providing a constructive definition. His notion was that a market in which prices always "fully reflect" available information is called "efficient," hence the term EMH. The logic is as follows. If prices are predictable, then there are opportunities for superior returns (i.e., a free lunch). These will be competed away quickly by traders seeking profit. Therefore, prices should be unpredictable. If a security is believed to be underpriced, buying pressure will lead to a jump up in the price level where it is no longer thought to be a bargain. If a security is believed to be overpriced, selling pressure would effect a jump down in the price to a level where the security was no longer thought too expensive. Market forces respond to news extremely fast and make prices the best available estimates of fundamental values, i.e., values justified by likely future cash flows and preferences of investors/consumers.

Example 18 *There is a famous joke about two Chicago economists who see a $10 bill on the sidewalk. One stops the other from picking it up by saying, "if this were real it would have already been picked up."*

It has been repeated so many times, it can't be funny any more, and of course we have all found money on the floor at some point. However, it does bring out the issue of timing and scale, which are important in the sequel. "Extremely fast" depends on the context. If it were a penny on the track of a monster truck racing track, then it could stay unpicked for a long time.

We distinguish among three forms of market efficiency depending on the information set with respect to which efficiency is defined:

1. **Weak form.** (1) Information from historical prices are fully reflected in the current price; and (2) one can't earn **abnormal profits** from trading strategies based on past prices alone.
2. **Semi-strong form.** (1) All public information (past prices, annual reports, quality of management, earnings forecasts, macroeconomic news, etc.) is fully reflected in current prices; and (2) one can't earn abnormal profits from trading strategies based on public information.
3. **Strong form.** (1) All private and public information is fully reflected in current prices; and (2) one can't earn abnormal profits from trading strategies based on all information including public and private.

We must define what is meant by abnormal profits. There must be a **risk premium** to compensate risk averse investors for taking risks, and this should feature in the benchmark against which to measure returns. The issue is that we need an asset pricing model to define the risk premium and hence "normal" security returns against which investor returns are measured. Different asset pricing models will deliver different risk premia. Suppose that R_t^* is the required or normal return over an interval $[t, T]$, for any $T > t$ (for notational simplicity we are dropping the dependence of R_t^* on T), that arrives from such an asset pricing model. This quantity is known at time t. Then under the null hypothesis ($\mathcal{F}_t$) it holds that the return R_T on any risky asset realized over the same time interval satisfies

$$E(R_T|\mathcal{F}_t) = R_t^*. \qquad (3.1)$$

You can't make more money on average than R_t^*. In the case where there is an observable risk free rate R_{ft} we can write $R_t^* = R_{ft} + \pi_t$, where π_t is the risk premium, which we may suppose is non-negative and determined by some model.

In the weak form case $\mathcal{F}_t$ contains only information on prices, in the semi-strong form case $\mathcal{F}_t$ contains additional publicly available information at time t, while in the strong form case $\mathcal{F}_t$ contains all information public and private available at time t. The words weak, semi-strong, and strong convey the meaning that, for example, the strong form hypothesis is stronger than the weak form hypothesis so that if the strong form hypothesis were true it would imply the weak form hypothesis' veracity. That is, if (3.1) holds and $\mathcal{F}_t' \subseteq \mathcal{F}_t$, then

$$E(R_T - R_t^*|\mathcal{F}_t') = 0, \qquad (3.2)$$

3.1 Efficient Markets Hypothesis

which is a consequence of the law of iterated expectations. There are further implications of this hypothesis. Recall that

$$\text{var}(r_{t+j}) = E\left(\text{var}(r_{t+j}|\mathcal{F}_t)\right) + \text{var}\left(E(r_{t+j}|\mathcal{F}_t)\right).$$

It follows that if $\mathcal{F}_t' \subseteq \mathcal{F}_t$, then $\text{var}(r_{t+j}|\mathcal{F}_t') \geq \text{var}(r_{t+j}|\mathcal{F}_t)$, so by increasing the information set we can reduce the conditional variance. However, this implies that $\text{var}(E(r_{t+j}|\mathcal{F}_t)) \geq \text{var}(E(r_{t+j}|\mathcal{F}_t'))$ so that as we increase the information set our forecasts become more variable.

The above formulation of the EMH is rather vague; it has many undefined terms. For example, the term "fully reflected" is usually taken to mean that all the information is instantly reflected in prices. The word instantly is really a fiction, since every action and reaction in our world are constrained by the speed of light, which is defined as 299792458 m/s in a vacuum. In practice one must decide what is a reasonable time frame to correspond to the essence of the theory. The precise definition of information sets is also subject to discussion. Typically, the weak form information set is taken to mean the historical prices of the security in question and recorded at the specific frequency that is the focus of study. For example, daily prices on IBM stock are used to predict daily prices on IBM stock. This does of course beg the question, which daily price? For daily data, the convention is the closing price, which is nowadays often obtained through an auction mechanism that delivers a single price at the close of trading. However, one could also include the opening price, the within day high price, the within day low price, and in fact the entire transaction record within each past day. One may additionally have transaction prices obtained from multiple exchanges or on slightly different partitions of the underlying security (like Berkshire Hathaway A and B shares). By historical prices, could one also mean the prices of other related securities? If the delineation is around the costs of acquiring the information and using it, it seems that past transaction prices of other assets are just as easy to obtain. The semi-strong form hypothesis allows for an even larger information set that includes publicly available information that is routinely released and other information that arrives spontaneously, not to a fixed schedule. The strong form allows for all information, whether publicly available or not. For example, some information is known only to insiders in the firm or regulatory authorities.

We next consider the notion that information is fully reflected in prices. We can in principle approach this issue without taking a stand on what the risk premium is.

Definition 49 *Suppose that $\mathcal{H}_t$ is the price history of a financial market at time $t \geq 0$, the set of all public information at time t can be denoted by $\mathcal{F}_t \supseteq \mathcal{H}_t$, and the set of all information can be denoted by $\mathcal{G}_t \supseteq \mathcal{F}_t$. Let P_T be the price of an asset at time $T > t$. Suppose that*

$$\Pr(P_T \leq x|\mathcal{F}_t) = \Pr(P_T \leq x|\mathcal{H}_t) \tag{3.3}$$

*for all $t \geq 0$ and $x \in \mathbb{R}$. A financial market where (3.3) is satisfied is said to **absorb** $\{\mathcal{F}_t\}$; Frahm (2013).*

All current information is present in current prices, but future prices may depend on additional sources of randomness, so that $P_t = f(P_{t-1}, P_{t-2}, \ldots, \varepsilon_t, \eta_t, \ldots)$ for random variables ε_t etc. that occur between $t-1$ and t. However, these represent unanticipated stochastic "shocks" to prices. The unanticipated component can be very large relative to the already known or forecastable part.

Equation (3.3) says that the price P_T is independent of $\mathcal{F}_t$ conditional on the price history $\mathcal{H}_t$. The conditional distribution of future asset returns might depend on the history $\mathcal{H}_t$ of asset prices but not on any additional information that is contained in $\mathcal{F}_t$. The relationship expressed by (3.3) can be interpreted as a probabilistic definition of the semi-strong and strong form hypotheses stated above. If the market is strong-form efficient, all private information, except for the price history $\mathcal{H}_t$, can be ignored because it is already "incorporated" in $\mathcal{H}_t$. Hence, if somebody aims at calculating the conditional distribution of P_T for any T with $0 \leq t < T$, the weaker information set $\mathcal{H}_t$ is as good as the stronger information set $\mathcal{F}_t$, i.e., all private information beyond the price history is simply useless. Likewise, if $\mathcal{F}_t$ is the set of all public information in addition to prices, then this definition corresponds to the notion of semi-strong form efficiency.

The condition (3.3) is rather hard to check in practice, and we should consider weaker conditions based around the first moments, specifically, $E(P_T|\mathcal{F}_t) = E(P_T|\mathcal{H}_t)$, which implies

$$E(R_T|\mathcal{F}_t) = E(R_T|\mathcal{H}_t), \tag{3.4}$$

provided the expectations exist. This implies that

$$E\left((R_T - E(R_T|\mathcal{H}_t))|\mathcal{F}_t\right) = 0, \tag{3.5}$$

by the law of iterated expectation. To test this hypothesis we need to make some restrictions on the expectations and information sets.

In the sequel we mostly focus on the very simple case where the required return is constant over time, that is, the risk free rate is constant and the risk premium is constant. If the marginal investor were risk neutral then the risk premium would be zero. The risk premium for a short horizon could be constant or small in a variety of other situations. The above formulae (3.1) and (3.3) implicitly allow that time is continuous, which Merton (1992) argued is a natural framework for financial analysis. We will mostly work with a discrete time framework in which time can take only integer values so that $T = t + k$ for $k = 1, 2, \ldots$ and $t = 1, 2, \ldots$.

3.1.1 Investment Strategies

We next discuss some well known investment strategies that are at odds with various versions of the EMH. First, **technical analysts** (or **chartists**) believe that prices contain patterns that are repeated and hence can be exploited by trading in the right direction. A well known example called **Head and Shoulders** is given in Figure 3.1. Once the rightmost "neckline" is breached, it is believed that prices will fall further, so this is a bearish prediction. Other tools they use look more like smoothing methods and time series trend analysis, but with different terminology.

3.1 Efficient Markets Hypothesis

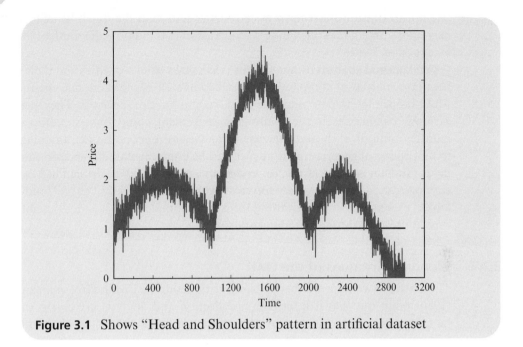

Figure 3.1 Shows "Head and Shoulders" pattern in artificial dataset

The weak form EMH implies that technical analysis is hopeless. If everybody used these trading rules profitably, wouldn't that invalidate them? Lo and Hasanhodzic (2010) connect the analysis of chartists to nonlinear time series analysis. They show how to convert observed price history into a numerical score that identifies, say, "head and shoulderness." They show that there is some statistical basis to their work, but provide the tools to replace them by automated systems.

Momentum investing looks for the market trend to move significantly in one direction on high volume and then trade in the same direction. Stock prices can surge suddenly, and continue to rise when people try to reap profit from the upward trend. This stems from the rate of information release and people's herd mentality. A simple example of this strategy is to buy a stock when the recent price is above a moving average of past prices and sell it when it's below the moving average; see Exercise 2. A simple strategy is to rank the sectors and buy the top stocks when their trailing moving average exceeds a threshold.

Mean reversion or **contrarian** investors assume that the price of the stock will over time revert back to its long-time moving average price. They use statistics to determine the trading bounds within which they operate. If the stock is trading significantly above the moving average, they will short sell it. On the other hand, if the stock is trending significantly below its moving average, they will buy it. For both momentum and mean reversion traders, the holding period is a key dimension along which the strategies operate. There are high frequency and low frequency momentum strategies for example. See Faber (2010) for a recent discussion of momentum and contrarian trading strategies. In Exercises 1–3 you are asked to investigate some of these.

Sentiment analysis trading derives from crowd psychology, where investors stay up-to-date on recent news, and purchase stocks predict the crowd's reaction. They attempt to capture short-term price changes and reap the quick benefits. Investors may monitor sources including Google search trends, media outlets, blogs and forums, and Twitter posts.

Fundamental analysts estimate future cash flows from securities and their riskiness, based on analysis of company-relevant data as well as the economic environment in which it operates, to determine the fundamental price of securities. They analyze past earnings, balance sheets, the quality of management, competitive standing within the industry, outlook of the industry, as well as the entire economy, i.e., anything relevant to the process of generating future profits. The fundamental reference on value investing is Graham and Dodd (1928). In a simple example, investors can use the price to earnings ratio (P/E) as a proxy for value where low P/E looks "cheap" and high P/E looks "expensive." The semistrong form EMH implies that it is not possible to earn superior returns through fundamental analysis.

3.1.2 Economic Critiques of the EMH

There are two well-known theory-based critiques of the EMH. Grossman and Stiglitz (1980) point out that if information collection and analysis are costly, there must be compensation for such activity in terms of extra risk-adjusted returns, otherwise rational investors would not incur such expenses. They propose a model where investors have to pay to acquire information that is relevant for the asset valuation. If the signal is perfectly informative about stock returns, the information is reflected in stock prices, and uninformed investors can observe the information through the informative price system; but then why should any investors purchase the information and bring it into prices? In their model markets cannot be fully informationally efficient; rather there can be equilibrium with less than perfectly informative prices. They establish that in equilibrium

$$1 - \rho_\theta^2 = \frac{\exp(\gamma c) - 1}{\sigma_\theta^2/\sigma_\varepsilon^2}, \qquad (3.6)$$

where the signal noise ratio is $\sigma_\theta^2/\sigma_\varepsilon^2$, the cost of signal acquisition is c, γ is the risk aversion of the investors, and ρ_θ^2 is the squared correlation between the signal and the equilibrium price, which measures the informativeness of the price system. A perfectly informative price system has $\rho_\theta^2 = 1$, which is not possible here when $c > 0$ and $\gamma > 0$. Their model allows some comparative statics regarding the change in efficiency that one might expect in a **big data** world with endless cheap data available on your smart phone. If $\sigma_\theta^2/\sigma_\varepsilon^2$ increases and c decreases then ρ_θ^2 should rise and the price system should be more informative. On the other hand, one might imagine a more complex scenario in which both the signal and the noise increase and where it is not the absolute cost of information (whatever that means) but the relative cost of information (cost per unit of σ_θ^2) that matters, in which case the predictions are ambiguous. In the real world, we don't expect investors to be faced with a one-shot game and we wouldn't expect prices to instantly reflect information; rather there may be a process of bringing information into prices. A model that has this feature is discussed in some detail in Chapter 5.

A further critique of market efficiency has come from Shleifer and Vishny (1997). They look at one of the cornerstones of finance: that **arbitrage opportunities** do not exist. They argue that this may not be true. Textbook arbitrage is a costless, riskless, and profitable trading opportunity. In practice it is usually both costly and risky. Also, it is conducted by a small number of highly specialized professionals in big banks using other people's capital. The agency relationship implies that the agents may have a short horizon. If mispricing temporarily worsens (the best opportunity to make long-term profit!), investors and clients may judge the manager as incompetent and not only refuse to provide additional capital (margin call) but make withdrawals, thus forcing him to liquidate positions at the worst time and/or on the worst terms. The consequence for the manager is no performance fees, and perhaps it could even be a career-ender. Therefore, a rational specialized arbitrageur stays away. Thus, there may be no easy way to make excess profit, and at the same time prices may deviate from fundamentals. In practice, some arbitrages are conducted by agents and some are by principals. In recent times, the time frame over which such arbitrages are conducted is often very short. For short-lived arbitrage opportunities – those lasting fractions of a second – attention costs and technological constraints are the main impediments to the imposition of the law of one price. These barriers are falling as high-frequency arbitrageurs massively invest to detect and exploit ever faster arbitrage opportunities. Some argue that returns on high-speed arbitrage are substantial because arbitrage opportunities are frequent at the timescale of milliseconds (see Budish, Cramton, and Shim 2015).

3.1.3 Econometric Issues with Testing

There are several econometric issues with testing the EMH. The main issue is that we require a specification of **normal returns** or the required risk premium against which to compute residuals. If we reject the hypothesis that investors can't achieve superior risk-adjusted returns, we don't know if markets are inefficient or if the underlying model for normal returns is misspecified (the **joint hypothesis problem**). Therefore, you can never truly reject the weak form EMH; it is not falsifiable. A further issue is that the information sets that feature in (3.1) are very large. Although this implies restrictions on the predictability given any smaller information set (3.2), if we do not reject the EMH given the smaller information set, this may be because we have chosen an uninformative information set. A further issue is exactly which properties of returns we should maintain for the purposes of constructing statistical hypothesis tests.

3.2 The Random Walk Model for Prices

The **random walk** model captures the notion of the absence of predictability. There are several different versions of this, which we will describe in turn. Consider the discrete time stochastic process X_t, $t = 1, 2, \ldots$, where

$$X_t = X_{t-1} + \varepsilon_t, \tag{3.7}$$

where ε_t is a shock or innovation process that usually has mean zero. This is called a random walk or a drunkard's walk because the position at time t is given by the position at the previous period plus a noisy increment that can be in any direction or magnitude. This captures the notion that the future position of the drunkard is hard to predict given his current location. A great deal has been written about the properties of random walks. Fortunately, the drunkard will get home, so long as he lives on the ground floor, but it may take a long time. The random walk plus drift satisfies

$$X_t = \mu + X_{t-1} + \varepsilon_t, \tag{3.8}$$

where the drift term μ may be non zero. In this case there is an overall direction of travel determined by μ but the position relative to that is as unpredictable as for a driftless random walk. In this case the drunkard will never get home if the drift takes him in the wrong direction. We have

$$X_t = \mu + \mu + X_{t-2} + \varepsilon_t + \varepsilon_{t-1} = \cdots = X_0 + t\mu + \sum_{s=1}^{t} \varepsilon_s. \tag{3.9}$$

From this we see that the position at time t is given by the accumulation of the random shocks $\sum_{s=1}^{t} \varepsilon_s$ and the cumulated drift as well as the starting point. In financial applications, $X_t = P_t$ or $X_t = p_t = \log P_t$. The drift μ is taken here to mean the required return or normal return in the market. Historically, μ was often assumed to be zero, and for high frequency data this may not be a bad assumption as we shall see. More generally, we may wish to allow μ to be time varying and to have arrived from some asset pricing model that specifies the evolution of the risk premium. That is, $X_t = \mu_t + X_{t-1} + \varepsilon_t$ with $\mu_t = \Psi(X_{t-1}, X_{t-2}, \ldots)$. For market returns, we might expect that

$$\mu_t = h(\text{var}(r_t|\mathcal{F}_{t-1})) \tag{3.10}$$

for some increasing function h, where the conditional variance $\text{var}(r_t|\mathcal{F}_{t-1})$ measures the risk of the asset. If the risk $(\text{var}(r_t|\mathcal{F}_{t-1}))$ is time varying, then so is return $(E(r_t|\mathcal{F}_{t-1}))$; see Chapter 11 for models that capture this.

For relatively high frequency data, such as daily or weekly, one may argue that the variation in μ_t should be small, and it is quite widespread practice to ignore its variation. In the next sections we will mostly be assuming that $\mu_t = \mu$ is constant or small. This may be reasonable for high frequency data but would not be appropriate, say, for quarterly or annual data. Allowing for time varying μ_t will complicate any type of empirical analysis, and we will defer treatment of this analysis to later.

3.2.1 Assumptions About Return Innovations

To make the random walk model precise we need to make explicit assumptions about the innovation ε_t, and there are a number of possibilities with varying strength and varying plausibility. Historically, as in Bachelier (1900), ε_t was considered to be i.i.d. and normally distributed with mean zero and constant variance. The properties of random walks generated by such increments are well understood. However, the normality assumption is not nowadays considered a good assumption empirically, even for quite low frequency data; see Chapter 14. For now we note that the

3.2 The Random Walk Model for Prices

normal distribution is not necessary for most statistical applications of the discrete time random walk. We next consider a number of different assumptions regarding the innovation process with a view to testing applications. The assumptions are:

rw1 ε_t are i.i.d. with $E(\varepsilon_t) = 0$.
rw2 ε_t are independent over time with $E(\varepsilon_t) = 0$.
rw3 ε_t is a martingale difference sequence (MDS) in the sense that for each t, $E(\varepsilon_t | \varepsilon_{t-1}, \varepsilon_{t-2}, \ldots) = 0$ with probability one.
rw4 For all k, t, we have $E(\varepsilon_t | \varepsilon_{t-k}) = 0$ with probability one.
rw5 For all k, t, we have $E(\varepsilon_t) = 0$, $\text{cov}(\varepsilon_t, \varepsilon_{t-k}) = 0$.

CLM used assumptions **rw1**, **rw2**, and **rw5**, and labelled them RW1–RW3. These assumptions are listed in apparent decreasing strength so that **rw1** is stronger than **rw2**, which is itself stronger than **rw3**, meaning that if **rw1** holds then perforce **rw2** holds and so **rw3** holds but not vice versa. Note that none of these assumptions is a complete specification of the stochastic process $\{X_t\}$. In **rw2** the distribution of ε_t can vary freely over time, which allows a wide range of specifications, many of which are not plausible and many of which would not allow any meaningful statistical analysis, so we would need additional restrictions in this case. Likewise in **rw5**, which has the further issue that the amount of dependence between ε_t and ε_s is completely unrestricted apart from the lack of correlation. None of these specifications restricts directly the moments of ε_t beyond the first moment except that **rw5** implicitly assumes that $E(\varepsilon_t^2) < \infty$. This last point shows that logically **rw2** is not necessarily stronger than **rw5** because **rw2** could contain random variables whose variance does not exist. It is possible to formulate **rw1** and **rw2** completely without the existence of moments, so that, for example, ε_t could be a Cauchy distributed random variable symmetric about zero. However, if we do assume that at least $E(\varepsilon_t^2) \leq C < \infty$ for **rw1–5**, then we can conclude that **rw$i$** is stronger than **rw$j$**, for $i < j$. Under **rw1**, the process X_t is nonstationary: if $\mu = 0$ the process wanders around the starting point making long sojourns but always coming back; if $\mu \neq 0$, then it makes those same sojourns but around a trend line formed by μ. If we look further though, the increment distribution matters: the properties of a random walk with normal increments and a random walk with Cauchy increments are quite different.

Fama (1970) argued that the most appropriate assumption is **rw3**, that ε_t is a **martingale difference sequence**, and we discuss this assumption further.

Definition 50 *A discrete time **martingale** is a time-series process $\{\widetilde{X}_t, t = 1, 2, \ldots\}$ obeying $E(|\widetilde{X}_t|) \leq C < \infty$ such that*

$$E\left(\widetilde{X}_{t+1} \mid \widetilde{X}_t, \widetilde{X}_{t-1}, \ldots\right) = \widetilde{X}_t \tag{3.11}$$

for each t. Equivalently, we call the process $\widetilde{X}_{t+1} - \widetilde{X}_t$ a martingale difference sequence if for all t

$$E\left(\widetilde{X}_{t+1} - \widetilde{X}_t \mid \widetilde{X}_{t-1}, \widetilde{X}_{t-2}, \ldots\right) = 0. \tag{3.12}$$

This framework requires the existence of first moments, which rules out Cauchy-valued increments for example, but is otherwise quite general: it does not require the process $\widetilde{X}_{t+1} - \widetilde{X}_t$ to be stationary or homoskedastic, and it says nothing about the dependence between $\widetilde{X}_{t+1} - \widetilde{X}_t$ and $\widetilde{X}_{s+1} - \widetilde{X}_s$ beyond (3.12). We take $\widetilde{X}_t = X_t - \mu$, which is consistent with our definition given above. The concept of martingale corresponds with the notion of a fair game: if you toss a coin against an opponent and bet successively at fair odds and you have initial capital X_0, then your current capital X_t at time t is a martingale. The martingale property implies that

$$\text{cov}(\varepsilon_t, g(X_{t-1}, X_{t-2}, \ldots)) = 0$$

for any (measurable) function g for which the relevant moments are defined, in particular for

$$g(X_{t-1}, X_{t-2}, \ldots) = X_{t-k} - X_{t-k-1} - \mu = \varepsilon_{t-k},$$

so it is a stronger condition in general than **rw5** but weaker than **rw2** (which implies additionally that $\text{cov}(h(\varepsilon_t), g(X_{t-1}, X_{t-2}, \ldots))$ for any measurable functions h, g). For a stationary Gaussian process **rw5** is equivalent to **rw3**, but not in general. The martingale hypothesis does not require stationarity of ε_t.

Condition **rw4** implies the somewhat weaker restriction that $\text{cov}(\varepsilon_t, g(\varepsilon_{t-k})) = 0$ for any (measurable) function g for which the relevant moments are defined.

Some have argued that real exchange rates should be stationary, for example Froot and Rogoff (1995). This is incompatible with **rw1**, but is compatible with **rw3**. We will discuss this issue further below in Chapter 11.

We will discuss the testing of these hypotheses from a sample of data. The test statistics will depend on which version of the hypothesis is maintained, and the stronger the hypothesis, the simpler the test statistic. On the other hand, bearing in mind Manski's **law of decreasing credibility**, we should ideally prefer the weakest assumptions when it comes to making inferences about data. Widespread practice, however, is to work implicitly with **rw1** as this leads to the simplest theory. We shall stick with this approach in the main part of the text but we shall return to this issue below.

3.2.2 Calendar Time or Trading Time

An issue of importance for empirical work is to decide whether our returns are determined in trading time or in calendar time. Under the trading time hypothesis, we might suppose that returns obey (3.8) with **rw3** for each unit of trading time, and then by construction we can directly work with the usual sample of daily returns. This is most convenient the and widespread practice.

On the other hand, suppose that returns are generated in calendar time and they obey (3.8) with **rw1** for each unit of calendar time. Furthermore, suppose that $E(r_t) = \mu$ and $\text{var}(r_t) = \sigma^2$. However, we only observe the trading time returns, that is, we may have some days where we do not observe a return since there is no trading then. In this case, if we use the observed returns we will go wrong. In particular, let D_t be the number of days since the last trading day (e.g., $D_t = 1$ for a standard Tuesday, ..., Friday, and $D_t = 3$ for a standard Monday). Then observed returns r_t^O will be independent but not identically distributed, and moreover

$$E(r_t^O) = \mu D_t; \quad \mathrm{var}(r_t^O) = \sigma^2 D_t.$$

The mean return is time varying. If we ignore this structure and work with $r_t^O - \mu^*$, where $\mu^* = \mu \times E(D_t)$, then our return series violates **rw3**, and we will likely reject our null hypothesis even though true returns are consistent with **rw3**. In this case, $r_t^O - \mu D_t$ will satisfy **rw3**, and we will see later how to modify our analysis to accommodate calendar time returns. In the sequel we shall implicitly assume the trading time hypothesis unless otherwise stated, since this is the most common approach.

3.3 Testing of Linear Weak Form Predictability

3.3.1 Autocorrelation Based Testing Under rw1

We next define some standard ways of measuring predictability in time series. We have to maintain further structure to implement statistical tests.

Definition 51 *The population **autocovariance** and **autocorrelation** functions of a (weakly) stationary series Y_t*

$$\gamma(j) = \mathrm{cov}(Y_t, Y_{t-j}) = E((Y_t - E(Y_t))(Y_{t-j} - E(Y_{t-j}))) \tag{3.13}$$

$$\rho(j) = \frac{\gamma(j)}{\gamma(0)} \tag{3.14}$$

for $j = 0, \pm 1, \pm 2, \ldots$. In the absence of stationarity we may define

$$\gamma(j) = \lim_{T \to \infty} \frac{1}{T} \sum_{t=j+1}^{T} \mathrm{cov}(Y_t, Y_{t-j}) \tag{3.15}$$

assuming that the limit exists, and $\rho(j) = \gamma(j)/\gamma(0)$.

These quantities require population second moments of Y_t to exist otherwise they are not well defined. Stationarity allows one to dispense with the time subscript t. It also implies that $\gamma(j) = \gamma(-j)$ for any j, which makes sense because correlation is a symmetric relationship between random variables. The autocovariance function and the autocorrelation function both measure the amount of linear dependence or predictability in the process Y_t with zero autocovariance or autocorrelation indicating absence of dependence. The autocorrelation function ρ has the advantage that it is numerically invariant under linear transformations, meaning for example if returns are measured in daily or annual units, the autocorrelation function is the same. Furthermore, the autocorrelation always lies between minus one and plus one. The best linear predictor of Y_t by Y_{t-j} is $E_L(Y_t|Y_{t-j}) = \mu + \rho(j)(Y_{t-j} - \mu)$, where $\mu = E(Y_t)$ is the mean of Y_t.

Stationary processes can have a variety of autocorrelations.

Example 19 *Suppose that $Y_t = Y$ for all t. This process is stationary. It has $\rho(j) = 1$ for all $j = 0, \pm 1, \ldots$, which means that the dependence between Y_t and Y_s never dies out no matter how far s and t are apart.*

We work with stationary processes for which $\gamma(j), \rho(j) \to 0$ as $j \to \infty$, and indeed for a **weakly dependent** process this convergence occurs rapidly, meaning that the behavior of Y_{t-j} has little influence on Y_t. This seems like a plausible scenario to maintain in general.

In practice we mostly work with stock returns or excess stock returns, i.e., $Y_t = r_t$ or R_t or $Y_t = R_t - R_{ft}$. The efficient markets hypothesis says in that case that $\gamma(j), \rho(j) = 0$ for all $j \neq 0$. The alternative hypothesis would have $\gamma(j), \rho(j) \neq 0$ for some $j \neq 0$, which implies some predictability from the past sequence of returns to the future values.

We next show how one can test the statistical implications of the EMH using a sample of data $\{Y_1, \ldots, Y_T\}$.

Definition 52 *The sample autocovariance and autocorrelation functions are computed as follows for $j = 0, 1, 2, \ldots, T-1$:*

$$\widehat{\gamma}(j) = \frac{1}{T} \sum_{t=j+1}^{T} (Y_t - \overline{Y})(Y_{t-j} - \overline{Y}); \quad \overline{Y} = \frac{1}{T} \sum_{t=1}^{T} Y_t$$

$$\widehat{\rho}(j) = \frac{\sum_{t=j+1}^{T} (Y_t - \overline{Y})(Y_{t-j} - \overline{Y})}{\sqrt{\sum_{t=j+1}^{T} (Y_t - \overline{Y})^2 \sum_{t=j+1}^{T} (Y_{t-j} - \overline{Y})^2}}.$$

These can be calculated in Eviews by the menu sequence Quick>Series Statistics>Correlogram (you have to enter series name and how many lags you wish to include). There are a number of slightly different variations on $\widehat{\gamma}(j)$ and $\widehat{\rho}(j)$ that can make a difference when T is small; we discuss some of these later. Note that $\widehat{\gamma}(j) = \widehat{\gamma}(-j)$ and $\widehat{\rho}(j) = \widehat{\rho}(-j)$. The sample autocorrelation function defined this way satisfies $-1 \leq \widehat{\rho}(j) \leq 1$, and is numerically invariant to linear transformations of the data. In practice we can compute $\widehat{\gamma}(j)$ for $j = 0, \pm 1, \ldots, \pm(T-1)$ although $\widehat{\gamma}(T-1) = (Y_T - \overline{Y})(Y_1 - \overline{Y})/T$ is evidently only based on one observation.

To provide a test of the EMH we will approximate the distribution of $\widehat{\rho}(j)$ using asymptotic approximations as the sample size T tends to infinity (large sample size). We give an outline argument in Exercise 6.

3.3 Testing of Linear Weak Form Predictability

Theorem 14 *Suppose that Y_t is i.i.d. with finite variance. Then for any p, as $T \to \infty$*

$$\sqrt{T}(\widehat{\rho}(1), \ldots, \widehat{\rho}(p))^\top \Longrightarrow N(0, I_p), \text{ i.e.,}$$

$$\Pr\left(\sqrt{T}\widehat{\rho}(1) \leq x_1, \ldots, \sqrt{T}\widehat{\rho}(p) \leq x_p\right) \to \Pr(Z_1 \leq x_1, \ldots, Z_p \leq x_p) = \prod_{j=1}^{p} \Phi(x_j),$$

where $Z = (Z_1, \ldots, Z_p)$ is a $N(0, I_p)$ random variable and Φ is the standard normal c.d.f. Here, I_p is the $p \times p$ identity matrix.

In particular, for any k

$$\sqrt{T}\widehat{\rho}(k) \Longrightarrow N(0, 1), \tag{3.16}$$

as $T \to \infty$. Therefore, we can test the null hypothesis **rw1** by comparing $\widehat{\rho}(k)$ with the so-called **Bartlett intervals**

$$[-z_{\alpha/2}/\sqrt{T}, z_{\alpha/2}/\sqrt{T}], \tag{3.17}$$

where z_α are normal critical values, i.e., $1 - \alpha = \Phi(z_\alpha)$, where Φ is the c.d.f. of the standard normal distribution. Values of $\widehat{\rho}(k)$ lying outside this interval are inconsistent with the null hypothesis. Literally, this is testing the hypothesis that $\rho(k) = 0$ versus $\rho(k) \neq 0$ for a given k (which is implied by, but not equivalent to, the EMH). This test is widely used. If $\rho(k) \neq 0$, then we may show that $\sqrt{T}\widehat{\rho}(k) \xrightarrow{P} \pm\infty$, which justifies this as a consistent test against the alternative that $\rho(k) \neq 0$.

Theorem 14 implies that the **Box–Pierce Q statistic**

$$Q = T \sum_{j=1}^{p} \widehat{\rho}^2(j) \tag{3.18}$$

satisfies $Q \Longrightarrow \chi_p^2$ under **rw1**. This can be used to test the joint hypothesis that $\rho(1) = 0, \ldots, \rho(p) = 0$ versus the general alternative that at least one $\rho(j) \neq 0$. The test is carried out by rejecting when $Q > \chi_p^2(\alpha)$ for an α-level test. This test will have power against all of the stated alternatives, in the sense that it will reject with probability tending to one whenever the null is violated, i.e., whenever $\rho(k) \neq 0$ for some $k = 1, \ldots, p$. These statistics are in widespread use.

We next consider an alternative way of delivering a test of the joint hypothesis that $\rho(k) = 0$ for all $k = 1, \ldots, p$. Note that the statistic Q can be inverted to obtain the acceptance region in the space where $\{\rho(1), \ldots, \rho(p)\}$ live. Under **rw1** this confidence region is a p-dimensional sphere, which is simple to define mathematically, but difficult to display graphically except when $p = 2$. This is why it is common practice to display the autocorrelations $\widehat{\rho}(1), \ldots, \widehat{\rho}(p)$ along with the Bartlett intervals for each lag. However, this will neglect the **multiple testing** issue, which is that by carrying out p

separate tests one generally ends up with a significance level of the tests that is different from the significance level of each individual test, and if one just selects the largest $\hat{\rho}(j)$ as evidence against the null hypothesis one would be abusing the interpretation of the testing process. Instead, one can present **uniform intervals** that are derived from the joint hypothesis and which have correct significance levels.

Uniform Confidence Intervals

An alternative to the Box–Pierce statistic for testing the joint hypothesis that $\rho(1) = 0, \ldots, \rho(p) = 0$ is to use **uniform confidence intervals**. By the **continuous mapping theorem**, we have that

$$\max_{1 \leq k \leq p} \sqrt{T} \hat{\rho}(k) \Longrightarrow \max_{1 \leq i \leq p} Z_i \equiv W_p,$$

where Z_i are i.i.d. standard normal random variables. The distribution of W_p is known (and not normal), in fact $\Pr(W_p \leq x) = \Phi(x)^p$. Therefore, the uniform confidence intervals are of the form

$$[-w_{\alpha/2}/\sqrt{T}, w_{\alpha/2}/\sqrt{T}] \times \cdots \times [-w_{\alpha/2}/\sqrt{T}, w_{\alpha/2}/\sqrt{T}], \qquad (3.19)$$

where $w_{\alpha/2}$ satisfies $(1 - 2\Phi(-w_{\alpha/2}))^p = 1 - \alpha$. For $\alpha = 0.05$, we have the following critical values for popular choices of p:

p	5	10	22
$w_{\alpha/2}$	2.57	2.80	3.04

This says that to construct a 95% interval for $p = 22$ we need to consider intervals $\pm 3.04/\sqrt{T}$ instead of the usual $\pm 1.96/\sqrt{T}$. These intervals are wider than the Bartlett intervals. Under the null hypothesis, the probability of an exceedance would be 5%. The idea is that one plots the autocorrelation function and then compares with these larger critical values for comparability with the Box–Pierce statistic. The test rejects whenever a single $\hat{\rho}(k)$ lies outside the permitted region.

Small Sample Improvements and Bias Reduction

In practice we may not have a very large sample. The sample autocorrelations are biased in finite samples, and the smaller the sample the larger the bias, and the tests based on the sample autocorrelations will tend to overreject or underreject relative to the nominal level α. This means that one may find evidence against something when it is not truly there.

There is one case where we can completely avoid the small sample issue. In the case that the data are i.i.d. and normally distributed, Anderson (1942) derived the exact

3.3 Testing of Linear Weak Form Predictability

distribution of the sample autocorrelation. For T odd

$$\Pr(\widehat{\rho} > \rho) = \sum_{i=1}^{m} \frac{(\lambda_i - \rho)^{\frac{T-3}{2}}}{\alpha_i}, \quad \lambda_{m+1} \leq \rho \leq \lambda_m$$

$$\alpha_i = \prod_{j=1, j \neq i}^{(T-1)/2} (\lambda_i - \lambda_j), \quad \lambda_k = \cos \frac{2\pi k}{T}, \ k = 1, \ldots, T-1.$$

For $T=9$, the 0.025 and 0.975 critical values are approximately -2.142 and 1.502 respectively, which shows the tendency of the sample autocorrelation to be biased downward.

Outside of the normal world, exact results are few and far between. We next discuss some standard approaches to improving the performance of the autocorrelation function and the test statistics associated with it when the sample size is small and we can't rely on normality. We are mostly concerned with the case where $\rho(j)$ is itself small and even zero, rather than the general case. The first issue here is that $\widehat{\gamma}(j)$ is biased in small samples, and in fact this bias is downwards so that $E(\widehat{\gamma}(j)) < \gamma(j)$.

Suppose we consider the slightly modified (dividing by $T-j$ rather than T) sample autocovariance

$$\widetilde{\gamma}(j) = \frac{1}{T-j} \sum_{t=j+1}^{T} (Y_t - \overline{Y}_j)(Y_{t-j} - \overline{Y}_j), \quad \overline{Y}_j = \frac{1}{T-j} \sum_{t=j+1}^{T} Y_t.$$

In that case, we show in Exercise 7 that

$$E(\widetilde{\gamma}(j)) = \gamma(j) - E(\mu - \overline{Y}_j)^2 < \gamma(j). \tag{3.20}$$

In general, $E((\mu - \overline{Y}_j)^2)$ is quite complicated, it is proportional to the long run variance of Y_t; see (2.17). However, in the case where $\gamma(j) = 0$ for all j, this formula simplifies and we have

$$E(\widetilde{\gamma}(j)) = \gamma(j) - \frac{\gamma(0)}{T-j}.$$

We may construct a **bias-corrected** estimator

$$\widetilde{\gamma}_{bc}(j) = \widetilde{\gamma}(j) + \frac{\widehat{\gamma}(0)}{T-j}, \tag{3.21}$$

which has the property that if $\gamma(j) = 0$, we have $E(\widetilde{\gamma}(j)) = 0$. This is not an unbiased estimator in the usual sense of the word but its bias is generally small when the $\gamma(j)$ are small.

Similar considerations apply to the sample autocorrelation in the sense that if the sample size T is small, $\widehat{\rho}(j)$ can be badly biased, and the direction of the bias is also negative. There are a number of ways of addressing this issue, but since $\rho(j)$ is defined as a ratio, it is only possible to find exactly unbiased estimators in special cases. One approach is to work with **higher order asymptotic approximations** such as defined in Campbell, Lo, and MacKinlay (1997, eq. 2.4.13). Define the bias corrected estimator

$$\widehat{\rho}^{bc}(j) = \widehat{\rho}(j) + \frac{T-j}{(T-1)^2} \times \left(1 - \widehat{\rho}^2(j)\right). \qquad (3.22)$$

Under some conditions one may show that the bias corrected estimator has bias that is of smaller order than that of the original estimator $\widehat{\rho}(j)$. In practice this method can work better for moderately sized T as shown in simulations. For T too small, e.g., $T=3$, then the correction can be very large so that when $\widehat{\rho}(j) = 0$, we have $\widehat{\rho}_1^{bc} = 1/2$.

The Box–Ljung test statistic, a modification of the Box–Pierce statistic, defined as

$$Q = T(T+2) \sum_{j=1}^{p} \frac{\widehat{\rho}^2(j)}{T-j} \qquad (3.23)$$

incorporates a bias correction (under the null hypothesis), and is known to have better finite sample performance than the Box–Pierce statistic under the null hypothesis.

The bias corrections require **rw1** for their justification, otherwise the expressions become much more difficult to employ and are more susceptible to the errors introduced by the bias correction itself. In any case, this simple theory is likely to be quite so misleading in practice because the conditions underlying the central limit theorem are unlikely to be met. Specifically, asset returns are unlikely to be i.i.d. If we maintain only **rw2**, say, we will generally obtain a different limiting distribution, and we may need further conditions in terms of moments and dependence in order to guarantee the CLT, and the form of the limiting variance will itself change. We will discuss this further below.

3.3.2 Cross Autocorrelations

The discussion so far has been focussed on the univariate case, but there are many asset prices observed every day, and it behooves us to consider the joint behavior of asset returns. The definition of the random walk and the assumptions **rw1**–**rw5** carry over directly to the multivariate case where we modify the information set in the case **rw3** to include all the lagged values of all the series. This is a larger information set than just including own past prices and so the statistical restrictions are stronger.

Definition 53 *The population **cross autocovariance** and **cross autocorrelation** functions of a (weakly) stationary vector series Y_t, X_t are defined for $j = 0, \pm 1, \pm 2, \ldots$ as*

$$\gamma_{YX}(j) = \operatorname{cov}(Y_t, X_{t-j}) = E\left((Y_t - E(Y_t))(X_{t-j} - E(X_{t-j}))\right) \qquad (3.24)$$

$$\rho_{YX}(j) = \frac{\gamma_{YX}(j)}{\sqrt{\gamma_Y(0)\gamma_X(0)}}. \qquad (3.25)$$

3.3 Testing of Linear Weak Form Predictability

We note that $\gamma_{YX}(-j)$ is not necessarily equal to $\gamma_{YX}(j)$, but $\gamma_{YX}(-j) = \gamma_{XY}(j)$. If $\gamma_{YX}(j) > \gamma_{XY}(j)$ for all $j > 0$, we say that X leads Y. The corresponding assumption **rw5** is that $\gamma_{YX}(j) = 0$ for all $j \neq 0$, which is in practice the easiest implication of the EMH to test.

Definition 54 *The sample versions of the cross-autocorrelation are for $j = 0, 1, 2, \ldots$*

$$\widehat{\gamma}_{YX}(j) = \frac{1}{T} \sum_{t=j+1}^{T} (Y_t - \overline{Y})(X_{t-j} - \overline{X})$$

$$\widehat{\rho}_{YX}(j) = \frac{\sum_{t=j+1}^{T} (Y_t - \overline{Y})(X_{t-j} - \overline{X})}{\sqrt{\sum_{t=j+1}^{T} (Y_t - \overline{Y})^2 \sum_{t=j+1}^{T} (X_{t-j} - \overline{X})^2}}. \tag{3.26}$$

We define $\widehat{\gamma}_{YX}(-j) = \widehat{\gamma}_{XY}(j)$.

Theorem 15 *Suppose that Y_t, X_t are i.i.d. with finite variance. Then for any p, as $T \to \infty$*

$$\sqrt{T}(\widehat{\rho}_{YX}(1), \ldots, \widehat{\rho}_{YX}(p))^\top \Longrightarrow N(0, I_p).$$

We can use this result to test the EMH. For example by comparing $\widehat{\rho}_{YX}(k)$ with the critical values $\pm z_{\alpha/2}/\sqrt{T}$, or by aggregating across lags like the Box–Pierce statistic.

The cross autocovariance is useful in understanding the different empirical results we obtain below for individual stocks and portfolios or indexes.

Example 20 *Consider a portfolio*

$$W_t = \omega X_t + (1 - \omega) Y_t,$$

where X_t, Y_t are two mean zero stock returns (and we use the approximation that portfolio returns are the average of individual stock returns) and ω is the portfolio weight value. Then

$$\text{cov}(W_t, W_{t-j}) = E((\omega X_t + (1-\omega) Y_t)(\omega X_{t-j} + (1-\omega) Y_{t-j}))$$
$$= \omega^2 \gamma_X(j) + (1-\omega)^2 \gamma_Y(j) + \omega(1-\omega)\gamma_{XY}(j) + \omega(1-\omega)\gamma_{YX}(j),$$

where $\gamma_{XY}(j) = E(X_t Y_{t-j})$ and $\gamma_{YX}(j) = E(Y_t X_{t-j})$. It follows that so long as the cross covariances $\gamma_{XY}(j), \gamma_{YX}(j) > 0$, we can have: $\text{cov}(W_t, W_{t-j}) > 0$, $\text{cov}(X_t, X_{t-j}) < 0$, and $\text{cov}(Y_t, Y_{t-j}) < 0$. In fact for large portfolios, cross correlations and cross-autocorrelations can be the dominant term.

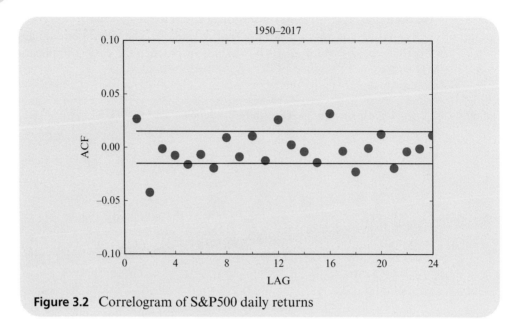

Figure 3.2 Correlogram of S&P500 daily returns

3.3.3 Empirical Evidence Regarding Linear Predictability

We first revisit the evidence presented in CLM, which concerns the period 19620703–19941230. They considered daily, weekly, and monthly data on the CRSP value weighted and equal weighted indexes, as well as a sample of 411 individual securities from the CRSP database. Their Table 2.4 shows a positive (first lag) autocorrelation for daily indexes (in the range 0.1–0.43), which are statistically significant using the Bartlett standard errors $1/\sqrt{T}$. They find that the Box–Pierce statistics Q_5 and Q_{10} are statistically significant. The evidence is weaker at the weekly and monthly horizon. The evidence is weaker for value weighted versus equal weighted. The results are not stable across subperiods. They find small negative autocorrelation for individual stocks at daily horizon. They argue based on the cross-autocorrelations that there is a lead lag relation between large and small stocks. We will consider their evidence again below, but for now we consider some other series.

In Figure 3.2 we show the correlogram of the daily return on the S&P500 index over the period 1950–2017. In Figure 3.3 we break this down into decade long subperiods. The sample sizes are very large here, with around 2500 observations per decade, so that the confidence intervals are narrow. Over the whole period, the index has positive and significant first order autocorrelation, but the second lag is negative and significant. Furthermore, the results are not robust over subperiods: in the most recent subperiod, the first two autocorrelations for the index are negative and significant.

In Figure 3.4 we also show the correlogram of the FTSE100 daily return series for comparison over the period 1984–2017. In this case the first lag is not significant, but the second and third lags are significant and negative. Figure 3.5 shows the correlogram for longer lags, out to a roughly annual horizon. There is not much evidence of significance at longer horizons.

3.3 Testing of Linear Weak Form Predictability

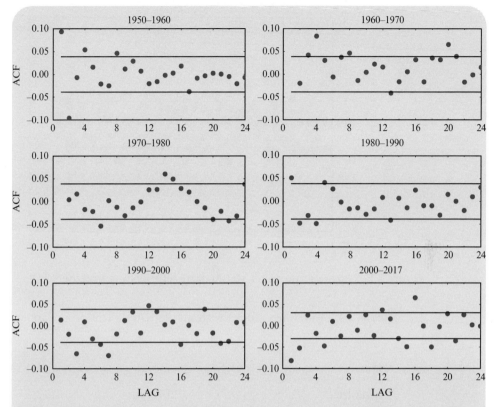

Figure 3.3 Correlogram of S&P500 daily returns by decade: (1,1) 1950–1960; (1,2) 1960–1970; (2,1) 1970–1980; (2,2) 1980–1990; (3,1) 1990–2000; (3,2) 2000–2017

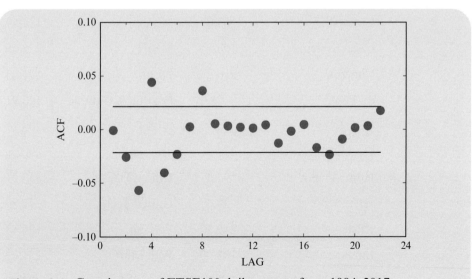

Figure 3.4 Correlogram of FTSE100 daily returns from 1984–2017

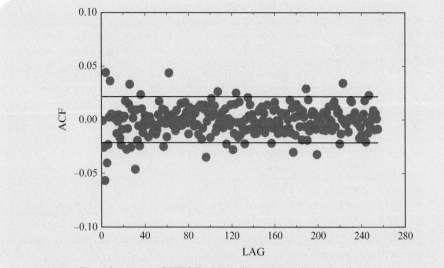

Figure 3.5 Correlogram of FTSE100 daily returns long horizon

Table 3.1 Dow Jones stocks with market capitalization in January 2013

Name	Cap US$b	Name	Cap US$b
Alcoa Inc.	9.88	JP Morgan	172.43
AmEx	66.71	Coke	168.91
Boeing	58.58	McD	90.21
Bank of America	130.52	MMM	65.99
Caterpillar	62.07	Merck	127.59
Cisco Systems	108.74	MSFT	225.06
Chevron	216.27	Pfizer	191.03
du Pont	42.64	Proctor & Gamble	188.91
Walt Disney	92.49	AT&T	200.11
General Electric	222.31	Travelers	28.25
Home Depot	94.47	United Health	53.21
HP	29.49	United Tech	77.89
IBM	219.20	Verizon	126.43
Intel	105.20	Wall Mart	231.02
Johnson & Johnson	198.28	Exxon Mobil	405.60

3.3 Testing of Linear Weak Form Predictability

We investigate the returns on stocks in the DJIA as of January 2013, whose names and market capitalization are given below in Table 3.1.

We report the value of $\widehat{\rho}_i(1)$ for each stock and plot them against the rank of market capitalization in Figure 3.6 to see if there is an association. There is a slightly negative ACF(1) on average and a slightly negative relation between size and the ACF(1), meaning large stocks tend to have less significant autocorrelations.

Cross-sectional Averages of Autocorrelations

Suppose that we have many individual stocks. How do we summarize their dependence properties? The EMH implies that $\rho_i(j) = 0$ for each j and for each stock i. We can't go through all the autocorrelation functions one by one, so how do we aggregate or present the combined information? One approach is to use a Wald statistic that forms a quadratic form in all the sample autocorrelations, like a generalized version of the Box–Pierce statistic. This is quite difficult to implement in practice when there are many stocks, and furthermore it doesn't yield information about the direction of departures from the null hypothesis; see Chapter 7. An alternative is to compute the cross-sectional average of the individual autocorrelations. Suppose that we compute $\widehat{\rho}_i(.)$ for a cross section of stocks and report the average estimated value

$$\overline{\rho}(j) = \frac{1}{n}\sum_{i=1}^{n} \widehat{\rho}_i(j). \tag{3.27}$$

This is quite common practice in applications; see for example CLM, Table 2.7. The EMH implies that $\sum_{i=1}^{n} \rho_i(j)/n = 0$ for each j, and so we can provide a formal test of

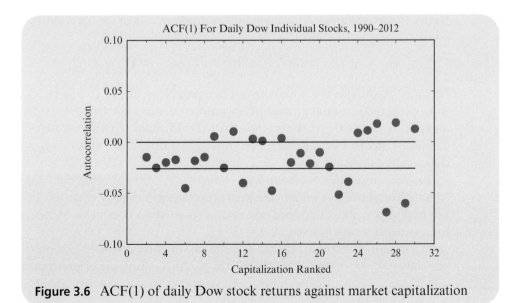

Figure 3.6 ACF(1) of daily Dow stock returns against market capitalization

EMH based on this statistic. What is the sampling distribution of this estimate under the null hypothesis?

We give the following result.

Theorem 16 *Suppose that Y_{it} are i.i.d. across t with finite variance, and let*

$$v^2 = \frac{1}{n^2}\left(n + \sum\sum_{i \neq j} \omega_{ij}^2\right),$$

where $\omega_{ij} = \mathrm{corr}(Y_{it}, Y_{jt})$. Then for any p, as $T \to \infty$

$$\sqrt{T}(\bar{\rho}(1), \ldots, \bar{\rho}(p))^\mathsf{T} \Longrightarrow N(0, v^2 I_p).$$

In Exercise 8 we explain this result. As discussed in CLM, Table 2.7, the cross sectional correlations are needed to conduct correct inference about the population value of $\bar{\rho}(j)$. Note that $v \leq 1$ with equality if and only if $\omega_{ij} = \pm 1$. If, for example, $\omega_{ij} = \omega$, then $v^2 = \omega^2 + \frac{1-\omega^2}{n}$. Typically, we might find $\omega \simeq 0.5$, so that the variance of the average autocorrelation is approximately one quarter the variance of any single autocorrelation, for large n. To carry out the statistical test we need to estimate v. We allow ω_{ij} to vary across i and j and to be unknown. Let

$$\hat{v}^2 = \frac{1}{n^2}\left(n + \sum\sum_{i \neq j} \hat{\omega}_{ij}^2\right), \quad \hat{\omega}_{ij} = \frac{\frac{1}{T}\sum_{t=1}^T (Y_{it} - \overline{Y}_i)(Y_{jt} - \overline{Y}_j)}{\sqrt{\frac{1}{T}\sum_{t=1}^T (Y_{it} - \overline{Y}_i)^2 \frac{1}{T}\sum_{t=1}^T (Y_{jt} - \overline{Y}_j)^2}}.$$

Under some conditions, $\hat{v}^2$ consistently estimates v^2. Then we reject the null hypothesis if

$$\bar{\rho}(j) \notin [-\hat{v} z_{\alpha/2}/\sqrt{T}, \hat{v} z_{\alpha/2}/\sqrt{T}] \tag{3.28}$$

for $j = 1, \ldots, P$. This is a pointwise test comparable to (3.17). In Figure 3.7 below we show the cross-sectional average of the autocorrelation function for the Dow stocks along with standard errors obtained from this formula. This shows that on average individual stocks have negative first order autocorrelation, and this is moderately statistically significant.

One may argue that there are $\rho_i(j) > 0$ and $\rho_j(j) < 0$ such that their sum is equal to zero. The above test would have zero power against such alternatives. Instead, one can look at test statistics based on sums of squared autocorrelations. You are asked to consider this case in Exercise 9.

CLM report in Table 2.7 the cross-sectional standard deviation as a measure of the variability of $\bar{\rho}(j)$ (in their case the related variance ratios, see below), that is,

$$\bar{s}^2 = \frac{1}{n}\sum_{i=1}^n (\hat{\rho}_i(j) - \bar{\rho}(j))^2.$$

3.3 Testing of Linear Weak Form Predictability

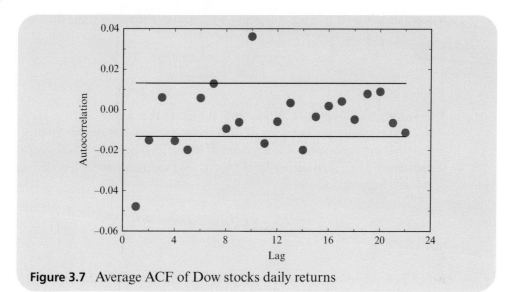

Figure 3.7 Average ACF of Dow stocks daily returns

Under the null hypothesis **rw1** we have as $T \to \infty$,

$$T\bar{s}^2 \Longrightarrow \frac{1}{n}\sum_{i=1}^{n} u_i^2 - \left(\frac{1}{n}\sum_{i=1}^{n} u_i\right)^2,$$

where u_i are normally distributed with mean zero and $\text{cov}(u_i, u_j) = \text{corr}(Y_{it}, Y_{jt})$. When n is large, the second term is of smaller order and $T\bar{s}^2$ converges in probability to one under some conditions on the cross-sectional correlation between the assets (we discuss this in Chapter 8). In this case, $\bar{s}^2$ provides an upper bound on the variance of $\bar{\rho}(j)$.

3.3.4 Variance Ratio Statistics

Variance ratio tests were introduced to finance by Poterba and Summers (1988) and Lo and MacKinlay (1988). They are closely related to autocorrelation and Box–Pierce tests. They are widely used in practice.

In this section we work with logarithmic returns r_t, which are assumed to follow a stationary process with finite variance. Define the q-period logarithmic returns

$$r_t(q) = p_{t+q} - p_t = r_{t+1} + \cdots + r_{t+q}, \quad (3.29)$$

for $q = 2, 3, \ldots$. The 2-period return, for example, satisfies

$$\text{var}(r_t(2)) = \text{var}(r_{t+2} + r_{t+1}) = \text{var}(r_{t+2}) + \text{var}(r_{t+1}) + 2\text{cov}(r_{t+2}, r_{t+1}).$$

Under the assumption that returns are uncorrelated (**rw5**), we have $\text{cov}(r_{t+2}, r_{t+1}) = 0$, and so $\text{var}(r_t(2)) = \text{var}(r_{t+2}) + \text{var}(r_{t+1})$. Therefore, we have

$$VR(2) = \frac{\text{var}(r_t(2))}{2\text{var}(r_t)} = 1.$$

In fact this holds for all horizons p under the null hypothesis. Define the variance ratio

$$VR(p) = \frac{\text{var}(r_t(p))}{p\text{var}(r_t)}. \qquad (3.30)$$

This is always non-negative. Under **rw5**, it follows that $VR(p) = 1$.

We next consider the behavior of the variance ratio when the return process is stationary but not necessarily uncorrelated. If the series is actually positively (negatively) serially correlated, then $\text{var}(r_t(2)) > < \text{var}(r_{t+2}) + \text{var}(r_{t+1})$. In fact, under stationarity of returns we have

$$VR(2) = \frac{\text{var}(r_t) + \text{var}(r_t) + 2\text{cov}(r_{t+2}, r_{t+1})}{2\text{var}(r_t)} = 1 + \rho(1),$$

which is equal to one if and only if $\rho(1) = 0$. For a general stationary process with ACF $\{\rho(j), j = 1, 2, \ldots\}$, we have by the same arguments that

$$VR(p) = 1 + 2\sum_{j=1}^{p-1} \left(1 - \frac{j}{p}\right) \rho(j). \qquad (3.31)$$

This shows that the variance ratio is a linear functional (an average) of the autocorrelation function of the high frequency data. The variance ratio has a sign relative to the central value one (unlike the Box–Pierce statistic, say), and this depends on all the first p autocorrelations and their relative magnitudes. By this we mean that if $\rho(j) > 0$ for all j, then $VR(p) > 1$, while if $\rho(j) < 0$ for all j, then $VR(p) < 1$. This captures an important idea that for the purposes of prediction it seems desirable to have a predictive relationship that is somewhat stable over lags, that is one suspects that a lag structure that flips signs repeatedly is unlikely to represent a stable or reliable phenomenon. The variance ratio statistics capture whether such a structure is present or not. The autocorrelation function itself does not, nor does the Box–Pierce statistic, which destroys the sign information by taking squares.

Example 21 *Suppose that*
$$r_t = \rho r_{t-1} + \varepsilon_t$$
with ε_t i.i.d. with mean zero and $|\rho| < 1$. It follows that

$$VR(p) = 1 + 2\sum_{j=1}^{p-1} \left(1 - \frac{j}{p}\right) \rho^j = 1 + 2\frac{\rho}{1-\rho} - \frac{2\rho(1-\rho^p)}{p(1-\rho)^2},$$

which is equal to one if and only if $\rho = 0$. For $\rho > 0$, $VR(p) > 1$.

3.3 Testing of Linear Weak Form Predictability

Example 22 *Consider the MA(2) process*

$$r_t = \varepsilon_t + \theta_1 \varepsilon_{t-1} + \theta_2 \varepsilon_{t-2}.$$

Then

$$VR(p) = 1 + 2\left(1 - \frac{1}{p}\right) \frac{\theta_1 + \theta_1 \theta_2}{1 + \theta_1^2 + \theta_2^2} + 2\left(1 - \frac{2}{p}\right) \frac{\theta_2}{1 + \theta_1^2 + \theta_2^2}.$$

This does not satisfy $\theta_1 = \theta_2 = 0$ if and only if $VR(p) = 1$. In fact, when p is large we see that if $\theta_2 = -\theta_1/(1+\theta_2)$ then $VR(p) = 1$. In fact,

$$\left\{ (\rho(1), \ldots, \rho(p-1)) : \sum_{j=1}^{p-1} \left(1 - \frac{j}{p}\right) \rho(j) = 0 \right\}$$

is a subspace of $\mathbb{R}^{p-1}$ with dimension $p - 2$, which means that it is testing a one dimensional restriction! In the case $p = 3$, we have

$$\{(\rho(1), \rho(2)) : 2\rho(1) + \rho(2) = 0\},$$

which is a line in $\mathbb{R}^2$.

Finally, we give an interpretation of the variance statistic as being proportional to the constant ρ in the autocorrelation model

$$\rho(j) = c_{j,p}\rho, \quad j = 1, \ldots, p.$$

This model corresponds to an $MA(p)$ process for r_t with declining coefficients $\theta_j = c_{j,p}\rho$. Consider the least squares objective functions

$$Q(\rho) = \sum_{j=1}^{p-1} (\rho(j) - c_{j,p}\rho)^2,$$

where $c_{j,p} = 1 - j/p$. Then the value ρ^* that minimizes $Q(\rho)$ is

$$\rho^* = \frac{\sum_{j=1}^{p-1} c_{j,p}\rho(j)}{\sum_{j=1}^{p-1} c_{j,p}^2}, \quad \sum_{j=1}^{p-1} c_{j,p}^2 = \frac{4}{6p}\left(2p^2 - 3p + 1\right).$$

We see that $VR(p) - 1$ is in a one to one relationship with this ρ^*, i.e.,

$$VR(p) - 1 = \rho^* \times 2 \sum_{j=1}^{p-1} c_{j,p}^2 \quad ; \quad \rho^* = \frac{VR(p) - 1}{2 \sum_{j=1}^{p-1} c_{j,p}^2}.$$

3.3.5 Variance Ratio Test

We next show how to compute the empirical variance ratios given a sample of $np + 1$ (log) price observations $p_0, \ldots, p_{np}$, from the highest frequency of observation. We calculate the return on each high frequency interval and on each lower frequency (length p) interval over the sample. We obtain a sample of high frequency returns $\{r_1, \ldots, r_T\}$, where $T = np$, a sample of non-overlapping low frequency returns $\{r_0(p), r_p(p), \ldots, r_{(n-1)p}(p)\}$ containing n observations of the form (3.29), and a sample of overlapping low frequency returns $\{r_0(p), r_1(p), \ldots, r_{(n-1)p}(p)\}$ containing $T - p + 1$ observations.

We calculate the variance of return for each sampling frame. Specifically, let:

$$\widehat{\sigma}_H^2 \equiv \frac{1}{T}\sum_{t=1}^{T}(r_t - \widehat{\mu})^2, \qquad \widehat{\mu} = \frac{1}{T}\sum_{t=1}^{T} r_t$$

$$\overbrace{\widehat{\sigma}_L^2(p)}^{\text{"Monday to Monday"}} \equiv \frac{1}{n}\sum_{t=0}^{n-1}(r_t(p) - p\widehat{\mu})^2$$

$$\widehat{VR}(p) = \frac{\widehat{\sigma}_L^2(p)}{p\widehat{\sigma}_H^2}.$$

Theorem 17 *Suppose that **rw1** holds and that p is fixed. Then we have*

$$\sqrt{T}\left(\widehat{VR}(p) - 1\right) \Longrightarrow N(0, 2(p-1)).$$

The main arguments for this result are explored in Exercise 10. This result suggests how to construct a formal test of the null hypothesis of no predictability. Define the test statistic

$$S = \frac{\sqrt{T}\left(\widehat{VR}(p) - 1\right)}{\sqrt{2(p-1)}}. \tag{3.32}$$

Reject the null hypothesis if $|S| > z_{\alpha/2}$, where z_α is the normal critical value at significance level α. If $S > 0$, this means positive autocorrelation (momentum) and if $S < 0$, this means negative autocorrelation (contrarian). This can be carried out for any p and typically results for a number of p are reported.

In this case we are using non-overlapping low frequency returns with a particular choice of starting period, we signify by "Friday." In fact we have a choice of how to construct the weekly returns: Monday to Monday, Tuesday to Tuesday, etc., so why not just work instead with the full set of overlapping weekly returns $\{r_0(p), r_1(p), \ldots, r_{T-p}(p)\}$ for $j = 1, \ldots, p$. Therefore, let

$$\widehat{\sigma}_{LO}^2(p) \equiv \frac{1}{T - p + 1}\sum_{t=0}^{T-p}(r_t(p) - p\widehat{\mu})^2.$$

3.3 Testing of Linear Weak Form Predictability

Specifically, this version of the variance ratio is asymptotically (i.e., in large samples) normal with asymptotic variance

$$\omega_1(p) = \frac{4}{6p}(2p-1)(p-1), \qquad (3.33)$$

which is smaller than $2(p-1)$. The asymptotic relative efficiency improvement is bounded by 2. This shows the modest improvement in efficiency obtained by using overlapping returns versus non-overlapping returns – it is equivalent to working with at best twice the sample size of the non-overlapping series.

An alternative estimator of the variance ratio is based on the correlogram (of the original highest frequency series) using the relation (3.31). This we can define as

$$\widetilde{VR}(p) = 1 + 2\sum_{j=1}^{p-1}\left(1 - \frac{j}{p}\right)\widehat{\rho}(j), \qquad (3.34)$$

where $\widehat{\rho}(j)$ is defined above using the highest frequency data. This is similar but not identical to $\widehat{VR}(p)$ (the same apart from end effects that are of magnitude T^{-1}): it shares the same asymptotic distribution as $\widehat{VR}(p)$ computed with overlapping data (i.e., it is asymptotically efficient (under Gaussianity) – suppose that $\widehat{\rho}(j)$, $j=1,\ldots,p-1$ were the maximum likelihood estimators of $\rho(j)$, $j=1,\ldots,p-1$, computed from some ARMA(p,q) model, say; then $\widetilde{VR}(p)$ would be the MLE of $VR(p)$).

Theorem 18 *Suppose that **rw1** holds. Then we have*

$$\sqrt{T}\left(\widetilde{VR}(p) - 1\right) \Longrightarrow N(0, \omega_1(p)).$$

This follows directly from Theorem 14, because $4\sum_{j=1}^{p-1}(1-\frac{j}{p})^2 = \omega_1(p)$. It is possible to correct the variance ratio test for heteroskedasticity of returns (**rw5** versus **rw1** null hypothesis) as we discuss below. There are also various possible small-sample adjustments to the variance ratio estimators. For example, we can bias adjust $\widetilde{VR}(p)$ by bias correcting each autocorrelation estimator as discussed above.

Time Aggregation and Overlapping Returns

The variance ratio statistics make use of temporal aggregation. We next consider how this temporal aggregation affects the relation between series. We are often interested in different horizons, daily, weekly, monthly, etc., and interested in how this horizon affects the predictability and the nature of this predictability. As we have seen, one can aggregate high frequency returns in two ways: overlapping and non overlapping. The non overlapping way presents no new problems except that one obtains a smaller sample size for the aggregated series. We consider the overlapping data case. Suppose that X_t, Y_t are two high frequency series, and let

$$X_t(p) = X_t + X_{t+1} + \cdots + X_{t+p-1}, \qquad Y_t(p) = Y_t + Y_{t+1} + \cdots + Y_{t+p-1}.$$

The series $X_t(p)$, $Y_t(p)$ represent p-period returns. If $p=5$ and X_t, Y_t are daily (log) returns, then $X_t(5)$, $Y_t(5)$ are weekly returns. The subset $\{X_{jp}(p)\}_{j=0}^{m}$ is a p-period return series with no overlap. It can be analyzed just as the one period return series. An issue arises however when one works with the **overlapping series** $\{X_t(p)\}_{t=1}^{n}$, since $X_t(p)$ and $X_{t+1}(p)$ contain an overlap of $p-1$ periods. Suppose that X_t, Y_t are mean zero processes. Then under stationarity we have

$$\begin{aligned}
\operatorname{cov}(X_t(p), Y_t(p)) &= E[(X_{t+1} + \cdots + X_{t+p})(Y_{t+1} + \cdots + Y_{t+p})] \\
&= pE(X_{t+1}Y_{t+1}) + 2(p-1)E(X_{t+1}Y_{t+2}) + 2(p-2)E(X_{t+1}Y_{t+3}) \\
&\quad + \cdots + 2E(X_{t+1}Y_{t+p}) \\
&= p\gamma_{XY}(0) + p\sum_{j=1}^{p-1}\left(1 - \frac{j}{p}\right)\gamma_{XY}(j).
\end{aligned}$$

Under the EMH we have that $\gamma_{XY}(j) = 0$, $\gamma_X(j) = 0$, and $\gamma_Y(j) = 0$ for $j \neq 0$. It follows that

$$\operatorname{corr}(X_t(p), Y_t(p)) = \frac{\operatorname{cov}(X_t(p), Y_t(p))}{\operatorname{sd}(X_t(p))\operatorname{sd}(Y_t(p))} = \operatorname{corr}(X_t, Y_t)$$

for all p. However, under the general stationary alternative we have

$$\begin{aligned}
\operatorname{corr}(X_t(p), Y_t(p)) &= \frac{p\gamma_{XY}(0) + p\sum_{j=1}^{p-1}\left(1 - \frac{j}{p}\right)\gamma_{XY}(j)}{\sqrt{\left(p\gamma_X(0) + p\sum_{j=1}^{p-1}\left(1 - \frac{j}{p}\right)\gamma_X(j)\right)\left(p\gamma_Y(0) + p\sum_{j=1}^{p-1}\left(1 - \frac{j}{p}\right)\gamma_Y(j)\right)}} \\
&= \frac{\rho_{XY}(0) + \sum_{j=1}^{p-1}\left(1 - \frac{j}{p}\right)\rho_{XY}(j)}{\sqrt{\left(1 + \sum_{j=1}^{p-1}\left(1 - \frac{j}{p}\right)\rho_X(j)\right)\left(1 + \sum_{j=1}^{p-1}\left(1 - \frac{j}{p}\right)\rho_Y(j)\right)}}.
\end{aligned}$$

The **Epps effect**, named after Epps (1979), is the phenomenon that the empirical correlation between the returns of two different stocks decreases as the sampling frequency of data increases. He calculated the contemporaneous correlations between different stocks using data from 10 minutes to 3 days in horizon, and found that the shortest horizon correlations were very small and this increased with the horizon examined. Our formula for $\operatorname{corr}(X_t(p), Y_t(p))$ shows how we could observe this increasing with p, which is consistent with the findings of Epps (1979), if cross autocovariances are

3.3 Testing of Linear Weak Form Predictability

positive and own autocovariances are negative. Other explanations for this are non-synchronous/asynchronous trading and discretization effects, which we will examine in Chapter 5.

3.3.6 Autoregression Tests

An alternative way of measuring predictability or modelling it is by regression or autoregression methods. Consider the pth order autoregression

$$Y_t = \mu + \beta_1 Y_{t-1} + \cdots + \beta_p Y_{t-p} + \varepsilon_t, \tag{3.35}$$

where ε_t satisfies the usual conditional moment restriction $E(\varepsilon_t|Y_{t-1},\ldots,Y_{t-p}) = 0$. This assumption essentially corresponds to (an extended version of) hypothesis **rw4**. The absence of predictability can be tested by the usual regression F-test (or rather Wald test) of the hypothesis $H_0: \beta_1 = \cdots = \beta_p = 0$ versus the general alternative that at least one of $\beta_j \neq 0$. For inference though it is natural to assume the stronger assumption **rw3** or even **rw1** for the error terms, and to assume that the process Y_t is strongly stationary and mixing.

Let X be the $(T-p-1) \times p + 1$ matrix whose first column consists of ones, whose second column consists of the observations $Y_{p+1},\ldots, Y_T$ etc. The Wald test statistic is

$$W = T\left(\widehat{\beta} - \beta\right)^\mathsf{T} \widehat{V}^{-1} \left(\widehat{\beta} - \beta\right), \tag{3.36}$$

where $\widehat{\beta} = (\widehat{\beta}_1,\ldots,\widehat{\beta}_p)^\mathsf{T}$ are the OLS estimates of β, and $\widehat{V}$ is a consistent estimate of the asymptotic variance of $\widehat{\beta}$. Under the i.i.d. assumption **rw1**, the asymptotic variance can be estimated by

$$\widehat{V} = \widehat{\sigma}_\varepsilon^2 (X^\mathsf{T} X/T)^{-1}.$$

Under the assumption **rw3**, the asymptotic variance can be estimated by the White's standard errors

$$\widehat{V}_W = (X^\mathsf{T} X/T)^{-1} (X^\mathsf{T} D X/T)(X^\mathsf{T} X/T)^{-1},$$

$$D = \text{diag}\{\widehat{\varepsilon}_{p+1}^2,\ldots,\widehat{\varepsilon}_T^2\},$$

where $\widehat{\varepsilon}_t = Y_t - \widehat{\mu} - \widehat{\beta}_1 Y_{t-1} - \cdots - \widehat{\beta}_p Y_{t-p}$. The test statistic W is compared with the critical value from the χ_p^2 distribution.

Table 3.2 shows the estimated coefficients for the AR(22) fit along with OLS standard errors and White's standard errors for the full sample of S&P500 index daily returns from 1950–2017.

The fitted coefficients are quite small and vary in sign, with 8 of the 22 significant at the 5% level using the OLS standard errors but only 4 of the 22 significant at the same level using the White's standard errors. The in-sample R^2 is 0.008, which is small.

Chapter 3 Return Predictability and the Efficient Markets Hypothesis

Table 3.2 Parameter estimates of AR(22) model

	1	2	3	4	5	6	7	8	9	10	11
$\widehat{\beta}$	0.0001	−0.0028	−0.0180	0.0110	0.0007	−0.0204	−0.0040	0.0316	−0.0140	0.0005	−0.0009
se_{OLS}	0.0000	0.0077	0.0077	0.0077	0.0077	0.0077	0.0077	0.0077	0.0077	0.0077	0.0077
se_W	0.0000	0.0125	0.0137	0.0130	0.0131	0.0132	0.0136	0.0133	0.0136	0.0123	0.0134
	12	13	14	15	16	17	18	19	20	21	22
$\widehat{\beta}$	0.0279	−0.0147	0.0144	−0.0110	0.0109	−0.0193	−0.0045	−0.0114	−0.0102	0.0029	−0.0426
se_{OLS}	0.0077	0.0077	0.0077	0.0077	0.0077	0.0077	0.0077	0.0077	0.0077	0.0077	0.0077
se_W	0.0140	0.0151	0.0142	0.0151	0.0154	0.0145	0.0151	0.0198	0.0151	0.0159	0.0193

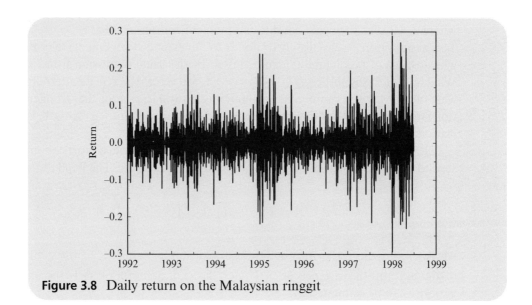

Figure 3.8 Daily return on the Malaysian ringgit

Example 23 *We fit an AR(4) to the daily Malaysian ringgit/US dollar currency return over the period 1992–1998. The time series is shown below in Figure 3.8. We obtain $\phi_1 = -1.97$, $\phi_2 = -2.08$, $\phi_3 = -1.39$, and $\phi_4 = -0.49$, which are surprisingly large values. The matrix*

$$A = \begin{pmatrix} -1.97 & -2.08 & -1.39 & -0.49 \\ 1 & 0 & 0 & 0 \\ 0 & 1 & 0 & 0 \\ 0 & 0 & 1 & 0 \end{pmatrix}$$

has eigenvalues: $-0.22307 + 0.81264i$, $-0.22307 - 0.81264i$, $-0.76193 + 0.33085i$, and $-0.76193 - 0.33085i$, where $i = \sqrt{-1}$. All eigenvalues of this matrix are complex valued! However, they have modulus less than one, which means that the process is stationary.

3.3 Testing of Linear Weak Form Predictability

We can compare the autoregression function with the autocorrelation function analytically. In the case $p=1$, the slope of the autoregression function is equal to $\rho(1)$, but when multiple lags are considered simultaneously, the relationship is more complicated.

Theorem 19 *The population best linear prediction of Y_t by a constant and X_t is $E_L(Y_t|X_t) = \mu + \beta^\top X_t$, where $\mu = E(Y_t) - \beta^\top E(X_t)$ and*

$$\beta = \mathrm{var}(X_t)^{-1}\mathrm{cov}(X_t, Y_t).$$

In this case, with $X_t = (Y_{t-1}, \ldots, Y_{t-p})^\top$, we have $\mu = E(Y_t)(1 - \beta^\top i_p)$ and

$$\begin{bmatrix} \beta_1 \\ \vdots \\ \vdots \\ \beta_p \end{bmatrix} = \begin{bmatrix} 1 & \rho(1) & & \rho(p-1) \\ & \ddots & \ddots & \\ & & \ddots & \rho(1) \\ & & & 1 \end{bmatrix}^{-1} \begin{bmatrix} \rho(1) \\ \vdots \\ \vdots \\ \rho(p) \end{bmatrix}. \tag{3.37}$$

There is a one to one relationship between the β and the ρ, which is nonlinear.

Example 24 *In the $p=2$ case we have:*

$$\beta_1 = \frac{\rho(1)(1-\rho(2))}{1-\rho^2(1)} \quad ; \quad \beta_2 = \frac{\rho(2)-\rho^2(1)}{1-\rho^2(1)}$$

$$\rho(1) = \frac{\beta_1}{1-\beta_2} \quad ; \quad \rho(2) = \beta_2 + \frac{\beta_1^2}{1-\beta_2}$$

so we see that $\beta_1 = \beta_2 = 0$ if and only if $\rho(1) = \rho(2) = 0$. This is not true in general.

The empirical version of the formula (3.37) is the set of **Yule–Walker** equations

$$\begin{bmatrix} \widehat{\beta}_1 \\ \vdots \\ \vdots \\ \widehat{\beta}_p \end{bmatrix} = \begin{bmatrix} 1 & \widehat{\rho}(1) & & \widehat{\rho}(p-1) \\ & \ddots & \ddots & \\ & & \ddots & \widehat{\rho}(1) \\ & & & 1 \end{bmatrix}^{-1} \begin{bmatrix} \widehat{\rho}(1) \\ \vdots \\ \vdots \\ \widehat{\rho}(p) \end{bmatrix}. \tag{3.38}$$

Autoregression based approaches have advantages and disadvantages. First, they are designed for the conditional moment hypothesis (Martingale hypothesis), whereas the autocorrelation approach is not specifically so tuned. Furthermore, the autoregression model can be explicitly used to deliver prediction of future values, whereas the autocorrelation approach does not. Specifically, the one-step ahead forecast is

$$\widehat{Y}_{T+1|T} = \widehat{\mu} + \widehat{\beta}_1 Y_T + \cdots + \widehat{\beta}_p Y_{T+1-p}.$$

This can be used to guide a trading strategy.

On the other hand, if p is large, the covariance matrix in OLS can be rank deficient, which makes the computation and analysis of $\widehat{\beta}$ difficult, whereas the sample autocorrelation can be defined for any lag $p < T - 1$. Furthermore, the regression approach is not so easy to encapsulate graphically.

3.4 Testing under More General Conditions than rw1

We next discuss the conditions that are needed to make the CLT for the sample autocorrelations and the variance ratio statistics to work, thereby allowing a statistical test of the EMH. We have mostly presented a theory based on **rw1**, which has the merit of being simple. Unfortunately, the data do not seem to conform well to this assumption. We shall discuss the theoretical issues and provide alternative standard errors and test statistics that are valid under weaker conditions such as **rw2–rw5**.

We first note that one needs to make stronger assumptions regarding the moments of the increments to the random walk in order to conduct inference than is strictly necessary for the random walk to be well-defined. Although normality of the return distribution is not needed for the above simple distribution theory for the **rw1** case, we do require at least $E(Y_t^2) < \infty$ in Theorem 14; see Brockwell and Davis (2006, Theorem 7.2.2.). It is quite notable that this result can be proved under only two moments, since the CLT for the autocovariance function requires four moments in any case. The autocorrelation function, since it is the ratio of the autocovariance function at j and 0 and is bounded between zero and one, somehow has less sensitivity, and the CLT holds under the weaker moment condition provided the series is stationary. However, under **rw3**, which does not require stationarity, the CLT for the sample autocorrelation function will require at least $E(Y_t^4) \leq C < \infty$. For daily data or higher frequency data more than four moments can be questionable, which motivates other approaches, as we will see in Chapter 4.

Second, we may have certain types of nonstationarity such as day of the week heteroskedasticity, which is consistent with **rw2** or **rw3** but not with **rw1**.

3.4.1 Testing under rw2

We consider the appropriate statistics when it is believed that **rw2** holds, i.e., returns are independent but not identically distributed over time (although they do have the same mean). We first give a simple example where this holds.

3.4 Testing under More General Conditions than rw1

Example 25 *Suppose that returns are generated in calendar time, so that Monday returns are the sum of Saturday, Sunday, and Monday returns (where weekend returns are not observed). We ignore holidays and other market closes. Suppose that (daily) returns Y_t are independent with mean zero (this restriction is important; see below) and variances $\sigma_j^2, j =$ Mon, Tue, Wed, Thur, Fri. Suppose that $\sigma_{Tue}^2 = \sigma_{Wed}^2 = \sigma_{Thur}^2 = \sigma_{Fri}^2 = \sigma^2$ and $\sigma_{Mon}^2 = 3\sigma^2$. Then for large T*

$$\operatorname{var}\left[\frac{1}{\sqrt{T}}\sum_{t=2}^{T} Y_t Y_{t-1}\right] = \frac{1}{T}\sum_{t=2}^{T} E(Y_t^2)E(Y_{t-1}^2)$$

$$\simeq \frac{1}{5}\left[\sigma_{Mon}^2\sigma_{Tue}^2 + \sigma_{Tue}^2\sigma_{Wed}^2 + \sigma_{Wed}^2\sigma_{Thur}^2 + \sigma_{Thur}^2\sigma_{Fri}^2 + \sigma_{Fri}^2\sigma_{Mon}^2\right]$$

$$\simeq \frac{9\sigma^4}{5}$$

$$\frac{1}{T}\sum_{t=1}^{T} E(Y_t^2) \simeq \frac{1}{5}\left[\sigma_{Mon}^2 + \sigma_{Tue}^2 + \sigma_{Wed}^2 + \sigma_{Thur}^2 + \sigma_{Fri}^2\right] = \frac{7\sigma^2}{5}.$$

In this case,

$$\sqrt{T}\hat{\rho}(1) \Longrightarrow N\left(0, \frac{45}{49}\right). \qquad (3.39)$$

*This sampling scheme is consistent with **rw2** and the CLM conditions given above. The limiting variance of $\sqrt{T}\hat{\rho}(1)$ is less than or equal to one and so the Bartlett confidence bands are conservative, meaning they reject less often than they should.*

We consider the general case where $\widetilde{Y}_t = Y_t - E(Y_t)$ is independent but not identically distributed. In this case, $\operatorname{cov}(Y_t, Y_{t-j}) = 0$ for all t, j, but $\operatorname{var}(Y_t)$ can change with t. Let:

$$\lambda_{jj} = \lim_{T\to\infty} \frac{1}{T}\sum_{t=j+1}^{T} E\left(\widetilde{Y}_{t-j}^2\right) E\left(\widetilde{Y}_t^2\right)$$

$$\gamma_0 = \lim_{T\to\infty} \frac{1}{T}\sum_{t=1}^{T} E\left(\widetilde{Y}_t^2\right)$$

$$\omega_2(p) = \sum_{j=1}^{p-1} c_{j,p}^2 V_{2;jj} \quad ; \quad c_{j,p} = 2\left(1 - \frac{j}{p}\right)$$

$$V_2(p) = \operatorname{diag}(V_{2;11}, \ldots, V_{2;pp}) \quad ; \quad V_{2;jj} = \frac{\lambda_{jj}}{\gamma_0^2}.$$

Here, $T = np$ is the total number of high frequency returns available, where p is fixed and n is large. The variance ratio statistic only uses the first $p-1$ sample autocorrelations, but we keep a common notation for simplicity.

Theorem 20 *Suppose that* **rw2** *holds and the following conditions hold:*

(1) $E(|Y_t|^{2+\delta}) \leq M$

(2) γ_0 and λ_{jj} exists, $j = 1, \ldots, p$.

Then, as $T \to \infty$

$$\sqrt{T}(\hat{\rho}(1), \ldots, \hat{\rho}(p))^\intercal \Longrightarrow N(0, V_2(p))$$

$$\sqrt{T}\left(\widetilde{VR}(p) - 1\right) \Longrightarrow N(0, \omega_2(p)).$$

In fact this CLT (i.e., this limiting variance) can continue to hold under weaker conditions than **rw2**; see below.

Define $\widehat{Y}_t = Y_t - \overline{Y}$ and for $j = 1, \ldots, p$

$$\hat{\lambda}_{jj} = \frac{1}{T}\sum_{t=1}^{T} \widehat{Y}_{t-j}^2 \widehat{Y}_t^2 \quad ; \quad \hat{\gamma}_0 = \frac{1}{T}\sum_{t=1}^{T} \widehat{Y}_t^2$$

$$\hat{\omega}_2(p) = 4\sum_{j=1}^{p-1}\left(1 - \frac{j}{p}\right)^2 \widehat{V}_{2;jj} \tag{3.40}$$

$$\widehat{V}_2(p) = \text{diag}(\widehat{V}_{2;11}, \ldots, \widehat{V}_{2;pp}) \quad ; \quad \widehat{V}_{2;jj} = \frac{\hat{\lambda}_{jj}}{\hat{\gamma}_0^2}. \tag{3.41}$$

Then as $T \to \infty$:

$$\frac{\sqrt{T}\hat{\rho}(j)}{\sqrt{\widehat{V}_{2;jj}}} \Longrightarrow N(0, 1) \tag{3.42}$$

$$\hat{\omega}_2(p)^{-1/2}\sqrt{T}\left(\widetilde{VR}(p) - 1\right) \Longrightarrow N(0, 1). \tag{3.43}$$

The standard Box–Pierce statistic is no longer asymptotically χ_p^2 under the null hypothesis and **rw2**. It needs to be modified in this case to reflect the different joint distribution of the sample autocorrelations. In particular, let

$$Q(p) = T\hat{\rho}^\intercal \widehat{V}_2(p)^{-1}\hat{\rho} = \sum_{j=1}^{p} \frac{\hat{\rho}(j)^2}{\widehat{V}_{2;jj}(p)}. \tag{3.44}$$

This statistic is asymptotically χ_p^2 under the null hypothesis.

The CLM Conditions

CLM give some conditions for asymptotic normality that are weaker than **rw2**; in fact they claim validity under **rw5**. Their conditions are very general because they do not require stationarity of the process and allow a wide range of weakly dependent processes (non-independent). We present their conditions here.

3.4 Testing under More General Conditions than rw1

CLM Assumption H *Let* $\widetilde{Y}_t = Y_t - E(Y_t)$

H1 For all t and for any $j \neq 0$, $E(\widetilde{Y}_t) = 0$ and $E(\widetilde{Y}_t \widetilde{Y}_{t-j}) = 0$.
H2 $\widetilde{Y}_t$ is strong mixing with coefficients $\alpha(m)$ of size $r/(r-1)$, where $r > 1$, such that for all t and for any $j \geq 0$, there exists some $\delta > 0$ for which $E(|\widetilde{Y}_t \widetilde{Y}_{t-j}|^{2(r+\delta)}) \leq C < \infty$.
H3 $\lim_{T \to \infty} \frac{1}{T} \sum_{t=1}^{T} E(\widetilde{Y}_t^2) = \sigma^2 < \infty$.
H4 For all t, $E(\widetilde{Y}_t^2 \widetilde{Y}_{t-j} \widetilde{Y}_{t-k}) = 0$ for any $j, k \neq 0$ with $j \neq k$.

They propose a bootstrap method for estimating the large sample variance of $\widehat{VR}(p)$. The result is basically an application of Herrndorf (1984) or White and Domowitz (1984) applied to the series $X_t = \widetilde{Y}_t \widetilde{Y}_{t-j}$. Conditions H1–H3 essentially yield a CLT for $\frac{1}{\sqrt{T}} \sum_{t=j+1}^{T} \widetilde{Y}_t \widetilde{Y}_{t-j}$, while condition H4 guarantees the simple form of the limiting variance (and by the way guarantees that it is positive). Condition H4 is weaker than independence, i.e., **rw2** implies this condition. However, it seems that their conditions are incomplete. For the form of the limiting variance they state, they need in addition to H1

H1+ *For all* $s \neq t$ *and all* $j, k = 1, \ldots, p$:

$$E\left(\widetilde{Y}_t \widetilde{Y}_{t-j} \widetilde{Y}_s \widetilde{Y}_{s-k}\right) = 0. \tag{3.45}$$

This condition does not follow from H1. In H4 they assume that $E(\widetilde{Y}_t^2 \widetilde{Y}_{t-j} \widetilde{Y}_{t-k}) = 0$ but that does not imply any restrictions on $E(\widetilde{Y}_t \widetilde{Y}_{t-j} \widetilde{Y}_s \widetilde{Y}_{s-k})$ except when $t = s$. Furthermore, for the inference result they state, one also needs to explicitly assume the following condition.

H5 *The following limit exists for* $j = 1, \ldots, p$

$$\lim_{T \to \infty} \frac{1}{T} \sum_{t=1}^{T} E(\widetilde{Y}_t^2 \widetilde{Y}_{t-j}^2) = \lambda_{jj} < \infty. \tag{3.46}$$

This does not automatically follow from the other conditions because no stationarity has been assumed. Suppose that H1, H1+, H2, H3, H4, and H5 hold, then the above CLT (3.43) continues to hold. In the absence of condition (3.45), the asymptotic distribution of the variance ratio statistic is much more complicated than they state – the asymptotic variance is an infinite sum; see below.

3.4.2 Testing under rw3

We next consider the more general setting **rw3** that may allow dependent data, which would violate **rw2**. For example, stock returns generated by a pure GARCH(1,1) process (see Chapter 11 below) can satisfy the martingale (difference) hypothesis but not be independent over time. We would also like to weaken condition H4, defined above. We call this the **no-leverage** assumption. It is weaker than independence, i.e., **rw2** implies this condition, but it is not implied by **rw3**. This condition is rather restrictive empirically. Essentially it is saying that we can't predict $\widetilde{Y}_t^2$ given $\widetilde{Y}_{t-j}, \widetilde{Y}_{t-k}$ or rather that whether $\widetilde{Y}_{t-j}, \widetilde{Y}_{t-k}$ are positive or negative should not matter in predicting $\widetilde{Y}_t^2$. This question has been the subject of much research in the time series literature, and Nelson (1991) introduced a model for financial volatility that violates this restriction. Best practice in that literature is to use models that are flexible with regard to their predictions about $E(\widetilde{Y}_t^2 \widetilde{Y}_{t-j} \widetilde{Y}_{t-k})$. We will visit this issue below in Chapter 11.

We make the following assumptions.

Assumption C Let $\widetilde{Y}_t = Y_t - E(Y_t)$.

C1 For all t, $E(\widetilde{Y}_t | \mathcal{F}_{t-1}) = 0$.
C2 $\widetilde{Y}_t$ is strong mixing with coefficients $\alpha(m)$ of size $r/(r-1)$, where $r > 1$, such that for all t and for any $j \geq 0$, there exists some $\delta > 0$ for which $E(|\widetilde{Y}_t \widetilde{Y}_{t-j}|^{2(r+\delta)}) \leq C < \infty$.
C3 The following limits exist and are finite for $j, k = 1, \ldots, p$:

$$\gamma_0 = \lim_{T \to \infty} \frac{1}{T} \sum_{t=1}^{T} E(\widetilde{Y}_t^2) \quad ; \quad \lambda_{jk} = \lim_{T \to \infty} \frac{1}{T} \sum_{t=1}^{T} E\left(\widetilde{Y}_t^2 \widetilde{Y}_{t-j} \widetilde{Y}_{t-k}\right).$$

These assumptions allow for nonstationarity as in the CLM conditions but strengthen their condition H1 (**rw5**) to the MDS assumption **rw3**. Let

$$\omega_3(p) = \sum_{j=1}^{p-1} \sum_{k=1}^{p-1} c_{j,p} c_{k,p} V_{3;jk} \tag{3.47}$$

$$V_3(p) = (V_{3;jk})_{j,k=1}^{p} \quad ; \quad V_{3;jk}(p) = \frac{\lambda_{jk}}{\gamma_0^2}.$$

Theorem 21 *Suppose that C1–C3 holds. Then, as $T \to \infty$*

$$\sqrt{T} (\widehat{\rho}(1), \ldots, \widehat{\rho}(p))^\top \Longrightarrow N(0, V_3(p))$$

$$\sqrt{T} \left(\widetilde{VR}(p) - 1\right) \Longrightarrow N(0, \omega_3(p)).$$

3.4 Testing under More General Conditions than rw1

Note that under these conditions, the sample autocorrelations are generally mutually correlated, that is,

$$\text{cov}(\sqrt{T}\hat{\rho}(j), \sqrt{T}\hat{\rho}(k)) \to \frac{\lambda_{jk}}{\gamma_0^2} \neq 0.$$

The CLM condition H4 implies that $\lambda_{jk} = 0$ for $j \neq k$, but the MDS assumption alone does not imply this. We discuss this issue further below.

Define

$$\hat{\omega}_3(p) = \sum_{j=1}^{p-1} \sum_{k=1}^{p-1} c_{j,p} c_{k,p} \hat{V}_{3;jk}. \tag{3.48}$$

The sample autocorrelation satisfies (3.42) under these weaker conditions, since $V_{2;jj} = V_{3;jj}$, i.e.,

$$\frac{\sqrt{T}\hat{\rho}(j)}{\sqrt{\hat{V}_{2;jj}}} \Longrightarrow N(0,1).$$

On the other hand the variance ratio statistic and the Box–Pierce statistic are affected. We have

$$\hat{\omega}_3(p)^{-1/2} \sqrt{T}\left(\widetilde{VR}(p) - 1\right) \Longrightarrow N(0,1). \tag{3.49}$$

The standard Box–Pierce statistic is no longer asymptotically χ_p^2 under the null hypothesis. In this case we modify the Box–Pierce statistic to

$$Q(p) = T\hat{\rho}^\intercal \hat{V}_3(p)^{-1}\hat{\rho}, \quad \hat{V}_3 = \left(\frac{\hat{\lambda}_{jk}}{\hat{\gamma}_0^2}\right)_{j,k}. \tag{3.50}$$

This statistic is asymptotically χ_p^2 under the null hypothesis. In conclusion, it is perfectly feasible to conduct tests that properly reflect the well documented features of asset returns.

3.4.3 Comparison of Asymptotic Variances

We give a discussion of the limiting variance $V_2(p)$ defined in comparison with the form assumed under the stronger **rw1** assumption. We first look at a single sample autocorrelation. We have

$$\sqrt{T}\hat{\rho}(1) = \frac{\frac{1}{\sqrt{T}}\sum_{t=2}^{T} \hat{Y}_t \hat{Y}_{t-1}}{\frac{1}{T}\sum_{t=1}^{T} \hat{Y}_t^2} \simeq \frac{\frac{1}{\sqrt{T}}\sum_{t=2}^{T} \tilde{Y}_t \tilde{Y}_{t-1}}{\frac{1}{T}\sum_{t=1}^{T} \tilde{Y}_t^2} \Longrightarrow N\left(0, \frac{E(\tilde{Y}_t^2 \tilde{Y}_{t-1}^2)}{E^2(\tilde{Y}_t^2)}\right). \tag{3.51}$$

Note that in general $E(\tilde{Y}_t^2 \tilde{Y}_{t-1}^2) \neq E(\tilde{Y}_t^2) E(\tilde{Y}_{t-1}^2)$ when dependent heteroskedasticity is allowed for. We next discuss how different the asymptotic variance can be from the **rw1** case under the more general sampling conditions. We can write

$$E(\tilde{Y}_t^2 \tilde{Y}_{t-1}^2) = E(\tilde{Y}_t^2) E(\tilde{Y}_{t-1}^2) + \text{cov}(\tilde{Y}_t^2, \tilde{Y}_{t-1}^2).$$

Therefore,

$$\frac{E(\tilde{Y}_t^2 \tilde{Y}_{t-1}^2)}{E^2(\tilde{Y}_t^2)} = 1 + \frac{\text{cov}(\tilde{Y}_t^2, \tilde{Y}_{t-1}^2)}{E^2(\tilde{Y}_t^2)}$$

$$= 1 + \frac{\text{var}(\tilde{Y}_t^2)}{E^2(\tilde{Y}_t^2)}\text{corr}(\tilde{Y}_t^2, \tilde{Y}_{t-1}^2)$$

$$= 1 + \underbrace{(\kappa_4(Y_t) - 1)}_{\text{heavy tails}} \times \underbrace{\text{corr}(\tilde{Y}_t^2, \tilde{Y}_{t-1}^2)}_{\text{dependent heteroskedasticity}} \quad (3.52)$$

$$\underset{\text{corr}\geq 0}{\leq} \kappa_4(Y_t),$$

where $\kappa_4(Y_t) = E(\tilde{Y}_t^4)/E^2(\tilde{Y}_t^2) \geq 1$ is the kurtosis of the series Y_t, and the last inequality follows if we assume $\text{corr}(\tilde{Y}_t^2, \tilde{Y}_{t-1}^2) \geq 0$, which empirically seems reasonable. In the third equality we use that $\text{var}(\tilde{Y}_t^2) = E(\tilde{Y}_t^4) - E^2(\tilde{Y}_t^2)$.

This says that the asymptotic variance of $\hat{\rho}(1)$ can be arbitrarily large depending on the kurtosis of Y_t. In principle, standard errors that allow for this dependence may be a lot wider than the Bartlett ones. In some cases they may be smaller. This shows that under (conditional) heteroskedasticity (models for which are discussed below in Chapter 11), the asymptotic variance of the sample autocorrelation could be arbitrarily far from the value predicted by the assumption **rw1**. We show in Table 3.3 some typical values of the quantities on the right hand side of (3.52).

We next give a discussion of the limiting variance $V_3(p)$ or rather the off-diagonal terms, which are proportional to

$$\lambda_{jk} = \lim_{T \to \infty} \frac{1}{T} \sum_{t=1}^{T} E\left(\tilde{Y}_t^2 \tilde{Y}_{t-j} \tilde{Y}_{t-k}\right)$$

for $j \neq k$. In Table 3.4 we give the sample estimates of λ_{jk} for $j, k = 1, \ldots, 5$ for the standardized daily S&P500 daily return. The off-diagonal terms tend to be positive but are much smaller than the diagonal terms, which represent the contribution of heteroskedasticity. Generally, $\omega_3(p) > \omega_2(p)$, and in some cases substantially so. In Tables 3.5, 3.6, and 3.7 we report t-statistics based on the three different standard errors $\hat{\omega}_j(p), j = 1, 2, 3$ for a different dataset.

3.4.4 Testing under rw5

We consider the properties of the test statistics we have considered when we only assume the weak condition of the absence of autocorrelation. In this case, the form of the asymptotic variance of the sample autocorrelations and of the variance ratio statistic is more complicated and involves an infinite sum. In particular,

$$\text{var}\left(\frac{1}{\sqrt{T}} \sum_{t=2}^{T} \tilde{Y}_t \tilde{Y}_{t-1}\right) = \frac{1}{T} \sum_{t=2}^{T} E(\tilde{Y}_t^2 \tilde{Y}_{t-1}^2) + \frac{1}{T} \sum_{t=2}^{T} \sum_{\substack{s=2 \\ s \neq t}}^{T} E(\tilde{Y}_t \tilde{Y}_{t-1} \tilde{Y}_s \tilde{Y}_{s-1}),$$

3.4 Testing under More General Conditions than rw1

Table 3.3 Autocorrelation of squared returns and kurtosis of returns

	$\rho_{Y^2}(1)$	$\kappa_4(Y)$
JP Morgan	0.1201	10.1305
Coke	0.3217	12.8566
McD	0.1801	7.4358
MMM	0.1137	7.4254
Merck	0.0412	22.9570
MSFT	0.1224	8.1895
Pfizer	0.1497	6.2051
P&G	0.0376	62.9128
AT&T	0.1494	8.0896
Travelers	0.3401	16.2173
United Health	0.0758	23.0302
United Tech	0.0365	21.5274
Verizon	0.2203	7.9266
Wall Mart	0.1841	6.1845
Exxon Mobil	0.2947	11.7152
S&P500	0.2101	11.4509

Table 3.4 The off-diagonal terms

j/k	1	2	3	4	5
1	4.893	0.106	0.177	0.129	0.133
2		6.835	0.622	0.188	1.420
3			4.173	0.185	−0.765
4				3.881	−0.189
5					5.882

Table 3.5 Variance ratios for weekly small-size portfolio returns

Sample period	# of obs	Lags			
		$p=2$	$p=4$	$p=8$	$p=16$
Portfolio of firms with market values in smallest CRSP quintile					
19620706–19780929	848	1.43	1.93	2.46	2.77
		(8.82)*	(8.49)*	(7.00)*	(5.59)*
		(8.82)*	(10.81)*	(11.00)*	(9.33)*
		(12.46)*	(14.47)*	(14.39)*	(11.70)*
19781006–19941223	847	1.43	1.98	2.65	3.19
		(6.20)*	(7.07)*	(7.37)*	(6.48)*
		(6.20)*	(8.62)*	(10.69)*	(10.70)*
		(12.52)*	(15.25)*	(16.26)*	(14.45)*
19941230–20131227	992	1.21	1.47	1.7	1.82
		(3.30)*	(3.58)*	(3.35)*	(2.50)*
		(3.30)*	(4.13)*	(4.15)*	(3.44)*
		(6.59)*	(7.91)*	(7.43)*	(5.82)*

where the second term is zero under **rw3** but not in general under **rw5**. When the series $\widetilde{Y}_t$ is also (weakly) stationary we saw in Chapter 2 that

$$\mathrm{var}\left(\frac{1}{\sqrt{T}}\sum_{t=2}^{T}\widetilde{Y}_t\widetilde{Y}_{t-1}\right) \to \gamma_X(0) + \sum_{j=1}^{\infty}\gamma_X(j),$$

where with $X_t = \widetilde{Y}_t\widetilde{Y}_{t-1}$, $\gamma_X(j) = \mathrm{cov}(X_t, X_{t-j}) = E(\widetilde{Y}_t\widetilde{Y}_{t-1}\widetilde{Y}_{t-j}\widetilde{Y}_{t-j-1}) \neq 0$. The asymptotic variance can be complicated! The limiting distribution of the variance ratio statistics and the Box–Pierce statistics can be obtained similarly, and they depend on the long run variance (2.17).

One can construct robust inference methods under **rw5** based on the general principles of long run variance estimation; see Newey and West (1987). Alternatively, Lobato (2001) shows how to test for the absence of serial correlation (i.e., under **rw5**) in a dependent process using **self normalized sums**. However, is it worth it? It seems that **rw5** is too weak an assumption and does not reflect fully the implications of the EMH. It results in a very complicated and potentially fragile inference method. The MDS assumption captures the essence of the EMH and leads to better inference procedures.

3.4 Testing under More General Conditions than rw1

Table 3.6 Variance ratios for weekly medium-size portfolio returns

Sample period	# of obs	Lags			
		$p=2$	$p=4$	$p=8$	$p=16$
Portfolio of firms with market values in central CRSP quintile					
62:07:06–78:09:29	848	1.25	1.54	1.79	1.91
		$(5.41)^*$	$(5.55)^*$	$(4.35)^*$	$(3.22)^*$
		$(5.41)^*$	$(6.41)^*$	$(5.93)^*$	$(4.69)^*$
		$(7.37)^*$	$(8.42)^*$	$(7.78)^*$	$(6.05)^*$
78:10:06–94:12:23	847	1.20	1.37	1.54	1.56
		$(3.29)^*$	$(3.35)^*$	$(3.18)^*$	$(2.14)^*$
		$(3.29)^*$	$(3.72)^*$	$(3.90)^*$	$(2.93)^*$
		$(5.73)^*$	$(5.80)^*$	$(5.36)^*$	$(3.74)^*$
94:12:30–13:12:27	992	0.99	1.05	1.02	0.89
		(-0.02)	(0.38)	(0.10)	(-0.38)
		(-0.02)	(0.43)	(0.11)	(-0.48)
		(-0.04)	(0.78)	(0.20)	(-0.78)

3.4.5 Trading Time Versus Calendar Time

Suppose that returns are generated in calendar time and prices obey **rw1** so that returns r_t are i.i.d. with mean μ and variance σ^2, but we only observe prices when the market is open. Then the observed return process is generally not consistent with even **rw5** because the mean return of the observed series will not be constant over time.

Let D_t be the number of days since the last trading day ($D_t=1$ for a standard Tuesday, ..., Friday, and $D_t=3$ for a standard Monday, and $D_t=4$ for a Tuesday following a holiday Monday). Then observed returns satisfy

$$r_t^O = \sum_{j=0}^{D_t-1} r_{t-j} = r_t(D_t),$$

which shows that observed returns are mutually independent but satisfy

$$E(r_t^O) = \mu D_t \quad ; \quad \text{var}(r_t^O) = \sigma^2 D_t, \tag{3.53}$$

which makes them non identically distributed, or nonstationary. Although non-stationarity in the variance does not affect the consistency arguments and can be

Table 3.7 Variance ratios for weekly large-size portfolio returns

		Lags			
Sample period	# of obs	$p=2$	$p=4$	$p=8$	$p=16$
Portfolio of firms with market values in largest CRSP quintile					
62:07:06–78:09:29	848	**1.05**	**1.15**	**1.21**	**1.19**
		(1.05)	(1.64)	(1.23)	(0.68)
		(1.05)	(1.54)	(1.32)	(0.84)
		(1.59)	(2.33)*	(2.06)*	(1.29)
78:10:06–94:12:23	847	**1.03**	**1.06**	**1.08**	**1.01**
		(0.63)	(0.61)	(0.54)	(0.03)
		(0.63)	(0.65)	(0.59)	(0.04)
		(0.95)	(0.91)	(0.75)	(0.04)
94:12:30–13:12:27	992	**0.93**	**0.94**	**0.89**	**0.81**
		(−0.99)	(−0.46)	(−0.53)	(−0.62)
		(−0.99)	(−0.52)	(−0.61)	(−0.77)
		(−2.05)*	(−1.01)	(−1.14)	(−1.35)

accounted for in the inference step, as we have seen, the presence of a time varying mean will bias the usual estimates of autocovariances as we now show.

For this process we have $\sum_{t=1}^{T} E(r_t^O) = \mu \sum_{t=1}^{T} D_t$. Note that since the typical number of trading days in a NYSE trading year is around 252 out of 365.25 calendar days, the average value of D is something like 1.5. Let's assume that every week has roughly the same structure, in which case for large T, $\frac{1}{T}\sum_{t=1}^{T} D_t = \frac{7}{5}$, and the sample mean of observed returns would overestimate the daily mean by around 40%. Furthermore, we have

$$\frac{1}{T}\sum_{t=1}^{T} E((r_t^O)^2) = \frac{1}{T}\sum_{t=1}^{T}(\mu^2 D_t^2 + \sigma^2 D_t) = \mu^2 \frac{1}{T}\sum_{t=1}^{T} D_t^2 + \sigma^2 \frac{1}{T}\sum_{t=1}^{T} D_t.$$

The sample variance of observed returns will converge to the same limit as the deterministic quantity

$$\gamma_T(0) = \frac{1}{T}\sum_{t=1}^{T} E(r_t^O)^2 - \left(\frac{1}{T}\sum_{t=1}^{T} E(r_t^O)\right)^2$$

$$= \mu^2\left[\frac{1}{T}\sum_{t=1}^{T} D_t^2 - \left(\frac{1}{T}\sum_{t=1}^{T} D_t\right)^2\right] + \sigma^2 \frac{1}{T}\sum_{t=1}^{T} D_t.$$

3.4 Testing under More General Conditions than rw1

Likewise, we have approximately

$$\gamma_T(k) = \mu^2 \left[\frac{1}{T} \sum_{t=1}^{T} D_t D_{t-k} - \left(\frac{1}{T} \sum_{t=1}^{T} D_t \right)^2 \right].$$

Furthermore the seasonal dummy variables satisfy:

$$\frac{1}{T} \sum_{t=1}^{T} D_t^2 = \frac{9+1+1+1+1}{5} = \frac{13}{5}$$

$$\frac{1}{T} \sum_{t=1}^{T} D_t D_{t-j} = \frac{3+1+1+1+3}{5} = \frac{9}{5}, \quad j = 1, 2, 3, 4$$

$$\frac{1}{T} \sum_{t=1}^{T} D_t D_{t-5} = \frac{1}{T} \sum_{t=1}^{T} D_t^2 = \frac{9+1+1+1+1}{5} = \frac{13}{5}.$$

Therefore,

$$\gamma^O(0) = \lim_{T\to\infty} \frac{1}{T} \sum_{t=1}^{T} \mathrm{var}(r_t^O) = \frac{16\mu^2 + 35\sigma^2}{25}$$

$$\gamma^O(k) = \lim_{T\to\infty} \frac{1}{T} \sum_{t=1}^{T} \mathrm{cov}(r_t^O, r_{t-k}^O) = \begin{cases} \frac{-4\mu^2}{25} & \text{if } k \neq 0 \bmod(5) \\ \frac{16\mu^2}{25} & \text{if } k = 0 \bmod(5) \end{cases}$$

$$\rho^O(k) = \frac{\gamma^O(k)}{\gamma^O(0)} = \begin{cases} \frac{-4\mu^2}{16\mu^2+35\sigma^2} & \text{if } k \neq 0 \bmod(5) \\ \frac{16\mu^2}{16\mu^2+35\sigma^2} & \text{if } k = 0 \bmod(5). \end{cases}$$

If $\mu/\sigma = 0.1$, then $\rho(k) = -0.04/(0.16+35) = -0.00114$, $k = 1, 2, 3, 4$ and $\rho(k) = 0.00456$. These are very small numbers. Note also that this predicts a negative sign for $\rho(k)$ for k not divisible by five but a positive sign and four times bigger value for those k divisible by five.

Suppose that we have two assets r_{it}, r_{jt} with $\mathrm{cov}(r_{it}, r_{jt}) = \sigma_{ij}$. Then

$$\frac{1}{T} \sum_{t=1}^{T} E(r_{it}^O r_{jt}^O) = \frac{1}{T} \sum_{t=1}^{T} (\mu_i \mu_j D_t^2 + \sigma_{ij} D_t) = \mu_i \mu_j \frac{1}{T} \sum_{t=1}^{T} D_t^2 + \sigma_{ij} \frac{1}{T} \sum_{t=1}^{T} D_t$$

$$\frac{1}{T} \sum_{t=1}^{T} \mathrm{cov}(r_{it}^O, r_{jt}^O) = \sigma_{ij} \frac{1}{T} \sum_{t=1}^{T} D_t + \mu_i \mu_j \left[\frac{1}{T} \sum_{t=1}^{T} D_t^2 - \left(\frac{1}{T} \sum_{t=1}^{T} D_t \right)^2 \right] \simeq \frac{16\mu_i\mu_j + 35\sigma_{ij}}{25},$$

which could go either way, although if $\mu_i, \mu_j \geq 0$, we have an upward bias.

There are several ways of dealing with this issue within the calendar time framework. First, we could estimate the mean of returns using the known structure (3.53). Specifically, let

$$\widehat{\gamma}(k) = \frac{1}{T} \sum_{t=1}^{T} \left(r_t^O - \widehat{\mu} D_t\right) \left(r_{t-k}^O - \widehat{\mu} D_{t-k}\right), \quad \widehat{\mu} = \frac{\sum_{t=1}^{T} D_t r_t^O}{\sum_{t=1}^{T} D_t^2}.$$

Here, $\widehat{\mu}$ is the OLS estimator of μ (we could use the GLS estimator here because of the known structure). This yields consistent estimators of the corresponding population quantities of true returns. This works also in the general case where $D_t \in \{1, 2, \ldots, m\}$ for some m (because of public holidays and other irregular closing events).

An alternative way of dealing with the missing data issue is to **impute** the **missing observations**, that is, to replace them by their best estimates given the available data.

Example 26 *Suppose we observe the log prices $p_1, \ldots, p_5, p_8$. Then*

$$p_6^I = p_5 + \frac{p_8 - p_5}{3} \quad ; \quad p_7^I = p_5 + \frac{2(p_8 - p_5)}{3}.$$

It follows that

$$r_6^I = p_6^I - p_5 = \frac{p_8 - p_5}{3} = \frac{r_6 + r_7 + r_8}{3}$$

$$r_7^I = p_7^I - p_6^I = \frac{p_8 - p_5}{3} = \frac{r_6 + r_7 + r_8}{3}$$

$$r_8^I = p_8 - p_7^I = p_8 - p_5 - \frac{2(p_8 - p_5)}{3} = \frac{r_6 + r_7 + r_8}{3}.$$

The daily return series is then $r_1, \ldots, r_5, r_6^I, r_7^I, r_8^I, r_9, \ldots$. This approach induces a positive autocorrelation in the imputed return series because of the overlap.

3.5 Some Alternative Hypotheses

In this section we discuss some leading alternatives to the EMH that may be detected by the testing strategies we have outlined.

3.5.1 Local Alternatives

We typically find that a certain test is consistent against a large class of global alternatives, meaning that with probability tending to one the test will reject when an alternative hypothesis is true. To compare tests one often has to consider local alternatives, which represent shrinking departures from the null hypothesis. Mathematically, this

3.5 Some Alternative Hypotheses

requires us to embed the problem in a **triangular array**, a sequence of distributions. For example, suppose that

$$r_{t,T} = \lambda_T r_{t-1} + \varepsilon_t$$

$$\lambda_T = \frac{\delta}{\sqrt{T}},$$

where δ is called the local parameter. In this case, the process $r_{t,T}$ is not stationary, but it has an autocorrelation function $\rho_T(k) = \lambda_T^k$, which decays to zero for each T as $k \to \infty$ and for each k as $T \to \infty$. In this case, we may show that

$$\sqrt{T} \hat{\rho}(k) \Longrightarrow N(\delta^k, 1).$$

Furthermore, the probability of rejecting the null hypothesis that $\rho(k) = 0$ is

$$\Pr\left(\left|\sqrt{T}\hat{\rho}(k)\right| \geq z_{\alpha/2}\right) \to \Pr\left(\left|Z + \delta^k\right| \geq z_{\alpha/2}\right) = 2\Phi(-z_{\alpha/2} - |\delta^k|) \in (\alpha, 1).$$

Therefore, the test has some power against any such local alternative, and the power increases with $|\delta|$. Faust (1992) considers the local power of variance ratio tests in more detail. He shows that variance ratio tests can be given a likelihood ratio interpretation and so are optimal against a certain local alternative.

3.5.2 Fads Model

We consider an alternative to the efficient market hypothesis, which allows for temporary mispricing through fads but assures that the rational price dominates in the long run.

Definition 55 *Suppose that observed log prices p_t satisfy:*

$$p_t^* = \mu + p_{t-1}^* + \varepsilon_t \tag{3.54}$$

$$p_t = p_t^* + \eta_t, \tag{3.55}$$

where ε_t is i.i.d. with mean zero and variance σ_ε^2, while η_t is a stationary weakly dependent process with unconditional variance σ_η^2, and the two processes are mutually independent.

This is the Muth (1960) model, which was employed by Poterba and Summers (1988). It follows that the observed return satisfies

$$r_t = p_t - p_{t-1} = \underbrace{\mu + \varepsilon_t}_{\text{i.i.d. fundamental return}} + \underbrace{\eta_t - \eta_{t-1}}_{\text{mean zero stationary fad}}.$$

It allows actual prices p to deviate from fundamental prices p^* but only in the short run through the fad process η_t. This process is a plausible alternative to the efficient markets hypothesis. If η_t were i.i.d., then r_t would be (to second order) an MA(1) process, which is a structure implied by a number of market microstructure issues; see Chapter 5 below. The covariance function of the sum of independent stochastic processes is the sum of the covariance functions. Therefore, with $u_t = \eta_t - \eta_{t-1}$, we have for $k \geq 1$,

$$\text{cov}(r_t, r_{t-k}) = \text{cov}(\varepsilon_t, \varepsilon_{t-k}) + \text{cov}(u_t, u_{t-k}) = \text{cov}(u_t, u_{t-k}),$$

so that there is short run autocorrelation in observed returns. Consider the horizon p returns, which are just the sum of the one period observed returns $r_t(p) = p\mu + \sum_{k=1}^{p} \varepsilon_{t-k} + \eta_t - \eta_{t-p}$. It follows that the variance of the horizon p returns is

$$\text{var}(r_t(p)) = p\text{var}(\varepsilon_t) + \text{var}(\eta_t - \eta_{t-p}).$$

Since the fad component is covariance stationary, for a large enough p the variance ratio of observed returns r is less than one. In fact, as $p \to \infty$

$$VR(p) = \frac{p\text{var}(\varepsilon_t) + \text{var}(\eta_t - \eta_{t-p})}{p\text{var}(\varepsilon_t) + p\text{var}(\eta_t - \eta_{t-1})} \to 1 - \frac{\text{var}(\Delta\eta)}{\text{var}(\Delta p)}.$$

This says that under this alternative hypothesis we should expect to see long horizon variance ratios less than one. Furthermore, the amount by which it is less than one tells us something about the magnitude of the fad component relative to the fundamental component. In fact, provided only that $\text{var}(\eta_t - \eta_{t-p}) < p\text{var}(\eta_t - \eta_{t-1})$, which is not a big ask, then $VR(p) < 1$.

CLM say that long horizon return tests often have low power and unreliable asymptotics. They explain this finding based on the case that $p/T \to (0, \infty)$, in which case the limiting distributions are no longer normal and so all bets are off. If $p/T \to 0$ fast enough then the test is asymptotically valid. We can write

$$VR(\infty) = 1 + 2 \lim_{p \to \infty} \sum_{j=1}^{p-1} \left(1 - \frac{j}{p}\right) \rho(j)$$

$$= 1 + 2 \sum_{j=1}^{\infty} \rho(j) - 2 \lim_{p \to \infty} \frac{1}{p} \sum_{j=1}^{p} j\rho(j)$$

$$= 1 + 2 \sum_{j=1}^{\infty} \rho(j) = \frac{\text{lrvar}(r)}{\text{var}(r)}$$

provided $\sum_{j=1}^{\infty} j|\rho(j)| < \infty$, which will be the case for many stationary and mixing processes. Under some conditions the lrvar is consistently estimable and asymptotic normal (at rate $n^{1/2}$ not $(np)^{1/2}$); see Andrews (1991).

3.5.3 Time Varying Expected Return

We briefly consider a simple statistical model for time varying expected return. This model could be consistent with rational pricing where the risk premium evolves slowly over time and has small variation relative to the shocks to risk adjusted returns.

3.6 Empirical Evidence regarding Linear Predictability based on Variance Ratios

Specifically, suppose that observed returns are composed of a slowly varying risk premium μ_t and an i.i.d. shock ε_t, i.e.,

$$r_t = \mu + \mu_t + \varepsilon_t, \qquad (3.56)$$

$$\mu_t = \mu_{t-1} + \eta_t, \qquad (3.57)$$

where $\mu_0 = 0$ and η_t is an i.i.d. mean zero shock that is small relative to ε_t. In this case observed returns are nonstationary so we must index populations by T. This specification is similar to that of equation 7.1.30 of CLM. We establish the following result.

Theorem 22 *Suppose that the model (3.56)–(3.57) holds with η_t independent and mean zero with $E(\eta_t^2) = \sigma_\eta^2/T > 0$ for each t, T, and ε_t i.i.d. mean zero with $E(\varepsilon_t^2) = \sigma_\varepsilon^2 > 0$. Then,*

$$\lim_{p \to \infty} \frac{1}{p} \lim_{T \to \infty} VR(p) = \frac{\frac{1}{2}\sigma_\eta^2}{\sigma_\eta^2 \frac{1}{2} + \sigma_\varepsilon^2} < 1.$$

This model predicts that the variance ratio should grow linearly with the horizon with a slope less than one.

3.5.4 Adaptive Markets Hypothesis

Suppose that $X_t = X_{t,T}$ can be approximated by a family of locally stationary processes $\{X_t(u), u \in [0, 1]\}$; see Dahlhaus (1997). For example, suppose that

$$X_t = \mu(t/T) + \varepsilon_t + \theta(t/T)\varepsilon_{t-1},$$

where $\mu(\cdot)$ and $\theta(\cdot)$ are smooth functions and ε_t is i.i.d. Here, $\mu(\cdot)$ can represent a slowly changing time varying risk premium, and $\theta(\cdot)$ captures predictability relative to that. This allows for zones of departure from the null hypothesis, say for $u \in U$, where U is a subinterval of $[0, 1]$, e.g., $\theta(u) \neq 0$ for $u \in U$. For example, during recessions the dependence structure may change and depart from efficient markets, but return to efficiency during normal times. This is consistent with the adaptive markets hypothesis of Lo (2004, 2005), whereby the amount of inefficiency can change over time depending on: "the number of competitors in the market, the magnitude of profit opportunities available, and the adaptability of the market participants." In this case, the variance ratio with a rolling window should detect the presence of predictability.

3.6 Empirical Evidence regarding Linear Predictability based on Variance Ratios

We consider again some of the evidence presented in CLM for the CRSP data over the periods 1962–1994 and the two subperiods 1962–1978 and 1978–1994. They used heteroskedasticity consistent standard errors, essentially those defined in (3.41). Their

Table 2.5 shows the variance ratios for weekly stock indexes (value weighted and equal weighted). They work with Wednesday to Wednesday returns, since this minimizes the number of missing observations due to holiday periods. They find that equal weighted indexes have strongly significant variance ratios, greater than one, and increasing with horizon. The evidence is stronger in the earlier period 1962–1978 compared with the second subperiod 1978–1994. The value weighted index shows a similar pattern but is not statistically significant. In their Table 2.6 they show the results for three size sorted portfolios. The smallest and medium sized portfolios are strongly significant, greater than one, and increasing with horizon. The largest portfolio variance ratios are not significant in the most recent period. In Table 2.7 they show the average of variance ratios for 411 individual stocks. They find these are less than one and declining with horizon.

We update their data to include the period 1994–2013. We divide the whole sample into three subsamples: 19620706–19780929 (848 weeks), 19781006–19941223 (847 weeks), and 19941230–20131227 (992 weeks). We first test for the absence of serial correlation in each of three weekly size-sorted equal-weighted portfolio returns (smallest quantile, central quantile, and largest quantile). We compare with the results reported in CLM (p71, Table 2.6). Table 3.5 reports the results for the portfolio of small-size firms, Table 3.6 reports the results for the portfolio of medium-size firms, and Table 3.7 reports the results for the portfolio of large-size firms. We examine $p = 2, 4, 8, 16$ as in CLM. We report in brackets below the VR itself the t statistics based on, respectively, $\widehat{V}_3(p)$, $\widehat{V}_2(p)$, and $\widehat{V}_1(p)$.

The results for the earlier sample periods are broadly similar to those in CLM (p71, Table 2.6) who compared the period 1962–1978 with the period 1978–1994 as well as the combined period 1962–1994. The variance ratios are greater than one and deviate further from one as the horizon lengthens. The departure from the random walk model is strongly statistically significant for the small and medium sized firms, but not so for the larger firms.

When we turn to the later period 1994–2013 we see that the variance ratios all reduce. For the smallest stocks the statistics are still significantly greater than one and increase with horizon. However, they are much closer to one at all horizons, and the statistical significance of the departures is substantially reduced. For medium sized firms, the variance ratios are reduced. They are in some cases below one and also no longer increasing with horizon. They are insignificantly different from one. For the largest firms, the ratios are all below one but are statistically inseparable from this value. One interpretation of these results is that the stock market (at the level of these portfolios) has become closer to the efficient benchmark. The biggest improvements seem to come in the most recent period, especially for the small stocks. The test statistics change quite a lot depending on which standard errors are used, and in some cases this could affect one's conclusions, for instance, for a large-size portfolio, test statistics based on the i.i.d. standard errors in some periods are statistically significant.

We also consider some long horizon variance ratios. In Figure 3.9 we show variance ratio statistics out to lags 252 for the S&P500 daily return series over the period 1950–2017. The ratio tends to around 0.55, which is quite low. On the other hand in Figure

3.6 Empirical Evidence regarding Linear Predictability based on Variance Ratios

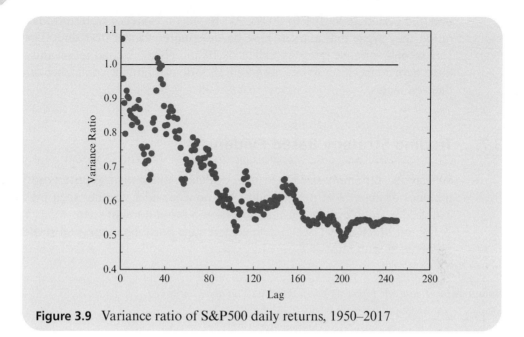

Figure 3.9 Variance ratio of S&P500 daily returns, 1950–2017

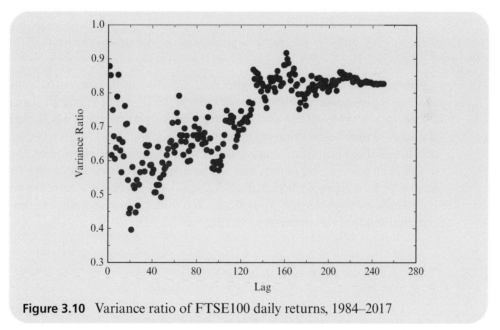

Figure 3.10 Variance ratio of FTSE100 daily returns, 1984–2017

3.10 we show the same results for the FTSE100 index over the period 1984–2017, which shows a somewhat different pattern and a larger limiting value, although still less than one.

There is a variety of evidence suggesting that high frequency price efficiency has improved with the growth of computer based trading. Castura, Litzenberger, and Gorelick (2010) investigate trends in market efficiency in Russell 1000/2000 stocks over the period 20060101 to 20091231. Based on evidence from intraday variance ratios, they argue that markets have become more efficient over time. They look at 10:1 second variance ratios as well as 60:10 and 600:60 second ratios and show that these have come closer to one, although they do not provide confidence intervals for their estimates.

3.7 Trading Strategy Based Evidence

We have so far emphasized statistical evidence and statistical criteria to judge the presence or absence of predictability. We now consider whether such predictability can yield a profit, on average, and how large a profit it might yield.

Lo and MacKinlay (1990a) defined arbitrage portfolio contrarian strategies on a set of assets with returns $X_{it}, i = 1, \ldots, d$.

Definition 56 *Define the following portfolio weights on asset i at time t,*

$$w_{it}(j) = \pm \frac{1}{d}\left(X_{i,t-j} - \overline{X}_{t-j}\right), \tag{3.58}$$

where $\overline{X}_t = \sum_{i=1}^{d} X_{it}/d$ is the equally weighted portfolio with return in period t.

The portfolio with positive weights (i.e., $\pm = +$) can be considered a momentum strategy: it puts positive weight on assets whose performance at time $t - j$ exceeded the average performance of the assets and negative weights on assets whose performance at time $t - j$ were below the average performance of the assets, i.e., it buys winners and sells losers. The contrarian strategy that Lo and MacKinlay considered corresponds to taking $\pm = -$, i.e., buy losers and sell winners. By construction with either $\pm = +$ or $\pm = -$ the weights (3.58) satisfy $\sum_{i=1}^{d} w_{it}(j) = 0$, so this is a zero net investment. For simplicity suppose that X_{it} is stationary with mean zero and variance one. The expected profit of the momentum strategy is

$$\begin{aligned}\pi_+(j) &= \sum_{i=1}^{d} E\left(w_{it}(j) X_{it}\right) \\ &= \frac{1}{d}\sum_{i=1}^{d} E\left(\left(X_{it-j} - \overline{X}_{it-j}\right) X_{it}\right) \\ &= \frac{1}{d}\sum_{i=1}^{d}\left(E(X_{it-j}X_{it}) - \frac{1}{d}\sum_{l=1}^{d} E(X_{l,t-j}X_{it})\right) \\ &= \frac{1}{d}\sum_{i=1}^{d} \rho_{ii}(j) - \frac{1}{d^2}\sum_{i=1}^{d}\sum_{l=1}^{d} \rho_{il}(j),\end{aligned}$$

3.7 Trading Strategy Based Evidence

where $\rho_{il}(j)$ is the cross-autocorrelation between asset i and l with lag j. Under the martingale hypothesis, $\pi_+(j) = 0$ for all j. One could test this hypothesis directly by calculating the realized performance of the strategy and testing whether it has mean zero – to do so one needs to apply this to a lot of periods $t, t-j$.

This approach has assumed one period, $t-j$, for portfolio formation and one period for evaluation. Consider instead the J, K method that forms portfolios using $X_{i,t-j} - \overline{X}_{t-j}, j = 1, \ldots, J$ and evaluates performance using periods $t, t+1, \ldots, t+K-1$. That is, let

$$w_{it}(1,\ldots,J) = \frac{1}{d}\sum_{j=1}^{J}(X_{i,t-j} - \overline{X}_{t-j}), \quad i=1,\ldots,d. \quad (3.59)$$

Then we have expected profits

$$\pi_+(j,K) = \sum_{i=1}^{d} E\left(w_{it}(1,\ldots,J)\sum_{k=0}^{K-1} X_{i,t+k}\right)$$

$$= \frac{1}{d}\sum_{i=1}^{d}\sum_{j=1}^{J} E\left((X_{it-j} - \overline{X}_{it-j})\sum_{k=0}^{K-1} X_{i,t+k}\right)$$

$$= \frac{1}{d}\sum_{i=1}^{d}\left(\sum_{j=1}^{J}\sum_{k=0}^{K-1} E(X_{i,t-j}X_{i,t+k}) - \frac{1}{d}\sum_{l=1}^{d}\sum_{j=1}^{J}\sum_{k=0}^{K-1} E(X_{l,t-j}X_{i,t+k})\right)$$

$$= \frac{1}{d}\sum_{i=1}^{d}\sum_{j=1}^{J}\sum_{k=0}^{K-1}\rho_{ii}(k+j) - \frac{1}{d^2}\sum_{i=1}^{d}\sum_{l=1}^{d}\sum_{j=1}^{J}\sum_{k=0}^{K-1}\rho_{il}(k+j)$$

$$= \frac{1}{d}\sum_{i=1}^{d}\sum_{j=1}^{J+K-1} c_{j,j+K}\rho_{ii}(j) - \frac{1}{d^2}\sum_{i=1}^{d}\sum_{l=1}^{d}\sum_{j=1}^{J+K-1} c_{j,j+K}\rho_{il}(j),$$

where $c_{j,p} = 1 - j/p$. This shows explicitly how expected profits depend on the cross-autocovariances.

Example 27 *A simple variation on this strategy is to put weight only on extreme winners and losers. Specifically, let*

$$i_{\max}(t-1:t-J) = \arg\max_{1 \leq i \leq n}\sum_{j=1}^{J}(X_{i,t-j} - \overline{X}_{t-j})$$

$$i_{\min}(t-1:t-J) = \arg\min_{1 \leq i \leq n}\sum_{j=1}^{J}(X_{i,t-j} - \overline{X}_{t-j}).$$

Then consider the portfolio that puts equal (in magnitude) weight on these two assets. The expected profit of this portfolio is

$$E\left(\sum_{k=0}^{K-1} \left(X_{i_{\max}(t-1:t-J),t+k} - X_{i_{\min}(t-1:t-J),t+k}\right)\right),$$

which is much harder to calculate analytically but can be approximated using sample information.

There are several well-known papers that use essentially this methodology. Jegadeesh and Titman (1993) found short-term momentum, i.e., good and bad recent performance (3–12 months), continues over time (which is consistent with positive autocorrelation and not zero as with a random walk). They considered NYSE and AMEX stocks over the period 1965 to 1989. They considered trading strategies that selected stocks based on their returns over the past J months and then evaluated their performance over a K month holding period. At the beginning of each month t the securities are ranked in ascending order on the basis of their returns in the past J months. Based on these rankings, ten decile portfolios were formed that equally weight the stocks contained in the top decile, the second decile, and so on. The top decile portfolio is called the losers decile and the bottom decile is called the winners decile. In each month t, the strategy buys the winner portfolio and sells the loser portfolio, holding this position for K months. The profits of the above strategies were calculated for a series of buy and hold portfolios. The returns of all the zero cost portfolios (i.e., the returns per dollar long in this portfolio) were positive. All these returns are statistically significant and strongly so even after accounting for the fact that many different tests were carried out. The most successful zero-cost strategy selects stocks based on their returns over the previous 12 months and then holds the portfolio for 3 months.

De Bondt and Thaler (1985, 1987) suggest on the other hand that stock prices overreact to information, suggesting that contrarian strategies (buying past losers and selling past winners) achieve abnormal returns over a longer horizon. De Bondt and Thaler (1985) consider monthly return data for NYSE common stocks for the period between January 1926 and December 1982. They show that over 3- to 5-year holding periods stocks that performed poorly over the previous 3 to 5 years achieve higher returns than stocks that performed well over the same period. Thus, prices might incorporate information only gradually and eventually overreact; this might subsequently be corrected as the market recognizes the error. In theory, however, risk aversion might be time varying and this might happen due to changes in rational risk premia. Some have argued that the De Bondt and Thaler results can be explained by the systematic risk of their contrarian portfolios and the size effect. In addition, since the long-term losers outperform the long-term winners only in Januaries, it is unclear whether their results can be attributed to overreaction. The Jegadeesh and Titman and De Bondt and Thaler results have become part of academic folklore. Khandani and Lo (2007) discuss the **Quant meltdown** of August 2007 where many momentum strategy funds collapsed as the relationships they believed in broke down.

3.7 Trading Strategy Based Evidence

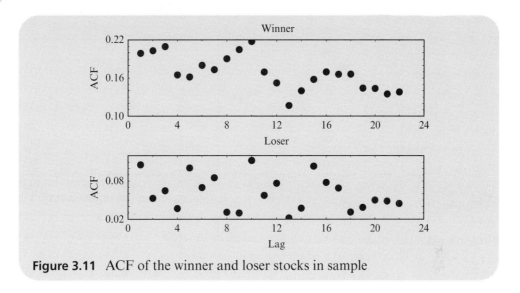

Figure 3.11 ACF of the winner and loser stocks in sample

We consider how a simplified version of this strategy works with daily data on the Dow Jones stocks. Specifically, we calculate the winner and loser every day. Let

$$r_{\max,t} = \max_{1 \leq i \leq n} r_{it} \quad ; \quad r_{\min,t} = \min_{1 \leq i \leq n} r_{it}.$$

In Figure 3.11 we show the autocorrelation function of the two series $r_{\max,t+1}$ and $r_{\min,t+1}$; under **rw1** these should also be i.i.d., but they are clearly not. Note that the winner stock is more positively autocorrelated than the loser one, but in both cases the autocorrelations are large relative to the usual null hypothesis critical bands.

We next consider whether the winner and loser stocks are predictable out of sample, that is, when we consider their next period return as the series of interest. That is, we let $i_{\max}(t) = \arg\max_{1 \leq i \leq n} r_{it}$ and consider $\text{cov}(r_{i_{\max}(t),t+1}, r_{i_{\max}(t-k,t+1-k)})$. The results are shown in Figure 3.12 (we also plot the standard Bartlett confidence bands). The apparent predictability is greatly reduced in magnitude, and the sign of the autocorrelations, at least the first ones, is swapped. Having been the winner or loser last period, you are likely to face a small short term reversal.

For random variables that satisfy **rw2**, $\text{acf}(\max_{1 \leq i \leq n}(X_{it})) = \max_{1 \leq i \leq n}(\text{acf}(X_{it})) = 0$, but for dependent processes (even uncorrelated ones) we may have any relation between these two population quantities, depending on the process governing returns. We compare the sample equivalents using the data on the Dow Jones stocks. In this case, clearly the ACF of the max return is much stronger than the max of the ACF's uniformly across lags.

The next example shows that this may not always be the case.

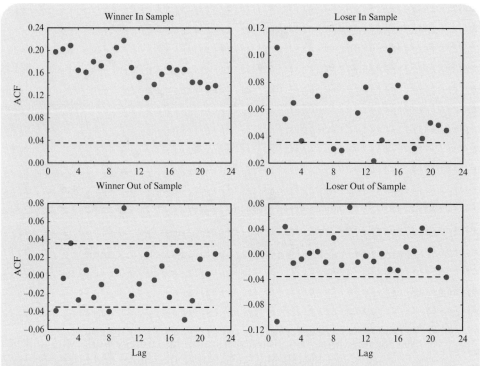

Figure 3.12 ACF of winner and loser stocks in and out of sample

Example 28 *Consider the process*

$$X_{it} = \beta_i f_t + \varepsilon_{it}$$

$$f_t = \rho f_{t-1} + \eta_t,$$

where: $\beta_i \sim U[0,1]$, $\varepsilon_{it} \sim N(0, \sigma_\varepsilon^2)$, *and* $\eta_t \sim N(0, \sigma_\eta^2)$ *and all random variables are mutually independent. In this case*

$$\gamma_{X_i}(k) = \begin{cases} \beta_i^2 \gamma_f(0) + \sigma_\varepsilon^2 & k=0 \\ \beta_i^2 \gamma_f(k) & k > 1. \end{cases}$$

This has maximal autocorrelation when $\beta_i = 1$ *in which case* $\max_{1 \leq i \leq n} \gamma_{X_i}(k) = \gamma_f(k)/(\gamma_f(0) + \sigma_\varepsilon^2) = \rho^k \sigma_\eta^2/(\sigma_\eta^2/(1-\rho^2) + \sigma_\varepsilon^2)$. *In the case where* $\sigma_\varepsilon^2 = 0$ *we have* $\max_{1 \leq i \leq n} \rho_{X_i}(k) = \rho_f(k) = \rho^k$. *Furthermore,*

$$\max_{1 \leq i \leq n}(X_{it}) = \begin{cases} f_t & \text{if } f_t > 0 \\ 0 & \text{if } f_t \leq 0. \end{cases}$$

The censoring of the process f_t *leads to an attenuated autocorrelation, Muthén (1990), so that* $\gamma_{\max X_i}(k) \leq \max \gamma_{X_i}(k)$. *Simulations confirm this, although the amount of attenuation is rather modest.*

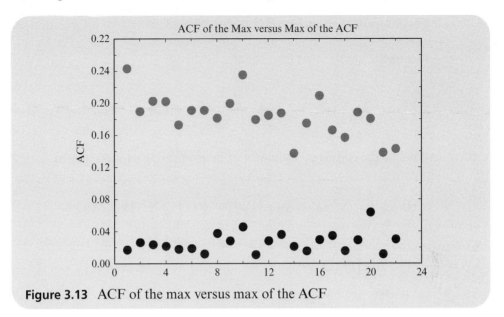

Figure 3.13 ACF of the max versus max of the ACF

The general methodology allows one to consider a longer period for determining winner/loser status and a longer period for determining the success of that portfolio. Variations on this include when portfolios are formed on the basis of some characteristic C such as market capitalization, which is time varying and cross sectionally varying.

3.8 Regression Based Tests of Semi-Strong Form Predictability

If stock returns are unforecastable given past price information this does not preclude them being forecastable given additional information, for example, by calendar effects, such as day of the week or month of the year effects, by accounting variables such as earnings, or by macroeconomic variables such as interest rates and inflation.

Suppose that

$$r_{t+1} = \mu + \beta^\top X_t + \varepsilon_t, \qquad (3.60)$$

where $X_t \in \mathbb{R}^d$ is observed (public information) at time t such as deterministic seasonal dummy variables. For example, $X_{it} = 1$ if t is in season i and $X_{it} = 0$ else. This specification includes the case where $X_t = (r_t, \ldots, r_{t-p}, z_{1t}, \ldots, z_{qt})^\top = (X_{1t}^\top, X_{2t}^\top)^\top$, where $p + q = d$ and z_{jt} are additional variables such as seasonal dummies or other variables. The EMH (along with constant mean or risk premium) says that $\beta_1 = \cdots = \beta_d = 0$. We may also be interested in testing whether given information contained in prices, $(r_t, \ldots, r_{t-p})^\top$, there is predictability coming from $(z_{1t}, \ldots, z_{qt})^\top$, in which case the null hypothesis is $\beta_{p+1} = \cdots = \beta_{p+q} = 0$. These hypotheses can be tested using a standard Wald statistic under the **rw1** assumption and relying on a large sample approximation. We shall consider a robust Wald statistic that is asymptotically valid under the regression version of **rw3** (that is, ε_t satisfies **rw3**, where the information set includes past r_t and X_t).

Define the unrestricted least squares estimator

$$\widehat{\beta} = \left(\sum_{t=1}^{T}(X_t - \overline{X})(X_t - \overline{X})^{\mathsf{T}}\right)^{-1}\left(\sum_{t=1}^{T}(X_t - \overline{X})r_t\right) = (\widehat{\beta}_1^{\mathsf{T}}, \widehat{\beta}_2^{\mathsf{T}})^{\mathsf{T}} \quad ; \quad \widehat{\mu} = \overline{r} - \widehat{\beta}^{\mathsf{T}}\overline{X},$$

and define the consistent estimator of its asymptotic variance matrix

$$\widehat{\Omega} = \left(\frac{1}{T}\sum_{t=1}^{T}(X_t - \overline{X})(X_t - \overline{X})^{\mathsf{T}}\right)^{-1}\left(\frac{1}{T}\sum_{t=1}^{T}(X_t - \overline{X})(X_t - \overline{X})^{\mathsf{T}}\widehat{\varepsilon}_t^2\right)$$
$$\times \left(\frac{1}{T}\sum_{t=1}^{T}(X_t - \overline{X})(X_t - \overline{X})^{\mathsf{T}}\right)^{-1}.$$

Then define the two robust Wald statistics

$$W_{rob} = T\widehat{\beta}^{\mathsf{T}}\widehat{\Omega}^{-1}\widehat{\beta} \quad ; \quad W_{2,rob} = T\widehat{\beta}_2^{\mathsf{T}}\widehat{\Omega}^{22}\widehat{\beta}_2, \tag{3.61}$$

where $\widehat{\Omega}^{22}$ is the corresponding block of $\widehat{\Omega}^{-1}$. These statistics are approximately distributed as χ_d^2 and χ_q^2, respectively, when the sample size is large.

For the day of the week hypothesis we may also want to test the calendar time version which says that expected returns on Monday should be three times the expected returns on any day. In Table 3.8 we show the results for the daily S&P500 return series where we drop the intercept. This shows that the mean return on Mondays was consistently negative and statistically significantly different from the returns on other days. The finding of negative returns on Monday is called the **Monday effect** because it is neither consistent with the calendar time or trading time interpretation of stock returns. In the most recent period, this discrepancy has reduced in terms of magnitude and statistical significance.

This is equivalent to the hypothesis that $\beta_{Mon} = 2\mu$ and $\beta_j = 0$ for $j =$ Tues, ..., Friday in the specification (3.60).

We may also consider longer horizon predictability, which involves the (predictive) regression

$$r_{t+j} = \mu + \beta^{\mathsf{T}}X_t + \varepsilon_{t+j},$$

where X_t is observed (public information) at time t, for example, price/earnings ratio effects, dividend rate, and so on. There is lots of evidence on this; see Chapter 9 for more discussion.

There is also a recent big data approach whereby the set of potential variables X_t is large. The least absolute selection and shrinkage operator (LASSO) method (see Tibshirani (1996)) is a popular way of estimating a regression model with many potential covariates (but a modest number of "active" ones. How to conduct formal statistical tests of the EMH in this context is not fully developed yet.

3.8 Regression Based Tests of Semi-Strong Form Predictability

Table 3.8 Day of the week effect

Sample period	Mon	Tue	Wed	Thur	Fri
1950–1960	−0.00128	0.00002	0.00109	0.00093	0.00188
	(3.969)	(0.075)	(3.463)	(2.926)	(5.859)
1960–1970	−0.00158	0.00013	0.00099	0.00053	0.00089
	(5.496)	(0.465)	(3.464)	(1.862)	(3.121)
1970–1980	−0.00120	−0.00013	0.00064	0.00039	0.00068
	(3.095)	(0.356)	(1.694)	(1.011)	(1.787)
1980–1990	−0.00107	0.00113	0.00141	0.00027	0.00088
	(−2.194)	(2.384)	(2.991)	(0.556)	(1.823)
1990–2000	0.00116	0.00062	0.00088	−0.00029	0.00062
	(2.882)	(1.601)	(2.247)	(−0.748)	(1.554)
2000–2016	−0.00013	0.00056	0.00010	0.00053	−0.00028
	(0.296)	(1.331)	(0.231)	(1.228)	(0.655)

3.8.1 Predictability in Prices

Suppose that we agree that there is predictability in stock returns. In practice, we are interested in predicting prices, so how are the two questions related? Suppose that

$$E(R_{t+1}|\mathcal{F}_t) = \mu(\mathcal{F}_t) = \mu_t.$$

What can we say about $E(P_{t+1}|\mathcal{F}_t)$? In this case, $E(P_{t+1}|\mathcal{F}_t) = P_t(1+\mu_t)$, and so we automatically obtain a forecast for prices. If $\widehat{\mu}_t$ is a conditionally unbiased forecast of μ_t, then

$$\widehat{E}(P_{t+1}|\mathcal{F}_t) = P_t(1+\widehat{\mu}_t)$$

is a conditionally unbiased forecast of P_{t+1}. On the other hand suppose that the model is for logarithmic returns, i.e.,

$$E(r_{t+1}|\mathcal{F}_t) = \mu_t,$$

where $r_{t+1} = \log P_{t+1} - \log P_t$. In this case, $E(\log P_{t+1}|\mathcal{F}_t) = \mu_t + \log P_t$. Unfortunately,

$$E(P_{t+1}|\mathcal{F}_t) \neq \exp(E(\log P_{t+1}|\mathcal{F}_t)) = P_t \exp(\mu_t),$$

so that we don't get a useful prediction for actual prices from the prediction of log returns. However, if μ_t is small (e.g., the data are high frequency), then the first equality is approximately true and we can forecast prices by $P_t \exp(\mu_t)$, that is, we let

$$\widehat{E}(P_{t+1}|\mathcal{F}_t) = P_t \exp(\widehat{\mu}_t),$$

where $\widehat{\mu}_t$ is our estimate of μ_t.

Pairs Trading

We consider an alternative approach to exploiting departures from the EMH. Suppose that P_t, Q_t are the prices of two assets whose logarithms obey the random walk hypothesis, i.e., **rw1**, so that they are nonstationary. Suppose however that for some β

$$\log(P_t) = \beta \log(Q_t) + \eta_t, \qquad (3.62)$$

where η_t is a stationary process with mean zero and finite variance. That is, there is a linear combination of the log prices of the two assets that is stationary, i.e., $\log(P_t)$ and $\log(Q_t)$ are **cointegrated**. Gatev, Goetzmann, and Rouwenhorst (2006) show how to construct a trading strategy that exploits this tendency of the two price series to move back towards each other after temporary separations.

Definition 57 *If $\log(P_t) - \widehat{\beta}\log(Q_t) > w_\alpha c_\eta$, then sell the stock P and buy Q, while if $\log(P_t) - \widehat{\beta}\log(Q_t) < -w_\alpha c_\eta$, then sell the stock Q and buy P. Here, $\widehat{\beta}$ is the least squares estimate of β, w_α is a critical value, and c_η is the square root of the variance or long run variance of η_t.*

Typical examples of paired stocks include Coca Cola and Pepsi Cola.

Example 29 *Suppose that η_t is i.i.d. and β were known. What is the expected profit of this strategy?*

Gatarek and Johansen (2014) consider the question of hedging portfolios in the cointegrated system (3.62) and in the more general multicointegration model.

3.9 Summary of Chapter

1. The efficient market hypothesis says that the stock market is informationally efficient and prices reflect all available information. It implies that returns are unpredictable relative to some required return.
2. Testing the EMH requires some specification of normal returns. When we can assume these are constant many statistical techniques are available for this task, such as autocorrelation based tests, variance ratios, and autoregression.
3. To carry out the statistical tests we need to impose further restrictions. First, we must choose the horizon over which we evaluate the evidence. Second, we need to

3.9 Summary of Chapter

specify some statistical assumptions to guarantee a CLT for the chosen statistic under the null hypothesis. We considered several versions of these. The Box–Pierce statistic does not distinguish between positive and negative correlations, and so although it aggregates the statistical evidence from many lags, it eradicates the direction of the evidence, whereas the variance ratio statistic cumulates the autocorrelations preserving the sign. The weakness is of course that it can fail to detect some alternatives.

4. The evidence mostly suggests that linear predictability in large stocks and stock indexes is rather small, and has reduced over time.

4 Robust Tests and Tests of Nonlinear Predictability of Returns

In this chapter we discuss nonlinear predictability and robust tests. The objective again is to provide methods that can be used to shed light on the validity of the EMH, but from a different angle. We also wrap up our discussion of predictability with some reflections on the evidence around predictability and explanations for the empirical findings.

4.1 Robust Tests

One issue with the tests we have considered so far is their vulnerability in the presence of extreme observations. The ACF and variance ratio tests we have considered require that $E(r_t^4) < \infty$ when returns are not i.i.d. (Mikosch and Starica (2000)), which may be hard to justify for some datasets; see Chapter 14. We will consider here tests that are robust to the existence of such moments. We first consider the more modest objective of trying to predict only the direction of travel of returns. We can decompose returns as

$$r_t = \text{sign}(r_t) \times |r_t|, \tag{4.1}$$

where

$$\text{sign}(x) = \begin{cases} 1 & \text{if } x > 0 \\ 0 & \text{if } x = 0 \\ -1 & \text{if } x < 0. \end{cases}$$

Under the EMH it should not be possible to predict the direction of returns, i.e., whether they go up or down (under the assumption of no risk premium). We now consider whether we can predict the direction of travel, $\text{sign}(r_t)$. We can think of this as like a coin toss. If the coin is unbiased, we expect heads and tails to be equally likely and to not occur in long sequences of heads and tails. Specifically, suppose that X_t is independent (**rw2**) and has median zero. Then

$$Y_t = \text{sign}(X_t) \tag{4.2}$$

is i.i.d. Bernoulli with a 50/50 chance of being ± 1 so $E(Y_t) = 0$ and $\text{var}(Y_t) = 1$. Furthermore, $E(Y_t Y_{t-j}) = 0$ for all j. We can apply standard testing procedures to the series Y_t. Dufour, Hallin, and Mizera (1998) establish various properties of sign tests under independent sampling.

4.1 Robust Tests

In fact, one doesn't need independence here to guarantee that $E(Y_t Y_{t-j}) = 0$ for all j, but just needs that for all j

$$\Pr(X_t > 0, X_{t-j} > 0) + \Pr(X_t < 0, X_{t-j} < 0) = \frac{1}{2}.$$

However, for statistical application we will need to apply a CLT and so may need to assume some additional independence or weak dependence condition. An example where this might be satisfied is stochastic volatility models; see Chapter 11.

The main issue for this approach is when X is not mean zero or rather not median zero, which is what we observe and should expect to observe in practice. Specifically, suppose that $X_t \sim N(\mu, \sigma^2)$. Then

$$\begin{aligned} E(Y_t) &= \Pr(X_t > 0) - \Pr(X_t < 0) \\ &= \Pr\left(\frac{X_t - \mu}{\sigma} > \frac{-\mu}{\sigma}\right) - \Pr\left(\frac{X_t - \mu}{\sigma} < \frac{-\mu}{\sigma}\right) \\ &= 1 - \Phi\left(\frac{-\mu}{\sigma}\right) - \Phi\left(\frac{-\mu}{\sigma}\right) \\ &= 1 - 2\Phi\left(\frac{-\mu}{\sigma}\right) \\ &= 2\left(\Phi\left(\frac{\mu}{\sigma}\right) - \frac{1}{2}\right) \neq 0. \end{aligned}$$

Furthermore, $\text{var}(Y_t) = 1 - E^2(Y_t) < 1$. If one assumes that the mean of returns is zero, then one may reject the null hypothesis incorrectly.

However, if we take account of the non zero mean then we can test the implication that $\text{cov}(Y_t, Y_{t-j}) = 0$. Dufour, Hallin, and Mizera (1998) consider the case where X_t has non zero median. When the median is known, they provide a test whose null distribution is known exactly; when the median is estimated, they show that this test is asymptotically distribution free under independent observations. To some extent this is about how one centers the returns. We will consider a similar test and show how to adjust it for non zero median/mean.

4.1.1 Cowles and Jones Test

Cowles and Jones (1937) proposed some tests of the efficient market hypothesis as then defined. These tests are robust to the existence of moments but do require the stronger independence assumption **rw2**. The idea is that we might be able to predict the direction of travel of the stock market rather than the actual level it achieves.

Definition 58 *For a stock return series r_t define continuations and reversals:*

$$\begin{aligned} \text{continuations}: & \quad r_t \times r_{t+1} > 0 \text{ or } \text{sign}(r_t) = \text{sign}(r_{t+1}) \\ \text{reversals}: & \quad r_t \times r_{t+1} < 0 \text{ or } \text{sign}(r_t) = -\text{sign}(r_{t+1}). \end{aligned}$$

This measures a natural tendency of returns to continue in the direction of travel or to reverse their direction of travel. This classification divides up the four quadrants of (r_t, r_{t+1}) space into two regions, such that $\Pr(r_t \times r_{t+1} < 0) = 1 - \Pr(r_t \times r_{t+1} > 0)$.

Definition 59 *The Cowles–Jones ratio is*

$$CJ = \frac{\Pr(continuation)}{\Pr(reversal)} = \frac{\Pr(r_t \times r_{t+1} > 0)}{1 - \Pr(r_t \times r_{t+1} > 0)}. \tag{4.3}$$

We suppose that returns are independent over time (**rw2**) and that $\pi = \Pr(r_t > 0)$ does not vary with time. It follows that

$$\begin{aligned}
\pi_c &= \Pr(continuation) \\
&= \Pr(r_t \times r_{t+1} > 0) \\
&= \Pr(r_t > 0, r_{t+1} > 0) + \Pr(r_t < 0, r_{t+1} < 0) \\
&= \Pr(r_t > 0)\Pr(r_{t+1} > 0) + \Pr(r_t < 0)\Pr(r_{t+1} < 0) \\
&= \pi^2 + (1 - \pi)^2,
\end{aligned}$$

which is constant over time but depends on π. Cowles and Jones assumed that $\pi = 1/2$ in which case $\pi_c = 1/2$ and hence $CJ = 1$. However, the testing idea can be applied to the case where $\pi \neq 1/2$ as we now show. In fact, assuming that $\pi = 1/2$ is hard to justify even for daily data: we do expect some more frequent up moves than down moves to reflect compensation for risk. Therefore, we shall not require that $\pi = 1/2$. We define the empirical version of CJ as follows.

Definition 60 *Suppose we observe a sample of returns $r_1, \ldots, r_T$. Let T_c be the number (#) of continuations, i.e., the number of times that $r_t \times r_{t+1} > 0$ and $T_r = \#$ of reversals, i.e., the number of times that $r_t \times r_{t+1} \leq 0$, and let*

$$\widehat{CJ} = \frac{T_c}{T_r} = \frac{T_c}{T - T_c} = \frac{\left(\frac{T_c}{T}\right)}{1 - \left(\frac{T_c}{T}\right)}. \tag{4.4}$$

Example 30 *We show the contingency table of the S&P500 daily stock return data over the period 1950–2014, that is, we record the frequency of up and down movements:*

	$r_{t+1} > 0$	$r_{t+1} < 0$
$r_t > 0$	0.297	0.227
$r_t < 0$	0.228	0.232

In this case, continuations $0.297 + 0.232 = 0.53$; reversals $0.228 + 0.227 = 0.46$, so that the CJ ratio is about 1.15.

4.1 Robust Tests

The Cowles–Jones statistic is a rational function of the sample mean of a binomial random variable. For large T, the sample mean of a binomial is approximately normal, using the CLT for sums of i.i.d. random variables. Let

$$m(\pi) = \frac{\pi_c}{1 - \pi_c} \quad ; \quad V(\pi) = \frac{\pi_c(1 - \pi_c) + 2\left(\pi^3 + (1-\pi)^3 - \pi_c^2\right)}{(1 - \pi_c)^4}.$$

Then $CJ = m(\pi)$. For example, when $X_t \sim N(\mu, \sigma^2)$, we have $\pi = 1 - \Phi\left(\frac{-\mu}{\sigma}\right)$.

Theorem 23 *Suppose that **rw2** holds and that $\pi = \Pr(r_t > 0)$ does not vary with time. Then,*

$$\sqrt{T}\left(\widehat{CJ} - m(\pi)\right) \Longrightarrow N(0, V(\pi)),$$

where $\pi_c = \pi^2 + (1 - \pi)^2$. If $\pi = 1/2$, then $\pi_c = 1/2$, $m(\pi) = 1$ and $V(\pi) = 4$.

Cowles and Jones assumed $\pi = 1/2$, in which case

$$S_{CJ} = \sqrt{T}\left(\frac{\widehat{CJ} - 1}{2}\right) \Longrightarrow N(0, 1). \tag{4.5}$$

In this case, the test is carried out by rejecting the null hypothesis (**rw1**) if $|S_{CJ}| > z_{\alpha/2}$.

CLM show this is not necessary to assume $\pi = 1/2$ to do the test. Instead, one can estimate π consistently by counting the number of positive returns in the sample, $\widehat{\pi}$. Define

$$S_{CJ*} = \sqrt{T}\left(\frac{\widehat{CJ} - m(\widehat{\pi})}{\sqrt{V(\widehat{\pi})}}\right).$$

Theorem 24 *Suppose that **rw2** holds and that $\pi = \Pr(r_t > 0)$ does not vary with time. Then as $T \to \infty$*

$$S_{CJ*} \Longrightarrow N(0, 1). \tag{4.6}$$

We may reject the null hypothesis (**rw1**) if $|S_{CJ*}| > z_{\alpha/2}$. If returns are not independent over time or if $\Pr(r_t > 0)$ does vary with time, then $CJ \neq m(\pi)$ and we expect the test to reject with probability tending to one.

An alternative approach for the case when $\pi \neq 1/2$ is to redefine continuations and reversals relative to the center of the return distribution rather than zero. In that case,

$$\begin{aligned}
\pi_c &= \Pr(continuation) \\
&= \Pr((r_t - \mu) \times (r_{t+1} - \mu) > 0) \\
&= \Pr(r_t - \mu > 0, \, r_{t+1} - \mu > 0) + \Pr(r_t - \mu < 0, \, r_{t+1} - \mu < 0) \\
&= \Pr(r_t - \mu > 0) \Pr(r_{t+1} - \mu > 0) + \Pr(r_t - \mu < 0) \Pr(r_{t+1} - \mu < 0) \\
&= \pi^2 + (1 - \pi)^2.
\end{aligned}$$

Definition 61 *Let*

$$\widehat{T}_c = \# \text{ of continuations}, \quad (r_t - \widehat{\mu}) \times (r_{t+1} - \widehat{\mu}) > 0$$
$$\widehat{T}_r = \# \text{ of reversals}, \quad (r_t - \widehat{\mu}) \times (r_{t+1} - \widehat{\mu}) \leq 0,$$

where $\widehat{\mu}$ is the sample median of returns. Then let

$$\widehat{CJ}_A = \frac{\widehat{T}_c}{\widehat{T}_r}.$$

In large samples (i.e., when $\widehat{\mu} \longrightarrow \mu$, where μ is the median of the return distribution) we may show that $\widehat{T}_c, \widehat{T}_r$ are well approximated by

$$T_c = \# \text{ of continuations}, \quad (r_t - \mu) \times (r_{t+1} - \mu) > 0$$
$$T_r = \# \text{ of reversals}, \quad (r_t - \mu) \times (r_{t+1} - \mu) \leq 0.$$

In this case, $\widehat{CJ}_A \to 1$ in probability. If $\widehat{\mu}$ were known, the critical values from the original Cowles–Jones test can be used, but in general the estimation of μ comes with a cost, i.e., the asymptotic variance is generally different from Theorem 23 and has to be estimated.

We next show that the adjusted Cowles–Jones ratio is related to the correlogram of the sign of returns. We have

$$\sum_{t=1}^{T-1} \text{sign}(r_t - \widehat{\mu}) \times \text{sign}(r_{t+1} - \widehat{\mu}) = T_c - T_r,$$

and since $T = T_c + T_r$, we may write

$$\widehat{T}_c = \frac{T}{2}\left(1 + \frac{1}{T}\sum_t \text{sign}(r_t - \widehat{\mu}) \times \text{sign}(r_{t+1} - \widehat{\mu})\right)$$

$$\widehat{T}_r = \frac{T}{2}\left(1 - \frac{1}{T}\sum_t \text{sign}(r_t - \widehat{\mu}) \times \text{sign}(r_{t+1} - \widehat{\mu})\right).$$

4.1 Robust Tests

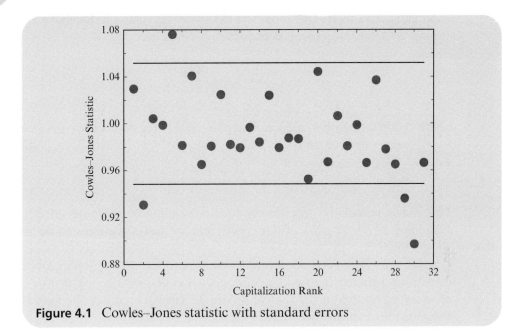

Figure 4.1 Cowles–Jones statistic with standard errors

Define the autocovariance function of the sign of returns $\gamma_{sign}(j) = E(\text{sign}(r_t - \mu) \times \text{sign}(r_{t+j} - \mu))$, and note that since $\text{sign}(x)^2 = 1$, $\gamma_{sign}(j)$ can be interpreted as an autocorrelation function, $\rho_{sign}(j)$. It follows that

$$\frac{\widehat{T}_c}{\widehat{T}_r} = \frac{1 + \widehat{\rho}_{sign}(1)}{1 - \widehat{\rho}_{sign}(1)}, \tag{4.7}$$

where $\widehat{\rho}_{sign}(1) = \sum_t \text{sign}(r_t - \widehat{\mu}) \times \text{sign}(r_{t+1} - \widehat{\mu})/T$. Linton and Whang (2007) show that $\widehat{\rho}_{sign}(1) \sim N(0, 1/T)$ under the i.i.d. null hypothesis, so that $\widehat{CJ}_A$ has asymptotic variance of 4 as for the unadjusted Cowles–Jones statistic. It follows that the **Fisher z-transform** of the sign autocorrelation

$$\frac{1}{2} \log\left(\widehat{CJ}_A\right) = \frac{1}{2} \log\left(\frac{1 + \widehat{\rho}_{sign}(1)}{1 - \widehat{\rho}_{sign}(1)}\right)$$

has asymptotic variance one. In other contexts, the Fisher z-transform is known to be better approximated by its normal limit than the raw correlations.

The Cowles–Jones statistic is nonparametric in the sense that it does not focus on the normal distribution like autocorrelations and linear regression models implicitly do. It is robust to heavy tailed distributions such as the Cauchy, which would fail all the tests of the previous chapter. However, the inference methods we have presented rely strongly upon the i.i.d. assumption. In practice, the Cowles–Jones test detects some departures from **rw1** but has limited power.

In Figure 4.1 below we show the Cowles–Jones test for the Dow stocks (assuming $\pi = 1/2$ confidence intervals).

Actually, the Cowles and Jones test has some similarities to **Kendall's Tau** derived around the same time. This is defined as

$$S = 1 - \frac{4T_D}{T(T-1)},$$

where T_D are the number of discordant pairs, that is, the number of pairs (r_t, r_{t+1}) and (r_s, r_{s+1}) that satisfy either $r_t < r_s$ and $r_{t+1} > r_{s+1}$, or $r_t > r_s$ and $r_{t+1} < r_{s+1}$. Under **rw1** we have

$$\sqrt{T}S \Longrightarrow N(0, 4/9).$$

This test is potentially much more time consuming to compute since it requires examining $T(T-1)$ pairs of data, but there are fast algorithms to do it in order $T \log T$ time.

Spearman's rank correlation is an alternative robust association measure. It is the correlation between the ranks of the series. Let $\mathcal{R}_t$, $t = 1, \ldots, T$ denote the ranks of observed returns in the sample of size T. Then $s(k)$ is just the autocorrelation computed from the integer valued time series $\mathcal{R}_t$, $t = 1, \ldots, T$.

4.1.2 Quantilogram

Define the α-quantile of a random variable Y by the number q_α that satisfies

$$\Pr(Y \leq q_\alpha) = \alpha,$$

assuming uniqueness (i.e., the c.d.f. of Y is strictly increasing). So if we take $\alpha = 1\%$, the number q_α is the value such that there is only a 1% chance of Y being less than q_α.

Suppose that Y has distribution F, where F is unknown. The standard estimator of the quantiles of a distribution is to use the sample quantity, $\widehat{q}_\alpha$, i.e., $\widehat{q}_\alpha$ is any number such that:

$$\widehat{F}(\widehat{q}_\alpha) = \alpha, \tag{4.8}$$

$$\widehat{F}(x) = \frac{1}{T} \sum_{t=1}^{T} 1(Y_t \leq x).$$

Here, $\widehat{F}(x)$ is the empirical c.d.f., the natural analogue of the population c.d.f. Because the empirical distribution function is not strictly increasing everywhere, $\widehat{q}_\alpha$ may not be unique even when q_α is, so we must apply some convention like take the midpoint of the values that solve the equation. For some purposes it is convenient to think of $\widehat{q}_\alpha$ as a minimizer of the sample function

$$Q_T(\mu) = \frac{1}{T} \sum_{t=1}^{T} \phi_\alpha(Y_t - \mu), \tag{4.9}$$

4.1 Robust Tests

where $\phi_\alpha(x) = x(\alpha - 1(x<0))$, or as a zero of the function

$$G_T(\mu) = \sum_{t=1}^{T} \psi_\alpha(Y_t - \mu), \qquad (4.10)$$

where $\psi_\alpha(x) = \text{sign}(x) - (1 - 2\alpha)$ is the check function. The sample quantile is consistent and asymptotically normal at rate square root sample size provided $f(q_\alpha) > 0$, where f is the density of Y. We discuss this further in Chapter 14.

We consider $\psi_\alpha(Y_t - q_\alpha)$ as a residual that is guaranteed to be mean zero by the definition of the quantile. Suppose that random variables $Y_1, Y_2, \ldots$ are from a stationary process whose marginal distribution has quantiles q_α for $0 < \alpha < 1$. Linton and Whang (2007) introduce the quantilogram, which checks whether the residual $\psi_\alpha(Y_t - q_\alpha)$ is autocorrelated.

Definition 62 *For $\alpha \in (0,1)$, define the quantilogram*

$$\rho_\alpha(k) = \frac{E\left(\psi_\alpha(Y_t - q_\alpha)\psi_\alpha(Y_{t+k} - q_\alpha)\right)}{E\left(\psi_\alpha^2(Y_t - q_\alpha)\right)}, \quad k = 1, 2, \ldots.$$

For a series that is i.i.d. $\rho_\alpha(k) = 0$ for all α and all $k \neq 1$. This can be (partially) true for some non-i.i.d. series, i.e., $\rho_\alpha(k) = 0$ for some α and all $k \neq 1$. Note that for $\alpha = 1/2$, $\psi_\alpha(Y_t - q_\alpha) = \text{sign}(Y_t - q_{1/2})$. The difference from the Cowles and Jones statistic is that we are centering the time series at its unconditional median or quantile.

We compute the empirical counterpart of $\rho_\alpha(.)$. We first estimate q_α by the sample quantile $\widehat{q}_\alpha$

$$\widehat{q}_\alpha = \arg\min_{\mu \in \mathbb{R}} \sum_{t=1}^{T} \phi_\alpha(Y_t - \mu).$$

Definition 63 *Then let*

$$\widehat{\rho}_\alpha(k) = \frac{\sum_{t=1}^{T-k} \psi_\alpha(Y_t - \widehat{q}_\alpha)\psi_\alpha(Y_{t+k} - \widehat{q}_\alpha)}{\sqrt{\sum_{t=1}^{T-k} \psi_\alpha^2(Y_t - \widehat{q}_\alpha)}\sqrt{\sum_{t=1}^{T-k} \psi_\alpha^2(Y_{t+k} - \widehat{q}_\alpha)}},$$

for $k = < T - 1$ and for any $\alpha \in (0,1)$.

Note that $-1 \leq \widehat{\rho}_\alpha(k) \leq 1$ for any α, k because this is just a sample correlation based on data $\psi_\alpha(Y_t - \widehat{q}_\alpha)$. Under the **rw1** assumption the large sample distribution is normal with mean zero and variance one. Linton and Whang (2007) establish the limiting variance of this statistic under a version of **rw3**.

In Figure 4.2 we show the quantilogram for S&P500 daily returns for $\alpha \in \{0.01, 0.05, 0.10, 0.25, 0.5, 0.75, 0.9, 0.95, 0.99\}$. This shows that there is much

Chapter 4 Robust Tests

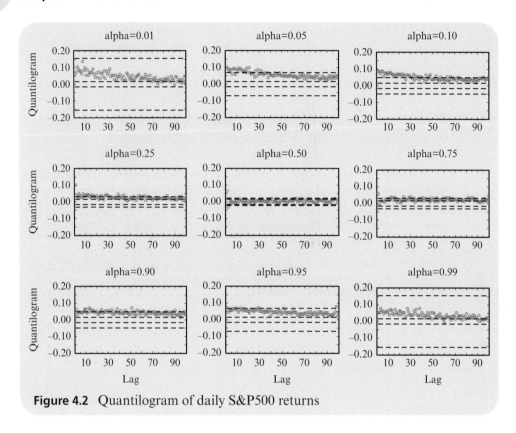

Figure 4.2 Quantilogram of daily S&P500 returns

stronger dependence at the lower quantiles than near the center of the distribution. Furthermore, this dependence is positive, meaning that positive values of the quantile residual tend to be followed by positive values and vice versa. An R-package called `quantilogram` is provided.

4.1.3 Cross-sectional Dimension

The count or ratio of the number of stocks on a given day whose prices went up compared with the number of stocks whose prices went down, for example

$$ad_t = \frac{\sum_{i=1}^{n} 1(r_{it} > 0)}{\sum_{i=1}^{n} 1(r_{it} < 0)},$$

can convey information about the direction of travel of the market. This is related to the sign statistics and the quantilogram except that the averaging is across stocks at a point in time. In practice, the fraction of positive returns

$$s_t = \frac{1}{n} \sum_{i=1}^{n} \operatorname{sign}(r_{it})$$

4.1 Robust Tests

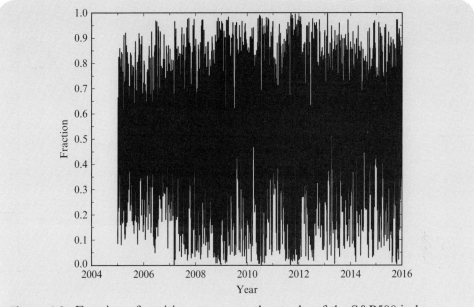

Figure 4.3 Fraction of positive returns on the stocks of the S&P500 index

contains equivalent information to ad_t. Under the i.i.d. hypothesis **rw1**, both s_t and ad_t should be uncorrelated over time.

In Figure 4.3 we show the time series s_t obtained from the S&P500 daily return series from 2005–2016. The series has mean 0.511 and standard deviation 0.263. It is highly volatile. There is some evidence of negative autocorrelation in the first few lags.

The so-called **advance-decline line** is defined as

$$AD_t = AD_{t-1} + \sum_{i=1}^{n} \text{sign}(r_{it}), \qquad (4.11)$$

where $i = 1, \ldots, n$ defines the universe of stocks and this quantity is defined relative to a starting point, perhaps the beginning of the year or the beginning of the week; see http://stockcharts.com/school/doku.php?id=chart_school:market_indicators:ad_line. It is an equally weighted indicator: it is a breadth indicator that reflects participation. Divergences between the price index and the AD line can help chartists identify potential reversals in the underlying index. Figure 4.4 shows the AD line cumulated since 2005 for the S&P500 index stocks, which shows the downward trend on the line over the period 2008–2009, which is consistent with what happened to the index level during this time.

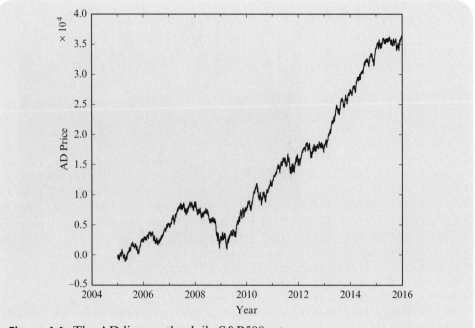

Figure 4.4 The AD line on the daily S&P500 return

4.1.4 Runs Tests

A run is a series of ups or downs. There is a formal (nonparametric) test of **rw1** using the predictions that this assumption makes about the number of runs, which is similar to the Cowles and Jones test.

We instead ask: how many days in a row should we see up markets or down markets? This is similar to the question of how many heads or tails in a row we should see in a coin tossing sequence or to how many reds or blacks in a row should we see at the roulette wheel. On August 18, 1913, the color black came up 26 times in a row at the casino in Monte Carlo. Incidentally, the casino made a fortune that day because so many people believed that "they were due a red" and bet overwhelmingly on the reversal. The probability of making 26 heads in a row is

$$\left(\frac{18}{37}\right)^{26} = 7.3087 \times 10^{-9},$$

which is low, but not impossible (and when you consider this to be the maximum of a long sequence of other such spins of the roulette wheel, then it is even less unlikely; see below). Many attempts have been made to beat the house odds at roulette, and one notable pioneer was **Joseph Jagger** (apparently no relation to Mick Jagger) at Monte Carlo in 1873. His scheme worked by identifying mechanically flawed wheels that gave more likelihood to numbers in certain regions of the wheel, which he identified by having his assistants count the relative frequencies for each of the wheels.

4.1 Robust Tests

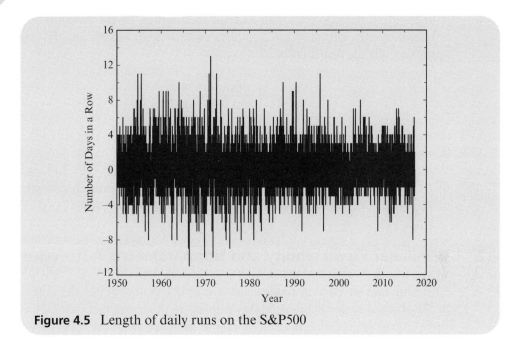

Figure 4.5 Length of daily runs on the S&P500

We next give a formal definition of a run.

Definition 64 *Define a time series run length*

1. $Z_0 = 0$.
2. Then,
   (a) If $\text{sign}(X_t) = \text{sign}(X_{t-1})$, we let $Z_t = Z_t + \text{sign}(X_t)$,
   (b) If $\text{sign}(X_t) = -\text{sign}(X_{t-1})$, we let $Z_t = 0$.
3. *Continue through the data.*

At any point in time, Z_t gives the length of the current run and its sign, so that if the last seven days have seen positive returns, $Z_t = 6$.

In Figure 4.5 we show the time series of the lengths of runs on the daily S&P500 return series. As you can see from the plot, there are both positive and negative runs through the data, and the longest run is of 10 days duration on both the positive and negative side. The mean of Z is 0.139 so that on balance we have more positive days than negative ones.

One can make a statistical test of the series of runs under some assumptions. Suppose that X_t is i.i.d. with a median of zero. Then $\text{sign}(X_t)$ is i.i.d. Bernoulli with probability one half of being ± 1. In that case one can verify by simulation that for $n = 5000$, the mean of the longest run length (i.e., $\max_{1 \leq t \leq 5000} Z_t$) is around 10.6

with standard deviation 1.87. In fact the distribution of the maximum run length is known exactly, although it is complicated to express analytically. For large n it is approximately the distribution of

$$Z_n = \left\lfloor \frac{W}{\log(2)} + \frac{\log(n)}{\log(2)} - 1 \right\rfloor, \quad (4.12)$$

where W is the random variable with distribution $\Pr(W \leq t) = \exp(-\exp(-t))$ and $\lfloor \ \rfloor$ denotes the greatest integer function. Let $c_\alpha(n)$ denote the largest integer such that $\Pr(Z_n \leq c_\alpha(n)) \leq 1 - \alpha$, then we find that for $n = 5000$, $c_\alpha(n) = 13$. In fact, the value of 10 is pretty close to the mean of the distribution and so entirely consistent with a coin tossing interpretation, that is, a run of length 10 is not unusual.

4.2 Nonlinear Predictability and Nonparametric Autoregression

One issue that we have not addressed is that one cannot reject the null hypothesis of the absence of linear predictability implicit in the above methodology even when there does exist predictability, but of a more complex nonlinear variety. For example, suppose that $X = \cos(\theta)$ and $Y = \sin(\theta)$, where θ is a random variable that is uniformly distributed on $[0, 2\pi]$. Then $Y^2 = 1 - X^2$ so the two random variables are functionally related and not independent, that is, if one knows one variable then one can predict the other one exactly with no error. However,

$$E(XY) = \int_0^{2\pi} \cos(\theta)\sin(\theta)d\theta = 0 = E(X) = \int_0^{2\pi} \cos(\theta)d\theta = E(Y) = \int_0^{2\pi} \sin(\theta)d\theta, \quad (4.13)$$

so that the two random variables are uncorrelated. The point is that even if there is no linear predictability (the best linear predictor here is also zero), there may be nonlinear predictability. The tests we have considered so far are not designed to detect nonlinear predictability, and although in many cases they might find some if it is there, in other cases they will not find it.

One approach is to look at $\gamma(g) = \mathrm{cov}(Y_t, g_{t-1})$ or $\mathrm{corr}(Y_t, g_{t-1})$, where $g_{t-1} = g(Y_{t-1}, \ldots)$ for particular functions g. For example, we may take $g_{t-1} = Y_{t-1}^2$, $g_{t-1} = \mathrm{sign}(Y_{t-1})$, or $g_{t-1} = \sum_{j=1}^p Y_{t-j}^2$. In this case we estimate the covariance using the sample covariance

$$\widehat{\gamma}(g) = \frac{1}{T} \sum_{t=2}^T (Y_t - \overline{Y})(g_{t-1} - \overline{g}) \quad ; \quad \overline{g} = \frac{1}{T} \sum_{t=1}^T g_t.$$

Under some conditions on the series Y_t, g_t we have an LLN and CLT for $\widehat{\gamma}(g)$, so that under the null hypothesis that $\gamma(g) = 0$

$$\sqrt{T}(\widehat{\gamma}(g) - \gamma(g)) \Longrightarrow N(0, V(g)), \quad V(g) = E\left[(Y_t - E(Y))^2(g_{t-1} - E(g_{t-1}))^2\right]$$

4.2 Nonlinear Predictability and Nonparametric Autoregression

and

$$\widehat{S}(g) = \widehat{V}(g)^{-1/2}\sqrt{T}(\widehat{\gamma}(g) - \gamma(g)) = \frac{\sqrt{T}(\widehat{\gamma}(g) - \gamma(g))}{\sqrt{\frac{1}{T}\sum_{t=1}^{T}(Y_t - \overline{Y})^2(g_{t-1} - \overline{g})^2}} \implies N(0,1). \tag{4.14}$$

This result can be used to provide a test of the absence of predictability: namely reject the null hypothesis if $|\widehat{S}(g)| > z_{\alpha/2}$. In general, we would like to consider a joint test over the set $\mathcal{G}$ of all functions g of all past information, which is an extremely large set. It includes all the linear ones. The question is how to test the joint hypothesis that $\gamma(g) = 0$ for all $g \in \mathcal{G}$, which requires some convenient indexing of $\mathcal{G}$. Note also that the covariance based tests are in some sense requiring some linearity in that they are allowing some potentially nonlinear function of the past information to affect returns in a linear way.

We now consider some nonlinear regression approaches. One approach is to use a model, a nonlinear parametric model. Threshold models allow for piecewise linear regressions according to different regimes. For example

$$Y_t = \mu + \rho Y_{t-1} 1(Y_{t-1} \geq \tau) + \gamma Y_{t-1} 1(Y_{t-1} < \tau) + \varepsilon_t, \tag{4.15}$$

where τ is an unobserved threshold. This piecewise linear predictability can fail to be detected by autocorrelation tests, especially if ρ and γ have different signs. The null hypothesis is still that $\rho = \gamma = 0$, and one way of testing this hypothesis is to estimate the parameters $(\mu, \rho, \gamma, \tau)$ of (4.15) and then use t-test or F-test of the null hypothesis. Unfortunately, there are some complications with this approach, since under the null hypothesis the parameter τ is not identified.

The nonparametric approach is to consider the (auto)regression function

$$m_k(y_1, \ldots, y_p) = E(Y_{t+k} | Y_t = y_1, \ldots, Y_{t+1-p} = y_p). \tag{4.16}$$

We may consistently estimate the functions $m_k(\cdot)$ using nonparametric methods without specifying the functional form except that we require some smoothness properties to hold on m_k. A popular estimator of m_k based on a sample of data $\{Y_1, \ldots, Y_T\}$ is the kernel estimator. Let K be a continuous function on $[-1, 1]$, symmetric about zero, called the kernel, which satisfies $\int K(u)du = 1$, and let the bandwidth $h > 0$. In the univariate case ($p = 1$), let

$$\widehat{m}_k(y) = \frac{\sum_{t=1}^{T-k} K\left(\frac{Y_t - y}{h}\right) Y_{t+k}}{\sum_{t=1}^{T-k} K\left(\frac{Y_t - y}{h}\right)}. \tag{4.17}$$

This forms a locally weighted average of the response using observations close to the predictor values. Silverman (1986) and Härdle and Linton (1994) discuss choices of bandwidth and kernel. Under some conditions, $\widehat{m}_k(y) \to m_k(y)$ and indeed satisfies a CLT.

Theorem 25 *Suppose that Y_t satisfies **rw3** and is stationary with finite variance so that $m_k(y) = \mu$ for all k, y. Suppose also that Y_t is continuously distributed with density f_Y that is differentiable and positive at y. Then provided $h \to 0$ and $Th \to \infty$*

$$\sqrt{Th}\left(\widehat{m}_k(y) - \widehat{\mu}\right) \Longrightarrow N(0, \omega), \qquad \omega = \int K(u)^2 du \frac{\sigma_k^2(y)}{f_Y(y)}$$

$$\widehat{S}_k(y) = \frac{\widehat{m}_k(y) - \widehat{\mu}}{\sqrt{\frac{\sum_{t=1}^{T-k} K^2\left(\frac{Y_t - y}{h}\right)(Y_{t+k} - \widehat{m}_k(y))^2}{\left(\sum_{t=1}^{T-k} K\left(\frac{Y_t - y}{h}\right)\right)^2}}} \Longrightarrow N(0, 1), \qquad (4.18)$$

where $\sigma_k^2(y) = \text{var}(Y_{t+k} | Y_t = y)$, while $f_Y(y)$ is the marginal density of the stationary process Y_t. Here, $\widehat{\mu} = \sum_{t=1}^{T} Y_t / T$. Furthermore, $\widehat{S}_k(y)$ and $\widehat{S}_l(y')$ are asymptotically independent when $y \neq y'$.

This can be used to test the null hypothesis of no predictability against nonlinear alternatives. Specifically, reject if $\sum_{k=1}^{K} \sum_{l=1}^{L} \widehat{S}_k(y_l)^2 > \chi_{KL}^2(\alpha)$, where $\{y_1, \ldots, y_L\}$ is a grid of distinct points in the support of Y_t. Sieve methods can also be used here; see Chen (2007).

Wolfgang Karl Härdle attained his Dr. rer. nat. in Mathematics at Universität Heidelberg in 1982, and in 1988 his habilitation at Universität Bonn. He is the head of the Ladislaus von Bortkiewicz Chair of Statistics at Humboldt-Universität zu Berlin and the director of the Sino German International Research Training Group IRTG1792 "High dimensional non stationary time series analysis," a joint project with WISE, Xiamen University. His research focuses on dimension reduction techniques, computational statistics and, quantitative finance. He has published over 30 books and more than 300 papers in top statistical, econometrics, and finance journals. He is highly ranked and cited on Google Scholar, REPEC, and SSRN. He has professional experience in financial engineering, smart (specific, measurable, achievable, relevant, timely) data analytics, machine learning, and cryptocurrency markets. He has created a financial risk meter, FRM hu.berlin/frm, and a cryptocurrency index, CRIX hu.berlin/crix.

4.3 Further Empirical Evidence on Semistrong and Strong Form EMH

Many studies have identified so-called "anomalies" that seem difficult to reconcile with the EMH. The following web site has a long list of examples of violations of the

4.3 Further Empirical Evidence on Semistrong and StrongForm EMH

EMH and explanations thereof www.behaviouralfinance.net/. At least some of these anomalies may be reinterpreted as rational rewards for risk if the asset pricing model that helps us adjust for risk is misspecified. Other examples like the Monday effect are harder to fit into a rational asset pricing paradigm.

Rashes (2001) presented an interesting anomaly. He examined the comovement of two stocks with similar ticker symbols MCI (large telecom, NASDAQ) and MCIC (a closed end mutual fund, NYSE). He found a significant correlation between returns, volume, and volatility at short frequencies. New information about MCI affects prices of MCIC and vice versa. Deviations from the fundamental value tend to be reversed within several days, although there is some evidence that the return comovement persists for longer horizons. Arbitrageurs appear to be limited in their ability to eliminate these deviations from fundamentals.

Thaler (2016) presented the example of a closed-end mutual fund (when the fund starts, a certain amount of money is raised and invested, and then the shares in the fund trade on organized secondary markets such as the NYSE, and are closed to new investors) that had the ticker symbol CUBA. It had around 70 percent of its holdings in US stocks with the rest in foreign stocks, but absolutely no exposure to Cuban securities, since it has been illegal for any US company to do business in Cuba since 1960. For the first few months of 2014 the share price was trading in the normal 10–15 percent discount range of the net asset value (the value of the shares it itself held). Then on December 18, 2014, President Obama announced his intention to relax the United States' diplomatic relations with Cuba. The price of CUBA shares jumped to a 70 percent premium. Although the value of the assets in the fund remained stable, the substantial premium lasted for several months, finally disappearing about a year later. The fund did declare its per share end year distribution of $0.635 on December 18th. Furthermore, its annual report says that

> "the Fund invests in issuers that are likely, in the Advisor's view, to benefit from economic, political, structural and technological developments in the countries in the Caribbean Basin, which include, among others, Cuba."

More recently, a number of firms with names overlapping with Bitcoin (but with no direct connection) have experienced substantial price appreciation.

Let us summarize anomaly characteristics or rather make some points of discussion.

1. They are small in terms of dollar amounts. For example, in the MCI Jr. versus MCI case. Maybe each one is small but many different anomalies could add up to some serious pocket money.
2. They are not scalable, i.e., the apparent profitability evident from historical data apparently implies an infinite amount of profitability by scaling up your bets, but this cannot be true, as your trades would change relative prices. Most pieces of evidence ignore this issue and just look for statistical significance without regard to impact. The next chapter will discuss price impact.
3. They are statistically suspect. At the simplest level, the standard errors are often calculated under unrealistic assumptions. Assuming that error terms are i.i.d. with

a normal distribution may seem like a good idea at the time, but it has been responsible for many catastrophes in financial markets; see Turner (2009, pp44–5). In addition, there is the **data mining** issue, which is nicely explained in White (2000). By trawling through many strategies one will end up with a best one that appears very good but the size of the trawling ocean will introduce a bias on the perceived benefits of a trading strategy based on the best case predictability found. White considers prediction of an index return using lagged values of the individual stock returns in combinations. If n is the total number of predictors, then there are a total of $2^n - 1$ different linear regression models that one could fit to the sample data. For example, if $n = 30$, there are over a billion different linear regression models that could be tried; if one just reports the best one of these then one is biasing the results. White (2000) describes how one might take account of the effect of data mining using **White's Reality Check** where you look at the maximum departure. One can typically observe the effects of data mining by evaluating the chosen strategy out of sample, i.e., for forecasting.

4. They are fleeting, i.e., they don't last long. For example, the small stock premium, the January effect, the Monday effect. These anomalies are known to have reduced or disappeared in later years. Ross called this the Heisenberg Principle of Finance: observing an anomaly brings about its own extinction because of trading. This is related to **Goodhart's law** of banking, which says that: any observed statistical regularity will tend to collapse once pressure is placed upon it for control purposes. Timmerman (2008) investigates the forecasting performance of a number of linear and nonlinear models and says: "Most of the time the forecasting models perform rather poorly, but there is evidence of relatively short-lived periods with modest return predictability. The short duration of the episodes where return predictability appears to be present and the relatively weak degree of predictability even during such periods makes predicting returns an extraordinarily challenging task."

5. They may not be realizable profit opportunities. There are transaction costs: commissions, exchange fees, and bid/ask spreads. These negatively impact the bottom line.

These are all important dimensions along which to evaluate anomalies. Rather than asking whether or not there is an anomaly, it might be more fruitful to measure the degree of inefficiency and make comparisons across markets or stocks or over time.

We don't expect markets to be strongly efficient. Trading on inside information is regulated and limited in many countries. Improper disclosure and misuse of information are kinds of insider dealing. Studies of trades by insiders (managers etc. who have to report such trades to the SEC) show they are able to make abnormal profits through their information.

4.4 Explanations for Predictability

We have found some modest evidence of predictability using a variety of tools. There are three potential sources for the predictability that is present in financial time series;

Boudoukh, Richardson, and Whitelaw (1994). First, market microstructure effects. That is to say details of how prices are set in financial markets can imply some predictability in observed returns. For example, nonsynchronous trading can cause: measured variances on single securities to overstate true variances; measured returns on single securities appear negatively serially correlated and leptokurtic relative to actual returns; market returns will be positively autocorrelated. We will consider this in the next chapter. The second source of predictability could come from the fact that the risk premium is time varying, while the results we have presented assumed a constant risk premium. That is to say, time variation of the risk premium could translate into autocorrelations in the observed return series. Changes over time in the required return on common stocks can generate mean-reverting stock price behavior. If required returns exhibit positive autocorrelation, then an innovation that raises required returns will cause share prices to fall. This will generate a holding period loss, followed by higher returns in subsequent periods. But it needs a great deal of variability in required returns to explain the degree of mean reversion. The final explanation is that investors are simply irrational and consequently any force that can reduce the predictability is swamped by bubbles or fads or other types of irrational behavior. This is the field of **behavioral finance**.

It would seem that there is a lot of evidence that microstructure effects should have reduced considerably over time. For example, it is hard to find even small cap stocks that do not trade now many times during a day. The microstructure explanation would imply that the long horizon daily or weekly variance ratios should return to unity, but this is not the case in our data even for the most recent period. There is also some evidence that the level of the market risk premium (and perhaps therefore its time variation) has reduced in recent years; see for example Hertzberg (2010).

Although the statistical magnitudes seem to have reduced, it is not clear whether the potential profit from exploiting linear predictability across the whole market has reduced, since the number of tradeable assets has increased and the transaction costs associated with any given trade seem to have reduced; Malkiel (2015).

4.5 Summary of Chapter

We described some robust tests for the EMH. We showed that these tests don't change the evidential picture very much and generally support the results based on the correlogram. We also summarized the anomaly characteristics and discussed some explanations for the predictability found in daily and weekly stock returns.

5 Empirical Market Microstructure

We are going to look in detail at three main topics in empirical market microstructure: stale prices, bid–ask spreads, and price and quantity discreteness. We will look at formal models that describe and explain the causes and consequences of these phenomena. These models predict that observed prices can depart from the perfect frictionless market that is often assumed in other contexts, and so provide some explanations for the findings of the previous two chapters. The study of market microstructure is relatively recent. The field is currently going through a big revolution due to recent developments in market organization and technology and we provide a short review of these recent developments.

5.1 Stale Prices

Stale prices are a very important issue in practice, i.e., for traders. A lot of trading strategies are designed to profit from stale prices by exploiting short-term arbitrage opportunities that arise from them. Academic empirical work also relies on prices being accurately measured. We consider some rather simple models of stale prices, the Scholes and Williams (1977) (SWnt) model and the Lo and MacKinlay (1990b) (LMnt) model of non trading. The general setting is that there is an underlying true price but it is only observed when a transaction occurs. The sequence of transaction prices generally behaves differently from the sequence of true prices, and we will look at how this is manifested in these two models.

In practice, stocks trade with different frequency, from Apple at one end (many times a millisecond) to **penny stocks** that may only trade once a week. Many empirical questions are concerned with the cross-section of returns, and nonsynchronous trading presents a big problem for the measurement of these returns with high frequency data. The SWnt and LMnt models are nominally about daily stock return data (in the 1980s this made sense because trading was less frequent). Now it is perhaps more relevant for intraday data, since all S&P500 stocks, say, trade every day many times now. The negative consequences of non trading can be mitigated at this horizon by using order book information (see below), in particular, quoted prices are generally always available. However, quotes can also be stale and the models and intuitions of the two models can be adapted to this situation.

5.1.1 Scholes and Williams

First, we consider a simplified version of the Scholes and Williams model. This says that the true stock price or economic value is generated continuously, and in

5.1 Stale Prices

particular, logarithmic returns are normally distributed with

$$\log P(t) - \log P(s) \sim N\left(\mu(t-s), \sigma^2(t-s)\right), \qquad (5.1)$$

and furthermore, returns from non-overlapping periods are independent; see Chapter 12 below for discussion of continuous time models. It says that returns are a linear function of the holding period, and the volatility also increases linearly with holding period. However, we don't observe the price continuously, only at discrete times, say when a transaction occurs.

Definition 65 *Consider the discrete times $1, 2, \ldots, T$, and suppose that in each interval $[t-1, t]$ there is a transaction at time $t-s$, where $s \in [0, 1]$ is a random variable and is independent of the stock price process. The value of s represents the staleness of the price relative to time t. Suppose that the staleness process s_t is i.i.d. with mean μ_s and variance σ_s^2. We suppose that the observed price sequence $P_1^O, \ldots, P_T^O$ to be given by*

$$P_t^O = P(t - s_t).$$

It follows that observed returns are

$$r_t^O = \log(P_t^O) - \log(P_{t-1}^O) = \log(P(t - s_t)) - \log(P(t - 1 - s_{t-1})).$$

This model captures well the situation concerning daily returns on relatively infrequently traded stocks or perhaps corporate bond data obtained from OTC markets. There may be other transactions in the interval $[t-1, t-s_t]$, but the transaction at $t - s_t$ is the last of the day and is taken as the closing price in practice. This was quite a common feature in the past, and led to concerns about biases in the estimation of key market parameters based on such censored data.

The model implies that conditionally on the staleness sequence $\{s_t\}_{t=1}^T$, $r_t^O \sim N(\mu(1 - s_t + s_{t-1}), \sigma^2(1 - s_t + s_{t-1}))$, that is, we can write

$$r_t^O = \mu(1 - s_t + s_{t-1}) + \sigma\sqrt{(1 - s_t + s_{t-1})} z_t, \qquad (5.2)$$

where z_t is an i.i.d. sequence of standard normal random variables. It follows from the law of iterated expectation that

$$E(r_t^O) = E(E(r_t^O | s_1, \ldots, s_T)) = \mu E(1 - s_t + s_{t-1}) = \mu$$

$$\mathrm{var}(r_t^O) = E\left(\mathrm{var}(r_t^O | s_1, \ldots, s_T)\right) + \mathrm{var}\left(E(r_t^O | s_1, \ldots, s_T)\right) = \sigma^2 + 2\mu^2 \sigma_s^2.$$

Furthermore,

$$E(r_t^O r_{t-1}^O) = E\left[\left(\mu(1 - s_t + s_{t-1}) + \sigma\sqrt{(1 - s_t + s_{t-1})}z_t\right)\right.$$
$$\left.\times \left(\mu(1 - s_{t-1} + s_{t-2}) + \sigma\sqrt{(1 - s_{t-1} + s_{t-2})}z_{t-1}\right)\right]$$
$$= \mu^2 E[(1 - s_t + s_{t-1})(1 - s_{t-1} + s_{t-2})]$$
$$= \mu^2 E[(1 - (s_t - \mu_s) + (s_{t-1} - \mu_s))(1 - (s_{t-1} - \mu_s) + (s_{t-2} - \mu_s))]$$
$$= \mu^2 (1 - \sigma_s^2),$$

so that

$$\text{cov}(r_t^O, r_{t-1}^O) = -\mu^2 \sigma_s^2. \qquad (5.3)$$

Furthermore, $\text{cov}(r_t^O, r_{t-j}^O) = 0$ for $j \geq 2$. This shows that observed returns may be negatively autocorrelated at lag one. This is consistent with observed returns following an MA(1) process.

Now what about covariances between returns on different firms? Suppose that the vector of true returns are generated by a vector version of (5.1), whereby they are normally distributed with independent increments and mean vector μ times holding period and variance covariance matrix $\Sigma = (\sigma_{ij})$ times the holding period, that is, they are contemporaneously correlated but not cross autocorrelated. Suppose also that the censoring times s_{it}, s_{jt} are i.i.d. over time and are mutually independent. We observe trading at times $\{t - s_{it}, t = 1, 2, \ldots\}$ for asset i and at times $\{t - s_{jt}, t = 1, 2, \ldots\}$ for asset j. The key observation is that the covariance between observed returns will be proportional to the overlap of the two time intervals $[t - 1 - s_{i,t-1}, t - s_{it}]$ and $[t - 1 - s_{j,t-1}, t - s_{jt}]$, which could be as little as zero or as much as one time unit. The overlap period is $[t - 1 - \min\{s_{i,t-1}, s_{j,t-1}\}, t - \max\{s_{it}, s_{jt}\}]$, which is of random length $O_{ijt} = 1 - (\max\{s_{it}, s_{jt}\} - \min\{s_{i,t-1}, s_{j,t-1}\}) \in [0, 1]$. It follows from the law of iterated expectation that

$$\text{cov}(r_{it}^O, r_{jt}^O) = E\left(\text{cov}(r_{it}^O, r_{jt}^O) | s_{i1}, \ldots, s_{iT}, s_{j1}, \ldots, s_{jT}\right)$$
$$+ \text{cov}\left(E\left(r_{it}^O | s_{i1}, \ldots, s_{iT}\right), E\left(r_{jt}^O | s_{j1}, \ldots, s_{jT}\right)\right)$$
$$= \sigma_{ij} E(O_{ijt}) + \text{cov}\left(1 - (s_{it} - s_{i,t-1}), 1 - (s_{jt} - s_{j,t-1})\right) \mu_i \mu_j,$$

and observed covariances will be less in magnitude than the true covariances σ_{ij} depending on the expected value of the overlap. It also follows that in general $\text{cov}(r_{it}^O, r_{j,t\pm 1}^O) \neq 0$.

Example 31 *Suppose that $s_{it} \sim U[0, 1]$ for all i, t and that s_{it} is independent of s_{jt}. Then*

$$E(O_{ijt}) = 1 - E(\max\{s_{it}, s_{jt}\}) + E(\min\{s_{i,t-1}, s_{j,t-1}\}) = \frac{2}{3}$$

$$\text{cov}\left(1 - (s_{it} - s_{i,t-1}), 1 - (s_{jt} - s_{j,t-1})\right) = 0.$$

Therefore, the observed covariance should be two-thirds the value of the true covariance.

5.1 Stale Prices

This model is quite easy to work with but it is a little arbitrary. It is assumed that for the given interval there is always a transaction between $t-1, t$, which can be rigged in practice but is a bit questionable from a theoretical point of view. The maximum time between transactions is two periods of time. We now consider the LMnt model which is a little more coherent with regard to whether transactions occur or not.

5.1.2 The LM Model

Suppose that the underlying price sequence P_t and hence returns r_t occur in Calendar time, i.e., r_t obeys **rw1** with $E(r_t) = \mu$ and $\text{var}(r_t) = \sigma^2$. Lo and MacKinlay (1990b) do not assume lognormality of prices as this is inessential to the questions they ask, so the true return could have a non-normal distribution. The key point is that the underlying price sequence is only observed when a trade occurs. The discrete time setting begs the question of which frequency, but we will defer discussion of that until later.

Define the random trading sequence

$$\delta_t = \begin{cases} 1 \text{ (no trade)} & \text{with probability } \pi \\ 0 \text{ (trade)} & \text{with probability } 1 - \pi, \end{cases}$$

which captures whether a given period has a trade or not. The frequency of trading is controlled by the parameter π. The trading time model we considered in Chapter 3 is a kind of special case of this with a deterministic δ_t that is equal to one on Saturdays, Sundays, and Exchange Holidays. We consider the case where this trade sequence is i.i.d. with the same probability of trading each period and independent of the past history.

Definition 66 *The observed log price p_t^o is given by*

$$p_t^o = \begin{cases} p_t & \text{if there is a trade at } t \text{ (i.e., } \delta_t = 0) \\ p_{t-1}^o & \text{if there is no trade at } t \text{ (i.e., } \delta_t = 1). \end{cases}$$

It follows that the observed return satisfies

$$r_t^o = p_{t-1}^o - p_{t-1}^o = \begin{cases} p_{t-1} - p_{t-1}^o & \text{if } \delta_t = 0 \\ 0 & \text{if } \delta_t = 1. \end{cases}$$

If $\delta_t = 0$ and $\delta_{t-1} = 0$, we have $r_t^o = r_t$, but otherwise the observed return is not equal to the actual return, and depends on the past history. Our objective is to establish the properties of the observed return series r_t^o. We have

$$p_t^o = p_t(1 - \delta_t) + p_{t-1}^o \delta_t = p_t(1 - \delta_t) + p_{t-1}\delta_t(1 - \delta_{t-1}) + p_{t-2}\delta_t\delta_{t-1}(1 - \delta_{t-2}) + \cdots .$$

Chapter 5 Empirical Market Microstructure

Taking differences of p_t^O we see that the observed return depends on the sequence of trade dummies $\delta_t, \delta_{t-1}, \ldots$. The key quantity is how stale is the price p_{t-1}^O, i.e., how many previous periods have we had no trading, i.e., how many j for which $\delta_{t-j} = 1$.

Definition 67 *The duration of non trading, denoted d_t, is*

$$d_t = \begin{cases} k & \text{if } \delta_t = 1, \ldots, \delta_{t-k+1} = 1, \delta_{t-k} = 0 \\ 0 & \text{if } \delta_t = 0. \end{cases}$$

It can take any value in $\{0, 1, 2, \ldots\}$.

Example 32 *The following concrete example shows a possible realization of the price and duration process*

true price	p_1	p_2	p_3	p_4	p_5		p_6		p_7	p_8	p_9	p_{10}
no trade?	0	0	0	1	1		0		0	0	1	1
observed price	p_1	p_2	p_3	p_3	p_3		p_6		p_7	p_8	p_8	p_8
true return		r_2	r_3	r_4	r_5		r_6		r_7	r_8	r_9	r_{10}
observed return		r_2	r_3	0	0		$r_4 + r_5 + r_6$		r_7	r_8	0	0
duration	0	0	0	1	2		0		0	0	1	2

CLM write the duration of non-trading as $d_t \equiv \sum_{k=1}^{\infty} \prod_{j=0}^{k} \delta_{t-j}$, which presupposes an infinite past and stationarity of d_t. This is a mathematical device to simplify the calculations. Perhaps easier to understand is the dynamic representation of d_t, which is

$$d_{t+1} = \begin{cases} d_t + 1 & \text{with probability } \pi \\ 0 & \text{with probability } 1 - \pi. \end{cases} \quad (5.4)$$

This is a stochastic process that evolves over time like an autoregression. Given the current state, the future state is either the current state plus one or it is zero. It is called a **Markov process**, because the next step only depends on where you currently are, not on how you arrived at where you are. It is also **time homogeneous** because the transition probabilities $\pi, 1 - \pi$, do not change over time. We have

$$E(d_{t+1}|\mathcal{F}_t) = \pi (d_t + 1) \text{ and } \mathrm{var}(d_{t+1}|\mathcal{F}_t) = (d_t + 1)^2 \pi(1 - \pi), \quad (5.5)$$

There is a lot of theory that governs the behavior of such processes. In particular, provided $\pi \in (0, 1)$ the process d is an **aperiodic positive recurrent process** (i.e., there

5.1 Stale Prices

is a positive probability of $d_{t+s} = k$ for some $s \geq 0$ given $d_t = i$ for any i, k), and hence it is a stationary ergodic process. The transition matrix ($\infty \times \infty$), which gives the probability of going from one state to another, can be written

$$T = \begin{pmatrix} 1-\pi & \pi & 0 & \cdots \\ 1-\pi & 0 & \pi & 0 & \cdots \\ 1-\pi & 0 & 0 & \ddots \\ \vdots & \vdots & \vdots & \ddots \end{pmatrix}.$$

It follows that the stationary distribution of d satisfies $\pi T = \pi$, where π is a row vector of probabilities of the stationary distribution. The unique solution to this equation is $\pi = \big((1-\pi), (1-\pi)\pi, (1-\pi)\pi^2, \ldots\big)$, which can easily be checked. This is the marginal distribution of d_t, which has the known form of a **geometric (type 2) distribution**. That is, for $k = 0, 1, 2, \ldots$

$$\Pr(d_t = k) = (1-\pi)\pi^k. \tag{5.6}$$

The moments of this distribution are easily calculated from the known marginal distribution:

$$E(d_t) = (1-\pi)\pi + 2(1-\pi)\pi^2 + \cdots = (1-\pi)\pi \sum_{j=0}^{\infty}(j+1)\pi^j = \frac{\pi}{1-\pi} \tag{5.7}$$

$$\mathrm{var}(d_t) = \frac{\pi}{(1-\pi)^2}. \tag{5.8}$$

This tells us how the mean and variance of the duration of non-trading episodes depends on the parameter π. The larger is π, the longer the expected length of non-trading intervals is. Other moments of d_t are given in the appendix to this chapter.

We return to analyze the observed (logarithmic) return series $r_t^o = p_t^o - p_{t-1}^o$. It is also a stationary process, and its marginal distribution satisfies

$$r_t^o = \begin{cases} 0 & \text{with prob } \pi \\ r_t & \text{with prob } (1-\pi)^2 \\ r_t + r_{t-1} & \text{with prob } (1-\pi)^2 \pi \\ \vdots & \vdots \end{cases} \tag{5.9}$$

There are two branches to this distribution, the zero branch and the non-zero branch, and we can write this compactly as

$$r_t^o = \begin{cases} 0 & \text{if } \delta_t = 1 \text{ (with prob } \pi\text{)} \\ \sum_{k=0}^{d_t} r_{t-k} & \text{if } \delta_t = 0 \text{ (with prob } 1-\pi\text{)}. \end{cases}$$

Chapter 5 Empirical Market Microstructure

This distribution is different from the distribution for true returns. For example, if r_t were normally distributed, then r_t^o is clearly not normally distributed, since it has a point mass at zero, i.e., a positive probability of being exactly zero, which is incompatible with any continuously distributed random variable. The second branch of observed returns (corresponding to $\sum_{k=0}^{d_t} r_{t-k}$) is continuously distributed and moreover if $r_t \sim N(\mu, \sigma^2)$, then $\sum_{k=0}^{d_t} r_{t-k}$ is normally distributed conditional on d_t, i.e.,

$$\sum_{k=0}^{d_t} r_{t-k} | d_t \sim N\left(\mu(d_t + 1), \sigma^2(d_t + 1)\right).$$

This is called a mixture distribution and in fact a normal mixture distribution. The unconditional distribution of this piece is not normal but its moments and other features are easy to obtain by the law of iterated expectation. In this calculation we are using the fact that true returns are independent of the entire trading history $\{\delta_s\}$.

Mean and Variance of Observed Returns

We first calculate the unconditional mean and variance of $\sum_{k=0}^{d_t} r_{t-k}$. By the law of iterated expectation, we have

$$E\left(\sum_{k=0}^{d_t} r_{t-k}\right) = E\left(E\left(\sum_{k=0}^{d_t} r_{t-k} | d_t\right)\right) = \mu E(d_t + 1) = \mu\left(\frac{\pi}{1-\pi} + 1\right) = \frac{\mu}{1-\pi}$$

$$\text{var}\left(\sum_{k=0}^{d_t} r_{t-k}\right) = E\left(\text{var}\left(\sum_{k=0}^{d_t} r_{t-k} | d_t\right)\right) + \text{var}\left(E\left(\sum_{k=0}^{d_t} r_{t-k} | d_t\right)\right)$$
$$= \sigma^2 E(d_t + 1) + \mu^2 \text{var}(d_t + 1)$$
$$= \sigma^2 \frac{1}{1-\pi} + \mu^2 \frac{\pi}{(1-\pi)^2}.$$

Finally, we should obtain the mean and variance of r_t^o by also conditioning on whether it is zero or not (i.e., whether $\delta_t = 1$ or $\delta_t = 0$). It follows that

$$E(r_t^o) = \pi E(r_t^o | \delta_t = 1) + (1-\pi) E(r_t^o | \delta_t = 0) = (1-\pi) E\left(E\left(\sum_{k=0}^{d_t} r_{t-k} | d_t\right)\right) = \mu.$$

Furthermore, we have $\text{var}(r_t^o) = E(\text{var}(r_t^o | \delta_t)) + \text{var}(E(r_t^o | \delta_t))$, where:

$$E(r_t^o | \delta_t) = \begin{cases} 0 & \text{if } \delta_t = 1 \\ \frac{\mu}{1-\pi} & \text{if } \delta_t = 0 \end{cases}$$

5.1 Stale Prices

$$\text{var}(r_t^O|\delta_t) = \begin{cases} 0 & \text{if } \delta_t = 1 \\ \sigma^2 \frac{1}{1-\pi} + \mu^2 \frac{\pi}{(1-\pi)^2} & \text{if } \delta_t = 0. \end{cases}$$

Combining these calculations we obtain that

$$\text{var}(r_t^O) = (1-\pi)\left(\sigma^2 \frac{1}{1-\pi} + \mu^2 \frac{\pi}{(1-\pi)^2}\right) + \frac{\mu^2}{(1-\pi)^2}\pi(1-\pi) = \sigma^2 + 2\mu^2 \frac{\pi}{1-\pi}. \tag{5.10}$$

This shows that the variance of observed returns is inflated relative to the variance of true returns. The long run variance of observed returns is equal to the variance of true returns.

Autocovariances of Observed Returns

By assumption, the true return series is not autocorrelated; we next calculate the autocorrelation of the observed return series. We calculate $E(r_t^O r_{t+1}^O)$ by first conditioning on the sequence $\{\delta_t\}$. If either $\delta_{t+1} = 1$ or $\delta_t = 1$, then $r_t^O r_{t+1}^O = 0$, so we only need consider the case that $\delta_{t+1} = \delta_t = 0$, which occurs with probability $(1-\pi)^2$, and in this case, $d_{t+1} = 0$. It follows that

$$E\left(r_t^O r_{t+1}^O\right) = (1-\pi)^2 E\left(r_{t+1} \sum_{i=0}^{d_t} r_{t-i}\right)$$
$$= (1-\pi)^2 \mu^2 E(d_t + 1)$$
$$= \mu^2 (1-\pi)^2 \left(\frac{\pi}{1-\pi} + 1\right)$$
$$= \mu^2 (1-\pi).$$

Therefore,
$$\text{cov}(r_t^O, r_{t+1}^O) = -\pi\mu^2. \tag{5.11}$$

In fact we may show the following.

Theorem 26 *For all $k = 1, 2, \ldots$*

$$\text{cov}(r_t^O, r_{t+k}^O) = -\mu^2 \pi^k \quad ; \quad \text{corr}(r_t^O, r_{t+k}^O) = \frac{-\mu^2}{\sigma^2 + \frac{2\pi}{1-\pi}\mu^2} \pi^k.$$

The implied autocovariance function of observed returns is consistent with them following an ARMA(1,1) process

$$r_t^O - \mu = \pi(r_{t-1}^O - \mu) + \eta_t + \theta\eta_{t-1}, \tag{5.12}$$

Chapter 5 Empirical Market Microstructure

where η_t is a MDS with variance σ_η^2, and θ is such that $|\theta| < 1$. That is, this process would give the same second order properties (autocovariance function) as in Theorem 26 for some such θ, σ_η^2. Specifically, the parameters σ_η^2 and θ must satisfy some restrictions; in particular, θ must solve the equation

$$\frac{-\mu^2 \pi}{\sigma^2 + \frac{2\pi}{1-\pi}\mu^2} = \frac{(\pi + \theta)(1 + \pi\theta)}{1 + 2\pi\theta + \theta^2}.$$

Essentially with $\theta < -\pi$ this delivers negative autocorrelations throughout. When π is small, the dominant term is the MA component, which delivers only first order autocorrelation.

Note that Theorem 26 implies that

$$\sum_{n=1}^{\infty} \text{cov}(r_t^O, r_{t+n}^O) = -\mu^2 \frac{\pi}{1-\pi}$$

so that

$$\text{lrvar}(r_t^O) = \text{var}(r_t^O) + \sum_{n=-\infty}^{\infty} \text{cov}(r_t^O, r_{t+n}^O) = \sigma^2. \tag{5.13}$$

This suggests that including autocovariances can reduce or in this case eliminate the effects of market microstructure noise. We will revisit this in Chapter 12.

Autocovariance of Observed Squared Returns

We consider the dynamics of observed squared returns. For the unobserved true return the autocorrelations are zero, since by construction returns are i.i.d. We work with the special case where $\mu = 0$, in which case

$$\begin{aligned}
\text{cov}((r_t^O)^2, (r_{t+1}^O)^2) &= E\left((r_t^O)^2 (r_{t+1}^O)^2\right) - E^2\left((r_t^O)^2\right) \\
&= E\left((r_t^O)^2 (r_{t+1}^O)^2\right) - \left(\text{var}(r_t^O)\right)^2 \\
&= (1-\pi)^2 E\left(r_{t+1}^2 \left(\sum_{i=0}^{d_t} r_{t-i}\right)^2\right) - \sigma^4 \\
&= (1-\pi)^2 \left(\sigma^2 \left(\frac{1}{1-\pi}\sigma^2\right)\right) - \sigma^4 \\
&= \left(\sigma^2 \left((1-\pi)\sigma^2\right)\right) - \sigma^4.
\end{aligned}$$

Therefore,

$$\text{cov}((r_t^O)^2, (r_{t+1}^O)^2) = -\pi\sigma^4. \tag{5.14}$$

5.1 Stale Prices

Suppose further that true returns are normally distributed, then $E(r_t^4) = 3\text{var}^2(r_t)$. In this case, we have

$$\text{var}((r_t^O)^2) = E((r_t^O)^4) - E^2((r_t^O)^2)$$

$$= (1-\pi)E\left(E\left(\left(\sum_{k=0}^{d_t} r_{t-k}\right)^4 |d_t\right)\right) - \sigma^4$$

$$= 3(1-\pi)\sigma^4 \left(\frac{1}{1-\pi}\right)^2 - \sigma^4$$

$$= \frac{\pi+2}{1-\pi}\sigma^4.$$

This is greater than the variance of true squared returns, $2\sigma^4$, for any $\pi > 0$. It follows that

$$\text{corr}((r_t^O)^2, (r_{t+1}^O)^2) = -\frac{\pi(1-\pi)}{\pi+2} < 0. \quad (5.15)$$

This shows that the non-trading model induces negative autocorrelation in squared returns also, and even when the mean of returns is zero (the mean of squared returns cannot be zero). In practice we find positively autocorrelated squared returns at all frequencies (see Chapter 11) so this model is over restrictive. One can relax the assumption of i.i.d. true returns and still obtain some implications. We will return to this later.

Main Implications for Single Stock

We summarize the main implications of this model for the observed time series:

1. Observed returns have the same expectation as true returns and are unaffected by stale prices.
2. The variance of observed returns is greater than true returns whenever $\mu \neq 0$.
3. The skewness of observed returns is

$$\kappa_3 + \frac{6\pi\mu}{(1-\pi)^2}\left((1-\pi)\sigma^2 + \pi\mu^2\right),$$

which is greater than the skewness of true returns κ_3 when $\mu > 0$. In the special case that $\mu = 0$, the excess kurtosis of observed returns is $2\pi/(1-\pi) \geq 0$ so that the observed return series will be heavier tailed than the true return series.
4. Observed returns are negatively autocorrelated whenever $\mu \neq 0$

$$\text{cov}(r_t^O, r_{t+k}^O) = -\mu^2 \pi^k.$$

5. Observed squared returns are negatively autocorrelated

$$\text{cov}((r_t^O)^2, (r_{t+k}^O)^2) = -\pi^k \sigma^4.$$

In practice the amount of autocorrelation this can predict is quite small, as discussed in CLM.

It is also of importance to understand the effect of non trading on the relationship between different return series, and the CLM framework allows us to calculate this.

Cross Covariances between Observed Return Series

We turn to the calculation of the autocovariances between two return series r_{it}, r_{jt}. We consider the bivariate case with mean zero returns and suppose that the true return series have some contemporaneous covariance

$$E(r_{it}r_{jt}) = \sigma_{ij}. \tag{5.16}$$

We suppose that the non-trading processes for each firm $\{\delta_{jt}\}$ and $\{\delta_{it}\}$ are mutually independent, and that the two series have non-trading probabilities π_i, π_j. This is clearly an unrealistic assumption, especially for stocks in similar industries, but we shall maintain this here for convenience. We calculate $E(r_{it}^O r_{jt}^O)$. If either $\delta_{jt} = 1$ or $\delta_{it} = 1$, then $r_{it}^O r_{jt}^O = 0$, so we only need consider the case that $\delta_{jt} = \delta_{it} = 0$, which has probability $(1 - \pi_i)(1 - \pi_j)$. In this case,

$$E\left(r_{it}^O r_{jt}^O\right) = (1 - \pi_i)(1 - \pi_j) E\left(\left(\sum_{l=0}^{d_{jt}} r_{jt-l}\right)\left(\sum_{l=0}^{d_{it}} r_{it-l}\right)\right),$$

and it remains to calculate the expectation of two aggregated return vectors of random lengths (d_{it} and d_{jt}). This calculation is a bit messy and we defer it to the appendix, where we show that

$$E\left(r_{it}^O r_{jt}^O\right) = \sigma_{ij} \frac{(1 - \pi_i)(1 - \pi_j)}{1 - \pi_i \pi_j} \leq \sigma_{ij}. \tag{5.17}$$

This can be extended to calculate $E(r_{it}^O r_{jt+n}^O)$ (see below) and to allow for a non-zero mean.

Theorem 27 *We have*

$$\gamma_{ij}(n) \equiv \mathrm{cov}(r_{it}^O, r_{jt+n}^O) = \begin{cases} -\mu_i^2 \pi_i^n & \text{for } i=j, n>0 \\ \frac{(1-\pi_i)(1-\pi_j)}{1-\pi_i\pi_j} \sigma_{ij} \pi_j^n & \text{for } i \neq j, n \geq 0. \end{cases}$$

The slope of the best linear predictor of $r_{it}^O | r_{jt}^O$ is

$$\frac{\mathrm{cov}(r_{it}^O, r_{jt}^O)}{\mathrm{var}(r_{jt}^O)} = \frac{\sigma_{ij}}{\sigma_j^2} \frac{(1-\pi_i)(1-\pi_j)}{1-\pi_i\pi_j} \leq \frac{\sigma_{ij}}{\sigma_j^2}.$$

5.1 Stale Prices

If we suppose that the market security is always traded ($\pi_m = 0$), then

$$\beta_i^O = \frac{\operatorname{cov}(r_{it}^O, r_{mt}^O)}{\operatorname{var}(r_{mt}^O)} = \frac{\sigma_{im}}{\sigma_m^2}(1 - \pi_i),$$

which shows that **betas** will be biased downward as a result of non trading. We may correct for the bias by including additional lags, in particular

$$\frac{\operatorname{cov}(r_{it}^O, r_{mt}^O) + \sum_{n=1}^{\infty} \operatorname{cov}(r_{it+n}^O, r_{mt}^O)}{\operatorname{var}(r_{mt}^O)} = \frac{\sum_{n=0}^{\infty} \operatorname{cov}(r_{it}^O, r_{m,t-n}^O)}{\operatorname{var}(r_{mt}^O)} = \frac{\sigma_{im}}{\sigma_m^2} = \beta_i. \quad (5.18)$$

Main Implications for Multiple Securities and Portfolios

Stale prices induce negative autocorrelation and positive cross-correlation. These formulae explain some patterns in autocorrelations and cross autocorrelations in individual stocks reported in CLM. The relative cross-covariances depend only upon the trading frequencies of the securities, that is, we have

$$\frac{\gamma_{ij}(n)}{\gamma_{ji}(n)} = \left(\frac{\pi_j}{\pi_i}\right)^n. \quad (5.19)$$

The more liquid stocks (small π) tend to lead more strongly and the less liquid stock tends to lag more strongly, but both lead and lag effects are present unless $\pi = 1$ for one stock. An always-traded asset $\pi = 1$ will have zero lead cross-covariances and positive lag cross-covariances with a not-always-traded asset.

The autocorrelation of portfolios is determined by the autocorrelation of the component stocks and the cross-autocorrelations. In some leading cases, the cross-autocorrelations can dominate and the portfolio has positive autocorrelation. This is consistent with the evidence presented in CLM and with some of our evidence in Chapter 3.

Epps Effect

Epps (1979) reported results showing that stock return cross correlations decrease as the sampling frequency of the data increases. Since his discovery the phenomenon has been detected in several studies of different stock markets and foreign exchange markets. The CLM model can partially explain this. Suppose that we shrink the trading interval to which the non trading model applies. In that case we expect the non-trading probability to increase for all stocks, and to go to one as the interval shrinks to zero. Suppose that $\pi_i = \pi_j = \pi$ and $\mu_i = \mu_j = 0$. Then

$$E(r_{it}^O r_{jt}^O) = \sigma_{ij} \times \frac{1 - \pi}{1 + \pi},$$

which goes to zero as $\pi \to 1$. As the time interval shrinks we may also expect that $\sigma_{ij} \to 0$ as the variances also shrink to zero. However, the correlation of the observed return series is $\rho_{ij}(1-\pi)/(1+\pi)$, and we may expect from the Brownian motion example (see Chapter 12) that ρ_{ij} is invariant as the time interval shrinks. It follows that $\operatorname{corr}(r_{it}, r_{jt}) = \rho_{ij}$ for any time interval but $\operatorname{corr}(r_{it}^O, r_{jt}^O) \to 0$ as the time interval shrinks under the natural assumption that $\pi \to 1$.

Conclusions

The non trading model gives some useful insights. However, as an econometric model it is oversimplified. It is only relevant now for intraday transactions data or the OTC market. In that case, π is surely not fixed within a day and depends on past trades and prices as well as on the time of day. Engle and Russell (1998) introduced the **autoregressive conditional duration** (ACD) model, which proposed a specific class of processes for the duration time series in the case where duration is a continuous state discrete time process. Also, order book information is relevant to trading, and an economic model is needed to understand how prices are determined in this environment.

5.2 Discrete Prices and Quantities

The models used to describe stock prices often assume that they are continuously distributed, but in fact prices are discrete. In the United States, for any stock over $1 in price level the minimum price increment or **tick size** is one cent (although there are exceptions – Berkshire Hathaway A priced currently at around $150,000.00 only takes $1 moves), and it follows that the values that stock prices can take lie on a grid $\{0, 0.01, 0.02, \ldots\}$. Prior to June 24, 1997 the tick size on the NYSE was $0.125 as per rule 62 of the NYSE Constitution and Rules. On June 24, 1997 the NYSE reduced the tick size from 1/8 of a dollar to 1/16 of a dollar (from an eighth to a sixteenth). The tick size was further reduced to one cent for most of the stocks on January 29, 2001. At the time of writing there is a regulatory discussion around the issue of **subpenny pricing**, whereby the tick size might be reduced to a tenth of a cent, for example. In the UK (and in many other countries except the USA), the tick size varies across stocks in bands according to the price level and market capitalization (generally speaking more liquid stocks have smaller tick sizes), but it is still the case that prices are discrete. One interesting consequence of the different systems used throughout the world is that the tick size can change for a given stock during trading as the price crosses the thresholds used to define the minimum price increment.

Example 33 *Table 5.1 gives the tick size for certain securities on the LSE in 2016.*

There are several issues raised by price discreteness. First, the standard models used to describe prices and return assume a continuous outcome or state, which makes analysis much easier. For example, the Black and Scholes (1973) option pricing

5.2 Discrete Prices and Quantities

Table 5.1 Tick size on the LSE

Price Range	Tick size in £
0–0.9999	0.0001
1–4.9995	0.0005
5–9.999	0.001
10–49.995	0.005
50–99.99	0.01
100–499.95	0.05
500–999.9	0.1
1000–4999.5	0.5
5000–9999	1
10000–∞	5

formula assumes continuous prices and also continuous time, and this specification leads to a simple formula. In practice, discrete prices can make observed data very far from the ideal situation hypothesized in many models. Just like the non-trading issue, we must ask what the consequences are for the properties we expect to find in stock returns. The higher the frequency of data the bigger the issue is, i.e., the bigger the difference between the observed prices series and what could have come from a continuous state model. We will consider some models that respect the discrete nature of prices. A second issue raised by price discreteness is broader, since the amount of discreteness has practical consequences for the functioning of markets and is determined by trading venues and regulators. One might ask why not make prices continuous and eliminate the tick size altogether? There are lots of reasons why this is not possible or desirable. The tick size puts a floor on the bid–ask spread (see below) and so provides some potential profits to market makers. In its absence, they might suffer because of miniscule undercutting (also called **sub-pennying**), which then inhibits them from participating in the market in the first place. There is also a bandwidth issue whereby if any price points are allowed, there will be many small changes in prices, and this imposes computational costs to keep track of all these price points. Finally, it is just not possible to engage the banking system with less than one penny, although with large orders this might imply subpenny ticks are viable, for example in the wholesale FX market where there is a minimum size of order. On the other hand, too large a tick size also has negative consequences: it increases the transaction costs for investors, especially small retail investors, and reduces competition between liquidity providers.

Chapter 5 Empirical Market Microstructure

One way to capture the discreteness in prices is through **rounding models**. These are discrete time series models in which the underlying true price is continuously distributed but the observed price is **rounded** to some grid of discrete values. We consider a simple version of this model.

Example 34 *Suppose that*
$$P_t = P_{t-1} + \varepsilon_t,$$
where ε_t is $N(0, \sigma^2)$, $P_0 = 10$ and $\sigma = 0.1$
$$P_t^R = round(P_t, 1/8),$$
where round (x, c) means to find the closest point in the grid $\{c, 2c, 3c, \ldots\}$ to the point $x \in \mathbb{R}$. Clearly, the observed price $P_t^R \in \{\frac{1}{8}, \frac{2}{8}, \ldots\}$, i.e., it is discrete.

In Figure 5.1 below we show a typical trajectory of the true price and the rounded price. It is quite difficult to give a full analysis of the properties of P_t^R. However, we can see that $P_t^R - P_{t-1}^R$ is negatively autocorrelated at first lag. Why? The logic is a bit like a non trading model except it also works even if there is no drift. If the rounding bites often, then many observed returns are zero and positive return is most likely followed by zero and negative return is most likely followed by zero. The round off

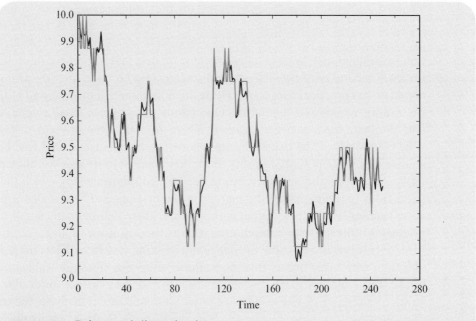

Figure 5.1 Prices and discretization

5.2 Discrete Prices and Quantities

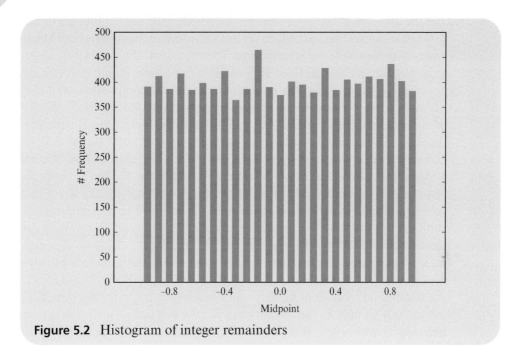

Figure 5.2 Histogram of integer remainders

errors are approximately uniformly distributed. We show the histogram of $16 * (P_t - P_t^R)$ in Figure 5.2.

An alternative approach is to model prices as a discrete time series, rather like discrete choice models in microeconometrics. That is, redefine price increments in terms of ticks. Let

$$x_t = \frac{P_t - P_{t-1}}{\text{tick size}} \in \{0, \pm 1, \pm 2, \ldots\}. \qquad (5.20)$$

This could be treated as a two sided Poisson random variable for example with

$$\Pr(x_t = k) = \frac{\exp(-\lambda)}{2 - \exp(-\lambda)} \frac{\lambda^{|k|}}{|k|!}.$$

We may specify some conditional distribution for the normalized price change in terms of past information $\mathcal{F}_{t-1}$, i.e.,

$$\Pr(x_t = k | \mathcal{F}_{t-1}) = H(k; \mathcal{F}_{t-1}, \theta),$$

where H is a known function, and θ are unknown parameters. One example is the Rydberg and Shephard (2003) model. It decomposes price changes into three components:

$$P_t - P_{t-1} = A_t D_t S_t.$$

The first component A_t is *activity*, i.e., whether the transaction price is different from the previous price. This variable takes the value zero or one. The second component

D_t is the *direction* of the price change, i.e., whether the price moves up or down, given that it has moved. This variable takes the value minus one or plus one. The third component S_t is the *size* of the price change. This represents the number of tick movements in the price, given its direction. This variable can be any strictly positive integer. The movement in price is then the product of these three components, and the joint distribution of price changes is modelled by sequentially modelling the distributions of the three components conditional on exogenous (predetermined) variables and past values. `Matlab` code is provided.

The Rydberg and Shephard model is particularly useful as it allows the inclusion of exogenous variables such as previous periods orders. So, one can estimate the effect of a particular exogenous variable on price activity and statistically test if that variable has a significant effect on price activity and direction. The included variables can be continuous or discrete. The main requirement is that they be exogenously determined, meaning effectively that they happen before the current price is determined. One can put in, for example, trading activity amounts or net amounts in a prior period and identify thereby whether these variables have a statistically significant effect on prices. One issue with this class of models is that it does not allow for the possibility that the tick size may change. In the UK for example, the tick size for a given stock may change dynamically within a trading day. In the US, the transition period from tick size of 1/8 to 1 cent is of interest in itself.

There is another source of discreteness, which is that quantities quoted or transacted are discrete. One can buy one share of a stock but not half a share. This can matter significantly for high price stocks such as Berkshire Hathaway A. In FX markets, there is often a minimum trade quantity such as US $1,000,000. The effect of this is to limit the participants to those who easily have that sum of money to work with. This aspect of discreteness has a non trivial impact on the so-called **triangular arbitrage** between currencies that should keep their prices in line: since if one converts US $1,000,000 into yen and then from yen to euros and from euros to US dollars one loses scale (and round numberedness) on each leg.

5.3 Bid, Ask, and Transaction Prices

We next consider how prices are actually determined in stock exchanges and other trading venues. The theoretical literature is mostly concerned with a stylized version of the dealer market. Market participants can be sorted into outsiders (everyone except the dealer or market maker) and insiders (the dealer). The dealer quotes buy (bid) and sell (ask or offer) prices, and the outsiders take it or leave it. She hopes to earn the spread, i.e., to buy at the bid and to sell at the ask. If she can make these two trades at exactly the same time, this is a money printing machine. Unfortunately for her, this is very difficult to achieve. In practice, quoted bid–ask spreads vary by stock, they vary over time within day for each stock, and they vary according to market conditions.

We next discuss the Roll (1984a) model, which traces out the consequences of a (fixed) bid–ask spread for observed transaction price changes.

5.3 Bid, Ask, and Transaction Prices

5.3.1 Roll Model

Suppose that the true price P^* is a random walk unrelated to order flow

$$P_t^* = \mu + P_{t-1}^* + \varepsilon_t, \qquad (5.21)$$

where ε_t is a mean zero and uncorrelated sequence (**rw5**). Buy and sell orders arrive randomly: Q_t is a trade direction indicator, $+1$ if the customer is buying and -1 if the customer is selling. The orders are matched, i.e., they result in a trade on one side each period at the observed transaction price P_t. The observed transaction price satisfies

$$P_t = P_t^* + Q_t \frac{s}{2}, \qquad (5.22)$$

where s is the full spread so the half-spread is $s/2$. That is, if the customer is buying he pays the price $P_t^A = P_t^* + s/2$ and if he is selling he receives $P_t^B = P_t^* - s/2$. Suppose that the order flow Q_t is an i.i.d. random variable (and independent of P_t^*) that takes the values $+1$ and -1 with equal probability. Then

$$X_t = P_t - P_{t-1} = P_t^* - P_{t-1}^* + (Q_t - Q_{t-1})\frac{s}{2} = \mu + \varepsilon_t + (Q_t - Q_{t-1})\frac{s}{2},$$

where $Q_t - Q_{t-1}$ is a random variable that can take the values $+2, 0, -2$ and is **1-dependent**, that is, $Q_t - Q_{t-1}$ and $Q_{t-1} - Q_{t-2}$ are correlated random variables.

It follows that

$$\mathrm{var}(X_t) = \sigma_\varepsilon^2 + \frac{s^2}{2} \qquad (5.23)$$

$$\mathrm{cov}(X_t, X_{t-1}) = -\frac{s^2}{4}, \qquad (5.24)$$

and $\mathrm{cov}(X_t, X_{t-j}) = 0$ for all $j > 1$. In the first equation the volatility of observed stock returns, $\mathrm{var}(X_t)$, is decomposed into the sum of fundamental volatility σ_ε^2 and transitory volatility $s^2/2$. The second equation says that we should see a negative autocorrelation in observed transaction prices, which is consistent with the evidence presented in Chapter 3. This is known as **bid ask bounce**: in the absence of the noise ε_t, if a transaction occurs at the ask price, then the subsequent price could be at the ask price or could decline to the bid price, which means the change in transaction price is non-negative, likewise a transaction at the bid price can only be followed by a weakly higher price – the bouncing between the upper and lower prices induces negative first order autocorrelation. In Figure 5.3 below we show a typical price sequence, which is consistent with this observation.

Higher order autocorrelations are zero in this model. Note that the autocorrelation magnitude depends positively on the ratio s^2/σ_ε^2. This can be interpreted as saying that less liquid stocks should have higher negative first order autocorrelation. The process of observed prices is equivalent to an MA(1) process because $\mathrm{cov}(X_t, X_{t-j}) = 0$ for all $j > 1$. The predictions of this model are similar in some respects to the non-trading model.

170 Chapter 5 Empirical Market Microstructure

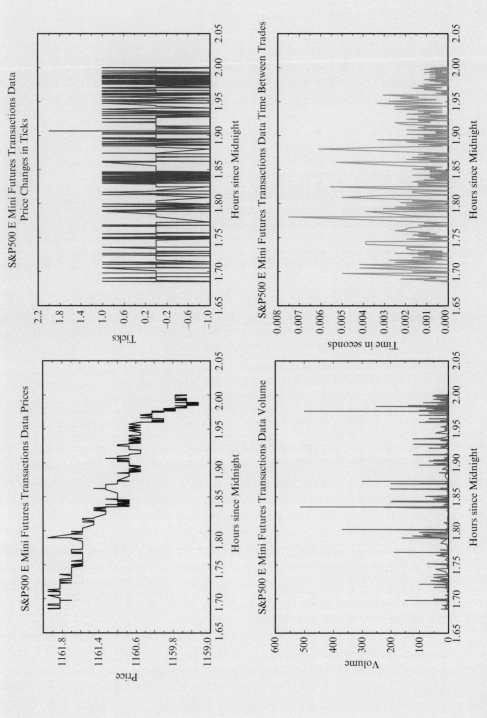

Figure 5.3 Price changes on S&P500 Emini contract

5.3 Bid, Ask, and Transaction Prices

Suppose that quoted prices were available, i.e., we observed the (best) ask price $P_t^A = P_t^* + s/2$ and the (best) bid price $P_t^B = P_t^* - s/2$. Then the **mid-quote**

$$P_t^M = \frac{P_t^A + P_t^B}{2} = P_t^* \tag{5.25}$$

is equal to the efficient price. Using the midquote instead of the transaction price can obviate the effects of bid–ask bounce, meaning that the mid-quote is serially uncorrelated and in fact equal to the efficient price. However, in many cases quote data is not available or is more expensive to acquire, and so we continue our narrative about the effects of this model on transaction prices.

A common application of the Roll model is to infer the value of s given only transaction price data. This is particularly useful when it is hard to obtain bid ask spreads directly. Specifically, by inverting the equation (5.24), we obtain the equation

$$s = 2\sqrt{-\mathrm{cov}(X_t, X_{t-1})}.$$

Furthermore, combining (5.23) and (5.24), the variance of the efficient price innovation is

$$\sigma_\varepsilon^2 = \mathrm{var}(X_t) + 2\mathrm{cov}(X_t, X_{t-1}).$$

This quantity is of interest in measuring **volatility**, which we will discuss in great detail in Chapters 11 and 12 below. It follows that both s and σ_ε^2 can be estimated from observed transaction price data. Suppose we have a sample of transaction prices $P_0, \ldots, P_T$ and price changes $X_t = P_t - P_{t-1}$, and let

$$\widehat{s} = \sqrt{-\frac{4}{T} \sum_{t=1}^{T} (X_t - \overline{X})(X_{t-1} - \overline{X})}, \qquad \overline{X} = \frac{1}{T} \sum_{t=1}^{T} X_t$$

$$\widehat{\sigma}_\varepsilon^2 = \frac{1}{T} \sum_{t=1}^{T} (X_t - \overline{X})^2 + \frac{2}{T} \sum_{t=1}^{T} (X_t - \overline{X})(X_{t-1} - \overline{X}). \tag{5.26}$$

Under the conditions of Chapter 3 these estimators are consistent and asymptotically normal. If ε_t satisfies **rw3**, then the limiting distribution of $\widehat{s}$ is tractable, and we can calculate confidence intervals for s, σ_ε^2. This is left as an exercise. We show in Figures 5.4 and 5.5 estimates of the autocovariance of the Dow stocks and the efficient price.

The model (5.21) is similar to but not the same as the random walk model of Chapter 3 because it is for price levels. If one uses returns (log prices) instead of price changes one obtains a percentage spread. Specifically, suppose that

$$p_t^* = \mu + p_{t-1}^* + \varepsilon_t$$

$$P_t = P_t^* \left(1 + Q_t \frac{s}{2}\right),$$

Chapter 5 Empirical Market Microstructure

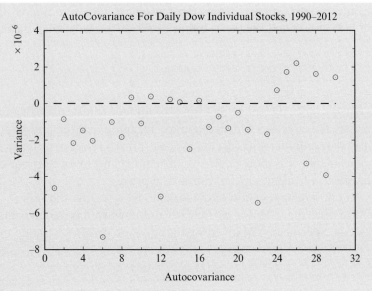

Figure 5.4 Autocovariance of daily Dow

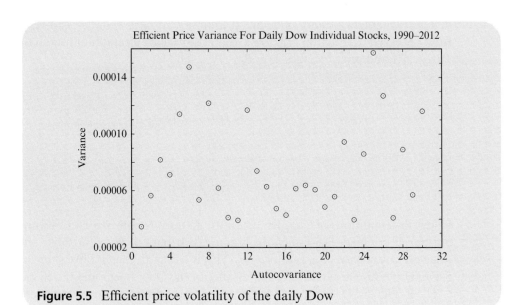

Figure 5.5 Efficient price volatility of the daily Dow

where Q, s, ε are as before. Then

$$r_t = \log P_t - \log P_{t-1} = \mu + \varepsilon_t + \log\left(1 + Q_t \frac{s}{2}\right) - \log\left(1 + Q_{t-1} \frac{s}{2}\right).$$

5.4 What Determines the Bid–Ask Spread?

In this case,

$$\text{var}(r_t) = \sigma_\varepsilon^2 + \frac{1}{2}\left(\log^2\left(1+\frac{s}{2}\right) + \log^2\left(1-\frac{s}{2}\right)\right) - \log\left(1-\frac{s^2}{4}\right)$$

$$\text{cov}(r_t, r_{t-1}) = -\frac{1}{2}\left(\log^2\left(1+\frac{s}{2}\right) + \log^2\left(1-\frac{s}{2}\right)\right).$$

We can invert these two nonlinear equations to obtain s as a function of $\text{cov}(r_t, r_{t-1})$ and σ_ε^2 as a function of $\text{var}(r_t)$ and $\text{cov}(r_t, r_{t-1})$. When s is small we can approximate $\log(1+s/2)$ by $s/2$ and $\log(1-s/2)$ by $-s/2$, in which case $\text{var}(r_t) \simeq \sigma_\varepsilon^2 + s^2/2$ and $\text{cov}(r_t, r_{t-1}) = -s^2/4$. In this case also, the mid-quote is free of the bid–ask bounce.

This model is a little oversimplified and is empirically problematic because one often finds that sample autocovariances are positive; see Chapter 3. A number of refinements to this method that address this issue have been suggested; see for example Hasbrouck (2009).

5.4 What Determines the Bid–Ask Spread?

We next ask, what determines the bid–ask spread? We discuss various models that are designed to explain the bid ask spread.

1. Inventory models. The bid–ask spread compensates the market maker for the risk of ruin through inventory explosion.
2. Asymmetric information/sequential trade models. Bid–ask spread is determined by adverse selection costs.
3. Models that combine both of the above explanations.

5.4.1 Inventory Models

An important role of market makers is to provide opportunity to trade at all times (**immediacy**). Market makers absorb temporary imbalances in order flow and will hold an inventory of assets that may deviate from their desired inventory position due to changes in demand and supply of their clients. The desired inventory position is usually zero, just like the fishmonger who intermediates between the fisherman and the housewife: he does not derive utility from the fish themselves but desires to unload his stock as soon as he can. Market makers, like fishmongers, require compensation for the service of providing immediacy.

Ho and Stoll (1981) develop the seminal model given here. There is one monopolistic dealer who sets bid and ask prices as a markup on his opinion of the true price p:

$$p_a = p + a, \quad p_b = p - b.$$

There is random arrival of buy and sell orders with possibly different rates. There is an elastic demand and supply so that the number of orders is declining in the markup, which limits the market power of the dealer. The market maker maximizes expected utility (he is risk averse) of final wealth W_T at time t, $E_t(U(W_T))$, where T is the terminal period. Here, E_t denotes expectation conditional on his current information. This model is solved using dynamic programming in continuous time. They derive an expansion (an approximation) for the bid and ask prices for the case where $\tau = T - t$ is small. In particular, they show that

$$p_a = p + \frac{1}{2}(A + \sigma^2 \theta \tau q) + O(\tau^2)$$

$$p_b = p - \frac{1}{2}(A + \sigma^2 \theta \tau q) + O(\tau^2),$$

where: A is a fixed measure of the monopoly power of the market maker; q is the trade size for which the quote is relevant; σ^2 is the volatility of true price; and θ is a measure of the risk aversion of the market maker. It follows that the bid–ask spread s is approximately independent of the current inventory level

$$s = A + \sigma^2 \theta \tau q + O(\tau^2). \tag{5.27}$$

If the market maker is risk neutral, then the spread is just A, which depends on the demand and supply elasticities of the traders. Under risk aversion ($\theta > 0$), the spread widens, and depends positively on risk aversion, volatility, order size, and remaining time horizon. In this model, trading affects future bid and ask prices. The market maker's guess of the price p is moved around by their inventory position (although the inventory position does not affect the spread). The quote midpoint p moves in the direction of trade, i.e., a buyer-initiated transaction pushes the price up, while a seller-initiated transaction pushes the price down, and the effect is proportional to trade size. In a later model Ho and Stoll (1983) introduce competition between dealers. Aït-Sahalia and Saglam (2013) have updated this literature.

5.4.2 Asymmetric Information Models

Adverse selection is an important issue for market makers. It is harder to price discriminate against anonymous traders as insurance companies do with their clients to mitigate risks. The bid–ask spread can be seen as fair compensation for the market maker's adverse selection costs. Copeland and Galai (1983) characterized the cost of supplying quotes, as writing a put and a call option to an information-motivated trader. In their static model, the bid–ask spread is a positive function of the price level and return variance, a negative function of measures of market activity, depth, and continuity, and negatively correlated with the degree of competition. Glosten and Milgrom (1985), Glosten (1987), and Glosten and Harris (1988) developed a class of sequential trade models that capture the effect of asymmetric information and adverse selection on the bid ask spread and the dynamic effect of trading. We present a simple version of their models; Hasbrouck (2007, pp. 44–6).

5.4 What Determines the Bid–Ask Spread?

A Simple Sequential Trade Model

The true value of the asset V is random, and is the outcome of the two-point distribution

$$V = \begin{cases} V_H & \text{with probability } 1 - \delta \\ V_L & \text{with probability } \delta, \end{cases}$$

where $V_H > V_L$. There are two types of traders: uninformed traders (noise traders) and informed traders. The type of investor that arrives in the market is random, and is the outcome of the two-point distribution

$$T = \begin{cases} I & \text{with probability } \mu \\ U & \text{with probability } 1 - \mu. \end{cases}$$

The market maker is uninformed about the value of the security (he knows the distribution of V but not its realization), he is risk-neutral, and he has no opportunity costs. He makes zero expected profits due to external competition that is not modelled. The market maker quotes bid and ask prices B and A, which are firm quotes to buy or sell one unit. The traders observe these quotes and take them or leave them (i.e., they submit market orders to trade).

The strategies of the participants are as follows. The informed traders, (I), will buy if the value is high V_H, and sell if the value is low V_L, that is, provided the market maker sets prices in the interval $[V_L, V_H]$. The uninformed traders, (U), buy or sell with probability 1/2 whatever. Let Q be the indicator of order direction, i.e., $Q = +1$ if the order is a buy order and $Q = -1$ if the order is a sell order. The order flow is described by the following probability table:

	$Q_t = +1$	$Q_t = +1$
$V = V_L$	$\frac{1}{2}(1-\mu)$	$\frac{1}{2}(1+\mu)$
$V = V_H$	$\frac{1}{2}(1+\mu)$	$\frac{1}{2}(1-\mu)$

where, for example, the probability of receiving a buy order when the asset has low value is $\Pr(Q_t = +1 | V = V_L) = (1 - \mu)/2$.

The dealer is allowed to set contingent prices B and A, one for incoming buy orders and one for incoming sell orders, and she sets these prices to make zero profits on ends side of the market (this is forced on her by external reasons) taking account of the value distribution and the distribution of trader types. Before trading starts, the dealer has valuation given from the prior distribution of V, and in particular,

$$E(V) = V_L \delta + V_H (1 - \delta)$$

$$\text{var}(V) = \delta (1 - \delta)(V_H - V_L)^2.$$

The variance is greatest when $\delta = 1/2$ and goes to zero as $\delta \to 0$ or $\delta \to 1$.

Chapter 5 Empirical Market Microstructure

If the dealer is forced to set just one price for both buy and sell orders, then presumably this would be $P = E(V) = V_L \delta + V_H(1 - \delta)$. This would consistently lose money unless $\mu = 0$, as we will see.

How should the dealer set different prices for buyers and sellers, B and A? The dealer uses **Bayesian updating** to adapt her value distribution conditional on a hypothetical signal that she might receive. The dealer reasons: if I receive a buy order how would I update my valuation. By Bayes rule, she would calculate the posterior distribution of the stock value given the order direction

$$\delta_+ = \overbrace{\Pr(V = V_L | Q = +1)}^{\text{posterior}} = \frac{\overbrace{\Pr(Q = +1 | V = V_L)}^{\text{likelihood}} \overbrace{\Pr(V = V_L)}^{\text{prior}}}{\Pr(Q = +1)} = \frac{\frac{1}{2}(1 - \mu) \times \delta}{(1 + \mu(1 - 2\delta))/2}$$

$$1 - \delta_+ = \Pr(V = V_H | Q = +1) = 1 - \Pr(V = V_L | Q = +1) = \frac{\frac{1}{2}(\mu + 1)(1 - \delta)}{(1 + \mu(1 - 2\delta))/2}.$$

Likewise, if she received a sell order she would calculate

$$1 - \delta_- = \overbrace{\Pr(V = V_H | Q = -1)}^{\text{posterior}} = \frac{\overbrace{\Pr(Q = -1 | V = V_H)}^{\text{likelihood}} \overbrace{\Pr(V = V_H)}^{\text{prior}}}{\Pr(Q = -1)}$$

$$= \frac{\frac{1}{2}(1 - \mu) \times (1 - \delta)}{(1 - \mu(1 - 2\delta))/2}$$

$$\delta_- = \Pr(V = V_L | Q = -1) = 1 - \Pr(V = V_H | Q = -1) = \frac{\frac{1}{2}\delta(1 + \mu)}{(1 - \mu(1 - 2\delta))/2}.$$

The revised distributions are two-pointers like the prior but with updated weights on the two possible outcomes reflecting the information in the buy/sell signal that would be received.

The profit of the dealer if she receives a buy order is $A - V$, and if she receives a sell order it is $V - B$. Therefore, she sets $A = E(V|Q = +1)$ and $B = E(V|Q = -1)$ to set expected profits to zero on each side, that is, she makes zero profits on the ask side of the market and zero profits on the bid side, on average. This implies that

$$A = \frac{V_L \delta(1 - \mu) + V_H(1 - \delta)(1 + \mu)}{1 + \mu(1 - 2\delta)}$$

$$B = \frac{V_L \delta(1 + \mu) + V_H(1 - \delta)(1 - \mu)}{1 - \mu(1 - 2\delta)}.$$

The bid–ask spread is

$$A - B = \frac{4\delta(1 - \delta)\mu}{1 - (1 - 2\delta)^2 \mu^2}(V_H - V_L). \tag{5.28}$$

5.4 What Determines the Bid–Ask Spread?

The spread is an increasing function of μ (more informed traders widen the spread) and an increasing function of $V_H - V_L$ (one could interpret this as volatility or uncertainty over the final value). It is also a decreasing function of $(\delta - 1/2)^2$, which also says that more uncertainty about the final valuation leads to higher spread. In the special case that $\delta = 1/2$, the bid–ask spread is

$$A - B = \mu(V_H - V_L),$$

and the midpoint of the quoted prices is

$$\frac{A + B}{2} = \frac{1}{2}V_L + \frac{1}{2}V_H = E(V).$$

The dealer gains from uninformed traders and loses to informed ones, and balances these two contributions to make zero profits. If the dealer were to set a single price $P = E(V)$, then she will lose money because $E(V|Q=+1) > E(V) > E(V|Q=-1)$.

Now consider this process evolving over time with constant value as determined in the first round but where in each period there arrives a different trader chosen from the same arrival distribution, i.e., $T = I$ with probability μ and $T = U$ otherwise. The dealer faces the same order flow distribution as before (although his information about the order flow improves over time). The dealer observes a transaction or order flow history at any time $t - 1$, i.e., he knows $\{Q_1, \ldots, Q_{t-1}\}$, for example $\{1, -1, -1, \ldots, 1\}$. Denote this history by $\mathcal{F}_{t-1}$. He updates his valuation of the security based on this history. Essentially he replaces his prior by the posterior distribution from the last round $\Pr(V = V_L | \mathcal{F}_{t-1})$, and then obtains the updated posterior $\Pr(V = V_L | Q_t = +1, \mathcal{F}_{t-1})$ conditional on the arrival of a sell order, or the updated posterior $\Pr(V = V_L | Q_t = -1, \mathcal{F}_{t-1})$ conditional on the arrival of a buy order. We may obtain the following dynamic equations for the dealers's updated value distribution

$$\Pr(V = V_L | Q_t = +1, \mathcal{F}_{t-1}) = \frac{(1 - \mu) \Pr(V = V_L | \mathcal{F}_{t-1})}{1 + \mu(1 - 2 \Pr(V = V_L | \mathcal{F}_{t-1}))}.$$

$$\Pr(V = V_L | Q_t = -1, \mathcal{F}_{t-1}) = \frac{\Pr(V = V_L | \mathcal{F}_{t-1})(1 + \mu)}{1 - \mu(1 - 2 \Pr(V = V_L | \mathcal{F}_{t-1}))}.$$

Likewise we can calculate the two complementary probabilities $\Pr(V = V_H | Q_t = +1, \mathcal{F}_{t-1})$ and $\Pr(V = V_H | Q_t = -1, \mathcal{F}_{t-1})$. The dealer sets the next period ask and bid price by the zero profit conditions, i.e.,

$$A_t = E(V | Q_t = +1, \mathcal{F}_{t-1}), \quad B_t = E(V | Q_t = -1, \mathcal{F}_{t-1}),$$

which depends on the history and can be calculated recursively using the above probabilities.

As time goes by, the quoted prices change in response to the received order flow. The order flow is serially correlated because informed investors always trade in the same direction. We may show this by computing the Cowles–Jones continuation/reversal ratio of the order direction variable. We have:

$$\Pr(Q_t = +1, Q_{t+1} = +1 | V = V_L) + \Pr(Q_t = -1, Q_{t+1} = -1 | V = V_L)$$
$$= \left(\frac{1}{2}(1-\mu)\right)^2 + \left(\frac{1}{2}(1+\mu)\right)^2 = \frac{1}{2}(1+\mu^2)$$

and likewise the other probabilities, so that we obtain

$$\frac{\Pr(Q_t = +1, Q_{t+1} = +1) + \Pr(Q_t = -1, Q_{t+1} = -1)}{\Pr(Q_t = +1, Q_{t+1} = -1) + \Pr(Q_t = -1, Q_{t+1} = +1)} = \frac{1+\mu^2}{1-\mu^2}, \quad (5.29)$$

which is greater than one if and only if μ is greater than zero. The dealer knows this, so that a long run of buy orders would tend to indicate a high valuation, and likewise a long run of sell orders would indicate a low valuation. This is reflected in the quotes he makes.

In this model, trades (or rather order flow) have price impact because they lead to an adjustment in the posterior distribution of the dealer. The transaction price each period is denoted by

$$P_t = \begin{cases} A_t & \text{if } Q_t = +1 \\ B_t & \text{if } Q_t = -1 \end{cases} = E(V|\mathcal{F}_t),$$

where $\mathcal{F}_t$ is the updated history $\mathcal{F}_t = \mathcal{F}_{t-1} \cup Q_t$. The transaction price will alternate between the bid and the ask price (bid–ask bounce) according to the direction of the order flow. However, the transaction price is a martingale with respect to $\mathcal{F}_t$ because

$$E(P_{t+1}|\mathcal{F}_t) = E(E(V|\mathcal{F}_{t+1})|\mathcal{F}_t) = E(V|\mathcal{F}_t) = P_t,$$

by the law of iterated expectation, since $\mathcal{F}_t \subset \mathcal{F}_{t+1}$. Therefore, this model cannot explain the observed patterns of serial correlation in transactions data.

In this model public information about the true value is improving over time. The dealer can consistently estimate the true (realized) value from the long sequence of orders he has received. Spreads get narrower over time. Transaction prices converge to the true value with probability one. There is a happy ending. It is a very simple model but it illustrates the process of **price discovery**, how fundamental information can be impacted in prices by the process of trading by informed traders. The process is not instantaneous, it takes a while to play out, but in this special case the transaction price itself remains a martingale throughout. The higher the proportion of informed traders, the faster the price discovery occurs, with the limiting case of only informed traders yielding instant value revelation. Although in that case $A = V_H$ and $B = V_L$ and the informed trader makes zero profits, so why he would trade is not clear. At the other extreme, if there were only uninformed traders, the true price would never

5.4 What Determines the Bid–Ask Spread?

be discovered, and the bid ask spread would be zero (through the exertion of competitive forces) and remain zero throughout sequential trading. Compare this with our discussion of the EMH in Chapter 3.

In the case where $\delta = 1/2$, the midpoint of the initially quoted values is $(V_H - V_L)/2$. The initial midquote is uncorrelated with the fundamental value in this case, sharing some features with the Grossman and Stiglitz (1980) model. However the process of trading brings the correlation up to one. In this model the informed trader makes profits (effectively from the noise trader) that might cover his costs of information acquisition, although eventually the profits are driven from the market.

The dealer does not explicitly care about his inventory. He is always able to buy or sell what he owns at the current price, presumably by selling to someone else. Suppose that he holds his position until the end when the true value is revealed. As time evolves the dealer's inventory increases or decreases without limit. The inventory process is an integer valued random walk plus drift, i.e., $X_{t+1} = X_t + \mu + \varepsilon_t$, where ε_t is mean zero, or $X_{t+1} = X_t - \mu + \varepsilon_t$ if the value is low. Clearly, this puts some limits on the interpretability here. For a fixed capital, the dealer will go bankrupt with probability going to one.

Example 35 *We show some simulated examples. We chose $V_H = 1$ and $V_L = 0$ and $V = V_H$. We considered two scenarios with regard to μ, δ: (1) $\mu = 0.5, \delta = 0.5$; (2) $\mu = 0.1, \delta = 0.5$. In Figure 5.6 below we show the trajectory of quoted prices as time goes by. In scenario one, after an initial wrong move the prices move rapidly toward the correct value. In the second scenario (5.7), there is much more volatility and reversal of direction, but ultimately we get a happy ending.*

This is obviously a very simple model, an unrealistic model to put it mildly. In Exercise (38) we consider the case where the value distribution V is uniformly distributed. However, the basic logic of this model is compelling and widely accepted.

5.4.3 Adverse Selection and Inventory

Glosten (1987) considers a more general model with both adverse selection and inventory costs. He supposes that prices can be decomposed into two parts:

$$P_a = P + Z_a + C_a \quad ; \quad P_b = P - Z_b - C_b$$

$$s = P_a - P_b = Z_a + C_a + Z_b + C_b,$$

where $C = C_a + C_b$ represents the inventory cost and $Z = Z_a + Z_b$ is the adverse selection component of the spread s. In the presence of the fixed inventory cost some of the strong conclusions of the pure asymmetric information model are weakened. Specifically, the transaction price is no longer a martingale. Under some model structure, he shows that

$$P_t = P_t^* + C_t Q_t, \tag{5.30}$$

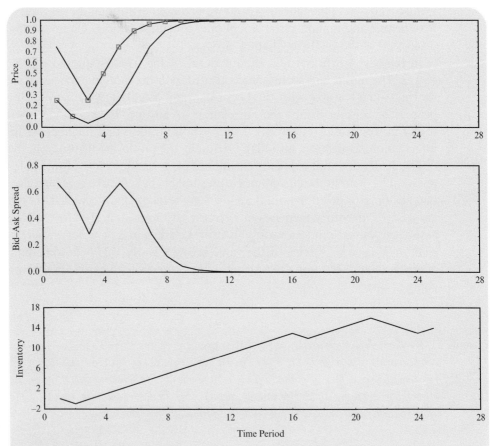

Figure 5.6 Price trajectory 1. In panel A, ask price, bid price, transaction price (grey squares); in panel B, bid–ask spread; in panel C, inventory of dealer.

where P_t is the observed transaction price and P_t^* is the common knowledge true price after the tth trade, Q_t is the trade direction indicator, while $C_t = C_A$ if it was a buy initiated trade and $C_t = C_B$ if it was a sell initiated trade. This equation is similar to the Roll model, except that P_t^* and Q_t are no longer independent. Under some conditions, the serial covariance of continuously compounded returns is approximately given by

$$\operatorname{cov}(r_t^O, r_{t-1}^O) = -\frac{\gamma s_p^2}{4}$$

$$\gamma = \frac{C}{C+Z} \quad ; \quad s_p = \frac{P_a - P_b}{(P_a + P_b)/2}.$$

In this case, the observed returns should be negatively autocorrelated consistent with the evidence presented in Chapter 3 and consistent with the Roll model. However, the precise magnitude of the negative autocovariance is affected by the adverse selection

5.4 What Determines the Bid–Ask Spread?

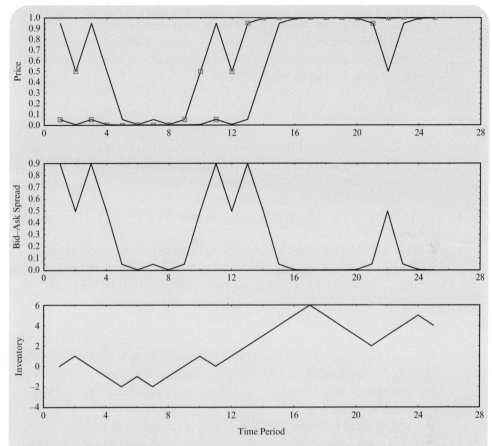

Figure 5.7 Price trajectory 2. In panel A, ask price, bid price, transaction price (grey squares); in panel B, bid–ask spread; in panel C, inventory of dealer.

component in such a way that the covariance estimator of the spread suggested by Roll may be a downwardly biased estimator of the quoted spread.

Note that these models assume that inventory costs are not affected by order flow, and that there is only a linear cumulative effect of order flow on implied asset value.

Glosten and Harris (1988) develop a model for empirical analysis that contains both permanent adverse selection effects and transitory inventory effects. They suppose that the fundamental price P^* is a random walk and the observed transaction price P_t satisfies

$$P_t^* = P_{t-1}^* + \varepsilon_t$$
$$P_t = P_t^* + Q_t C_t + Q_t Z_t,$$

where Q_t is the trade direction indicator. The true price innovations are of two types. The first term ε_t is due to the arrival of public information, while the second term $Q_t Z_t$, the adverse-selection spread, is due to the revision in expectations conditional on an order arrival. Assuming that Z_t is positive, buy orders cause "true" prices to rise by Z_t while sell orders cause them to fall by Z_t. The adverse-selection spread has a permanent effect on prices since it is due to a change in expectations. They assume that Q_t is i.i.d. with equal probability of $+1$ and -1 and unrelated to P_t^*. Here, Z_t is the unobserved adverse selection spread component, C_t is the unobserved transitory spread component and they are determined as:

$$Z_t = \alpha_0 + \alpha_1 V_t \quad ; \quad C_t = \beta_0 + \beta_1 V_t,$$

where V_t is the observed number of shares traded in transaction t. They work with transaction time, so that $t = 1, 2, \ldots$ indexes transactions, rather than calendar time. The observed time between transactions $t-1$ and t is T_t, and ε_t is normal with mean $f_1(T_t)$ and variance $f_2(T_t)$ for some functions f_1 and f_2. They also have a rounding component $r_t = P_t^R - P_t$, whereby $P_t^R = round(P_t, 1/8)$, but we shall ignore this in the sequel.

There is an identification issue here and to solve this they set $\beta_1 = \alpha_0 = 0$. The information model we saw above predicts that $\alpha_0 = 0$ and $\alpha_1 > 0$. Models of inventory may predict $\beta_1 > 0$ or $\beta_1 < 0$, but they all agree that $\beta_0 > 0$. It follows that

$$P_t - P_{t-1} = \beta_0 (Q_t - Q_{t-1}) + \alpha_1 Q_t V_t + \varepsilon_t.$$

This says that returns (or rather price differences) are serially correlated. They estimate β_0 and α_1 from data $\{P_t, V_t, T_t\}_{t=1}^T$ using some additional assumptions. They find that the adverse selection component α_1 is significant and is related to relevant stock characteristics.

A simple version of this model is as follows. Suppose that

$$p_t = p_t^* + Q_t \frac{s_0}{2}, \quad p_t^* = p_{t-1}^* + \delta Q_t + \varepsilon_t,$$

see Equation (5.4) in Foucault, Pagano, and Röell (2013). Here, δ measures the contribution of adverse selection, i.e., the effect of a market order on the latent true efficient price, or rather the best guess of the efficient price. This implies that

$$\Delta p_t = \varepsilon_t + \alpha_0 Q_t - \beta_0 Q_{t-1}, \quad \text{with } \alpha_0 \equiv \frac{s_0}{2} + \delta, \quad \beta_0 \equiv \frac{s_0}{2}. \tag{5.31}$$

This model has both adverse selection and the usual Roll effect: returns would be serially correlated, although the magnitude of the first order autocorrelation is different from the standard Roll model, which means that the Roll and Hasbrouck spread estimators would be inconsistent (i.e., they are biased even as the sample size goes to infinity). If $\{p_t\}$ is the only observable, even assuming $\varepsilon_t \sim$ i.i.d. $N(0, \sigma_\varepsilon^2)$, the parameter vector $(\alpha_0, \beta_0, \sigma_\varepsilon^2)$ is still not jointly identified. When one is able to observe the trade direction indicators, one can estimate all these parameters.

Easley and O'Hara (1987, 1992) extend these models in various directions. They allow that: (1) informed investors do not know precisely the value of stock but observe private signals (information events); (2) not all trading periods have information events; (3) trades can be of different sizes and the price impact can depend on the trade size; and (4) participants may or may not trade so that the time between trades is variable and this may have information content.

5.5 Strategic Trade Models

The final issue we consider is that of **price impact** or **strategic trading**. Price impact refers to the association between an incoming order (to buy or to sell) and the subsequent price change. *Ceteris paribus*, a buy trade should push the price up and a sell trade should push the price down. This is an issue for traders for whom price impact is tantamount to a cost: their second buy trade is on average more expensive than the first because of their own impact (and vice-versa for sells). Monitoring and controlling impact is therefore of key importance for market participants. The Kyle (1985) model is a classic treatment of this issue, which we will now outline.

Albert S. (Pete) Kyle has been the Charles E. Smith Chair Professor of Finance at the University of Maryland's Robert H. Smith School of Business since 2006. He earned his BS degree in mathematics from Davidson College (1974), studied philosophy and economics at Oxford University as a Rhodes Scholar from Texas (Merton College, 1974–1976, and Nuffield College, 1976–1977), and completed his PhD in economics at the University of Chicago in 1981. He has been a professor at Princeton University (1981–1987), the University of California Berkeley (1987–1992), and Duke University (1992–2006). He has made pioneering contributions to the study of market microstructure through his dynamic equilibrium model of trading in Kyle (1985). He works on topics such as high frequency trading, informed speculative trading, market manipulation, price volatility, the informational content of market prices, market liquidity, and contagion. He has combined an outstanding academic career with an engagement in and contribution to public policy. He was a staff member of the Presidential Task Force on Market Mechanisms (Brady Commission, 1987), a consultant to the SEC (Office of Inspector General), Commodities Futures Trading Commission (CFTC), and US Department of Justice, a member of NASDAQ's economic advisory board (2004–2007), a member of the FINRA economic advisory board (2010–2014), and a member of the CFTC's Technology Advisory Committee (2010–2012).

There is a single informed trader who knows the impact of their own trades and therefore chooses how to trade strategically. The single security has value $v \sim N(\mu_v, \sigma_v^2)$. The informed trader knows v and submits demand x to maximize his expected profit. There are also noise traders who submit random order flow $u \sim N(0, \sigma_u^2)$. The risk neutral market maker observes total demand

$$y = x + u \qquad (5.32)$$

and then sets a price p to make zero profits in expectation. Equilibrium can be characterized by the pricing rule of the market maker $p(y)$ and the informed trader demand function $x(v)$, which are mutually consistent. There is no formal bid ask spread here, but the market maker is submitting an entire schedule whose prices depend on the total quantity of orders submitted.

The solution is obtained by conjecturing that the market maker uses a linear rule

$$p = \lambda y + \mu \qquad (5.33)$$

for some coefficients λ, μ. This seems like a reasonable conjecture given the normally distributed shocks in the system.

The informed trader chooses $x(v)$ to maximize expected profit

$$\begin{aligned} E(\pi|x) &= E(x(v-p)|x) \\ &= E(x(v-\lambda y)|x) - x\mu \\ &= x(v-\mu) - \lambda x^2, \end{aligned}$$

using (5.32) and (5.33) and the fact that $E(u|x) = 0$. The first order condition guarantees that $x = (v - \mu)/2\lambda$, which can be written as

$$x = \alpha + \beta v \quad ; \quad \alpha = \frac{-\mu}{2\lambda} \quad ; \quad \beta = \frac{1}{2\lambda}. \qquad (5.34)$$

So if the market maker uses a linear pricing rule in y then the informed trader uses a linear demand schedule in v.

We determine λ, μ from the constraint that the market maker sets $p = E(v|y)$, which ensures zero profits for him. From the properties of the bivariate normal distribution we have

$$E(v|y) = \mu_v + \frac{\beta(y - \alpha - \beta\mu_v)\sigma_v^2}{\sigma_u^2 + \beta^2\sigma_v^2}.$$

Substituting in the expressions for α, β we obtain:

$$\mu = \mu_v + \frac{-\alpha\beta\sigma_v^2 + \sigma_u^2\mu_v}{\sigma_u^2 + \beta^2\sigma_v^2} = \mu_v \quad ; \quad \lambda = \frac{\beta\sigma_v^2}{\sigma_u^2 + \beta^2\sigma_v^2} = \frac{1}{2\varphi}$$

$$p = \mu_v + \frac{1}{2\varphi}y \qquad (5.35)$$

$$x = -\mu_v\varphi + \varphi v, \qquad (5.36)$$

where $\varphi = \sqrt{\sigma_u^2/\sigma_v^2}$, which verifies the conjecture of the linear pricing rule (5.33). The informed trader and the market maker both care about the ratio σ_u^2/σ_v^2, which is the noise to signal ratio. The informed trader's demand schedule satisfies $\partial x/\partial \varphi = v - \mu$.

The expected profit of the informed trader is $(v - \mu_v)^2 \varphi/2 \geq 0$, which comes at the expense of the noise traders who lose the same amount of money. The market maker

5.5 Strategic Trade Models

breaks even in the long run. Half of the informed trader's information is impounded in price in the sense that

$$\text{var}(v|y) = \frac{\sigma_u^2 \sigma_v^2}{\sigma_u^2 + \beta^2 \sigma_v^2} = \frac{\sigma_v^2}{2}.$$

This model suggests the **Kyle's Lambda** (λ) as a measure of the **liquidity** of market or price impact. We see that λ is the amount that the market maker raises the price when the total order flow, $(u + x)$, goes up by 1 unit. Hence, the amount of order flow necessary to raise the price by $1 equals

$$1/\lambda = 2\varphi, \tag{5.37}$$

which is a measure of the depth of the market or market liquidity. The higher is the proportion of noise trading to the value of insider information, the deeper or more liquid is the market. Intuitively, the more noise traders there are, the less the market maker needs to adjust the price in response to a given order, since the likelihood of the order being that of a noise trader, rather than an insider, is greater. Although there is not an explicit bid ask spread in this model, one might think of $p(\varepsilon) - p(-\varepsilon)$ as the bid ask spread for small buy or sell quantities. We have

$$\lambda = \frac{dp}{dy} = \lim_{\varepsilon \to 0} \frac{p(\varepsilon) - p(-\varepsilon)}{\varepsilon}.$$

Measurement of Liquidity

Liquidity is an important concept but hard to define, and it means different things to different people. Indeed, O'Hara (1995) drew on an analogy with pornography: it is hard to define, but you know it when you see it. She went on to argue that a liquid market is one in which buyers and sellers can trade into and out of positions quickly and without having large price effects. There are a number of ways of measuring this type of liquidity from low frequency data suggested by this literature; see Goyenko, Holden, and Trzcinka (2009) for a review. A common approach is to estimate (5.33) by regressing price on order flow. Amihud (2002) develops a liquidity measure that captures the daily price response associated with one unit of trading volume. Suppose that we observe intraperiod returns R_{t_j} and trading volume V_{t_j}, where $j = 1, \ldots, n_t$ and n_t is the number of intraperiod returns. Then, let

$$Am_t = \frac{1}{n_t} \sum_{j=1}^{n_t} \ell_{t_j}, \quad \ell_{t_j} = \frac{|R_{t_j}|}{V_{t_j}}. \tag{5.38}$$

The idea is to average the daily value of ℓ_{t_j} over a longer period like a week or a month because the daily series ℓ_{t_j} can be quite noisy, especially when there are very quiet days with little trading volume. Large values of Am_t indicate an illiquid market where small amounts of volume can generate big price moves. It is considered a good proxy for the theoretically founded Kyle's price impact coefficient. In Figure 5.8 we show the raw

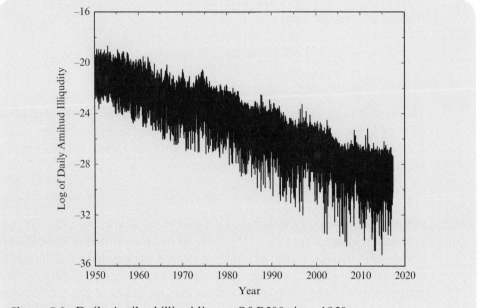

Figure 5.8 Daily Amihud illiquidity on S&P500 since 1950

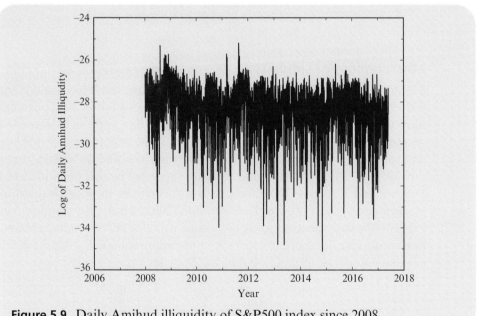

Figure 5.9 Daily Amihud illiquidity of S&P500 index since 2008

Amihud series (ℓ_{t_j}) for daily data on the S&P500 index. Actually, we show the log of the series (deleting all cases where the raw series is exactly zero). This series has shown a sustained downward trend indicating an improvement in liquidity on the S&P500, at least until around 2008 when this series seems to have lost its mojo; see Figure 5.9.

5.6 Electronic Markets

In the Kyle model the impact of an order is linear in volume and permanent, and there is a perfect correlation between order flow and prices. This seems a bit strong for empirical use. There are a number of models that allow for nonlinear impact of volume on prices and allow a distinction between permanent and transitory impact, i.e., allow a dynamic model.

5.6 Electronic Markets

Some markets (or segments of markets) still have dealers (e.g., small stocks on the LSE) who quote buy and sell prices with some manual element, and people who want to buy or sell either take it or leave it. Other markets are fully electronic, perhaps with an (electronic) auction phase to open and close the market and to reopen the market after a break, but otherwise trading takes place continuously through the limit order book with anonymity of buyers and sellers. In this case, the role of the dealer is carried out by multiple competing firms who **supply liquidity**, i.e., post limit orders to the market on one or both sides of the market and on multiple competing venues. In fact the larger **electronic market makers** such as Knight Capital (KCG) (a listed (NYSE) electronic market maker) are operating on many securities simultaneously, and so they are effectively able to diversify across stocks as well as across venues. They operate tight inventory control, i.e., they target zero inventory within the day (because overnight positions are subject to interest rates and margin payments etc.) and even within each hour. Stock market participants may use limit orders or market orders. In fact these are just broad categories; actually, there are many types of orders that traders may use such as: pegged orders, iceberg orders, fill-or-kill, hide not slide, intermarket sweep orders, etc. Harris (2003) is an excellent guide to the purpose and usage of different order types in trading strategies. In Table 5.2 below we show a typical limit order book. The bid–ask spread is 0.01, which is the difference between the best bid price and the best ask price. The term **the touch** is sometimes used to describe the spread between the best bid and ask prices. The quantities displayed at different prices could come from one limit order or multiple limit orders and we have aggregated across the unitary orders to obtain total quantity at the stated price.

This information is available to participants who have access to the market. Market orders or marketable limit orders (that cross the spread) will execute against **the book**. Suppose there arrives a market buy order for 15000. The first 7000 shares get executed at 15.72, the next 3000 at 15.73, the next 4000 at 15.74, and the last 1000 at 15.75. This order was so large that it had to walk down the book. Notice that there is not a single price for this transaction, but several. In this case we may speak about the **volume weighted average price** (VWAP) of the order, which in this case is

$$VWAP = \frac{7}{15} \times 15.72 + \frac{3}{15} \times 15.73 + \frac{4}{15} \times 15.74 + \frac{1}{15} \times 15.75 = 15.729.$$

After this transaction (or these transactions) has concluded, the new order book is as follows.

Table 5.2a Limit order book example

Bid		Ask	
Price	Quantity	Price	Quantity
15.71	2000	15.72	7000
15.70	4500	15.73	3000
15.69	5000	15.74	4000
15.68	10000	15.75	2000
15.67	15000	15.76	12000

Table 5.2b Limit order book example

Bid		Ask	
Price	Quantity	Price	Quantity
15.71	2000	15.75	1000
15.70	4500	15.76	12000
15.69	5000		
15.68	10000		
15.67	15000		

The bid–ask spread is now 0.04. This is called **market impact**, because the price was moved by the trade.

If the market buy order were for a smaller quantity 2000, then the order would be executed at $15.72. In this case the best ask price would not change; only the quantity available at that price would reduce. Whose limit orders are filled? The order executes by **price-time priority** – the limit order at the best price that was placed first chronologically gets filled first and then the second etc. In fact, there is often an additional complication due to so-called iceberg orders and hidden orders, which are *ceteris paribus* given lower priority than orders that are made visible at the same price. Frequently, hidden orders are placed between the touch, so that an incoming market order would execute against them first.

How do electronic markets compare with dealer markets?

5.6.1 Measuring Market Quality

There are some obvious comparisons to make between how markets work now and how they worked ten or twenty years ago. First, the average size of each transaction has decreased with the advent of electronic trading. Second, the total number

5.6 Electronic Markets

of transactions has increased for individual stocks over time. Third, the number of stocks traded has increased and the number of synthetic instruments like ETFs that mimic indexes has increased. Fourth, the average holding period per share has decreased considerably (this is partly due to portfolio rebalancing: for example ETFs require daily rebalancing; HFTs constantly try to drive their inventory to zero). These background facts are not perhaps of first order importance when it comes to evaluating trading costs. We next discuss how to measure the bid–ask spread in electronic markets.

In the current market structure, the quoted bid ask spreads may not properly reflect the transaction cost or price impact faced by investors because:

1. Quotes are located on many different exchanges in different physical locations, and frequently cancelled and updated. The National Market System (NMS) of the US stock market integrates the top of the book across venues to give the National Best Bid and Offer (NBBO).
2. Transactions could occur within the spread because of hidden liquidity.
3. Even moderate sized transactions could have to walk down the book (take worse prices than at the touch) because the volume of liquidity at the touch is small nowadays.

The desire for spreads that better reflect economic costs has led to the concept of **effective and realized spreads**, which make use of the order book and the transaction record, so are grounded in real transaction events. The SEC mandates that trading venues in the US report and disseminate these measures monthly through their Rule 605 so that investors may gauge the quality of the marketplace they host. This rule states:

> The term average effective spread shall mean the share-weighted average of effective spreads for order executions calculated, for buy orders, as double the amount of difference between the execution price and the midpoint of the consolidated best bid and offer at the time of order receipt and, for sell orders, as double the amount of difference between the midpoint of the consolidated best bid and offer at the time of order receipt and the execution price.

The effective spread measures the transaction cost in dollars for market orders. The same rule specifies:

> The term average realized spread shall mean the share-weighted average of realized spreads for order executions calculated, for buy orders, as double the amount of difference between the execution price and the midpoint of the consolidated best bid and offer five minutes after the time of order execution and, for sell orders, as double the amount of difference between the execution price and the midpoint of the consolidated best bid and offer five minutes after the time of order execution.

This impounds price movements after the trade (including the price impact due to the information in the trade). This cost can be interpreted as the profit realized by the other side of the trade, assuming they could lay off the position at the new quote midpoint. The explicit use of five minutes in this definition seems like a hostage to fortune, and Bandi, Lian, and Russell (2012) have investigated whether alternative time delay factors might improve measurement. The realized spread and effective spread can be combined to measure the **price impact**, specifically:

$$Priceimpact = (effectivespread - realisedspread)/2. \quad (5.39)$$

The **depth** of the market is another measure of market quality or liquidity. Depth can be measured by the amount of visible volume at **the touch**, that is, the sum of visible buy and sell orders at the best prices. One can also include volume close to the touch or weight the volume according to how far it is from the touch. One issue with this is the presence of hidden liquidity due to **iceberg** orders on lit venues and orders that sit inside dark pools.

There is some evidence that effective and realized spreads have improved over the last ten years and certainly over the last twenty years. Depth is harder to gauge. Perhaps visible depth at the touch has decreased; see Foresight (2012) for more discussion.

5.6.2 Flash Crashes

There is some evidence that information is much faster impacted into prices in electronic markets. In Figure 5.10 above we show minute by minute returns on the FTSE100 index over a period which includes a positive US nonfarm payroll day in 2012. The price level seems to jump up instantaneously to reflect the good news that was announced on that day.

On the other hand a feature of current markets that is not so attractive is rapid movements in price, sometimes called **flash crashes**, that do not appear to be related to fundamentals. The best known such event occurred in the US stock market on May 6th, 2010. In Figure 5.11 we show the trajectory of the E-mini futures price in the hour containing the peak declines and rise. The lower panel shows the price changes between consecutive transactions in terms of ticks (5.20).

The price level dropped nearly 10% in under 5 minutes only to rebound. The market drop transmitted from the futures market to the stock market in New York with similar drops across the board and a collapse of the market structure. The SEC/CFTC report on the 2010 Flash Crash provided some possible explanations for the initiation and promulgation of the flash crash in terms of the activities of algorithmic traders and high frequency traders. They argued that a starting point was a large parent sell order by the asset management firm Waddell and Reed for 75,000 E-mini contracts that was fed into the market by a price insensitive algorithm but tuned according to the volume found in the market. This lead to a nonlinear feedback loop (Shin (2010)), which is described in Figure 5.12. At some point, the high frequency traders "withdrew liquidity" and the market crashed.

5.6 Electronic Markets

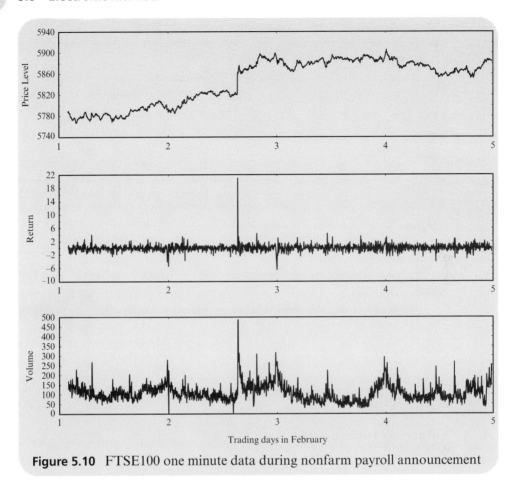

Figure 5.10 FTSE100 one minute data during nonfarm payroll announcement

There have been a number of other recent technology related disasters and crashes including: the Facebook IPO (#Faceplant) May 18, 2012 (the opening on NASDAQ was delayed by 30 minutes); the BATS IPO (the stock opened at $15.5 but traded down to a penny in 1.4 seconds); and the Google (#Pending Larry) mistaken early earnings announcement in 2012 (price went down 10% in 8 minutes). There was also the so-called **Knightmare on Wall Street** on August 1, 2012: Knight Capital was a market maker and HFT firm, listed on the NYSE. A trading error caused widespread disruption on NYSE, and the firm lost $450m in a few minutes. Subsequently, they were bought out by another firm. The so-called **Hash Crash** occurred on April 23rd, 2013. There was a hack of a Reuters twitter account and a story tweeted about a bomb at the White House, which resulted in a rapid drop in the Dow Jones index, and subsequent rapid recovery when the hack was uncovered. The **Treasury Flash Rally** of October 15, 2014 was a very big event without apparent macroeconomic cause.

Flash crashes are short and relatively deep events that are not apparently driven by fundamentals, or rather the movement in prices is in excess of what would be warranted based on fundamentals, according to hindsight anyway. Unlike some other

Chapter 5 Empirical Market Microstructure

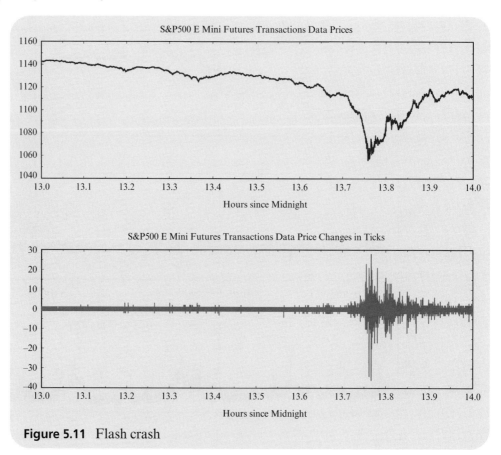

Figure 5.11 Flash crash

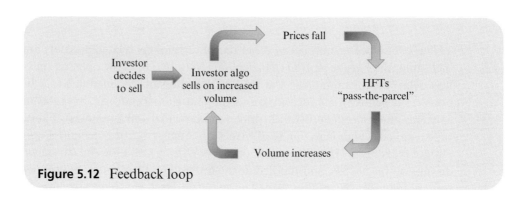

Figure 5.12 Feedback loop

market crashes (e.g., 1929 and 1987), one may not easily identify a prior period where the market was dominated by bubbles. Also, in many cases they are not contagioned globally, unlike say the 1929 and 1987 crashes, which were worldwide phenomena. The fact they invariably end in a rebound as fast as the crash could be interpreted

5.6 Electronic Markets

Table 5.3 Round trip speed of the LSE

System	Implementation	Latency 10^{-6}
SETS	<2000	600000
SETS1	Nov 2001	250000
SETS2	Jan 2003	100000
SETS3	Oct 2005	55000
TradElect	June 18, 2007	15000
TradElect 2	October 31, 2007	11000
TradElect 3	September 1, 2008	6000
TradElect 4	May 2, 2009	5000
TradElect 4.1	July 20, 2009	3700
TradElect 5	March 20, 2010	3000
Millenium	February 14, 2011	113

as a sign of market resilience. Funny stuff can happen at the quantum level but one doesn't need to be concerned about that when investing for the long term.

5.6.3 Speed

Modern markets are very fast. Since this speed is the subject of much criticism, we review its role in financial markets. In fact, speed has always mattered. A well known example is the legend of Nathan Mayer Rothschild profiting from news of the defeat of Napoleon Bonaparte at the battle of Waterloo. He had news about the outcome of the battle ahead of the British government (the fantasy version has it that he received the information by carrier pigeons, other versions say by personal couriers). There are also two versions of how he made money. In the first, he just bought bonds on news of victory, while the second version of events has that he **spoofed** the market by himself publicly selling bonds while having his agents buy them and keep on buying as the prices dropped. In any case, he made lots of money from having the key information 24 hours before other participants. He was the high frequency trader of his day.

Table 5.3 illustrates the evolution of speed in the last 20 years as measured by the **system latency** of the LSE's matching engine. This has roughly followed **Moore's law** with the end result that the trading system itself can handle many more messages and deliver execution very much faster than it could prior to 2000 and of course much faster than any human intermediated system as was common prior to the 1980s. For comparison, the time for light to travel round trip from London to New York is around 37200 microseconds.

This shows that the trading infrastructure has become faster or alternatively the costs of delivering a given speed have declined. The benefits of speed are small but not zero. Posting limit orders give options to trade to other traders, since they have the right but not the obligation to execute against you; Copeland and Galai (1983). The Black and Scholes (1973) call option price can be used to value the option. In a standard notation, the price of the call option C in terms of the underlying price S is

$$C(S, K, \tau, r_f, \sigma) = S \times \Phi(d_+) - K \times e^{-r_f \tau} \times \Phi(d_-), \qquad (5.40)$$

$$d_\pm = \frac{\log \frac{S}{K} + \left(r_f \pm \frac{\sigma^2}{2}\right) \times \tau}{\sigma \times \sqrt{\tau}},$$

where Φ is the standard normal c.d.f., while r_f is the interest rate, τ is the time to maturity, K is the strike price, and σ is volatility. A competitive limit order can be considered as **at the money**, i.e., $S = K$; furthermore, intraday interest rates r_f are zero. Letting $\tau \to 0$, we obtain the approximation

$$\frac{C(S, K, \tau, r_f, \sigma)}{S} = \frac{1}{\sqrt{2\pi}} \sigma \sqrt{\tau} + O(\tau^{3/2}). \qquad (5.41)$$

This says that there is a positive albeit small value in a single order that sits for a small time (if there are many orders, then the total value being given away can be large). Also, (5.41) says that the value being given away increases with volatility measured by the parameter σ, so that in times of market stress the value per unit time increases. The posters of limit orders should be compensated for their service to the market by, for example, the bid–ask spread. The faster the limit order poster is at updating his quotes, the less compensation he would require for posting them in the first place. Note that the value in (5.41) scales with the square root of time: millisecond to microsecond value shrinks by 1/30 not by 1/1000.

The classical Glosten and Milgrom (1985) microstructure model of a dealer market explains the bid–ask spread in terms of adverse selection. The dealer is uninformed, and he posts limit orders for informed and uninformed traders who arrive randomly. In this model if the dealer can update quotes in response to new information faster than the informed trader can act, then he will be able to set narrower spreads than if he is slow and keeps stale quotes on the table. On the other hand, if the informed trader is faster than the dealer, then the dealer will have to set wider spreads to protect himself. In the classical dealer market, the dealer alone knew the order flow and they alone set prices; investors had to take it or leave it. They earned a profit from providing the service of immediacy. In the new system, that advantage has been eliminated, and without some advantage the human and physical capital that would have gone to the market making activity would find better employ elsewhere. Speed of action and superior information about order flow from datafeeds are the advantage that the new market makers seek; Menkveld (2013).

5.6 Electronic Markets

The classic inventory models, e.g., Ho and Stoll (1981, 1983), explain the spread as the cost of providing immediacy to impatient investors. In their models, the spread varies positively with the degree of the monopoly power of the dealer, the volatility of the asset price, the trade size, and the horizon of the dealer. This class of models predicts that competition between dealers would lead to small spreads. It also predicts that small sized orders would require small spreads whereas larger orders would face wider spreads, *ceteris paribus*. Also, if the dealer has a short horizon, then spreads should be narrow. All these suggest that dealers who can quote and cancel faster will benefit the market. Aït-Sahalia and Saglam (2013) extend this literature to analyze the consequences for liquidity provision of competing market makers operating at high frequency. They find that competition increases overall liquidity and deters the fast market maker's use of order flow signals. They show that the market maker provides more liquidity as he gets faster but shies away as volatility increases.

Maureen O'Hara is Purcell Professor of Finance at the Johnson Graduate School of Management, Cornell University and also Professor of Finance at UTS (Sydney). A citizen of both Ireland and the US, she received her doctorate from Northwestern University and Honorary Doctorates from Facultés Universitaires Catholiques à Mons (FUCAM), Universität Bern, and University College Dublin. Professor O'Hara is an expert on market microstructure, and she publishes widely in banking and financial intermediaries, law and finance, and experimental economics. She is the author of numerous journal articles as well as the books *Market Microstructure Theory* (Blackwell, 1995) and *High Frequency Trading: New Realities for Traders, Markets, and Regulators* (Risk Books, 2013). Her most recent book, *Something for Nothing: Arbitrage and Ethics on Wall Street*, was published in 2016 by Norton Books. A past President of the American Finance Association, the Western Finance Association, and the Financial Management Association, she was Executive Editor of the Review of Financial Studies. Professor O'Hara has served on a variety of corporate boards including NewStar Financial, Teachers Insurance and Annuity Association (TIAA), and Investment Technology Group, Inc. (ITG), where she was Chairwoman of the Board. She was named in Institutional Investors Trading Technology Top 40 and she is currently an Advisor to Symbiont, a company focussing on blockchain and smart securities. A member of the CFTC–SEC Emerging Regulatory Issues Task Force (the "flash crash" committee), she has also served on the Global Advisory Board of the Securities Exchange Board of India (SEBI), the Advisory Board of the Office of Financial Research, US Treasury, and the SEC Equity Market Structure Advisory Committee. She has published extensively on the topic of high frequency trading and its impacts, including its role in flash crashes.

5.7 Summary of Chapter

We considered some of the models developed by economists to describe observed high frequency intraday prices: the stale price model, the Roll model, the Kyle model, and the model of asymmetric information. We discussed recent developments in financial markets.

5.8 Appendix

5.8.1 Non-trading Model

We derive some additional properties of the observed returns in the non-trading model. The characteristic function of observed r_t^O is

$$\varphi_O(u) = E\left(\exp(iur_t^O)\right)$$

$$= \pi + (1-\pi)E\left(E\left(\exp\left(iu\sum_{k=0}^{d_t} r_{t-k} \mid d_t\right)\right)\right)$$

$$= \pi + (1-\pi)E\left[\varphi(u)^{d_t+1}\right]$$

$$= \pi + (1-\pi)^2 \sum_{k=0}^{\infty} \pi^k \varphi(u)^{k+1}$$

$$= \pi + \frac{(1-\pi)^2 \varphi(u)}{1-\pi\varphi(u)},$$

where $\varphi(u) = E(\exp(iur_t))$ is the characteristic function of true returns.

From this we can obtain the cumulants of r_t^O by differentiating at $u=0$. For example,

$$\varphi_O'(u) = (1-\pi)^2 \varphi'(u) (1-\pi\varphi(u))^{-1} + (1-\pi)^2 \varphi(u) (1-\pi\varphi(u))^{-2} \pi \varphi'(u)$$

for $u=0$, $\varphi_O'(u) = \varphi'(0) = \mu$. Furthermore

$$\varphi_O''(u) = (1-\pi)^2 \varphi''(u) (1-\pi\varphi(u))^{-1} + (1-\pi)^2 \varphi'(u) (1-\pi\varphi(u))^{-2} \pi \varphi'(u)$$
$$+ (1-\pi)^2 \varphi'(u) (1-\pi\varphi(u))^{-2} \pi \varphi'(u)$$
$$+ 2(1-\pi)^2 \varphi(u) (1-\pi\varphi(u))^{-3} \pi^2 \varphi'(u)^2$$
$$+ (1-\pi)^2 \varphi(u) (1-\pi\varphi(u))^{-2} \pi \varphi''(u),$$

which has $\varphi_O''(u) = \varphi''(0) + 2(\pi + (1-\pi)^{-1}\pi^2)\varphi'(0)^2 = \varphi''(0) + 2\pi (1-\pi)^{-1} \varphi'(0)^2$, which is exactly as shown in (5.10).

5.8 Appendix

Furthermore, we have

$$\varphi_O'''(u) = (1-\pi)^2 \varphi'''(u)(1-\pi\varphi(u))^{-1} + (1-\pi)^2 \varphi''(u)(1-\pi\varphi(u))^{-2}\pi\varphi'(u)$$
$$+ (1-\pi)^2 \varphi''(u)(1-\pi\varphi(u))^{-2}\pi\varphi'(u)$$
$$+ 2(1-\pi)^2 \varphi'(u)(1-\pi\varphi(u))^{-3}\pi^2\varphi'(u)^2$$
$$+ (1-\pi)^2 \varphi'(u)(1-\pi\varphi(u))^{-2}\pi\varphi''(u)$$
$$+ (1-\pi)^2 \varphi''(u)(1-\pi\varphi(u))^{-2}\pi\varphi'(u)$$
$$+ 2(1-\pi)^2 \varphi'(u)(1-\pi\varphi(u))^{-3}\pi^2\varphi'(u)^2$$
$$+ (1-\pi)^2 \varphi'(u)(1-\pi\varphi(u))^{-2}\pi\varphi''(u)$$
$$+ 2(1-\pi)^2 \varphi'(u)(1-\pi\varphi(u))^{-3}\pi^2\varphi'(u)^2$$
$$+ 6(1-\pi)^2 \varphi(u)(1-\pi\varphi(u))^{-4}\pi^3\varphi'(u)^3$$
$$+ 4(1-\pi)^2 \varphi(u)(1-\pi\varphi(u))^{-3}\pi^2\varphi'(u)\varphi''(u)$$
$$+ (1-\pi)^2 \varphi'(u)(1-\pi\varphi(u))^{-2}\pi\varphi''(u)$$
$$+ 2(1-\pi)^2 \varphi(u)(1-\pi\varphi(u))^{-3}\pi^2\varphi''(u)\varphi'(u)$$
$$+ (1-\pi)^2 \varphi(u)(1-\pi\varphi(u))^{-2}\pi\varphi'''(u),$$

which has

$$\varphi_O'''(0) = \varphi'''(0) + \left[6\pi + 4(1-\pi)^{-1}\pi^2 + 2(1-\pi)^{-1}\pi^2\right]\varphi''(0)\varphi'(0)$$
$$+ \left[2(1-\pi)^{-1}\pi^2 + 2(1-\pi)^{-1}\pi^2 + 6(1-\pi)^{-2}\pi^3 + 2(1-\pi)^{-1}\pi^2\right]\varphi'(0)^3$$
$$= \varphi'''(0) + \frac{6\pi}{1-\pi}\varphi''(0)\varphi'(0) + \frac{6\pi^2}{(1-\pi)^2}\varphi'(0)^3$$
$$= \kappa_3 + \frac{6\pi\mu}{(1-\pi)^2}\left((1-\pi)\sigma^2 + \pi\mu^2\right),$$

where $\kappa_3 = \varphi'''(0)$ is the skewness of true returns. Therefore, when $\mu > 0$, the non-trading process induces a positive skewness to observed returns, which may not be surprising.

We show the result that

$$E\left(r_{it}^O r_{jt}^O\right) = \sigma_{ij}\frac{(1-\pi_i)(1-\pi_j)}{1-\pi_i\pi_j}. \tag{5.42}$$

We have

$$E\left(r_{it}^O r_{jt}^O\right) = (1-\pi_i)(1-\pi_j)E\left(\left(\sum_{l=0}^{d_{jt}} r_{jt-l}\right)\left(\sum_{l=0}^{d_{it}} r_{it-l}\right)\right)$$
$$= (1-\pi_i)(1-\pi_j)\sigma_{ij}E\left(\min\{d_{jt}+1, d_{it}+1\}\right)$$
$$= (1-\pi_i)(1-\pi_j)\frac{\sigma_{ij}}{2}\left(2 + E\left(d_{jt}+d_{it}\right) - E\left(|d_{jt}-d_{it}|\right)\right),$$

where we have used $\min\{x+1, y+1\} = 1 + \min\{x, y\}$ and $2\min\{x, y\} = (x + y) - |x - y|$. We have

$$E(|d_{jt} - d_{it}|) = (1 - \pi_i)(1 - \pi_j) \sum_{k_j=0}^{\infty} \sum_{k_i=0}^{\infty} \pi_i^{k_i} \pi_j^{k_j} |k_j - k_i|$$

$$= (1 - \pi_i)(1 - \pi_j) \left[\sum_{k_i=0}^{\infty} \sum_{k_j=k_i+1}^{\infty} \pi_i^{k_i} \pi_j^{k_j} (k_j - k_i) + \sum_{k_j=0}^{\infty} \sum_{k_i=k_j+1}^{\infty} \pi_i^{k_i} \pi_j^{k_j} (k_i - k_j) \right]$$

$$= (1 - \pi_i)(1 - \pi_j) \left[\sum_{k_i=0}^{\infty} \sum_{\Delta=1}^{\infty} \pi_i^{k_i} \pi_j^{k_i+\Delta} \Delta + \sum_{k_j=0}^{\infty} \sum_{\Delta=1}^{\infty} \pi_j^{k_j} \pi_i^{k_j+\Delta} \Delta \right]$$

$$= (1 - \pi_i)(1 - \pi_j) \sum_{k_i=0}^{\infty} (\pi_i \pi_j)^{k_i} \left[\sum_{\Delta=1}^{\infty} \pi_j^{\Delta} \Delta + \sum_{\Delta=1}^{\infty} \pi_i^{\Delta} \Delta \right]$$

$$= \frac{(1 - \pi_i)(1 - \pi_j)}{1 - \pi_i \pi_j} \left[\frac{\pi_j}{(1 - \pi_j)^2} + \frac{\pi_i}{(1 - \pi_i)^2} \right].$$

$$E\left(r_{it}^O r_{jt}^O\right) = (1 - \pi_i)(1 - \pi_j) \frac{\sigma_{ij}}{2} \left(2 + \frac{\pi_i}{1 - \pi_i} + \frac{\pi_j}{1 - \pi_j} - \frac{(1 - \pi_i)(1 - \pi_j)}{1 - \pi_i \pi_j} \right.$$

$$\left. \times \left[\frac{\pi_j}{(1 - \pi_j)^2} + \frac{\pi_i}{(1 - \pi_i)^2} \right] \right)$$

$$= \frac{\sigma_{ij}}{2} \left(2(1 - \pi_i)(1 - \pi_j) + (1 - \pi_j)\pi_i + \pi_j(1 - \pi_i) \right.$$

$$\left. - \frac{(1 - \pi_i)^2 (1 - \pi_j)^2}{1 - \pi_i \pi_j} \left(\frac{\pi_j}{(1 - \pi_j)^2} + \frac{\pi_i}{(1 - \pi_i)^2} \right) \right)$$

$$= \sigma_{ij} \frac{(1 - \pi_i)(1 - \pi_j)}{1 - \pi_i \pi_j}.$$

We give an alternative way of calculating (5.8). Define

$$I_k(\pi) = 1 + \pi + 2^k \pi^2 + 3^k \pi^3 + \cdots = \sum_{j=0}^{\infty} \pi^j j^k.$$

We have

$$I_{k+1}(\pi) + I_k(\pi) = \sum_{j=0}^{\infty} (j+1) j^k \pi^j = \frac{\partial}{\partial \pi} \sum_{j=0}^{\infty} j^k \pi^{j+1} = \frac{\partial}{\partial \pi} \pi \sum_{j=0}^{\infty} j^k \pi^j = I_k(\pi) + \pi I'_k(\pi).$$

Therefore, $I_{k+1}(\pi) = \pi I'_k(\pi)$. We have: $I_0(\pi) = \frac{1}{1-\pi}$, $I_1(\pi) = \frac{\pi}{(1-\pi)^2}$, $I_2(\pi) = \frac{\pi}{(1-\pi)^2} + \frac{2\pi^2}{(1-\pi)^3} = \frac{\pi(1+\pi)}{(1-\pi)^3}$, etc. Therefore $I_1(\pi) + I_0(\pi) = \frac{1}{(1-\pi)^2}$ and $I_2(\pi) + 2I_1(\pi) + I_0(\pi) = \frac{1+\pi}{(1-\pi)^3}$. Therefore,

$$E(d_t) = (1 - \pi)\pi \left[I_1(\pi) + I_0(\pi) \right] = (1 - \pi)\pi \left(\frac{1}{1 - \pi} + \frac{\pi}{(1 - \pi)^2} \right) = \frac{\pi}{1 - \pi}$$

5.8 Appendix

$$E(d_t^2) = (1-\pi)\pi \sum_{j=0}^{\infty} (j+1)^2 \pi^j = (1-\pi)\pi \left[I_2(\pi) + 2I_1(\pi) + I_0(\pi) \right]$$

$$= (1-\pi)\pi \left(\frac{\pi(1+\pi)}{(1-\pi)^3} + 2\frac{\pi}{(1-\pi)^2} + \frac{1}{1-\pi} \right)$$

$$= \pi \left(\frac{\pi(1+\pi) + 2\pi(1-\pi) + (1-\pi)^2}{(1-\pi)^2} \right)$$

$$= \pi \frac{(1+\pi)}{(1-\pi)^2}.$$

Non-trading Model with Serially Correlated True Returns

Suppose that

$$r_t = \varepsilon_t - \theta \varepsilon_{t-1},$$

where ε_t is i.i.d. $N(0, \sigma_\varepsilon^2)$. That is, the true returns are mean zero serially correlated, which is incompatible with the EMH but is consistent with various behavioral theories of stock markets. Then we may write $r_t = (1 - \theta L)\varepsilon_t$. Suppose that observed returns are generated exactly as before. In this case, conditional on d_t we may write

$$\sum_{k=0}^{d_t} r_{t-k} = \left(1 + \cdots + L^{d_t}\right)(1-\theta L)\varepsilon_t = \left(1 + (1-\theta)L + \cdots + (1-\theta)L^{d_t} - \theta L^{d_t+1}\right)\varepsilon_t.$$

In this case $E(r_t^O) = 0$. Furthermore, conditional on d_t,

$$\text{var}\left(\sum_{k=0}^{d_t} r_{t-k} | d_t \right) = (1 + d_t(1-\theta)^2 + \theta^2)\sigma_\varepsilon^2.$$

Therefore, the variance of observed returns is

$$\text{var}(r_t^O) = (1-\pi)\sigma_\varepsilon^2 \left(1 + \theta^2 - 2\pi\theta\right) \leq \text{var}(r_t) = \sigma_\varepsilon^2(1+\theta^2)$$

when $\theta > 0$. We also show

$$E\left(r_t^O r_{t+1}^O\right) = (1-\pi)^2 E\left(r_{t+1} \sum_{i=0}^{d_t} r_{t-i}\right)$$

$$= (1-\pi)^2 E(r_{t+1} r_t)$$

$$= -\sigma_\varepsilon^2 (1-\pi)^2 \theta.$$

The autocovariance of the observed return series is less than the autocovariance of the true return series. In fact

$$\rho^O(1) = -\frac{(1-\pi)^2 \theta}{(1+\theta^2 - 2\pi\theta)}.$$

The ratio of the autocorrelation of the observed series to the underlying series is given by

$$\frac{(1-\pi)^2(1+\theta^2)}{(1+\theta^2-2\pi\theta)} = \frac{(1-\pi)^2}{1-\frac{2\pi\theta}{1+\theta^2}} \leq 1,$$

where the inequality is true for all $\theta \in [-1, 1]$ and $\pi \in [0, 1]$. In this case, non-trading can decrease the variance and the autocorrelation of the observed series. This is also true when there is a non-zero mean. In practice, how does one tell whether the observed autocorrelation is coming from an autocorrelated series or not?

6 Event Study Analysis

A common objective in social science is to try to measure the impact of an event E on an outcome variable Y. This may be a policy event or some corporate decision, and the outcome variable may measure some attribute under question. Specifically, we may want to compare the outcomes before and after the event, which we denote symbolically $Y_{after} - Y_{before}$. Unfortunately, in many cases this approach may lead to biased inferences because there may be many things happening at the same time as the putative event that affect the outcomes. A general strategy for dealing with this issue is to use a **control group** against which to measure the change in outcomes. Thus we divide into treated and control and compute the **diff in diffs**,

$$\left(Y^{treated}_{after} - Y^{treated}_{before}\right) - \left(Y^{control}_{after} - Y^{control}_{before}\right); \tag{6.1}$$

see Imbens and Wooldridge (2009). We look at how the treated group changed in comparison with how the control group changed. This is a general principle applied in many fields from labor economics to medicine etc. The questions that need to be addressed are: how is Y measured; what is the treatment and control group; and what is before and after? The terminology originates from medical experiments in which one group of patients receives a drug and a control group does not. Assignment to the treatment and control groups is random. Extending the analogy, a treatment can be any random intervention into the life of an economic agent. Economics and finance are, for the most part, not experimental sciences. We therefore look for instances in which history provides random assignment. These instances are called **natural experiments**. In finance researchers have been using a standard methodology to evaluate the effects of firm specific events on stock returns based on the so-called **market model** since the 1960s. This creates a control group by regression extrapolation. We review the standard approach and show how it can be adapted to deal with common events, heteroskedasticity, and other nuisances. We also consider approaches designed for other outcome variables, which have arisen from the microeconometric and biostatistical literatures.

6.1 Some Applications

We first discuss some particular financial applications. There are many firm specific events whose effects are in contention such as: stock splits; reverse splits; share repurchase; mergers and acquisitions; earnings announcements; seasoned equity offering; inclusion in stock index; short selling restrictions; insider trading; circuit breaker;

and macroeconomic announcements. Often, the main outcome of interest is the valuation of the firm in question, and whether the response to the event is consistent with the EMH. In other cases we may be interested in outcome variables like volatility and bid–ask spreads to capture whether a market is functioning well. In some cases hypothesis testing is central to the activity whereas in other cases quantifying the effect, i.e., estimation, is of interest.

Example 36 *Stock Splits. Firm splits stock 2:1 means that it doubles the number of shares (allocating them pro rata to original owners) and halfs the price level. Rational expectations theory says that workers and consumers correctly calculate the payoffs they face and act on them correctly. According to the present value calculations, the stock split should have no effect on the valuation of the firm and on the return on holding the stock.*

Example 37 *Earnings example. Under strong-form efficiency, earnings announcements have no effect on the firm's stock price. Under semi-strong form, the anticipated part of the announcement should have no effect on stock prices, but the unanticipated part of the announcement should have an immediate effect. The difficulty here is in defining what is anticipated and what is new news. There are various forecasts around that can be used to benchmark this. One may define the earnings surprise as the announced earnings minus the previous day's Institutional Brokers' Estimate System(I/B/E/S) mean forecast: "good news" if earnings surprise $> .025 \times I/B/E/S$ mean, and "bad news" if earnings surprise $< -.025 \times I/B/E/S$ mean. https://financial.thomsonreuters.com/en/products/data-analytics/company-data/ibes-estimates.html.*

Example 38 *In August 2016 the Bank of England's (BoE) Monetary Policy Committee voted to introduce a Corporate Bond Purchase Scheme (CBPS), which mandated the BoE to purchase up to £10 billion of sterling investment grade, non financial corporate bonds over a period of 18 months. The purpose of the CBPS was to impart monetary stimulus by: (i) lowering the yields on corporate bonds, thereby reducing the cost of borrowing for companies, (ii) triggering portfolio rebalancing, and (iii) by stimulating new issuance. The question is what was the effect of this scheme on the liquidity, volatility, and other market outcomes for the individual bonds included in the scheme.*

6.2 Basic Structure of an Event Study

The basic structure of an event study is laid out in Fama, Fisher, Jensen, and Roll (1969) and Brown and Warner (1980). The approach taken will depend on whether the event is firm specific, like a stock split, or whether it is a common effect like a macroeconomics announcement. It may also depend on whether the focus is on a single event or on a generic class of events. However, many of the issues are common to both cases. We will require the following assumption in the sequel.

6.2 Basic Structure of an Event Study

Key assumption. *The event occurrence is exogenous to earlier changes in the outcome variable.*

This makes the event a natural experiment that can be used to identify its effect on the outcome variable. We will discuss the validity of this assumption later. We will work under the null hypothesis that the event has no effect on the stock returns. We later consider which alternatives we can address.

6.2.1 Event Definition

There are many practical questions around the definition of the event and how it unfolds in time and how we are to measure the effect of interest. The usual issues that need to be addressed are:

1. *Sampling interval of the data.* This choice determines the purpose of the study. Typically daily, weekly, or monthly have been used, but nowadays there are a number of high frequency event studies that use intraday data. If the sampling frequency is too low, one may miss the fine structure of the event effect and suffer from endogeneity problems. This has become a substantial issue now, since market reactions can be so fast that a daily sampling interval may already be too long in some cases.
2. *Event window.* This is chosen to define where the main effect of the event is likely to be found. It depends on the purpose of the study, and what one's view of the EMH is. A common choice might be $-1, 0, +1$ days. The Fama *et al.* study chose a very long event window of ± 30 months for specific reasons.
3. *Estimation window.* The risk adjustment or control group measurement requires parameter estimation and we need data to do that accurately. Ideally, one would like to take as long a sample as possible but there may be an issue about stationarity, which limits the useable horizon here. For example, $-250, \ldots, -21$ days is quite common. The Fama *et al.* study also used post event window observations in some cases.

6.2.2 The Null Hypothesis: The Market Model

The econometric issue here is to specify a model for the outcome variable that holds during the period of interest and under the null hypothesis of no effect. This model defines what is meant by normal; the alternative hypothesis will bring departures from this model that can be quantified. When the outcome of interest is stock returns, the so-called market model is a commonly accepted setting. This is a panel regression model for stock returns with a single observed common factor, the return on the market, i.e., a broad stock market index, and heterogeneous response to that factor.

Definition 68 *The **market model** relates asset i return to the market return in a linear fashion*

$$R_{it} = \alpha_i + \beta_i R_{mt} + \varepsilon_{it}, \tag{6.2}$$

where the error terms satisfy for all $T, t \neq s$

$$E(\varepsilon_{it}|R_{m1},\ldots,R_{mT}) = 0 \quad ; \quad E(\varepsilon_{it}\varepsilon_{is}|R_{m1},\ldots,R_{mT}) = \begin{cases} \sigma_{\varepsilon i}^2 & \text{if } t = s \\ 0 & \text{if } t \neq s \end{cases}. \quad (6.3)$$

These are precisely the standard assumptions made for linear regression for each i; see (2.3) and (2.5). For some arguments we may further assume that the errors are normally distributed.

Definition 69 *For each i, ε_{it} are normally distributed independently of market returns, that is*

$$(\varepsilon_{i1},\ldots,\varepsilon_{iT})^\intercal \sim N(0, \sigma_{\varepsilon i}^2 I_T). \quad (6.4)$$

The classical version of the market model derives from an assumption that asset returns are jointly normal, in which case one obtains linear regression relationships between all the return variables. Of course one may arrive at the same destination without recourse to joint normality. This model does not allow time series heteroskedasticity but does allow the **idiosyncratic error** variance to differ across i as seems reasonable. We give more discussion about these assumptions in Chapter 7.

This model allows for common components that are not affected by the event under consideration (or at least that is the assumption); controlling for these common components has the consequence of reducing the measured abnormal return variance in a reasonable way. Economic models such as the CAPM restrict $\alpha_i = 0$ (see Chapter 7), and one can impose these restrictions, although usual practice is not to do so. For short event windows, this adds little value since volatilities dominate unconditional means over short sample intervals. For long event windows, the CAPM is subject to known anomalies which can bias the findings. Latest research uses characteristic-based multifactor benchmarks (see Chapter 8) again without the intercept restrictions implied by economic theory.

In some cases, data is limited, and it is not feasible to use the market model. For example, in studying IPOs, one does not have any historical time series to fit the market model to. In this case it is common practice to set $\alpha_i = 0$ and $\beta_i = 1$, i.e., just use the market return as a proxy for the normal return. One could instead use a matching type approach (see below) where the return on similar firms or industry returns is used to provide a benchmark relative to which the abnormal returns are calculated.

6.2.3 The Alternative Hypothesis

The null hypothesis is that the data are generated by (6.2) throughout the whole period (estimation window plus event window). The simplest alternative of interest is that during the event window the mean of returns changes. That is, the mean $\mu_{is} = E(\varepsilon_{is})$ of

6.2 Basic Structure of an Event Study

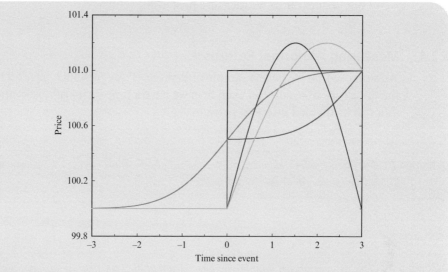

Figure 6.1 Shows examples of price trajectories around event at time 0

the residuals $\varepsilon_{is} = R_{is} - \alpha_i - \beta_i R_{ms}$ is not necessarily zero for all s in the event window. A general hypothesis would allow μ_{is} to vary with s in the event window and with the stock i.

We consider some different ways that outcomes could adjust to an event. First, there could be an immediate effect or a gradual effect. For an immediate effect we should see the price immediately jump to a new level that reflects the event. In that case, only one return is large and the rest should be small, i.e., consistent with the null distribution. If the effect is gradual then the price may take several periods to adjust to the new level, and one may see several returns that are anomalous. In practice the choices made about event window and sampling interval discussed above are related to what we take to be immediate or gradual. Second, the effect of the event on prices could be **Permanent** or it could be **Temporary**, that is, prices could increase permanently consistent with a positive total return over the specified period, or they could increase and subsequently decrease back to base camp, which would be consistent with positive returns followed by negative returns with zero effect on returns overall. Third, we may consider the trajectory of dynamic adjustment. There could be **underreaction**, i.e., gradual adjustment of prices to a new level, or **overreaction**, rapid adjustment that overshoots the new level and then returns to it. In Figure 6.1 we show examples of price trajectories around the event time.

Example 39 *Dubow and Monteiro (2006) develop a measure of "market cleanliness." The measure of market cleanliness was based on the extent to which "informed price movements" were observed ahead of "significant" (i.e. price-sensitive) regulatory announcements made by issuers to the market. These price movements could indicate insider trading. In that case they were looking for movements of prices before the announcement date 0.*

We discuss next how we measure post event returns.

6.2.4 Measuring Abnormal Returns

We suppose that we have a sample of firms $i = 1, \ldots, n$, which have all been subject to the same type of event. For each firm we have a time series of observations on returns that can be divided into two periods.

Definition 70 *The event window (denoted $\mathcal{E}_i$) consists of observations $t = t_{0i} - k_i, \ldots, t_{0i} + k_i$, and let $\tau_i = 2k_i + 1$ be the length of the event window.*

Definition 71 *The estimation or sample window (denoted $\mathcal{S}_i$) consists of observation points $t = t_{0i} - h_i, t_{0i} - 1, \ldots, t_{0i} - h_i - T_i + 1$, where $h_i > k_i$.*

We define our origins (or event dates) to be $t_{01}, \ldots, t_{0n}$, and the estimation window and event windows are defined relative to that. The estimation window may be quarantined from the event window, when $h_i > k_i$, which can be desirable in some cases.

In practice it is quite common to choose the same sized estimation window (although Fama et al. (1969) allowed for different sized windows) and the same size event window across firms. In this case, the only difference between the firms is that the chronological time when the event occurs may differ across firms. For notational convenience we drop the i subscript on T_i, h_i, k_i, τ_i in the sequel. We denote the event window residuals with an asterisk to distinguish them from the estimation window and normal time observations, i.e., we define the event window errors (abnormal returns) to be

$$\varepsilon_{is}^* = R_{is} - \alpha_i - \beta_i R_{ms}, \quad s = t_{0i} - k, \ldots, t_{0i} + k. \tag{6.5}$$

Under the null hypothesis these random variables should be mean zero and have variance $\sigma_{\varepsilon_i}^2$. If ε_{is}^* is also normally distributed, then $\varepsilon_{is}^*/\sigma_{\varepsilon_i} \sim N(0, 1)$. Suppose that we observed ε_{is}^* itself for a given i and s. Then each such abnormal return could be used to test the null hypothesis by comparing $\varepsilon_{is}^*/\sigma_{\varepsilon_i}$ with the critical values z_α or $z_{\alpha/2}$ for each i and each s. This allows one to identify the precise trajectory of stock returns through the event window, and to say when the largest impact was etc. The time indexing of ε_{is}^* by calendar time is not very helpful for some purposes; it is better sometimes to interpret time relative to the origin t_{0i}, which is called **event time**, and to denote this we will write $\varepsilon_{i;e}^*$, $e = -k, \ldots, k$. This suppresses the dependence of the time argument on i. The semicolon does not have a big role to play in history but I will use it throughout this chapter to distinguish event time from calendar time in the subscripting.

6.2 Basic Structure of an Event Study

Definition 72 *The cumulative abnormal return, CAR, is $CAR_{it} = \sum_{t=t_{0i}-k}^{t} \varepsilon_{it}^{*}$ for $t \in \mathcal{E}_i$ or rather*

$$CAR_{i;l} = \sum_{e=-k}^{l} \varepsilon_{i;e}^{*}, \qquad l = -k+1, \ldots, k.$$

We define this generically as $CAR_i(c) = \sum_{e=-k}^{k} c_e \varepsilon_{i;e}^{*}$, where c_e, $e = -k, \ldots, k$ are weights. If one takes $c \in \mathbb{R}^{\tau}$ equal to $(1, 1, \ldots, 1, 0, \ldots, 0)^{\top}$ with the first zero in position $k+l+2$, then $CAR_i(c)$ is the sum of abnormal returns up to the period $l \leq k$.

If returns were logarithmic and $\alpha_i = \beta_i = 0$, then the CAR represents the cumulative (log) price increase over the event window and one can compare with Figure 6.1. By considering $CAR_{i;k}$, for example, one benefits statistically from the reduction in variance achieved by the averaging of the errors over the entire window. This approach can be interpreted as comparing the mean of abnormal returns with the mean of the errors before the event (which is zero by construction). This can be interpreted in the diff in diff framework, (6.1), where the control group here is synthetically generated from the risk model

$$\left(Y_{after}^{treated} - Y_{before}^{treated}\right) - \left(Y_{after}^{control} - Y_{before}^{control}\right)$$
$$= \left(Y_{after}^{treated} - Y_{after}^{control}\right) - \left(Y_{before}^{treated} - Y_{before}^{control}\right)$$
$$= \text{mean of abnormal returns} - \text{mean of in sample errors}$$
$$= \text{mean of abnormal returns}.$$

Under the null hypothesis and (6.4), $CAR_{i;l}/\sqrt{(l+k)\sigma_{\varepsilon_i}^2} \sim N(0, 1)$. Likewise, the standardized cumulated squared abnormal return $CSAR_i = \sum_{e=-k}^{k} (\varepsilon_{i;e}^{*})^2 / (\tau \sigma_{\varepsilon_i}^2)$ is distributed as χ_{τ}^2 under the null hypothesis and (6.4).

Estimated Parameters

In practice, we do not observe α_i, β_i so it is not feasible to compute ε_{it}^{*} and hence CAR. We next extend the discussion to include estimation of the unknown parameters using the estimation window data; this is a standard econometric exercise in linear regression methods, with the main difficulty coming from the notation that we must employ.

We assume that the data are generated (under the null hypothesis) from the market model (6.2) for all i, t. We can write the estimation window data in matrix form (R_i is $(T \times 1)$)

$$R_i = X_i \theta_i + \varepsilon_i$$

$$R_i = \begin{pmatrix} R_{i,t_{0i}-h-T+1} \\ \vdots \\ R_{i,t_{0i}-h} \end{pmatrix} \; ; \; X_i = \begin{pmatrix} 1 & R_{m,t_{0i}-m-T+1} \\ \vdots & \vdots \\ 1 & R_{m,t_{0i}-h} \end{pmatrix} \; ; \; \theta_i = \begin{pmatrix} \alpha_i \\ \beta_i \end{pmatrix}.$$

The OLS estimator of the market model, its (estimation window) residuals, and its estimated error variance are:

$$\widehat{\theta}_i = \left(X_i^\top X_i\right)^{-1} X_i^\top R_i$$

$$\widehat{\sigma}_{\varepsilon i}^2 = \frac{\widehat{\varepsilon}_i^\top \widehat{\varepsilon}_i}{T-2}, \qquad \widehat{\varepsilon}_i = R_i - X_i \widehat{\theta}_i. \tag{6.6}$$

If (6.4) holds, $\widehat{\theta}_i \sim N(\theta_i, \sigma_{\varepsilon i}^2 \left(X_i^\top X_i\right)^{-1})$. Under more general conditions discussed in Chapter 2, $\widehat{\theta}_i$ is approximately normal with the same mean and variance provided the sample size T is large.

We use these estimated parameters to compute the event window residuals using the event window data R_i^* ($\tau \times 1$), X_i^* ($\tau \times 2$):

$$R_i^* = \begin{pmatrix} R_{i,t_{0i}-k} \\ \vdots \\ R_{i,t_{0i}+k} \end{pmatrix}, \qquad X_i^* = \begin{pmatrix} 1 & R_{m,t_{0i}-k} \\ \vdots & \vdots \\ 1 & R_{m,t_{0i}+k} \end{pmatrix}$$

$$\widehat{\varepsilon}_i^* = R_i^* - X_i^* \widehat{\theta}_i.$$

The residuals are sometimes called the (estimated) abnormal returns and denoted $\widehat{AR}_i \equiv \widehat{\varepsilon}_i^* = (\widehat{AR}_{i,t_{0i}-k}, \ldots, \widehat{AR}_{i,t_{0i}+k})^\top$. They are estimates of the true underlying error vector $\varepsilon_i^* = (\varepsilon_{it_{0i}-k}^*, \ldots, \varepsilon_{it_{0i}+k}^*) = R_i^* - X_i^* \theta_i$, which we discussed above.

We next discuss how to test statistically whether these event window returns are consistent with the null hypothesis that the market model holds throughout the whole period. We have

$$\widehat{\varepsilon}_i^* = R_i^* - X_i^* \theta_i - X_i^* (\widehat{\theta}_i - \theta_i) = \varepsilon_i^* - X_i^* (X_i^\top X_i)^{-1} X_i^\top \varepsilon_i.$$

Let $\mathcal{X}_i$ denote all the covariate data, i.e., $\mathcal{X}_i = (X_i, X_i^*)$. Under the null hypothesis of no effect, i.e., that the market model assumptions remain valid in the event window, we have

$$E\left(\widehat{\varepsilon}_i^* \mid \mathcal{X}_i\right) = 0. \tag{6.7}$$

This means that abnormal returns are expected to have mean zero and be uncorrelated with the market return over the event window. Furthermore, we have

$$E\left(\widehat{\varepsilon}_i^* \widehat{\varepsilon}_i^{*\top} \mid \mathcal{X}_i\right) = \sigma_{\varepsilon i}^2 \left(I_\tau + X_i^* \left(X_i^\top X_i\right)^{-1} X_i^{*\top}\right) \equiv \Omega_{T;ii}. \tag{6.8}$$

6.2 Basic Structure of an Event Study

If we assume (6.4), then $\widehat{\varepsilon}_i^*$ is a $(\tau \times 1)$ vector of normal random variables with mean zero and $\tau \times \tau$ covariance matrix $\Omega_{T;ii}$ conditional on the covariates, i.e., $\widehat{\varepsilon}_i^* \sim N(0, \Omega_{T;ii})$. If T is large, $\Omega_{T;ii} \simeq \Omega_{T;ii}^\infty = \sigma_{\varepsilon_i}^2 I_\tau$.

Definition 73 *Let*

$$\widehat{CAR}_i(c_i) = c_i^\top \widehat{\varepsilon}_i^* = \sum_{e=-k}^{k} c_{i;e} \widehat{\varepsilon}_{i;e}^*. \tag{6.9}$$

Under the null hypothesis $\widehat{CAR}_i(c_i)$ has conditional mean zero and $\text{var}(\widehat{CAR}_i(c_i)|\mathcal{X}_i) = c_i^\top \Omega_{T;ii} c_i$. In practice we estimate $\Omega_{T;ii}$ by either

$$\widehat{\Omega}_{T;ii} = \widehat{\sigma}_{\varepsilon_i}^2 \left(I_\tau + X_i^* \left(X_i^\top X_i \right)^{-1} X_i^{*\top} \right) \quad ; \quad \widehat{\Omega}_{T;ii}^\infty = \widehat{\sigma}_{\varepsilon_i}^2 I_\tau,$$

where the ∞ superscript is used to convey the fact that this quantity is based on a large sample (large T) approximation.

Definition 74 *We form the standardized quantities based on either exact or approximate standard errors:*

$$\widehat{SCAR}_i(c_i) = \frac{\widehat{CAR}_i(c_i)}{\widehat{\sigma}_i} \quad ; \quad \widehat{SCAR}_i^\infty(c_i) = \frac{\widehat{CAR}_i(c_i)}{\widehat{\sigma}_i^\infty} \tag{6.10}$$

$$\widehat{\sigma}_i^2 = c_i^\top \widehat{\Omega}_{T;ii} c_i = \widehat{\sigma}_{\varepsilon_i}^2 c_i^\top \left(I + X_i^* \left(X_i^\top X_i \right)^{-1} X_i^{*\top} \right) c_i \quad ; \quad \widehat{\sigma}_i^\infty = c_i^\top \widehat{\Omega}_{T;ii}^\infty c_i = \widehat{\sigma}_{\varepsilon_i}^2 c_i^\top c_i.$$

If (6.4) holds, then under the null hypothesis

$$\widehat{SCAR}_i(c_i) \sim t(T-2),$$

which allows one to carry out a standard t-test. As $T \to \infty$ this distribution approaches the standard normal distribution. For typical estimation window sizes, like 250+, the standard normal approximation is indistinguishable from the t-distribution. As the estimation sample increases, whether or not we maintain (6.4), the estimation error disappears, and so $\widehat{SCAR}_i(c_i)$ and $\widehat{SCAR}_i^\infty(c_i)$ are approximately equal. Note that $\widehat{\sigma}_i^2 \geq \widehat{\sigma}_i^\infty$ for all possible values of X_i^*, X_i, so that $|\widehat{SCAR}_i(c_i)| \leq |\widehat{SCAR}_i^\infty(c_i)|$.

An alternative testing strategy can be based on aggregating the squared errors.

Definition 75 *Let*

$$\widehat{CSAR}_i = \sum_{e=-k}^{k} \frac{(\widehat{\varepsilon}_{i;e}^*)^2}{\tau \widehat{\sigma}_{\varepsilon_i}^2} = \frac{\widehat{\varepsilon}_i^{*\top} \widehat{\varepsilon}_i^* / \tau}{\widehat{\varepsilon}_i^\top \widehat{\varepsilon}_i / (T-2)}. \tag{6.11}$$

This statistic is asymptotically (as $T \to \infty$) χ_τ^2 under the null hypothesis of no effect and (6.4). In fact, this can be seen as a standard F-test and has $F_{\tau, T-2}$ distribution under the null hypothesis and (6.4).

If the error distribution is not specified, i.e., we do not maintain (6.4), then the distributions of $\widehat{CAR}_i$, $\widehat{SCAR}_i$, and $\widehat{CSAR}_i$ are unknown. This is because the event window is of finite length and so we can't invoke a CLT for averaging over the event window. The event window is the primary source of sampling variation, not the estimation of the parameters, so even in the case where the parameters are known we do not have a test in general unless we specify the error distribution or estimate it using the estimation window data.

6.2.5 Aggregating Abnormal Returns for Statistical Power and Validity in the Absence of Normality

We next consider how to aggregate evidence across firms that are subject to similar events. In this case we have a number of parameters of interest or, rather, ways of measuring departures from the null hypothesis. In some cases we are interested in the mean of abnormal returns $\mu_{it} = E(\varepsilon_{it}^*)$, their cross-sectional sum $\mu_e = \sum_{i=1}^n \mu_{i;e}$ for point $e \in \{-k, \ldots, k\}$, or their temporal and cross-sectional sum $\sum_{i=1}^n \sum_{e=-k}^k \mu_{i;e}$. As we have seen, if we uphold the normality assumption we can obtain a valid test under normality provided the estimation sample is large ($T \to \infty$) for any number of firms. In the absence of the normality assumption the test is not valid. By considering the number of firms to be large ($n \to \infty$), we obtain a valid test. In this case, we can apply a CLT that exploits the independence or approximate independence across firms. An alternative is to require a long event window, i.e., $\tau \to \infty$, but this is seldom invoked in practice because of the issues around constancy of effect etc. One of the key issues we face in the sequel is the correlation between the idiosyncratic error terms.

We suppose that the event times may be different across firms and are ordered so that $t_{01} \leq t_{02} \leq \cdots \leq t_{0n}$. We collect all the covariate information into a large array $\mathcal{X} = \{\mathcal{X}_i, i = 1, \ldots, n\}$. We now need to consider the joint behavior of the idiosyncratic terms across firms. We suppose that the market model conditions (6.2) hold for each i and that (we drop the ε subscript)

$$\text{cov}(\varepsilon_{it}^*, \varepsilon_{js}^* | \mathcal{X}) = \begin{cases} \sigma_{ij} & \text{if } s = t \\ 0 & \text{if } s \neq t. \end{cases} \quad (6.12)$$

That is, errors may be contemporaneously correlated but there is not cross-autocorrelation. This assumption is consistent with the EMH. Define the $n\tau \times 1$ vector of abnormal returns and their estimated counterparts

$$\varepsilon^* = \left(\varepsilon_{1, t_{01}-k}^*, \ldots, \varepsilon_{1, t_{01}+k}^*, \varepsilon_{2, t_{02}-k}^*, \ldots \varepsilon_{n, t_{0n}+k}^*\right)^\mathsf{T},$$
$$\widehat{\varepsilon}^* = \left(\widehat{\varepsilon}_{1, t_{01}-k}^*, \ldots, \widehat{\varepsilon}_{1, t_{01}+k}^*, \widehat{\varepsilon}_{2, t_{02}-k}^*, \ldots \widehat{\varepsilon}_{n, t_{0n}+k}^*\right)^\mathsf{T}. \quad (6.13)$$

6.2 Basic Structure of an Event Study

This represents all the potential and actual information we have about the effect of the event. The abnormal return vector ε^* is conditional mean zero and has $n\tau \times n\tau$ conditional covariance matrix

$$\Omega^* = E(\varepsilon^* \varepsilon^{*\top}|\mathcal{X}) \tag{6.14}$$

with $\tau \times \tau$ blocks Ω_{ij}^*. The estimated abnormal return vector $\widehat{\varepsilon}^*$ under the null hypothesis is conditional mean zero and has $n\tau \times n\tau$ conditional covariance matrix

$$\Omega_T = E(\widehat{\varepsilon}^* \widehat{\varepsilon}^{*\top}|\mathcal{X}), \tag{6.15}$$

with $\tau \times \tau$ blocks $\Omega_{T;ij}$, which we will specify below. The structure of Ω^* (and hence Ω_T) is the main consideration in the sequel.

Common practice is to work with event time and to relabel the times t_{0i} to be all equal, and equal to one. This simplifies the notation and discussion. If $\sigma_{ij}=0$ for all i,j, then there is no loss of generality in taking this approach. But this may be a strong assumption to make and so we will allow the possibility that the error terms of the market model are contemporaneously correlated. In this case, it is important to know about the overlap of the windows, both estimation and event, across firms. We consider several scenarios with regard to the overlap of the estimation and event windows. In some applications there is little or no overlap, while in others there may be full overlap, i.e., the events take place at the same time. We accommodate both possibilities.

Windows not Overlapping

We first consider the case where there is no overlap whatsoever.

Assumption No Overlap *For all $i \neq j$*

$$(\mathcal{S}_i \cup \mathcal{E}_i) \cap (\mathcal{S}_j \cup \mathcal{E}_j) = \emptyset.$$

This says that neither the estimation windows nor the event window overlap at all. In this case, the $n\tau \times n\tau$ covariance matrix Ω^* of the abnormal returns is block diagonal with diagonal blocks $\Omega_{T;ii}$ given above in (6.8). In this case, the distribution theory is simple because $\text{cov}(CAR_i, CAR_j) = \text{cov}(\widehat{CAR}_i, \widehat{CAR}_j) = 0$ for all $i \neq j$. Let $c = (c_1^\top, \ldots, c_n^\top)^\top$, where $c_i = (c_{i1}, \ldots, c_{i\tau})$, be some weighting scheme. Let

$$\widehat{CAR}(c) = c^\top \widehat{\varepsilon}^* = \sum_{i=1}^{n} \sum_{e=-k}^{k} c_{i;e} \widehat{\varepsilon}_{i;e}^*.$$

Under the null hypothesis $\widehat{CAR}(c)$ has conditional mean zero and $\text{var}(\widehat{CAR}(c)|\mathcal{X}_i) = c^\top \Omega_T c = \sum_{i=1}^{n} c_i^\top \Omega_{T;ii} c_i$.

Definition 76 *We form the standardized quantities based on either exact or approximate standard errors:*

$$\widehat{SCAR}(c) = \frac{\widehat{CAR}(c)}{\widehat{\sigma}} \quad ; \quad \widehat{SCAR}_\infty(c) = \frac{\widehat{CAR}(c)}{\widehat{\sigma}_\infty} \qquad (6.16)$$

$$\widehat{\sigma}^2 = \sum_{i=1}^n c_i^\top \widehat{\Omega}_{T;ii} c_i = \sum_{i=1}^n \widehat{\sigma}_{\varepsilon i}^2 c_i^\top (I + X_i^* (X_i^\top X_i)^{-1} X_i^{*\top}) c_i \ ; \ \widehat{\sigma}_\infty = \sum_{i=1}^n \widehat{\sigma}_{\varepsilon i}^2 c_i^\top c_i.$$

Suppose that the error terms are normally distributed (6.4), then $\widehat{CAR} \sim N(0, c^\top \Omega_T c)$. In this case it is possible to obtain the exact distribution of $\widehat{SCAR}(c)$ by simulation methods and to perform an exact test. If T is large, then both $\widehat{SCAR}(c)$ and $\widehat{SCAR}_\infty(c)$ are approximately standard normally distributed. Again, $\widehat{SCAR}(c) \leq \widehat{SCAR}^\infty(c)$ for all values of the covariates.

We next consider the case where the error terms are not necessarily normally distributed but the cross-section is large. Under some conditions we have that $\widehat{\theta}_i \xrightarrow{P} \theta_i$ and $\widehat{\sigma}_{\varepsilon i}^2 \xrightarrow{P} \sigma_{\varepsilon i}^2$ for each i as $T \to \infty$. With some further arguments we can show that

$$\widehat{SCAR}(c), \widehat{SCAR}_\infty(c) \simeq SCAR(c) = \frac{\sum_{i=1}^n \sum_{e=-k}^k c_{i;e} \varepsilon_{i;e}^*}{\sqrt{\sum_{i=1}^n \sigma_{\varepsilon i}^2 c_i^\top c_i}} \qquad (6.17)$$

for large T. We may apply **Lindeberg's CLT** to the right hand side of (6.17). That is, let $U_i = \sum_{e=-k}^k c_{i;e} \varepsilon_{i;e}^*$, $i = 1, \ldots, n$. Then U_i, $i = 1, \ldots, n$, are independent with $E(U_i) = 0$ and $\sigma_{U_i}^2 = \text{var}(U_i) = \sigma_{\varepsilon i}^2 \sum_{e=-k}^k c_{is}^2 = \sigma_{\varepsilon i}^2 c_i^\top c_i$.

Theorem 28 *Suppose that Lindeberg's condition holds for $\sum_{i=1}^n U_i$ as $n \to \infty$. Then*

$$SCAR(c) = \frac{\sum_{i=1}^n U_i}{\sqrt{\sum_{i=1}^n \sigma_{U_i}^2}} \Longrightarrow N(0,1).$$

The Lindeberg condition is satisfied provided: (1) $\sigma_{\varepsilon i}^2$ are bounded away from zero and infinity; (2) some higher moment exists for the error terms; and (3) the weighting sequence c_{is} is "spread out" enough (for example equal weighting satisfies this requirement). This means that for large n, T we may approximate the distribution of $\widehat{SCAR}(c)$ or $\widehat{SCAR}^\infty(c)$ by that of a standard normal.

However, it is hard to imagine that both estimation windows and event windows for firm i do not overlap with any estimation window or event window for firm j. Specifically, since we require the sample windows $\mathcal{S}_i$ to be large to deliver good estimators of θ_i it is hard to imagine that there is no overlap of these windows at all.

6.2 Basic Structure of an Event Study

Overlapping Windows

We next consider the case where the windows overlap: either the estimation windows, the event windows, or both. This is quite likely to be the case in practice. The bigger issue arises with overlap of the event windows, since the estimation effects can usually be assumed to be small.

Definition 77 For $i, j = 1, \ldots, n$, let E_{ij} be the $\tau \times \tau$ matrix of ones and zeros where ones indicate that the calendar times of observations in the two event windows $\mathcal{E}_i$ and $\mathcal{E}_j$ coincide. Let

$$\tau_{ij} = \sum_{t=t_{01}-k}^{t_{0n}+k} 1\left(t \in \mathcal{E}_i \cap \mathcal{E}_j\right) = i_\tau^\top E_{ij} i_\tau, \tag{6.18}$$

that is, τ_{ij} is the number of common calendar time periods in the event window of firm i and the event window of firm j.

If the event windows do not overlap, $\mathcal{E}_i \cap \mathcal{E}_j = \emptyset$, then the matrix E_{ij} consists entirely of zeros, but if $\mathcal{E}_i \cap \mathcal{E}_j \neq \emptyset$, then E_{ij} contains some ones. Clearly, $0 \leq \tau_{ij} \leq \tau$. If the windows do not overlap at all then $\tau_{ij} = 0$, while if they perfectly coincide then $\tau_{ij} = \tau$. Note that given the ordering of event times $t_{01} \leq \cdots \leq t_{0n}$, $\tau_{ij} = g(|i - j|)$, where $g(u) = 0$ for all $u > \tau$. With this notation we have

$$\Omega_{ij}^* = E(\varepsilon_i^* \varepsilon_j^{*\top} | \mathcal{X}) = \sigma_{ij} E_{ij}. \tag{6.19}$$

We define the $n\tau \times n\tau$ covariance matrix of the true event window errors $\Omega^* = (\Omega_{ij}^*)$. This covariance matrix may have non zero elements in different locations due to the non-synchronization of the events in calendar time. We give an example to illustrate this.

Example 40 Suppose that $n = 3$ and $\tau = 3$, and that $t_{01} = t_{02} = t_{03} = 0$ so that the events all occur at the same time. Then

$$\Omega^* = \begin{pmatrix} \sigma_{11} & 0 & 0 & \sigma_{12} & 0 & 0 & \sigma_{13} & 0 & 0 \\ 0 & \sigma_{11} & 0 & 0 & \sigma_{12} & 0 & 0 & \sigma_{13} & 0 \\ 0 & 0 & \sigma_{11} & 0 & 0 & \sigma_{12} & 0 & 0 & \sigma_{13} \\ \sigma_{12} & 0 & 0 & \sigma_{22} & 0 & 0 & \sigma_{23} & 0 & 0 \\ 0 & \sigma_{12} & 0 & 0 & \sigma_{22} & 0 & 0 & \sigma_{23} & 0 \\ 0 & 0 & \sigma_{12} & 0 & 0 & \sigma_{22} & 0 & 0 & \sigma_{23} \\ \sigma_{13} & 0 & 0 & \sigma_{23} & 0 & 0 & \sigma_{33} & 0 & 0 \\ 0 & \sigma_{13} & 0 & 0 & \sigma_{23} & 0 & 0 & \sigma_{33} & 0 \\ 0 & 0 & \sigma_{13} & 0 & 0 & \sigma_{23} & 0 & 0 & \sigma_{33} \end{pmatrix}.$$

Now suppose that $t_{01} = 0$, $t_{02} = 1$, and $t_{03} = 4$. Then

$$\Omega^* = \begin{pmatrix} \sigma_{11} & 0 & 0 & 0 & 0 & 0 & 0 & 0 & 0 \\ 0 & \sigma_{11} & 0 & \sigma_{12} & 0 & 0 & 0 & 0 & 0 \\ 0 & 0 & \sigma_{11} & 0 & \sigma_{13} & 0 & 0 & 0 & 0 \\ 0 & \sigma_{12} & 0 & \sigma_{22} & 0 & 0 & 0 & 0 & 0 \\ 0 & 0 & \sigma_{13} & 0 & \sigma_{22} & 0 & 0 & 0 & 0 \\ 0 & 0 & 0 & 0 & 0 & \sigma_{22} & 0 & 0 & 0 \\ 0 & 0 & 0 & 0 & 0 & 0 & \sigma_{33} & 0 & 0 \\ 0 & 0 & 0 & 0 & 0 & 0 & 0 & \sigma_{33} & 0 \\ 0 & 0 & 0 & 0 & 0 & 0 & 0 & 0 & \sigma_{33} \end{pmatrix}.$$

We now turn to the behavior of the estimated abnormal returns or residuals. The conditional covariance matrix Ω_T of the event window residuals consists of blocks $\Omega_{T;ij}$ of dimension $\tau \times \tau$, where $\Omega_{T;ii}$ was defined above, and

$$\Omega_{T;ij} = E\left(\widehat{\varepsilon}_i^* \widehat{\varepsilon}_j^{*\top} \mid \mathcal{X}\right)$$

$$= E\left(\left(\varepsilon_i^* - X_i^*(X_i^\top X_i)^{-1} X_i^\top \varepsilon_i\right)\left(\varepsilon_j^* - X_j^*(X_j^\top X_j)^{-1} X_j^\top \varepsilon_j\right)^\top \mid \mathcal{X}\right)$$

$$= E\left(\varepsilon_i^* \varepsilon_j^{*\top} \mid \mathcal{X}\right) - E\left(\varepsilon_i^* \varepsilon_j^\top \mid \mathcal{X}\right) X_j (X_j^\top X_j)^{-1} X_j^{*\top}$$

$$- X_i^*(X_i^\top X_i)^{-1} X_i^\top E\left(\varepsilon_i \varepsilon_j^{*\top} \mid \mathcal{X}\right)$$

$$+ X_i^*(X_i^\top X_i)^{-1} X_i^\top E\left(\varepsilon_i \varepsilon_j^\top \mid \mathcal{X}\right) X_j (X_j^\top X_j)^{-1} X_j^{*\top}.$$

The main issue is that the matrices $E(\varepsilon_i^* \varepsilon_j^{*\top} \mid \mathcal{X})$, $E(\varepsilon_i^* \varepsilon_j^\top \mid \mathcal{X})$, and $E(\varepsilon_i \varepsilon_j^\top \mid \mathcal{X})$ depend on the overlap of the event windows and the estimation windows. To complete the formula for $\Omega_{T;ij}$ we need to define the $\tau \times T$ and $T \times T$ elementary matrices (contain just zeros and ones) E_{ij}^* and E_{ij}^{**} that indicate the presence of overlap between $\mathcal{E}_i$ and $\mathcal{S}_j$ and $\mathcal{S}_i$ and $\mathcal{S}_j$ respectively. It follows that

$$\Omega_{T;ij} = \sigma_{ij} O_{ij},$$

$$O_{ij} = E_{ij} - E_{ij}^* X_j (X_j^\top X_j)^{-1} X_j^{*\top} - X_i^*(X_i^\top X_i)^{-1} X_i^\top E_{ij}^{*\top} + X_i^*(X_i^\top X_i)^{-1} X_i^\top E_{ij}^{**} X_j (X_j^\top X_j)^{-1} X_j^{*\top},$$

where the overlap matrix O_{ij} just depends on the covariate data and the event times, i.e., it is a known known. Yes, this is a lot of work. If the sample size T is large, we may work with the approximation $\Omega_{T;ij} \simeq \sigma_{ij} E_{ij} \simeq \Omega_{ij}^*$.

6.2 Basic Structure of an Event Study

We return to the question of testing based on the quantity $\widehat{CAR} = c^\mathsf{T} \widehat{\varepsilon}^*$. The problem facing us is how to **studentize** this statistic to reflect the overlap of the event and estimation windows, that is, we need to estimate consistently its standard deviation when the error terms are correlated. We consider several approaches. We first consider the case where (6.4) holds. In this case, we have the exact result $c^\mathsf{T} \widehat{\varepsilon}^* / \sqrt{c^\mathsf{T} \Omega_T c} \sim N(0,1)$. We may estimate σ_{ij} consistently provided we have a long time series of observations on $\widehat{\varepsilon}_{it}, \widehat{\varepsilon}_{jt}$. Let $T_{ij} = \sum_{t=t_{01}-h-T}^{t_{0n}+k} 1(t \in \mathcal{S}_i \cap \mathcal{S}_j)$ for $i,j = 1, \ldots, n$, that is, T_{ij} is the number of elements in the overlap of the estimation windows, i.e., $\mathcal{S}_i \cap \mathcal{S}_j$, which is assumed to be large. In practice this can be achieved by expanding the estimation windows so that they overlap. Then let

$$\widehat{\sigma}_{ij} = \frac{1}{T_{ij}} \sum_{t \in \mathcal{S}_i \cap \mathcal{S}_j} \widehat{\varepsilon}_{it} \widehat{\varepsilon}_{jt},$$

and let $\widehat{\Omega}_T = (\widehat{\sigma}_{ij} O_{ij})$ be the $n\tau \times n\tau$ estimated covariance matrix.

Definition 78 *We form the standardized quantities based on either exact or approximate standard errors:*

$$\widehat{SCAR}^O(c) = \frac{c^\mathsf{T} \widehat{\varepsilon}^*}{\sqrt{c^\mathsf{T} \widehat{\Omega}_T c}} \quad ; \quad \widehat{SCAR}^O_\infty(c) = \frac{c^\mathsf{T} \widehat{\varepsilon}^*}{\sqrt{c^\mathsf{T} \widehat{\Omega}_T^\infty c}}, \qquad (6.20)$$

where $\widehat{\Omega}_{T;ij} = \widehat{\sigma}_{ij} O_{ij}$ *and* $\widehat{\Omega}_{T;ij}^\infty = \widehat{\sigma}_{ij} E_{ij}$.

Suppose that the error terms are normally distributed, then $\widehat{CAR}(c)$ is normally distributed, and both $\widehat{SCAR}^O(c)$ and $\widehat{SCAR}^O_\infty(c)$ are approximately normally distributed for large T. In this case there is no necessary ranking of $\widehat{SCAR}^O(c)$ and $\widehat{SCAR}^O_\infty(c)$. Using simulation methods one can compute the exact distribution of both these statistics.

We next dispense with the normality assumption. In this case, we shall have to rely on a CLT for $\widehat{CAR}$ holding as $n \to \infty$. This will hold if the errors are uncorrelated or independent but will not hold otherwise unless one restricts the cross-sectional dependence of the event window observations. We will see some such assumptions in Chapter 8. For now we just state that under some conditions on this cross-sectional dependence and the relative size of n and T, $\widehat{SCAR}^O(c)$ and $\widehat{SCAR}^O_\infty(c)$ are asymptotically standard normal under the null hypothesis. That is, reject the null hypothesis if $|\widehat{SCAR}^O(c)| > z_{\alpha/2}$ or if $|\widehat{SCAR}^O_\infty(c)| > z_{\alpha/2}$.

We next consider a special case, which is quite common in practice in which one can actually ignore the cross-sectional dependence issue. We make the following assumption.

Assumption Sparse Overlap. *For τ_{ij} defined in Definition 77 as $n \to \infty$*

$$\frac{\sum\sum_{i \neq j} \tau_{ij}\sigma_{ij}}{\sum_{i=1}^{n} \sigma_{ii}} \to 0. \qquad (6.21)$$

This can be satisfied either:

1. If the event windows do not coincide very much regardless of σ_{ij}.
2. If the idiosyncratic error term is weakly dependent across firms, meaning that many of the σ_{ij} are small, or even zero, regardless of τ_{ij}.

Suppose that $0 < a \leq |\sigma_{ij}| \leq a^{-1}$, which corresponds to case (1), then to satisfy (6.21) we should require that the event windows do not coincide much, i.e., $\sum\sum_{i \neq j}\tau_{ij}/n \to 0$.

On the other hand if the event windows coincide a lot, then we should require that $\sum\sum_{i \neq j}\sigma_{ij}/\sum_{i=1}^{n}\sigma_{ii} \to 0$, i.e., the matrix Ω^* is **sparse** in the sense that the number of non zero off-diagonal elements is a small fraction of the total number of diagonal elements. Note that this condition is about the overlap of the event windows; the overlap of the estimation windows is of no consequence when we rely on large T approximations, since the estimation error disappears in that scenario.

Provided the sparse overlap condition (6.21) holds for the event windows, we have

$$\text{var}\left(\sum_{i=1}^{n}\sum_{s=t_{0i}-k}^{t_{0i}+k} c_{is}\varepsilon_{is}^*\right) = \sum_{i=1}^{n}\sum_{s=t_{0i}-k}^{t_{0i}+k} c_{is}^2\sigma_{\varepsilon i}^2 + \sum\sum_{i \neq j} E\left(\left(\sum_{s=t_{0i}-k}^{t_{0i}+k} c_{is}\varepsilon_{is}^*\right)\left(\sum_{s=t_{0j}-k}^{t_{0j}+k} c_{js}\varepsilon_{js}^*\right)\right)$$

$$= \sum_{i=1}^{n}\sum_{s=t_{0i}-k}^{t_{0i}+k} c_{is}^2\sigma_{\varepsilon i}^2 + \sum\sum_{i \neq j}\sigma_{ij}c_i^\intercal E_{ij}c_j$$

$$\simeq \sum_{i=1}^{n}\sum_{s=t_{0i}-k}^{t_{0i}+k} c_{is}^2\sigma_{\varepsilon i}^2.$$

Note that $|c_i^\intercal E_{ij}c_j| \leq \tau_{ij}\max|c_{il}c_{jk}|$ and so $c_i^\intercal E_{ij}c_j$ is like a constant times τ_{ij}. Under the sparse overlap condition (6.21), the second term in the middle expression is of smaller order and can be ignored. Under some additional minor conditions the CLT holds for $\sum_{i=1}^{n}\sum_{s=t_{0i}-k}^{t_{0i}+k} c_{is}\varepsilon_{is}^* / \sqrt{\text{var}\left(\sum_{i=1}^{n}\sum_{s=t_{0i}-k}^{t_{0i}+k} c_{is}\varepsilon_{is}^*\right)}$ as $n \to \infty$. Provided the estimation sample size also increases, the estimation error when we replace ε_{is}^* by $\widehat{\varepsilon}_{is}^*$ is of smaller order and can be ignored.

We consider in this case the statistic $\widehat{SCAR}_\infty(c)$ defined in (6.16) that ignores the overlap.

6.2 Basic Structure of an Event Study

Theorem 29 *Suppose that the sparse overlap condition holds and that some other technical conditions hold. Then under the null hypothesis*

$$\widehat{SCAR}_\infty(c) \Longrightarrow N(0,1).$$

We may test the null hypothesis by comparing $\widehat{SCAR}_\infty(c)$ with critical values from a standard normal distribution, that is, reject the null hypothesis if $|\widehat{SCAR}_\infty(c)| > z_{\alpha/2}$.

It also can be shown that $\widehat{SCAR}^o_\infty(c)$, $\widehat{SCAR}^o(c)$ are approximately standard normal, and one could base a test on them. In theory, there may be some power gains from this approach. However, they of course require we estimate the error covariances and execute the painful algebra necessary to construct them.

6.2.6 A Discussion on Power

Choice of Weights

We will return to the no overlap case for simplicity and discuss the power of $\widehat{SCAR}_\infty$. The power of this test statistic depends on the alternative hypothesis and on the weighting vector c. Specifically, suppose that

$$E(\varepsilon_i^*) = \delta_i i_\tau,$$

where $\delta_i \neq 0$ and i_τ is a $\tau \times 1$ vector of ones. Then we have approximately as $n, T \to \infty$

$$\widehat{SCAR}_\infty(c) \Longrightarrow N(\mu_\delta(c), 1), \qquad \mu_\delta(c) = \lim_{n \to \infty} \frac{\sum_{i=1}^n \delta_i c_i^\top i_\tau}{\sqrt{\sum_{i=1}^n \sigma_{\varepsilon i}^2 c_i^\top c_i}}.$$

The power of the test will increase with the magnitude of μ_δ, which itself depends on the sequences δ, c, and $\{\sigma_{\varepsilon i}^2\}$. By choosing c we may improve the power by making μ_δ large, although this will require knowledge of the alternative and the error variances. The optimal choice of c depends on $\{\sigma_{\varepsilon i}^2\}$.

One approach is to think of the problem as one where we have $n\tau$ observations $\widehat{\varepsilon}^*$ on a population with scalar mean μ (=0 under null) but where the covariance matrix of the observations is Ω_T. The GLS estimator of μ is $(i_{n\tau}^\top \Omega_T^{-1} i_{n\tau})^{-1} i_{n\tau}^\top \Omega_T^{-1} \widehat{\varepsilon}^*$, where $i_{n\tau}$ is a $n\tau \times 1$ vector of ones, which has variance $(i_{n\tau}^\top \Omega_T^{-1} i_{n\tau})^{-1}$. This suggests use of the GLS weighted test statistic

$$\widehat{SCAR}_{GLS}(c) = \frac{\sum_{i=1}^n i_\tau^\top \widehat{\Omega}_T^{-1} \widehat{\varepsilon}_i^*}{\sqrt{\sum_{i=1}^n i_\tau^\top \widehat{\Omega}_T^{-1} i_\tau}},$$

which can be interpreted as a normalized averaged CAR. A simple version appropriate for large T replaces $\widehat{\Omega}_{T;ij}$ by $\widehat{\sigma}_{\varepsilon i}^2 I_\tau$ if $i=j$. Under the null hypothesis, $\widehat{SCAR}_{GLS}(c)$ is asymptotically standard normal. Under the alternative hypothesis, we have approximately as $n \to \infty$,

$$\widehat{SCAR}_{GLS}(c) \sim N\left(\mu_\delta^{GLS}, 1\right), \quad \mu_\delta^{GLS} = \frac{\sum_{i=1}^n \delta_i \sigma_{\varepsilon i}^{-2} i_\tau^\mathsf{T} i_\tau}{\sqrt{\sum_{i=1}^n \sigma_{\varepsilon i}^{-2} i_\tau^\mathsf{T} i_\tau}} = \sqrt{\tau} \frac{\sum_{i=1}^n \delta_i \sigma_{\varepsilon i}^{-2}}{\sqrt{\sum_{i=1}^n \sigma_{\varepsilon i}^{-2}}} \to \infty.$$

Suppose that $\delta_i = \delta$, then $\mu_\delta^{GLS} \geq \mu_\delta(c)$ for any choice of c, which follows by the **Cauchy–Schwarz** inequality (essentially by the efficiency of GLS).

Tests based on CAR or $SCAR$ can detect shifts in the mean need to be modified to detect under reaction and overreaction, for example.

6.3 Regression Framework

An alternative way of looking at event studies is as a panel (or seemingly unrelated) regression model that allows a certain type of structural change where the null hypothesis corresponds to no change. This corresponds to certain linear restrictions on the model parameters. We can express the general model as

$$R_{it} = \begin{cases} \alpha_i + \beta_i R_{mt} + \varepsilon_{it} & \text{for } t < t_{0i} - k \\ \alpha_i + \gamma_{i1} + \beta_i R_{mt} + \varepsilon_{it} & \text{for } t = t_{0i} - k \\ \vdots & \vdots \\ \alpha_i + \gamma_{i\tau} + \beta_i R_{mt} + \varepsilon_{it} & \text{for } t = t_{0i} + k, \end{cases}$$

where we make the same assumptions about the error terms ε_{it} as before. We can write this model compactly as a linear regression for each firm using indicator variables (they also used to be known as dummy variables):

$$R_{it} = \alpha_i + \beta_i R_{mt} + \sum_{e=-k}^{k} \gamma_{i;e} D_{e,it} + \varepsilon_{it} \qquad (6.22)$$

$$D_{e,it} = \begin{cases} 1 & \text{if time } t \text{ corresponds to the } e^{th} \text{ time of firm } i \text{ event window} \\ 0 & \text{else} \end{cases}.$$

Here, $\gamma_{i;e} = \gamma_{is}$ with $s = e + k$. We may use data from both the estimation window and the event window, i.e., $t = t_{0i} - h - T + 1, \ldots, t_{0i} - h$ and $t = t_{0i} - k, \ldots, t_{0i} + k$, so the total sample size is $(T + \tau)n$ and the total number of unknown parameters (in the mean) is $n(\tau + 2)$.

The parameter vector $\gamma = (\gamma_{11}, \ldots, \gamma_{n\tau})^\mathsf{T}$ captures the departure from the null hypothesis. The (population) mean cumulative abnormal return for stock i is $CAR_i = \sum_{s=t_{0i}-k}^{t_{0i}+k} \gamma_{is}$, and so the above methodology is effectively testing whether $\sum_{s=t_{0i}-k}^{t_{0i}+k} \gamma_{is} = 0$ versus the alternative that $\sum_{s=t_{0i}-k}^{t_{0i}+k} \gamma_{is} \neq 0$ either for a single i or for all $i = 1, \ldots, n$. Note that there are many values of $\gamma_{i1}, \ldots, \gamma_{i\tau}$ that are individually

6.3 Regression Framework

non zero but for which $\sum_{s=t_{0i}-k}^{t_{0i}+k} \gamma_{is} = 0$. An alternative null hypothesis of interest is that $\gamma_{is} = 0$ for all i, s, which says that the event has no effect on the outcome after controlling for risk through the market return R_{mt}.

For this to be a valid regression model we require that

$$E\left(\varepsilon_{it} \mid R_{mt}, D_{-k,it}, \ldots, D_{k,it}\right) = 0. \tag{6.23}$$

This rules out **endogeneity** between the event and the current error term, although the event can depend on previous error terms.

Example 41 *Suppose that*

$$D_{0,it} = 1\left(\sum_{s=a}^{t-b} \varepsilon_{is} > \lambda\right),$$

so that the event occurs in firm i at time t when the cumulative error term passes a certain threshold $\lambda > 0$ at time $t - b$. In this case, (6.23) is satisfied provided $k < b$, but is not satisfied if $k \geq b$.

In practice whether (6.23) is satisfied depends on the data frequency relative to the organic frequency of the event. It also depends on whether the estimation window is sufficiently quarantined from the event window and whether the event window itself is short enough to avoid the contamination.

To parallel exactly the approach taken above we shall use estimation window data $\{R_{it}, R_{mt}, t = t_{0i} - h, t_{0i} - 1, \ldots, t_{0i} - h - T + 1, i = 1, \ldots, n\}$ and event window data $\{R_{it}, R_{mt}, t = t_{0i} - k, \ldots, t_{0i} + k, i = 1, \ldots, n\}$, which makes this **unbalanced** in common panel regression parlance. For notational simplicity we assume that $h = k + 1$ so that the time series for each firm contains contiguous observations. We can estimate the parameters α, β, γ by OLS (asset by asset) using all the data. Let

$$R_i^+ = \begin{pmatrix} R_{i,t_{0i}-h-T+1} \\ \vdots \\ R_{i,t_{0i}+k} \end{pmatrix}; X_i^+ = \begin{pmatrix} 1 & R_{m,t_{0i}-h-T+1} & D_{1,i,t_{0i}-h-T+1} & \cdots & D_{\tau,i,t_{0i}-h-T+1} \\ \vdots & \vdots & \vdots & & \vdots \\ 1 & R_{m,t_{0i}+k} & D_{1,i,t_{0i}+k} & \cdots & D_{\tau,i,t_{0i}+k} \end{pmatrix};$$

$$\theta_i^+ = \begin{pmatrix} \alpha_i \\ \beta_i \\ \gamma_i \end{pmatrix},$$

where $\gamma_i = (\gamma_{i;-k}, \ldots, \gamma_{i;k})^\top \in \mathbb{R}^\tau$, then let:

$$\widehat{\theta}_i = \left(X_i^{+\top} X_i^+\right)^{-1} X_i^{+\top} R_i^+$$

$$\widehat{\sigma}_{\varepsilon i}^2 = \frac{\widehat{\varepsilon}_i^\top \widehat{\varepsilon}_i}{T - 2}, \quad \widehat{\varepsilon}_i = R_i^+ - X_i^+ \widehat{\theta}_i^+.$$

By the algebra of OLS, we can show that $\widehat{\gamma}_{is} = \widehat{AR}_{is}$ and $\widehat{CAR}_i = \sum_{s=t_{0i}-k}^{t_{0i}+k} \widehat{\gamma}_{is}$; see the appendix. This shows the equivalence with the method we described earlier.

We can apply standard regression testing results here. We first consider the case where normality is assumed. Under normality,

$$\widehat{\theta}_i \sim N(\theta_i, W_i),$$

where $W_i = \sigma_{ii}(X_i^{+\tau} X_i^+)^{-1}$. The single equation F-test for the hypothesis that $\gamma_{i;e} = 0$ for all $e = -k, \ldots, k$ is

$$F_i = \frac{\left(\sum_{t=t_{0i}-h-T+1}^{t_{0i}+k} \widehat{\varepsilon}_{it}^2 - \sum_{t=t_{0i}-h-T+1}^{t_{0i}+k} \widetilde{\varepsilon}_{it}^2\right)/\tau}{\sum_{t=t_{0i}-h-T+1}^{t_{0i}+\tau} \widehat{\varepsilon}_{it}^2/(T-2)}, \qquad (6.24)$$

which is exactly $F_{\tau, T-2}$ distributed under the null hypothesis with normal errors. Here, $\widetilde{\varepsilon}_{it} = R_{it} - \widetilde{\alpha}_i - \widetilde{\beta}_i R_{mt}$, where $\widetilde{\alpha}_i, \widetilde{\beta}_i$ are the restricted least squares estimators that use the entire sample $\{t_{0i} - h - T + 1, \ldots, t_{0i} + k\}$. The likelihood ratio test of the hypothesis that $\gamma_{i;e} = 0$ for $i = 1, \ldots, n$ and $e = -k, \ldots, k$ is based on the ratio $\det(\widehat{\Omega})/\det(\widetilde{\Omega})$, where $\widehat{\Omega}$ and $\widetilde{\Omega}$ are the unrestricted and restricted MLE's of the error covariance matrix respectively. Anderson (1984) gives the exact distribution of this statistic, which nowadays would be calculated by simulation methods. We discuss these testing problems more in the next two chapters.

The regression framework is quite convenient. We can also allow for multiple splits for the same firm by introducing additional dummy variables in (6.22): in effect it is like just having another firm since in that case the overlap of the event windows is zero.

The above distribution theory works for the case where the errors are normally distributed. In the absence of this condition, the test statistics no longer have the stated distributions even when the estimation sample size is large. In that case we must rely on large n approximations, but the problem with that is that the number of restrictions being tested is large and conventional asymptotics do not accommodate such situations. Recent work has suggested how to obtain large n approximations based on the idea that $(\chi_n^2 - n)/\sqrt{n}$ is approximately normal; see Pesaran and Yamagata (2012). Instead, the literature has focussed on testing the implication of the null hypothesis that $\sum_{i=1}^n \sum_{s=t_{0i}+1}^{t_{0i}+\tau} c_{is} \gamma_{is} = 0$ versus the general alternative. This can be tested by a t-test.

M. Hashem Pesaran is the John Elliot Distinguished Chair in Economics at the University of Southern California, Emeritus Professor of Economics at the University of Cambridge, and a Fellow of Trinity College. He also holds the directorship of Dornsife Institute for New Economic Thinking and the Center for Applied Financial Economics, both at USC. Previously he was the head of the Economic Research Department of the Central Bank of Iran, the Under-Secretary of the Ministry of Education in Iran, Professor of Economics at the University of California in Los Angeles, and a Vice President at the Tudor Investment Corporation. He is a fellow of the British Academy, Econometric Society, Journal of Econometrics, and International Association for Applied Econometrics. He was awarded Honorary Doctorates by the University of Salford (1993), the University of Goethe, Frankfurt (2008), Maastricht University (2013), and the University of Economics in Prague (2016). He is the recipient of the 1992 Royal Economic Society Prize. He is the founding editor of the Journal of Applied Econometrics (1986–2014) and a co-developer of Microfit, an econometric software package published by Oxford University Press. He was named one of "The World's Most Influential Scientific Minds" by Thomson Reuters in 2014 and 2015. He has over 200 publications published in leading scientific journals and edited volumes in the areas of econometrics, empirical finance, macroeconomics, and the Iranian economy. His most recent book *Time Series and Panel Data Econometrics* was published by Oxford University Press in 2015. He is an expert in the economics of oil and the Middle East and his research has been cited over 70000 times according to Google Scholar. He holds a PhD in economics from Cambridge University. He is currently the Chair of the Board of Directors of the International Association for Applied Econometrics (IAAE).

Econometric methods can be used that allow for and account for heteroskedasticity – specifically, the event window error variances can be allowed to be different from the pre-event error variances. This may be important in practice – event induced heteroskedasticity is a commonly reported issue. We treat this issue below.

6.4 Nonparametric and Robust Tests

Nonparametric tests can be used as a complement to the standard parametric tests. The motivation for considering nonparametric tests is that they are less sensitive to extreme observations and outliers. They guard against a few outliers dominating the results. Specifically, one typically does not need to assume that the error has a variance or indeed any moments. On the other hand, they usually require stronger assumptions such as independence in order to conduct inference.

We first consider the sign test. Suppose that the null hypothesis is that the abnormal returns, $\varepsilon_{it}^* = R_{it}^* - X_{it}^* \theta_i$, are independent with median zero conditional on the covariates in the estimation window and the event window

$$\Pr(\varepsilon_{it}^* > 0 | \mathcal{F}_t) = \frac{1}{2}. \tag{6.25}$$

We can test the hypothesis by comparing the fraction of positive abnormal returns with the fraction of negative abnormal returns (estimated as before by least squares).

Cowan (1992) defined the sign test as follows. Let

$$S_{np} = \frac{\hat{p}(c) - 0.5}{\sqrt{0.25 \times \sum_{i=1}^{n} \sum_{e=-k}^{k} c_{i;e}^2}}, \quad \hat{p}(c) = \sum_{i=1}^{n} \sum_{e=-k}^{k} c_{i;e} 1(\hat{\varepsilon}_{i;e}^* > 0) \quad (6.26)$$

for some weighting sequence $c \in \mathbb{R}^\tau$. Then under the null hypothesis this statistic is approximately standard normal provided n and T are large. This is assuming that both the mean and the median of the error terms are zero, which is weaker than symmetry but in practice it is similar to it. If we further assume that ε_{it} (ε_{it}^*) are symmetrically distributed about zero conditional on the covariates, then one can perform an exact test, since in that case under the null hypothesis $\hat{\varepsilon}_{it}^*$ are symmetrically distributed about zero conditional on the covariates. Cowan (1992) defined a generalized sign test, which is robust to the assumption of symmetric errors. Rank tests are similar but guard also against asymmetric return distributions.

An alternative approach is, instead of estimating the parameters of the market model by OLS, to estimate them by quantile regression, for example least absolute deviation (median regression), for each equation. In this case we are assuming that the errors are median zero, say, conditional on the market returns under the null hypothesis, and do not require the errors or residuals to be mean zero. We estimate the parameters α_i, β_i by the median regression using the estimation window data and construct $\hat{\varepsilon}_{it}^* = R_{it} - \hat{\alpha}_i - \hat{\beta}_i R_{mt}$ for $t = t_{0i} - k, \ldots, t_{0i} + k$. We then obtain the median residuals $\psi_{1/2}(\hat{\varepsilon}_{it}^*), t = t_{0i} - k, \ldots, t_{0i} + k$, where $\psi_{1/2}(x) = \text{sign}(x)$. Provided T is large these are close to $\psi_{1/2}(\varepsilon_{it}^*), t = t_{0i} - k, \ldots, t_{0i} + k$, which are mean zero under the null hypothesis. Let

$$\widehat{CAR}_{robust} = \sum_{i=1}^{n} \sum_{e=-k}^{k} c_{i;e} \psi_{1/2}(\hat{\varepsilon}_{i;e}^*).$$

If we suppose that $E(\psi_{1/2}(\varepsilon_{it}^*)\psi_{1/2}(\varepsilon_{js}^*)) = 0$ unless $i = j$ and $t = s$ (for example, if ε_{it}^* are i.i.d.), then under some regularity conditions (see Koenker (2005)) we have

$$\frac{\widehat{CAR}_{robust}}{\sqrt{\sum_{i=1}^{n} \sum_{e=-k}^{k} c_{i;e}^2}} \Longrightarrow N(0, 1) \quad (6.27)$$

provided $n, T \to \infty$.

6.5 Cross-sectional Regressions

This is a commonly used method to describe the relationship between the effect of the event and security characteristics. Specifically, cross-sectional regressions relate the size of the abnormal return to cross-sectional characteristics. Is there a sensible relationship?

6.6 Time Series Heteroskedasticity

Example 42 *Abnormal return due to a new stock offering is linearly related to the size of the offering (as a % of total equity). Y = cross-section of CARs; X = cross-section of offering sizes*

A common approach is to fit a linear regression of the n CARs on the covariates. It is often argued that one needs to use heteroskedasticity-adjusted standard errors since this is a cross-sectional regression.

We may relate this approach to the work in the previous section. That is, suppose that $\gamma_{i1} = \cdots = \gamma_{i\tau} = \overline{\gamma}_i$ for each i and

$$\overline{\gamma}_i = \delta_i^\mathsf{T} Z_i + u_i,$$

where Z_i are firm specific characteristics and u_i is an error term that is mean zero given Z_i. Then we have

$$R_{it} = \alpha_i + \beta_i R_{mt} + \delta_i^\mathsf{T} Z_i D_{it} + \varepsilon_{it} + u_i D_{it},$$

where $D_{it} = \sum_{s=1}^{\tau} D_{s,it}$. This regression can be estimated by OLS provided T is large.

But there are many other issues with the interpretation of these regressions. Suppose the events (or announcements) are voluntary, i.e., endogenous. The fact that the firm chooses to announce at a particular time conveys information. Presumably, they are going to announce only at a time most favorable. This introduces a truncation bias. When events are modelled accounting for the firm's choice to announce some event, the resulting specifications are typically nonlinear rather than linear.

6.6 Time Series Heteroskedasticity

Suppose that the market model (6.2) holds but that

$$\operatorname{var}(\varepsilon_{it} | R_{m1}, \ldots, R_{mT}) = \sigma_{it}^2,$$

where σ_{it}^2 varies over firm and time. The OLS estimators $\widehat{\theta}_i$ of θ_i are still consistent as $T \to \infty$. Heteroskedasticity is widely acknowledged as an important problem in many fields of economics that affects the validity of conventional standard errors. Nevertheless, since Eicker (1967) and White (1980) there have been simple effective remedies for the standard errors in regression under the presence of heteroskedasticity.

Suppose that we observed the $n\tau \times 1$ vector of abnormal returns ε^* defined in (6.13), and that our event window small overlap condition (6.21) holds. Then

$$\operatorname{var}\left(\sum_{i=1}^{n} \sum_{j=-k}^{k} c_{ij} \varepsilon_{i,t_{0i}+j}^* \right) \simeq \sum_{i=1}^{n} \sum_{e=-k}^{k} c_{i;e}^2 \sigma_{i;e}^2,$$

and furthermore under the Lindeberg CLT

$$\frac{\sum_{i=1}^{n}\sum_{e=-k}^{k}c_{i;e}\varepsilon_{i;e}^{*}}{\sqrt{\sum_{i=1}^{n}\sum_{e=-k}^{k}c_{i;e}^{2}\sigma_{i;e}^{2}}}\Longrightarrow N(0,1).$$

We just require essentially that $\sum_{i=1}^{n}\sum_{e=-k}^{k}\sigma_{i;e}^{2}/n\tau$ converges to a finite positive limit, which it would do under homoskedasticity, but this is also true under a variety of heteroskedastic models.

In practice, we consider the test statistic

$$S=\frac{\sum_{i=1}^{n}\sum_{e=-k}^{k}c_{i;e}\widehat{\varepsilon}_{i;e}^{*}}{\sqrt{\sum_{i=1}^{n}\sum_{e=-k}^{k}c_{i;e}^{2}\widehat{\varepsilon}_{i;e}^{*2}}}, \tag{6.28}$$

which is also asymptotically standard normal under the null hypothesis. Note that although we cannot estimate $\sigma_{i,e}^{2}$ consistently without further structure, we can estimate its average, which is all that is needed for the test statistic.

This is consistent with the view that the null hypothesis concerns the mean of returns and not its variance. We may also want to test specifically whether the event has had an effect on the volatility of abnormal returns. This can be effected by comparing event window average variance with pre-event variance. The average pre- and post-event volatility could be measured by

$$V_{pre}=\frac{1}{nT}\sum_{i=1}^{n}\sum_{s=1}^{T}\widehat{\varepsilon}_{i,t_{0i}-s}^{*2} \quad ; \quad V_{post}=\frac{1}{n\tau}\sum_{i=1}^{n}\sum_{e=-k}^{k}\widehat{\varepsilon}_{i;e}^{*2}.$$

See Ohlson and Penman (1985).

6.7 Panel Regression for Estimating Treatment Effects

We next consider the **fixed effect** model, which is a mainstay of applied microeconometrics; see Angrist and Pischke (2009). This model can be used to evaluate the effects of some event or treatment on a general outcome. Suppose that we observe outcomes y_{it} for individual i at time t. We suppose that

$$y_{it}=\beta x_{it}+\alpha_{i}+\gamma_{t}+\varepsilon_{it}=\beta x_{it}+u_{it}, \tag{6.29}$$

where x_{it} is the covariate including the treatment effect. For example, $x_{it}=1$ if unit i received a treatment in period t and $x_{it}=0$ otherwise. Alternatively, x_{it} could measure the magnitude of the treatment in the case where this varies across units that receive the treatment (the non-treated units have $x_{it}=0$). The error term ε_{it} is assumed to be mean zero and moreover satisfy at least $E(\varepsilon_{it}|x_{it})=0$ so if $\alpha_{i}+\gamma_{t}$ were observed this would be a valid regression. The parameter β measures the treatment effect, while

6.7 Panel Regression for Estimating Treatment Effects

the α_i and γ_t are firm specific and time specific variables that are not observed and are **nuisance parameters**. One can think of them as random variables but it is not necessary to assume that they are i.i.d., that is, α_i could be a function of $x_{i1}, \ldots, x_{iT}$ and γ_t could be a function of $x_{1t}, \ldots, x_{nt}$, so that $E(u_{it}|x_{it}) \neq 0$. The model says that in the absence of the treatment by x_{it}, the time series outcomes of different firms would follow the same common trend γ_t, and at each time period the cross-section of outcomes would follow the same α_i. This is sometimes called the **parallel trends assumption**; see Bertrand, Duflo, and Mullainathan (2004). The model is non-nested with the market model (6.2) because the time effect γ_t is not restricted to be related to the return or excess return on the market portfolio; but on the other hand the model has a homogeneous effect β of the treatment on the outcome.

In the presence of $\alpha_i + \gamma_t$ the regression of y_{it} on x_{it} would be biased and inconsistent because $E(u_{it}|x_{it}) \neq 0$, so we eliminate these terms. For any i, j, t, s we have

$$\begin{aligned}
\Delta\Delta y_{it;js} &= (y_{it} - y_{is}) - (y_{jt} - y_{js}) \\
&= y_{it} + y_{js} - y_{is} - y_{jt} \\
&= \alpha_i + \gamma_t + \beta x_{it} + \varepsilon_{it} + \alpha_j + \gamma_s + \beta x_{js} + \varepsilon_{js} \\
&\quad - \alpha_i - \gamma_s - \beta x_{is} - \varepsilon_{is} - \alpha_j + \gamma_t - \beta x_{jt} - \varepsilon_{jt} \\
&= \beta(x_{it} + x_{js} - x_{is} - x_{jt}) + \varepsilon_{it} + \varepsilon_{js} - \varepsilon_{is} - \varepsilon_{jt} \\
&= \beta(\Delta\Delta x_{it;js}) + \Delta\Delta\varepsilon_{it;js}.
\end{aligned} \quad (6.30)$$

The double differencing has eliminated the nuisance parameters α_i, γ_t; the error term we assume satisfies $E(\Delta\Delta\varepsilon_{it;js}|\Delta\Delta x_{it;js}) = 0$ for this particular i, j, t, s. Therefore we can estimate β by the regression of $\Delta\Delta y_{it;js}$ on $\Delta\Delta x_{it;js}$ using all the double paired observations i, j, t, s for which $\Delta\Delta x_{it;js}$ is not identically zero. Let us just look at the classical setting where there are two groups of firms, treated (I) and untreated (J). Suppose that treated firms $i \in I$ are treated in period(s) t_i (i.e., $x_{it_i} > 0$) but are not treated in period(s) s_i (i.e., $x_{is_i} = 0$), time periods which could vary across firms or could be the same. In the previous terminology, $t_i \in \mathcal{E}_i$ but $s_i \notin \mathcal{E}_i$. We let $\ell = 1, \ldots, L$ denote all the cases (i, t_i, j, s_i), where $i \in I$ and $j \in J$. Then we can rewrite the regression model as

$$y_\ell^* = \beta x_\ell^* + \varepsilon_\ell^*,$$

where $\ell = 1, \ldots, L$ and $y_\ell^* = \Delta\Delta y_{it_i;js_i}$, $x_\ell^* = \Delta\Delta x_{it_i;js_i}$, and $\varepsilon_\ell^* = \Delta\Delta \varepsilon_{it_i;js_i}$. The error term ε_ℓ^* satisfies $E(\varepsilon_\ell^*|x_\ell^*) = 0$. We may estimate β and its asymptotic variance by

$$\widehat{\beta} = \left(\sum_{\ell=1}^L x_\ell^{*2}\right)^{-1} \sum_{\ell=1}^L x_\ell^* y_\ell^* \quad (6.31)$$

$$\widehat{V} = \left(\sum_{\ell=1}^L x_\ell^{*2}\right)^{-1} \sum_{\ell=1}^L x_\ell^{*2} \widehat{\varepsilon}_\ell^{*2} \left(\sum_{\ell=1}^L x_\ell^{*2}\right)^{-1}, \quad (6.32)$$

where $\widehat{\varepsilon}_\ell^* = y_\ell^* - \widehat{\beta} x_\ell^*$ are the least squares residuals. Provided $E(\varepsilon_\ell^* \varepsilon_{\ell'}^*) = 0$ for $\ell \neq \ell'$ the White's standard errors are consistent even allowing for heteroskedasticity in ε_ℓ^*, which could arrive from time series heteroskedasticity in ε_{it} or just cross-sectional heteroskedasticity across firms. In the special case where x_{it} is a binary treatment, we can write the estimator (6.31) as

$$\widehat{\beta} = \bar{y}_{11} - \bar{y}_{10} - (\bar{y}_{01} - \bar{y}_{00}), \tag{6.33}$$

where $\bar{y}_{11}$ is the mean of the post-treatment treatment group, $\bar{y}_{10}$ is the mean of the pre-treatment treatment group, $\bar{y}_{01}$ is the mean of the post-treatment control group, and $\bar{y}_{00}$ is the mean of the pre-treatment control group.

If the treatment effect occurs at the same time t for all firms, then it may be hard to argue that all error terms ε_ℓ^* are uncorrelated if the original error terms ε_{it} are contemporaneously correlated. We may be willing to assume, however, that errors from different time periods are uncorrelated. Suppose that $\mathrm{cov}(\varepsilon_{it}, \varepsilon_{js}) = 0$ unless $t = s$, then

$$E(\varepsilon_\ell^* \varepsilon_{\ell'}^*) = E(\Delta\Delta\varepsilon_{it;js} \times \Delta\Delta\varepsilon_{i't;j's'}) = E(\varepsilon_{it}\varepsilon_{i't}) + E(\varepsilon_{jt}\varepsilon_{j't}) \neq 0.$$

In this case, one needs some restriction on the cross-sectional dependence to justify consistency and inference. One approach to standard error construction in this case is based on **clustering techniques**, in which one assumes that there are groups G_g, $g = 1, \ldots, M$ of the observations $\ell = 1, \ldots, L$ such that within the groups there is correlation of unknown form but if observations ℓ, ℓ' lie in two different groups then they are uncorrelated. The estimated asymptotic variance of $\widehat{\beta}$ in this case is given by

$$\widehat{V}_{cluster} = \left(\sum_{\ell=1}^{L} x_\ell^{*2}\right)^{-1} \sum_{g=1}^{M} \left(\sum_{\ell \in G_g} x_\ell^* \widehat{\varepsilon}_\ell^*\right)^2 \left(\sum_{\ell=1}^{L} x_\ell^{*2}\right)^{-1},$$

which collapses to $\widehat{V}$ when there are L clusters (that is blocks of correlated observations). The differences in differences (DID) argument works well under the parallel trend assumption embodied in (6.29), but if there are interaction effects then it fails to eliminate the nuisance parameters. A more general framework is given in Imbens and Wooldridge (2009).

Example 43 *Jovanovic and Menkveld (2012) conducted an empirical study of the entry of a HFT liquidity provider into the market for Euronext Amsterdam listed Dutch index stocks that were also on the Chi-X stock exchange in London in 2007/2008. A simple before and after analysis would be confounded because 2007/2008 was the beginning of the macro financial crisis. Therefore, they compare the Dutch stocks with Belgian stocks that had no such HFT entry but were affected similarly by the macro financial crisis. They consider a binary treatment effect, and compute*

$$\left(Dutch_{after} - Dutch_{before}\right) - \left(Belgian_{after} - Belgian_{before}\right), \tag{6.34}$$

where $Dutch_j$, $Belgian_j$ are Dutch and Belgian outcomes respectively averaged over the time period $j \in \{before, after\}$, i.e., (6.33). Their results show improved market quality metrics, reduced adverse selection components, and more trading due to the treatment.

6.7 Panel Regression for Estimating Treatment Effects

An alternative method for estimating (6.29) is to use the fixed effects estimator. Let $\bar{y}_i, \bar{x}_i$ be the time series average for firm i, let $\bar{y}_t, \bar{x}_t$ be the cross-sectional average at time t, and let $\bar{\bar{y}}, \bar{\bar{x}}$ be the average of y_{it}, x_{it} over both time and cross-section. Then

$$y_{it} - \bar{y}_i - \bar{y}_t + \bar{\bar{y}} = \beta \left(x_{it} - \bar{x}_i - \bar{x}_t + \bar{\bar{x}} \right) + v_{it},$$

where $v_{it} = \varepsilon_{it} - \bar{\varepsilon}_i - \bar{\varepsilon}_t + \bar{\bar{\varepsilon}}$. The fixed effect transform has also eliminated the nuisance parameters $\alpha_i + \gamma_t$, and one can estimate β by the OLS estimator of $y_{it} - \bar{y}_i - \bar{y}_t + \bar{\bar{y}}$ on $x_{it} - \bar{x}_i - \bar{x}_t + \bar{\bar{x}}$. In this case, the error structure v_{it} will have some correlation over the cross section and over time.

We conclude with a discussion of the construction of standard errors in a general panel data regression model where the errors may be correlated across time and across individuals. Suppose that

$$y_{it} = \beta^\mathsf{T} x_{it} + u_{it}, \quad i = 1, \ldots, n, \ t = 1, \ldots, T$$

where $x_{it} \in \mathbb{R}^K$ are observed regressors and $E(u_{it}|x_{it}) = 0$. However, the error terms are not i.i.d. and satisfy $E(u_{it}u_{is}|x_{it}, x_{is}) \neq 0$ and $E(u_{it}u_{jt}|x_{it}, x_{jt}) \neq 0$. In this case the OLS estimator of y_{it} on x_{it}

$$\widehat{\beta} = \left(X^\mathsf{T} X \right)^{-1} X^\mathsf{T} y$$

is consistent, where X is the $nT \times K$ matrix containing the regressors x_{it} and y is the $nT \times 1$ vector containing y_{it}. However, its asymptotic variance may be quite complicated.

Example 44 *Suppose that*

$$u_{it} = \theta_i^\mathsf{T} f_t + \varepsilon_{it},$$

where ε_{it} is i.i.d. and θ_i and f_t are mutually independent random variables with mean zero (but $E(\theta_i\theta_j) = \omega_{ij} \neq 0$ and $E(f_t f_s) = g_{st} \neq 0$, say). In this case, double differencing will not eliminate the nuisance parameters θ_i, f_t. On the other hand the OLS estimator of y_{it} on x_{it} is consistent but its limiting variance depends on all the ω_{ij} and g_{st}.

The main issue in this model is how to obtain valid standard errors for $\widehat{\beta}$. Thompson (2011) proposes the following robust **clustered** covariance matrix estimator

$$\widehat{V} = \widehat{V}_{firm} + \widehat{V}_{time} - \widehat{V}_{White}$$

$$\widehat{V}_{firm} = \left(X^\mathsf{T} X \right)^{-1} \sum_{i=1}^n \left(\sum_{t=1}^T x_{it}\widehat{u}_{it} \right) \left(\sum_{t=1}^T x_{it}\widehat{u}_{it} \right)^\mathsf{T} \left(X^\mathsf{T} X \right)^{-1}$$

$$\widehat{V}_{time} = \left(X^\mathsf{T} X \right)^{-1} \sum_{t=1}^T \left(\sum_{i=1}^n x_{it}\widehat{u}_{it} \right) \left(\sum_{i=1}^n x_{it}\widehat{u}_{it} \right)^\mathsf{T} \left(X^\mathsf{T} X \right)^{-1}$$

$$\widehat{V}_{White} = \left(X^\top X\right)^{-1} \sum_{t=1}^{T}\left(\sum_{i=1}^{n} x_{it} x_{it}^\top \widehat{u}_{it}^2\right)\left(X^\top X\right)^{-1},$$

where $\widehat{u}_{it}$ are the OLS residuals. He shows that this is consistent in the case where both n and T are large and of similar size. In Exercise 54 you are asked to explore this further.

6.8 Matching Approach

This method is widely used in observational studies and applied microeconomics. The method is to compare average changes in outcomes for the matched firms (control) with the event firms (treated) over the event window relative to the estimation window. This can be particularly advantageous when the outcome of interest is not returns but, for example, trading volume or bid–ask spreads. The matching is done as follows. For a sample of firms E that have an event, find firms M that did not have the event and then match according to some observed vector of characteristics C. That is, for each $i \in E$ find the firm $j \in M$ that is closest to i, that is,

$$j(i) = \arg\min_{j \in M} ||C_i - C_j||, \qquad (6.35)$$

where $||.||$ is some norm, for example $||x|| = (\sum_j x_j^2)^{1/2}$. For example C could be market capitalization, prior earnings growth, etc. Recently, interest has focussed on **synthetic controls** where one forms portfolios of firms that do not have the event to better match on observed characteristics with the event firm. **Propensity score** matching involves fitting a logit or probit model to the pooled sample of event firms and nonevent firms using the characteristics as predictors of whether they had the event or not and then matching the event firms with nonevent firms according to the estimated probabilities $\widehat{p}_i$ of the firms.

One compares the average changes in outcomes for the matched firms (control) with the event firms (treated) over the event window relative to the estimation window, that is, we compute

$$\widehat{\lambda} = \sum_{i \in E}\left(\Delta Y_i - \Delta Y_{j(i)}\right),$$

where

$$\Delta Y_i = \frac{1}{\tau}\sum_{t=T+1}^{T+\tau} Y_{it} - \frac{1}{T}\sum_{t=1}^{T} Y_{it}.$$

The usual assumption is that $u_i = \Delta Y_i - \Delta Y_{j(i)}$, $i = 1, \ldots, n$ are i.i.d. from some population. The null hypothesis is that the mean of u_i is zero. This can be tested by a standard difference in mean outcomes. Let

$$S = \frac{\widehat{\lambda}}{\sqrt{\sum_{i \in E}\left(\Delta Y_i - \Delta Y_{j(i)}\right)^2}}.$$

6.9 Stock Splits

If u_i are normally distributed, then under the null hypothesis $S \sim t(n-1)$. If u_i are merely i.i.d. with mean zero and finite variance, S is approximately standard normal when n is large.

In this literature it is quite common to use nonparametric tests, in particular the **Wilcoxon signed rank test**, which replaces the mean by the median. This is a nonparametric test of the hypothesis that median outcomes in the treatment and control groups are the same. For a set of real numbers $x_1, \ldots, x_n$, let $\text{Rank}(x_1), \ldots, \text{Rank}(x_n)$ denote their (integer valued) rank order. The Wilcoxon test is carried out as follows

1. Let
$$W = \left| \sum_{i \in E} \text{sign}(\Delta Y_i - \Delta Y_{j(i)}) \text{Rank}(|\Delta Y_i - \Delta Y_{j(i)}|) \right|,$$
where the set E contains n_E firms.

2. Let
$$S_W = \frac{W - 0.5}{\sqrt{n_E(n_E + 1)(2n_E + 1)/6}}$$
and compare S_W with standard normal critical values.

Example 45 *O'Hara and Ye (2011) compare the execution quality statistics of stocks that have **fragmented trading** with that of stocks with consolidated trading. Each fragmented stock was matched with a stock that was listed on the same exchange (NYSE or NASDAQ) also using market capitalization and their closing price on January 2, 2008. They compare the stocks according to bid–ask spread, effective spread, and realized spread. They used both the t-test and the Wilcoxon test, and found significant effects in some cases.*

Barber and Lyon (1997) argue that for long term event studies based on returns one should use buy and hold returns relative to a matched control firm $j(i)$, that is compare, $\mathcal{R}_i - \mathcal{R}_{j(i)}$ instead of the CAR, where

$$\mathcal{R}_i = \prod_{t=T+1}^{T+\tau} (1 + R_{it}), \qquad \mathcal{R}_{j(i)} = \prod_{t=T+1}^{T+\tau} (1 + R_{j(i)t}). \tag{6.36}$$

They argue that the use of the market model to generate abnormal (log) returns has three biases: (1) new listing bias (sampled firms have longer history than firms included in the index); (2) rebalancing bias (index frequently rebalances whereas individual stock does not); and (3) skewness bias (long run AR has a positive skewness).

6.9 Stock Splits

We consider stock splits in some detail. The announcement date (when the split is announced) is generally between one and two months prior to the ex-date (when the

Table 6.1 Recent stock splits on NASDAQ

Stock	Ratio	Ex-date	Announcement
Shore Community Bank (NJ) (SHRC)	10.000%	20180418	20170302
Heico Corporation (HEI)	5 : 4	20180118	N/A
Security National Financial Corporation (SNFCA)	5.000%	20180111	20171201
Dommo Energia SA ADR (Sponsored) (DMMOY)	1 : 100	20180108	20180108
United States Natural Gas Fund LP (UNG)	1 : 4	20180105	20180105
Life Healthcare Group Holdings Ltd ADR (LTGHY)	1.825%	20180104	N/A
Mahindra & Mahindra Ltd. Global Dep Rcpt 144A (MAHDY)	2 : 1	20180104	N/A
Mahindra & Mahindra Ltd. Global Dep. Rcpt. Reg. S (MAHMF)	2 : 1	20180104	N/A
Citizens National Bancshares of Bossier Inc (CNBL)	5.000%	20180103	20171221

split is made, which is during non-trading hours). Fama et al. (1969) found an average difference of 44.5 days. Table 6.1 shows recently upcoming splits on NASDAQ, which shows quite considerable variation in both the split amount and the timings.

One approach is to choose the event window starting with the announcement day and end after the split day. Alternatively, we may take the event window just around the split day.

Most splits are 2:1, but there are exceptions: Berkshire Hathaway in January 2010 did a 50:1 split. Empirically, stock splits are procyclical – that is, many stock splits come at the end of a bull market.

There is considerable literature on the effect of stock splits, starting with Dolley (1933) who studied stock splits between 1921 and 1931 and found price increases at the time of the split. This study used short windows. Fama et al. (1969) introduced a new methodology. They argue that Dolley and other studies did not control for the price appreciation trend established prior to the split. They computed CARs with a

6.9 Stock Splits

simple market model and monthly data, and chose windows ±30 months around the ex-date for 940 splits between 1927 and 1959.

1. They found that CAR increased linearly up to the split date and then stayed constant, that is, abnormal returns prior to the split date but not afterwards.
2. They argued that the results are consistent with the semi-strong form of efficiency. Most of the effect occurs a long time before the actual split date. This illustrates the **sample selection** issue that is endemic to these studies: firms that split tend to have had a period of high price appreciation prior to split decision. Splits tend to happen more during bull markets than bear markets.
3. They argue that a split announcement signals that dividends may increase in the future. They provide some evidence on this by dividing the sample into a high dividend after ex group and a low dividend after ex group. They found that the high group had positive ARs after ex-date, while the low group had negative ARs upto a year after the ex-date.

Other studies have found significant short term and long term effects. Ikenberry, Rankine, and Stice (1996) found a significant post-split excess return of 7.93 percent in the first year and 12.15 percent in the first three years for a sample of 1275 two-for-one stock splits. These excess returns followed an announcement return of 3.38 percent, indicating that the market underreacts to split announcements. Ohlson and Penman (1985) examine stock return volatilities prior to and subsequent to the ex-dates of stock splits. They find an increase of approximately 30% in the return standard deviations following the ex-date. This holds for daily and weekly returns and persists for a long while.

6.9.1 Dow Stocks

We consider stock splits that occurred on the Dow stocks during our sample. There was a total of 167 splits for the Dow stocks over the sample period, which in some cases go back to 1960. In Figure 6.2 below we show the time frame of when the splits occurred. Typically, there are multiple splits per year. The estimation windows and event windows in some cases overlap considerably, although the average degree of overlap appears to be consistent with our sparse overlap assumption. We can also see from this graphic that there were many splits at the end of the 1990s as stock prices rose during the tech boom. There have been relatively few stock splits in the period 2004–2009.

In Table 6.2 we show the distribution of the size of the stock splits in our sample. Most splits are 2:1, with a split of 1.5 being the next most common.

For Exxon, there were 5 splits during the sample period. In Figure 6.3 below we show the ±52 day CAR for each split. In Figure 6.4 we average the CAR across these five events for Exxon. This shows a fairly strong trend increase during the event window starting somewhat before the split date.

In Figures 6.5 and 6.6 we carry out the same exercise for the longer event window of ±252 days. This shows that some of the increase occurs before the split date.

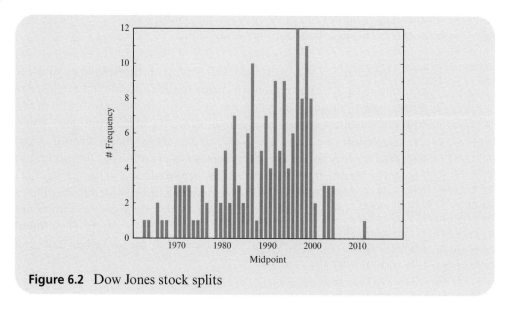

Figure 6.2 Dow Jones stock splits

Table 6.2 Dow stocks split size distribution

Size	Number
<1.5	3
1.5	30
2	119
3	12
4	3

Finally, in Figures 6.7 and 6.8 we show the short event window ±5 days, which shows average positive performance.

Some Explanations of Stock Splits

Why do firms split their stocks? Some general reasons for stock splits from management surveys (i.e., as stated by industry participants) are as follows. (a) The stock split announcement draws attention to a company's success. This results in increased buying and higher prices. (b) Companies will often report high earnings and raise dividends at the same time they announce a stock split. The synergy of these events can drive the price of the stock up even more. (c) The reduced price per share after companies split a stock attracts many smaller investors. For example: Berkshire Hathaway (Warren Buffet's company) stock prices soar above $200,000 (*Financial Times*, 20140815): "Shareholder eugenics might appear to be a hopeless undertaking. However, were we to split the stock or take other actions focussing on stock price rather

6.9 Stock Splits

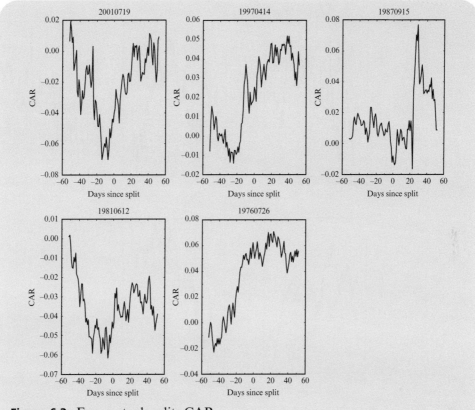

Figure 6.3 Exxon stock splits CAR

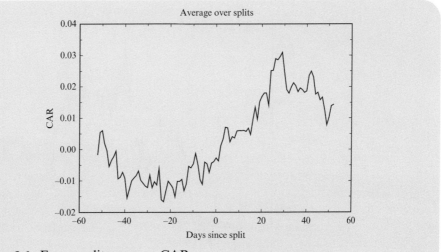

Figure 6.4 Exxon splits average CAR

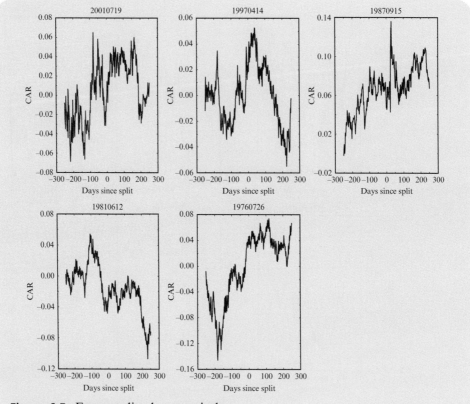

Figure 6.5 Exxon splits, longer window

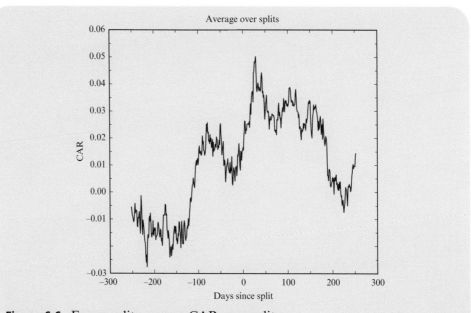

Figure 6.6 Exxon splits average CAR over splits

6.9 Stock Splits

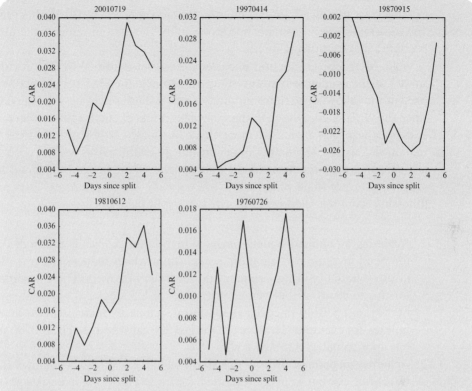

Figure 6.7 Exxon splits, shorter window

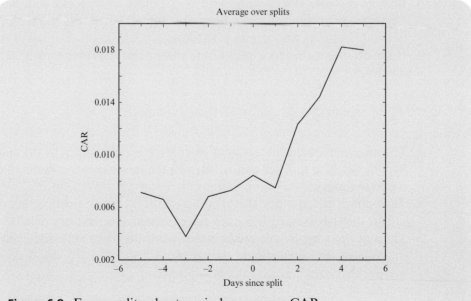

Figure 6.8 Exxon splits, shorter window average CAR

than business value, we would attract an entering class of buyers inferior to the existing class of sellers" (WB, Letter to shareholders 1983). (d) With so many news and information services reporting stock splits, the announcements themselves have become a market-moving force.

Weld, Michaely, Thaler, and Benartzi (2009), henceforth WMTB, provide a recent survey of the academic literature that tries to explain why stock splits occur. They present the following background facts: (1) US share prices have remained constant since the Great Depression (so there have been lots of stock splits), whereas the general price level has increased more than 10 times; (2) IPO share prices have also remained constant; (3) maintaining constant prices increases trading costs (real bid–ask spreads); (4) large firms tend to have higher share prices than small firms; and (5) the pattern of share prices varies dramatically across countries. They present the following economic theories as to why firms split their stock:

1. **Achieving an optimal trading range.** When the price level is too high it is harder for retail investor participation, so splitting, which reduces the price level, may improve liquidity in the primary and secondary markets. There is some evidence for this in terms of the average price level of US stocks: it has kept pretty much in the range $30–40 since the 1930s despite a massive increase in the level of stock indexes, for example. (However, in the last 30 years the average S&P500 stock price has approximately doubled.)
2. **Achieving an optimal effective tick size.** The bid–ask spread is bounded from below by the tick size but what really matters for traders is the relative tick size (and hence the relative bid–ask spread), which is determined largely by the price level. Too small a tick size discourages liquidity provision; too large a tick size is costly for retail investors especially.
3. Brokers promote stocks with lower prices more than stocks with higher prices (relative commissions higher). Some institutional investors are prohibited from investing in stocks whose price is too low.
4. It signals that managers expect future sustainable improvements in earnings and dividends.

WMTB present evidence against all these hypotheses:

1. The long term decline in real stock price levels is not justified by the marketability hypothesis. Also it is not consistent with the increase in institutional ownership (pension funds).
2. The optimal effective tick size hypothesis fails because it predicts that if tick sizes fall (as they have since 1999), then prices should also fall (they didn't).
3. The signalling hypothesis predicts that when the cost of the signal changes, the intensity of the signal should change. So, we should have seen a decline in the average share price, which has not happened.

WMTB fall back on "tradition" as the explanation!

6.10 Summary of Chapter

We described the standard methodology used in financial event studies based on the market model. We discussed the issue of overlap of event and estimation windows and gave some conditions under which these issues can be avoided. We also discussed other panel data regression models and methods for identifying the effect of an event on outcome variables.

6.11 Appendix

6.11.1 Equivalence of the Regression Formulation

Here, we show the equivalence of the CAR and the regression formulation, in a special case. Consider the bivariate regression

$$y_t = \beta x_t + \gamma D_{t,T} + \varepsilon_t,$$

where $D_{t,T}$ is a dummy variable with $D_{t,T}=1$ if $t=T$ and $D_{t,T}=0$ otherwise. The OLS estimator of γ satisfies

$$\hat{\gamma} = e_2^{\top} \begin{bmatrix} \sum_{t=1}^T x_t^2 & x_T \\ x_T & 1 \end{bmatrix}^{-1} \begin{bmatrix} \sum_{t=1}^T x_t y_t \\ y_T \end{bmatrix}$$

$$= \frac{1}{\sum_{t=1}^T x_t^2 - x_T^2} e_2^{\top} \begin{bmatrix} 1 & -x_T \\ -x_T & \sum_{t=1}^T x_t^2 \end{bmatrix} \begin{bmatrix} \sum_{t=1}^T x_t y_t \\ y_T \end{bmatrix}$$

$$= \frac{1}{\sum_{t=1}^{T-1} x_t^2} \begin{bmatrix} -x_T & \sum_{t=1}^T x_t^2 \end{bmatrix} \begin{bmatrix} \sum_{t=1}^T x_t y_t \\ y_T \end{bmatrix}$$

$$= \frac{1}{\sum_{t=1}^{T-1} x_t^2} \left[y_T \sum_{t=1}^T x_t^2 - x_T \sum_{t=1}^T x_t y_t \right]$$

$$= \frac{1}{\sum_{t=1}^{T-1} x_t^2} \left[y_T \sum_{t=1}^{T-1} x_t^2 + y_T x_T^2 - x_T \sum_{t=1}^{T-1} x_t y_t - x_T^2 y_T \right]$$

$$= y_T - x_T \frac{\sum_{t=1}^{T-1} x_t y_t}{\sum_{t=1}^{T-1} x_t^2} = y_T - \hat{\beta}_{T-1} x_T,$$

where $\hat{\beta}_{T-1}$ is the OLS estimator computed from the sample $t=1,\ldots,T-1$. This is exactly the abnormal return in this case. Note that $\hat{\varepsilon}_T = y_T - \hat{\beta} x_T - \hat{\gamma} D_{T,T} = 0$.

7 Portfolio Choice and Testing the Capital Asset Pricing Model

In this chapter we cover the implementation of mean variance portfolio choice based on a panel dataset of stock returns. We also outline the popular methods for testing the CAPM. We cover the method of portfolio grouping, which is popular in practice. Finally, we discuss some of the evidence about this theory, which points us in the direction of the subsequent chapter, which is about the multifactor extensions.

7.1 Portfolio Choice

We first consider the practical problem of implementing portfolio choice. We first consider the situation with just risky assets collected in the vector $R_t \in \mathbb{R}^N$, $t=1,\ldots,T$, where the population mean and covariance matrix is denoted by μ, Σ. We want to minimize the portfolio variance subject to achieving a mean return of m. The optimal population weighting vector is given in (1.19).

Let

$$\widehat{\mu} = \frac{1}{T}\sum_{t=1}^{T} R_t, \qquad \widehat{\Sigma} = \frac{1}{T}\sum_{t=1}^{T}(R_t - \widehat{\mu})(R_t - \widehat{\mu})^\top,$$

which are consistent estimates of the population quantities as $T \to \infty$ for fixed N. Let also $\widehat{A} = i^\top \widehat{\Sigma}^{-1} i$, $\widehat{B} = i^\top \widehat{\Sigma}^{-1} \widehat{\mu}$, $\widehat{C} = \widehat{\mu}^\top \widehat{\Sigma}^{-1} \widehat{\mu}$, $\widehat{\Delta} = \widehat{A}\widehat{C} - \widehat{B}^2$, and

$$\widehat{\lambda} = \frac{\widehat{C} - \widehat{B}m}{\widehat{\Delta}} \quad ; \quad \widehat{\gamma} = \frac{\widehat{A}m - \widehat{B}}{\widehat{\Delta}}.$$

Then

$$\widehat{w}_{opt}(m) = \widehat{\lambda}\widehat{\Sigma}^{-1} i + \widehat{\gamma}\widehat{\Sigma}^{-1}\widehat{\mu}, \tag{7.1}$$

is the optimal sample weighting scheme. The corresponding estimated variance is

$$\widehat{\sigma}^2(m) = \widehat{w}_{opt}(m)^\top \widehat{\Sigma} \widehat{w}_{opt}(m) = \left(\widehat{\lambda} i + \widehat{\gamma}\widehat{\mu}\right)^\top \widehat{\Sigma}^{-1}\left(\widehat{\lambda} i + \widehat{\gamma}\widehat{\mu}\right)$$

and the estimated efficient frontier is the set $\{m, \widehat{\sigma}(m) : m \geq 0\}$.

The GMV portfolio has weights that are proportional to $w_{GMV} \propto \Sigma^{-1} i$, which can be estimated by

$$\widehat{w}_{GMV} = \frac{\widehat{\Sigma}^{-1} i}{i^\top \widehat{\Sigma}^{-1} i}. \tag{7.2}$$

7.1 Portfolio Choice

The weights vector is a complicated nonlinear function of the $N \times 1$ vector $\widehat{\mu}$ and the $N \times N$ covariance matrix $\widehat{\Sigma}$.

Suppose that we also include the risk free asset R_f in the portfolio calculations, and let $Z_t = R_t - R_{ft}$, and let

$$\widehat{\mu}_Z = \frac{1}{T}\sum_{t=1}^{T} Z_t, \quad \widehat{\Sigma}_Z = \frac{1}{T}\sum_{t=1}^{T}(Z_t - \widehat{\mu}_Z)(Z_t - \widehat{\mu}_Z)^\top. \quad (7.3)$$

The optimal weights w_{TP} on the risky assets of the so-called tangency portfolio are proportional to $\Sigma^{-1}(\mu - R_f i)$. We can estimate these by

$$\widehat{w}_{TP} = \frac{\widehat{\Sigma}^{-1}\widehat{\mu}_Z}{i^\top \widehat{\Sigma}^{-1}\widehat{\mu}_Z}. \quad (7.4)$$

We present estimated versions of these portfolio weights below for the Dow Jones stocks along with annualized returns and standard deviations in Table 7.1.

In Figure 7.1 we show the empirical mean variance efficient frontier for the Dow stocks along with the individual means and standard deviations.

A necessary condition for $\widehat{\Sigma}$ to be full rank (and hence invertible) is that $T > N$, which limits the application of this method. In practice, there are many thousands of assets that are available for purchase. One approach is to start with a large universe of n assets and then divide them into $N < T$ portfolios based on some characteristic, and then finally find the optimal weights on these portfolios. Another approach, which is quite popular, is based on modifying the covariance matrix estimate to guarantee that it is strictly positive definite. The Ledoit and Wolf (2003) shrinkage method replaces $\widehat{\Sigma}$ by

$$\widetilde{\Sigma} = \alpha \widehat{\Sigma} + (1 - \alpha)\widehat{D},$$

where $\widehat{D}$ is the diagonal matrix of $\widehat{\Sigma}$ and α is a tuning parameter. For $\alpha \in (0, 1)$ the matrix $\widetilde{\Sigma}$ is of full rank and invertible regardless of the relative sizes of N and T. Basically, the matrix $\widehat{\Sigma}$ has non-negative eigenvalues and adding a positive diagonal matrix bounds the eigenvalues of $\widetilde{\Sigma}$ strictly away from zero. Anderson (1984, Corollary 7.7.2) says that if X is normally distributed with mean μ and covariance matrix Σ and that Σ has a prior distribution that is inverse **Wishart** with parameters Ψ, m, where Ψ is a positive definite matrix and m is an integer, then

$$\widetilde{\Sigma}_B = \frac{1}{T + m - N - 2}\left(T\widehat{\Sigma} + \Psi\right)$$

is the Bayes estimator of Σ with respect to a quadratic loss function (assuming that $T + m > N + 2$). For a suitable choice of Ψ we can include the Ledoit and Wolf (2003) estimator.

There is now an active area of research proposing new methods for estimating optimal portfolios when the number of assets is large. Essentially there should be some structure on Σ that reduces the number of unknown quantities from $N(N+1)/2$ to

Chapter 7 Portfolio Choice

Table 7.1 Tangency and GMV portfolio weights for Dow Stocks

	μ	σ	$\widehat{w}_{GMV}$	$\widehat{w}_{TP}$
Alcoa Inc.	−0.0665	0.2151	−0.0665	−0.0346
AmEx	0.0009	0.1851	−0.2475	−0.2482
Boeing	−0.0852	0.2350	−0.0006	0.0097
Bank of America	−0.0093	0.1540	0.0369	0.0377
Caterpillar	0.0375	0.1595	0.0715	0.0103
Cisco Systems	0.0140	0.1103	−0.0584	−0.0732
Chevron	−0.0110	0.2151	−0.1038	−0.1016
du Pont	−0.0088	0.1504	0.1374	0.1503
Walt Disney	−0.0046	0.1408	0.0820	0.0799
General Electric	−0.0782	0.2267	−0.0187	−0.0223
Home Depot	−0.0803	0.2200	0.0092	0.0220
HP	−0.0137	0.1937	−0.0978	−0.0907
IBM	−0.0296	0.2014	−0.0147	0.0065
Intel	0.0282	0.1318	0.1769	0.1463
Johnson&Johnson	−0.0512	0.2268	0.0706	0.0679
JP Morgan	0.0129	0.0991	0.1868	0.1834
Coke	−0.0577	0.2234	0.0353	0.0331
McD	−0.0188	0.1491	0.1096	0.1087
MMM	−0.0094	0.1458	0.0773	0.0903
Merck	−0.0754	0.1972	−0.0087	0.0127
MSFT	−0.0579	0.2092	−0.0332	−0.0227
Pfizer	−0.1093	0.2057	−0.0176	0.0062
Proctor & Gamble	0.0309	0.1045	0.3443	0.2999
AT&T	−0.0042	0.1327	0.0237	0.0364
Travelers	−0.0234	0.1703	−0.0043	0.0092
United Health	0.0190	0.2132	0.0342	0.0326
United Tech	−0.0054	0.1852	−0.0092	−0.0197
Verizon	0.0075	0.1280	0.0299	0.0071
Wall Mart	0.0179	0.1268	0.1877	0.1858
Exxon Mobil	0.0070	0.1470	0.0680	0.0770

7.2 Testing the Capital Asset Pricing Model

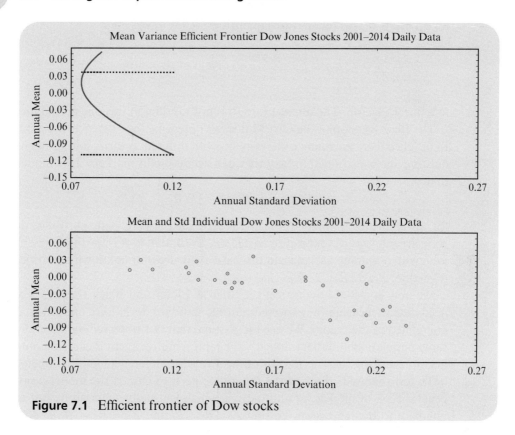

Figure 7.1 Efficient frontier of Dow stocks

some smaller quantity. The market model and factor models are well established ways of doing this and we will visit them shortly.

7.2 Testing the Capital Asset Pricing Model

We next present the most common methods for testing the CAPM. This model is very widely taught and used and is of central importance in finance. Therefore, its testing is of importance. Of course, it has been done many many times using many many different methods, but there are still disagreements about the interpretation of the results and the appropriate methods.

The theory is about a one period representative agent who chooses investments optimally subject to a budget constraint. It concerns a universe of risky assets whose random payoffs we shall denote by R_i, $i = 1, \ldots, N$, the market return R_m, which is determined by the underlying universe of assets, and possibly a risk free asset whose non-random payoff is denoted by R_f. We record here the main implications of the theory. The Sharpe–Lintner version with a riskless asset (borrowing or lending) says that

$$E(R_i) = R_f + \beta_i(E(R_m) - R_f) \tag{7.5}$$

for all assets i. This relates three quantities:

$$\mu_i = E(R_i) - R_f \quad ; \quad \mu_m = E(R_m) - R_f \quad ; \quad \beta_i = \frac{\text{cov}(R_i, R_m)}{\text{var}(R_m)}$$

in a linear fashion. The interest rate R_f is not random in this one period model.

The Black version of the CAPM does not presume the existence of a riskless asset. In practice there are many riskless assets with different returns, and different individuals may be constrained in how they can access credit markets. This model predicts that

$$E(R_i) = E(R_0) + \beta_i(E(R_m) - E(R_0)) \tag{7.6}$$

for all securities i, where R_0 is the return on the zero beta portfolio, defined as $R_0 = \arg\min \text{var}(R_x)$ subject to $\text{cov}(R_x, R_m) = 0$, that is, it is the market neutral portfolio with minimum variance. In this case, there are four unknown quantities that are related: $E(R_i)$, $E(R_0)$, $E(R_m)$, and β_i.

We now consider how to test the Sharpe–Linter and Black theories. For this we need data to measure the expected returns, standard deviations, and covariances that are related in the theory. We need to assume that our observed data comes from the same population and that implicitly the representative agent is carrying out the same decision-making each period with time invariant preferences and parameters.

To test the predictions of the theory we need to embed the model inside a more general class of models, or alternatives, which is usually achieved by just adding an intercept to the regression. Letting $Z_i = R_i - R_f$ and $Z_m = R_m - R_f$, in the Sharpe–Lintner case we write

$$E(Z_i) = \alpha_i + \beta_i E(Z_m).$$

The theory predicts that $\alpha_i = 0$ for all i. The alternative here is that $\alpha_i \neq 0$ for at least one i. In the Black case, we suppose that

$$E(R_i) = \alpha_i + \beta_i E(R_m).$$

The theory predicts that $\alpha_i = (1 - \beta_i)E(R_0)$ for all i, with the alternative hypothesis that $\alpha_i \neq (1 - \beta_i)E(R_0)$ for some i. In both cases, α_i (alpha) captures an excess reward to security i.

7.2.1 Market Model

Suppose that we have a sample of observations on each asset R_{it}, $t = 1, \ldots, T$, that are random draws from the same distribution and we likewise observe market returns R_{mt}, $t = 1, \ldots, T$, which may be composed of assets not contained in $i = 1, \ldots, N$. We may or may not also observe a sample of risk free rates R_{ft} (which may vary over time but are known for the one period that they correspond to).

We suppose that the market model (6.2) and (6.3) holds with $Z_{it} = R_{it}$ or $Z_{it} = R_{it} - R_{ft}$ and $Z_{mt} = R_{mt}$ or $Z_{mt} = R_{mt} - R_{ft}$. Estimation of the parameters $\alpha_i, \beta_i, \sigma^2_{\varepsilon i}$ from a time series sample is well understood. It is not necessary that ε_{it} be i.i.d. over time

7.2 Testing the Capital Asset Pricing Model

and it is certainly not necessary that it be normally distributed (6.4), but a lot of the literature starts from this stronger assumption.

So far we have only specified the behavior of the error terms for each stock. In fact, it is later necessary to specify the joint behavior of $\varepsilon_{1t}, \ldots, \varepsilon_{Nt}$. For the most part we will suppose the following

Assumption For $\varepsilon_t = (\varepsilon_{1t}, \ldots, \varepsilon_{Nt})^\top$ and $Z_t = (Z_{mt}, \ldots, Z_{mt})^\top$

$$E(\varepsilon_t \varepsilon_s^\top | Z_1, \ldots, Z_T) = \begin{cases} \Omega_\varepsilon & \text{if } t = s \\ 0 & \text{if } t \neq s. \end{cases} \quad (7.7)$$

This says that the errors are contemporaneously correlated as represented by the time invariant covariance matrix Ω_ε, but are not cross autocorrelated. The simplest assumption is that they are uncorrelated across i, in which case Σ is a diagonal matrix. The consequence of this assumption depends on whether N is large or not. We shall assume that N is fixed (i.e., small) unless otherwise stated. In some cases we make use of the following stronger assumption.

Assumption The vector ε_t is independent of $Z_1, \ldots, Z_T$ and i.i.d. with

$$\varepsilon_t \sim N(0, \Omega_\varepsilon). \quad (7.8)$$

Regardless of the error correlation structure, because the regressors are the same $(1, Z_{mt})$ for each asset, we can estimate the parameters of this seemingly unrelated regression model efficiently by OLS using the time series of observations for each asset. These estimates are unbiased and consistent under heteroskedasticity and non-normality, as is well known. However, the classical theory here emphasizes the normal assumption, in which case these estimates are efficient and more importantly perhaps one can obtain exact test statistics. We next review some evidence around this assumption.

7.2.2 Evidence for Normality

Normality is not necessary for the CAPM (see Ingersoll (1987) for discussion) but much of the literature assumes it and takes advantage of its properties. There is a famous quote about the ubiquity of the normal distribution (Whittaker and Robinson (1967))

> *Everybody believes in the exponential law of errors (i.e., the Normal distribution): the experimenters, because they think it can be proved by mathematics; and the mathematicians, because they believe it has been established by observation.*

Fama (1965) compares the empirical stock return distributions (log returns) with normal distribution for daily prices for each of the thirty stocks in the Dow Jones average over the period 1957–1962. He adjusted for dividends and stock splits. He found the frequency distribution to be highly leptokurtic. Fama argues that a better characterization of return distributions is the stable Paretian, which we will discuss below in Chapter 14.

We consider here whether stock returns are well approximated by a normal distribution by looking at cumulants. Standard measures of non-normality include the higher cumulants such as skewness and excess kurtosis defined in (1.7). For a normal distribution $\kappa_3, \kappa_4 = 0$. The **Jarque–Bera** (1980) statistic combines empirical estimates of these cumulants into a test statistic. Daily stock returns typically have large negative skewness and large positive kurtosis and fail normality tests. However, Fama argues that monthly returns are closer to normality. We investigate this with recent data. Let us state the following result.

Theorem 30 *Aggregation of (logarithmic) returns.* Let A be the aggregation (e.g., weekly, monthly) level such that

$$r_A = r_1 + \cdots + r_A.$$

Then under **rw1**

$$E(r_A) = AE(r) \quad ; \quad \text{var}(r_A) = A\text{var}(r)$$

$$\kappa_3(r_A) = \frac{1}{\sqrt{A}}\kappa_3(r) \quad ; \quad \kappa_4(r_A) = \frac{1}{A}\kappa_4(r).$$

Proof. The mean and variance results are obvious and have been shown already in Chapter 3. Let $\tilde{r}_j = r_j - E(r_j)$. We have

$$E\left[(\tilde{r}_1 + \cdots + \tilde{r}_A)^3\right] = \sum_{j=1}^{A} E(\tilde{r}_j^3) + \sum_{j=1}^{A}\sum_{\substack{k=1 \\ j \neq k}}^{A} E(\tilde{r}_j^2 \tilde{r}_k) + \sum_{j=1}^{A}\sum_{k=1}^{A}\sum_{\substack{l=1 \\ j \neq k, k \neq l, j \neq l}}^{A} E(\tilde{r}_j \tilde{r}_k \tilde{r}_l)$$

$$= AE(\tilde{r}_j^3),$$

because $E(\tilde{r}_j \tilde{r}_k \tilde{r}_l) = 0$ and $E(\tilde{r}_j^2 \tilde{r}_k) = 0$, whenever $j \neq k \neq l$. Then dividing by the 3/2 power of the variance gives the desired result. A similar calculation holds for the kurtosis. □

This says that as you aggregate more ($A \to \infty$), returns become more normal, i.e., $\kappa_3(r_A) \to 0$ and $\kappa_4(r_A) \to 0$ as $A \to \infty$, which is consistent with the CLT.

Is this true empirically? In Figures 7.2 and 7.3 we show the skewness and kurtosis of the Dow Jones stocks plotted against the aggregation factor A. The skewness and kurtosis decline with aggregation horizon as predicted, but the monthly horizon ($A = 22$) still has substantial skewness and kurtosis.

We may also consider the case where returns are not i.i.d., e.g., they satisfy **rw3**. In that case, the formulae are more complex, but the result is fundamentally similar: aggregation should reduce the magnitude of skewness and kurtosis. We give such a result in the appendix.

We next ask how the market model aggregates.

7.2 Testing the Capital Asset Pricing Model

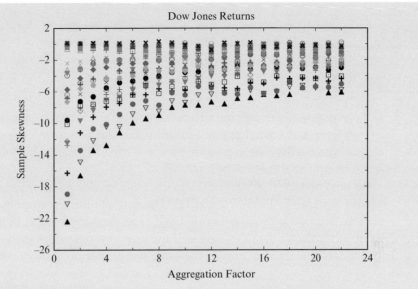

Figure 7.2 Skewness of Dow Jones returns by sampling frequency

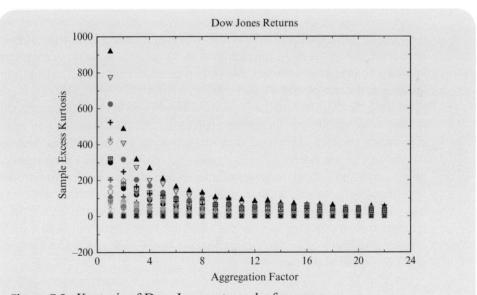

Figure 7.3 Kurtosis of Dow Jones returns by frequency

Theorem 31 *Suppose that the market model (6.2) holds for the highest frequency of excess return data Z_{it}. Then for integer A, let $Z_{it}^A = \sum_{s=t-A}^{t} Z_{is}$ and $Z_{mt}^A = \sum_{s=t-A}^{t} Z_{mt}$. Then we have*

$$Z_{it}^A = \alpha_i^A + \beta_i^A Z_{mt}^A + \varepsilon_{it}^A, \tag{7.9}$$

where $\alpha_i^A = K\alpha_i$ and $\beta_i^A = \beta_i$, where $\varepsilon_{it}^A = \sum_{s=t-A}^{t} \varepsilon_{it}$ satisfies $E(\varepsilon_{it}^A | Z_{m1}, \ldots, Z_{mT}) = 0$ and $\text{var}(\varepsilon_{it}^A | Z_{m1}, \ldots, Z_{mT}) = K\text{var}(\varepsilon_{it} | Z_{m1}, \ldots, Z_{mT})$. If ε_{it} are i.i.d., then $\kappa_3(\varepsilon_{it}^A) = \kappa_3(\varepsilon_{it})/\sqrt{A}$ and $\kappa_4(\varepsilon_{it}^A) = \kappa_4(\varepsilon_{it}^A)/A$, so that the aggregated error terms should be closer to normality than the high frequency returns.

If one uses non-overlapping returns $t = 1, A+1, \ldots$, then ε_{it}^A are also serially uncorrelated. This is the usual practice in this particular literature. However, if one works with all the overlapping returns for $t = A+1, \ldots, T$, then $\varepsilon_{it}^A, \varepsilon_{is}^A$ will be correlated, which complicates the task of inference.

We next present estimates of the market model along with standard errors for different frequency in Tables 7.2 and 7.3. The market return is given by the S&P500 and the risk free rate is the one month T-bill. The sample period is 1990–2013. The first set of tables shows the daily data and the second set of tables shows the same results for the monthly data series over the same sample period.

7.3 Maximum Likelihood Estimation and Testing

The classical approach to testing for the CAPM is based on the theory of likelihood based tests under a normality assumption. In this setting one can obtain simple exact tests of the null hypothesis of interest, which is a desirable feature. However, as we have seen, the evidence of normality of stock returns is weak, even at the monthly horizon. In that case, however, the tests presented here are all asymptotically valid (as long as the sample size T gets large) under some weaker conditions on the process generating the data. We shall discuss these conditions later. In this approach, it is customary to condition on the market returns, i.e., to treat them as non random. This is common practice in regression models with no dynamic effects, and is based on the concept of **ancillarity** – the marginal distribution of market returns contains no information about the parameters of interest, which describe the conditional distribution of individual asset returns conditional on the market return. This may run into logical problems since the market returns are themselves composed of the individual returns, and so conditioning on the weighted sum of the outcome variables seems especially dubious when the chosen assets are themselves a relatively large fraction (e.g. portfolios) of the market portfolio.

Let $Z_t = (Z_{1t}, \ldots, Z_{Nt})^\top$ be the $N \times 1$ vector of excess returns and rewrite the market model (6.2) in vector notation

$$Z_t = \alpha + \beta Z_{mt} + \varepsilon_t, \tag{7.10}$$

where ε_t satisfies (7.7). Here, $\alpha = (\alpha_1, \ldots, \alpha_N)^\top$ and $\beta = (\beta_1, \ldots, \beta_N)^\top$ are vectors of unknown parameters. The $N \times N$ covariance matrix Ω_ε is not restricted, i.e., we allow correlation between the idiosyncratic errors. It is common practice in other settings to assume that the idiosyncratic terms are uncorrelated, in which case Ω_ε is diagonal, but it is not necessary for the theory, and it is not common practice here. In some

7.3 Maximum Likelihood Estimation and Testing

Table 7.2 Market model estimates of Dow Stocks, daily data

	α	se(α)	β	se(β)	se$_W$(β)	R^2
Alcoa Inc.	−0.0531	0.0480	1.3598	0.0305	0.0400	0.3929
AmEx	0.0170	0.0332	1.4543	0.0211	0.0317	0.6073
Boeing	−0.0644	0.0492	1.6177	0.0312	0.0667	0.4661
Bank of America	−0.0066	0.0341	0.9847	0.0217	0.0301	0.4020
Caterpillar	0.0433	0.0335	1.0950	0.0213	0.0263	0.4635
Cisco Systems	0.0060	0.0264	0.6098	0.0168	0.0272	0.3005
Chevron	0.0017	0.0486	1.3360	0.0308	0.0401	0.3793
du Pont	−0.0101	0.0358	0.8422	0.0228	0.0314	0.3084
Walt Disney	−0.0009	0.0281	1.0198	0.0178	0.0241	0.5159
General Electric	−0.0732	0.0575	1.0650	0.0365	0.0296	0.2169
Home Depot	−0.0720	0.0534	1.1815	0.0339	0.0418	0.2835
HP	−0.0083	0.0462	1.0797	0.0294	0.0315	0.3055
IBM	−0.0232	0.0483	1.1107	0.0307	0.0349	0.2992
Intel	0.0282	0.0280	0.8903	0.0178	0.0242	0.4490
Johnson&Johnson	−0.0400	0.0539	1.2803	0.0342	0.0360	0.3132
JP Morgan	0.0041	0.0231	0.5810	0.0147	0.0226	0.3382
Coke	−0.0380	0.0456	1.5811	0.0290	0.0595	0.4927
McD	−0.0270	0.0392	0.6012	0.0249	0.0237	0.1598
MMM	−0.0117	0.0349	0.8093	0.0221	0.0228	0.3032
Merck	−0.0783	0.0519	0.7893	0.0329	0.0268	0.1575
MSFT	−0.0536	0.0521	1.0427	0.0331	0.0408	0.2444
Pfizer	−0.1121	0.0545	0.7912	0.0346	0.0256	0.1455
Proctor & Gamble	0.0221	0.0250	0.5804	0.0159	0.0230	0.3036
AT&T	−0.0065	0.0303	0.8076	0.0193	0.0264	0.3643
Travelers	−0.0209	0.0402	0.9750	0.0255	0.0410	0.3224
United Health	0.0173	0.0564	0.8278	0.0358	0.0563	0.1482
United Tech	−0.0029	0.0452	0.9779	0.0287	0.0298	0.2742
Verizon	0.0039	0.0296	0.7606	0.0188	0.0238	0.3472
Wall Mart	0.0141	0.0293	0.7555	0.0186	0.0249	0.3495
Exxon Mobil	0.0053	0.0349	0.8290	0.0222	0.0292	0.3128

Chapter 7 Portfolio Choice

Table 7.3 Market model estimates of Dow Stocks, monthly data

	α	se(α)	β	se(β)	se$_W$(β)	R^2
Alcoa Inc.	−0.0528	0.0499	1.4107	0.1759	0.2185	0.3179
AmEx	0.0193	0.0254	1.5594	0.0895	0.1045	0.6875
Boeing	−0.0675	0.0499	1.6575	0.1759	0.2298	0.3915
Bank of America	0.0018	0.0311	1.2767	0.1097	0.1232	0.4955
Caterpillar	0.0454	0.0306	1.2212	0.1078	0.1300	0.4819
Cisco Systems	0.0134	0.0243	0.8418	0.0856	0.1114	0.4122
Chevron	0.0034	0.0472	1.4240	0.1663	0.1505	0.3468
du Pont	−0.0191	0.0318	0.5596	0.1120	0.1082	0.1532
Walt Disney	−0.0048	0.0258	0.9200	0.0908	0.0867	0.4264
General Electric	−0.0630	0.0568	1.4430	0.2002	0.2678	0.2734
Home Depot	−0.0715	0.0532	1.1787	0.1877	0.0992	0.2222
HP	0.0048	0.0435	1.4292	0.1535	0.1651	0.3858
IBM	−0.0270	0.0451	1.1867	0.1592	0.1274	0.2872
Intel	0.0299	0.0247	0.9255	0.0872	0.0935	0.4492
Johnson&Johnson	−0.0449	0.0550	1.1815	0.1940	0.1281	0.2118
JP Morgan	0.0046	0.0201	0.5912	0.0708	0.0738	0.3358
Coke	−0.0435	0.0428	1.4111	0.1508	0.1590	0.3880
McD	−0.0262	0.0364	0.6342	0.1284	0.1218	0.1501
MMM	−0.0129	0.0317	0.7906	0.1116	0.0814	0.2665
Merck	−0.0716	0.0506	0.9235	0.1784	0.2365	0.1627
MSFT	−0.0529	0.0571	1.1163	0.2012	0.1862	0.1824
Pfizer	−0.1135	0.0586	0.7811	0.2066	0.1804	0.0938
Proctor & Gamble	0.0255	0.0223	0.6626	0.0786	0.0838	0.3401
AT&T	−0.0111	0.0305	0.6668	0.1075	0.1069	0.2182
Travelers	−0.0255	0.0361	0.8321	0.1272	0.0977	0.2367
United Health	0.0186	0.0566	0.8556	0.1997	0.1751	0.1174
United Tech	−0.0047	0.0455	0.9470	0.1606	0.1620	0.2012
Verizon	0.0004	0.0277	0.6265	0.0976	0.1200	0.2301
Wall Mart	0.0142	0.0255	0.6802	0.0899	0.1026	0.2933
Exxon Mobil	0.0025	0.0339	0.7124	0.1196	0.1020	0.2044

7.3 Maximum Likelihood Estimation and Testing

applications authors use overlapping portfolios constructed from some anomaly and this would induce correlation in the error terms even when the underlying assets are idiosyncratically uncorrelated. So we allow a general Ω_ε. This is a seemingly unrelated regression in the terminology of Zellner (1962) with the same regressors, an intercept, and the market return or excess return.

Suppose that (7.8) were satisfied, then the Gaussian log likelihood for the observed vectors $Z_1, \ldots, Z_T$ conditional on the market returns $Z_{m1}, \ldots, Z_{mT}$ is

$$\ell(\alpha, \beta, \Omega_\varepsilon) = c - \frac{T}{2} \log \det(\Omega_\varepsilon) - \frac{1}{2} \sum_{t=1}^{T} (Z_t - \alpha - \beta Z_{mt})^\mathsf{T} \Omega_\varepsilon^{-1} (Z_t - \alpha - \beta Z_{mt}), \tag{7.11}$$

where c is a constant not depending on unknown parameters. The maximum likelihood estimates $\widehat{\alpha}, \widehat{\beta}$ are the equation-by-equation time-series OLS estimates, because the market return and the intercept are common regressors in each equation. For $i = 1, \ldots, N$ we have

$$\widehat{\alpha}_i = \frac{1}{T} \sum_{t=1}^{T} Z_{it} - \widehat{\beta}_i \overline{Z}_m$$

$$\widehat{\beta}_i = \frac{\sum_{t=1}^{T} (Z_{mt} - \overline{Z}_m) Z_{it}}{\sum_{t=1}^{T} (Z_{mt} - \overline{Z}_m)^2} = \frac{1}{\widehat{\sigma}_m^2} \frac{1}{T} \sum_{t=1}^{T} (Z_{mt} - \overline{Z}_m) Z_{it}$$

$$\overline{Z}_m = \frac{1}{T} \sum_{t=1}^{T} Z_{mt} \quad ; \quad \widehat{\sigma}_m^2 = \frac{1}{T} \sum_{t=1}^{T} (Z_{mt} - \overline{Z}_m)^2.$$

This can be written in vector form:

$$\widehat{\alpha} = \frac{1}{T} \sum_{t=1}^{T} Z_t - \widehat{\beta} \overline{Z}_m \tag{7.12}$$

$$\widehat{\beta} = \left(\sum_{t=1}^{T} (Z_{mt} - \overline{Z}_m)^2 \right)^{-1} \sum_{t=1}^{T} (Z_{mt} - \overline{Z}_m) Z_t = \frac{1}{\widehat{\sigma}_m^2} \frac{1}{T} \sum_{t=1}^{T} (Z_{mt} - \overline{Z}_m) Z_t. \tag{7.13}$$

The maximum likelihood estimate of Ω_ε is

$$\widehat{\Omega}_\varepsilon = (\widehat{\sigma}_{ij}) = \frac{1}{T} \widehat{\varepsilon} \widehat{\varepsilon}^\mathsf{T} = \left(\frac{1}{T} \sum_{t=1}^{T} \widehat{\varepsilon}_{it} \widehat{\varepsilon}_{jt} \right)_{i,j},$$

where $\widehat{\varepsilon}_t = Z_t - \widehat{\alpha} - \widehat{\beta} Z_{mt}$.

We may alternatively collect parameters equation by equation, that is, let $\Theta = (\theta_1, \ldots, \theta_N)^\mathsf{T}$ be the $N \times 2$ matrix with $\theta_i = (\alpha_i, \beta_i)^\mathsf{T}$, let X be the $T \times 2$ matrix containing a column of ones and a column of observations on Z_{mt}, and let Z_i be the $T \times 1$

vector containing the observations on Z_{it}. Then $\widehat{\theta}_i = (\widehat{\alpha}_i, \widehat{\beta}_i)^\mathsf{T} = (X^\mathsf{T} X)^{-1} X^\mathsf{T} Z_i$, and we can write in system notation the estimator of Θ as

$$\widehat{\Theta} = Z^\mathsf{T} X \left(X^\mathsf{T} X \right)^{-1} \tag{7.14}$$

$$X^\mathsf{T} X = \begin{bmatrix} T & \sum_{t=1}^T Z_{mt} \\ \sum_{t=1}^T Z_{mt} & \sum_{t=1}^T Z_{mt}^2 \end{bmatrix} \; ; \; Z^\mathsf{T} X = \begin{bmatrix} \sum_{t=1}^T Z_{1t} & \sum_{t=1}^T Z_{mt} Z_{1t} \\ \vdots & \vdots \\ \sum_{t=1}^T Z_{Nt} & \sum_{t=1}^T Z_{mt} Z_{Nt} \end{bmatrix},$$

where $Z^\mathsf{T} = (Z_1, \ldots, Z_N)$ is the $T \times N$ matrix containing the returns or excess returns on the individual assets. We also let $\theta = \text{vec}(\Theta)$ and $\widehat{\theta} = \text{vec}(\widehat{\Theta})$ be the stacked $2N \times 1$ vector of parameters and estimators in this ordering. The representations, (7.12), (7.13), and (7.14) are useful.

Under the normality assumption, we have, conditional on excess market returns, the exact distributions:

$$\widehat{\alpha} \sim N\left(\alpha, \frac{1}{T}\left(1 + \frac{\overline{Z}_m^2}{\widehat{\sigma}_m^2}\right) \Omega_\varepsilon \right) \tag{7.15}$$

$$\widehat{\beta} \sim N\left(\beta, \frac{1}{T}\frac{1}{\widehat{\sigma}_m^2} \Omega_\varepsilon \right), \tag{7.16}$$

because the estimators are linear combinations of the normally distributed error terms. Note also that $\widehat{\alpha}$ and $\widehat{\beta}$ are correlated with $\text{cov}(\widehat{\alpha}, \widehat{\beta}) = -(\overline{Z}_m/T\widehat{\sigma}_m^2)\Omega_\varepsilon$, which is negative when the mean excess return on the market is positive. The distribution of $T\widehat{\sigma}_{ii}$ is $t(T-2)$ and the matrix random variable $T\widehat{\Omega}_\varepsilon$ follows the Wishart distribution. We may write $\widehat{\theta} \sim N(\theta, \Omega_\varepsilon \otimes (X^\mathsf{T} X)^{-1})$, where $\otimes$ denotes the **Kronecker product** of two matrices, that is, $\Omega_\varepsilon \otimes (X^\mathsf{T} X)^{-1} = (\sigma_{\varepsilon ij}(X^\mathsf{T} X)^{-1})_{i,j}$.

In some cases we may be interested in the intercept from a single equation. We may test the hypothesis that $\alpha_i = 0$ versus $\alpha_i \neq 0$ using the t-test

$$t_i = \frac{\sqrt{T}\widehat{\alpha}_i}{\sqrt{\left(1 + \frac{\overline{Z}_m^2}{\widehat{\sigma}_m^2}\right) \widehat{\sigma}_{ii}}}, \tag{7.17}$$

where $\widehat{\sigma}_{ii}$ is the corresponding diagonal element of $\widehat{\Omega}_\varepsilon$. We may compare t_i with the critical value from the $t(T-2)$ distribution if the normal distribution is believed or with the standard normal distribution if one is relying on large samples. This test will have power against the alternative that $\alpha_i \neq 0$. We could calculate all the t statistics, $t_1, \ldots, t_N$ but we then need to adjust the significance level for the multiple testing issue. Instead we look at standard tests of the joint hypothesis based on the likelihood trilogy.

7.3 Maximum Likelihood Estimation and Testing

The Wald test statistic for testing the joint null hypothesis that $\alpha = 0$ is

$$W = T\left(1 + \frac{\bar{Z}_m^2}{\hat{\sigma}_m^2}\right)^{-1} \hat{\alpha}^{\mathsf{T}} \hat{\Omega}_\varepsilon^{-1} \hat{\alpha}, \qquad (7.18)$$

where $\hat{\Omega}_\varepsilon$ is the MLE defined above. In large samples (i.e., $T \to \infty$), W is approximately distributed as χ_N^2 under the null hypothesis. Under the alternative hypothesis $W \to \infty$, so the test has power against all alternatives.

The Lagrange multiplier or score test is based on the residual vector of the restricted estimators. The restricted MLE of β is the no intercept OLS estimator

$$\tilde{\beta} = \left(\sum_{t=1}^T Z_{mt}^2\right)^{-1} \sum_{t=1}^T Z_{mt} Z_t = \frac{1}{\hat{\mu}_{m2}} \frac{1}{T} \sum_{t=1}^T Z_{mt} Z_t$$

$$\hat{\mu}_{m2} = \frac{1}{T} \sum_{t=1}^T Z_{mt}^2,$$

while the restricted maximum likelihood estimate of Ω_ε is

$$\tilde{\Omega}_\varepsilon = \frac{1}{T} \tilde{\varepsilon} \tilde{\varepsilon}^{\mathsf{T}} = \left(\frac{1}{T} \sum_{t=1}^T \tilde{\varepsilon}_{it} \tilde{\varepsilon}_{jt}\right)_{i,j},$$

where $\tilde{\varepsilon}_t = Z_t - \tilde{\beta} Z_{mt}$. The null hypothesis in this case only restricts the intercept parameters so the score test can be expressed purely in terms of the score with respect to these parameters. The score function is proportional to the vector of restricted residuals

$$\frac{\partial \ell}{\partial \alpha}(\tilde{\theta}) = \frac{1}{T} \sum_{t=1}^T \tilde{\varepsilon}_t = \frac{1}{T} \sum_{t=1}^T \varepsilon_t - (\tilde{\beta} - \beta) \frac{1}{T} \sum_{t=1}^T Z_{mt}$$

$$= \frac{1}{T} \sum_{t=1}^T \left(1 - \frac{\bar{Z}_m}{\hat{\mu}_{m2}} Z_{mt}\right) \varepsilon_t,$$

which is normally distributed with mean zero and variance that you can calculate. The LM test statistic is

$$LM = T\left(1 + \frac{\bar{Z}_m^2}{\hat{\mu}_{m2}}\right)^{-1} \overline{\tilde{\varepsilon}}^{\mathsf{T}} \tilde{\Omega}_\varepsilon^{-1} \overline{\tilde{\varepsilon}}, \qquad (7.19)$$

where $\overline{\tilde{\varepsilon}} = \sum_{t=1}^T \tilde{\varepsilon}_t / T$. Under the null hypothesis, this is approximately χ_N^2 as $T \to \infty$.

The likelihood ratio (LR) test is based on a comparison of the residual sum of squares from the restricted model with the unrestricted model

$$LR = 2(\log \ell_{restricted} - \log \ell_{unrestricted}) = T\left(\log \det(\tilde{\Omega}_\varepsilon) - \log \det(\hat{\Omega}_\varepsilon)\right). \qquad (7.20)$$

This statistic will also be χ_N^2 in large samples under the null hypothesis.

Berndt and Savin (1977) show that $LM \leq LR \leq W$ in finite samples, which says that the Wald statistic is the most likely to reject, while the LM test is least likely to reject, when using the asymptotic χ_N^2 critical values. In fact, under the normality assumption there is an exact finite-sample variant of the Wald test statistic using the known Wishart distribution of $\widehat{\Omega}_\varepsilon$. Specifically, let

$$F = \frac{(T-N-1)}{N}\left(\left(1 + \frac{\overline{Z}_m^2}{\widehat{\sigma}_m^2}\right)^{-1}\right) \times \widehat{\alpha}^\top \widehat{\Omega}_\varepsilon^{-1} \widehat{\alpha}.$$

This is exactly distributed as $F_{N,T-N-1}$, and so we can carry out the test using the critical value from this distribution. This test is often called the GRS test after Gibbons, Ross, and Shanken (1989). Note that this is like the standard F-test from regression except that there we test whether the covariates are jointly significant; here we are testing whether a vector of intercepts is jointly significant. We can express it in terms of the goodness of fit of the restricted model versus the goodness of fit of the unrestricted model.

Another important insight due to GRS is that F is proportional to the difference in the Sharpe ratios of the market portfolio m and the ex-post efficient tangency portfolio TP, i.e.,

$$F = \frac{T-N-1}{N}\left(\frac{\frac{\widehat{\mu}_{TP}^2}{\widehat{\sigma}_{TP}^2} - \frac{\overline{Z}_m^2}{\widehat{\sigma}_m^2}}{1 + \frac{\overline{Z}_m^2}{\widehat{\sigma}_m^2}}\right).$$

This has a useful graphical interpretation in terms of investment theory.

The three likelihood tests have an exact one to one relationship with the F statistic. For example, we may write

$$F = \frac{T-N-1}{N}\left(\exp(\frac{LR}{T}) - 1\right) \quad ; \quad LR = \log\left(F\frac{N}{T-N-1} + 1\right).$$

It follows that if $F_{N,T-N-1}(\alpha)$ is the level α critical value for F, then if one used $\log(F_{N,T-N-1}(\alpha)\frac{N}{T-N-1} + 1)$ as critical value for LR, then one would obtain exactly the same results. The difference between the tests only arises when asymptotic approximate critical values are used. All three tests are consistent against all alternatives for which $\alpha \neq 0$.

Example 46 *For the Dow stocks: Daily $W = 22.943222$ (p-value $= 0.81758677$); Monthly $W = 33.615606$ (p-value $= 0.29645969$).*

Note that all these joint tests require that $N < T$, because otherwise the matrix $\widehat{\Omega}_\varepsilon$ is not invertible. The CRSP database contains many thousands of individual stocks, so this methodology cannot be applied to all of the stocks directly. Alternatively, one may work with subsets of assets and do joint tests of the intercepts from the subsets. Since

7.3 Maximum Likelihood Estimation and Testing

the estimation is equation by equation, there is no bad consequence from ignoring the other equations, except that potentially one is losing power. Instead, typically authors work with portfolios of stocks that are constructed using various criteria; see below for a discussion of **portfolio grouping** methodology. The approach widely used in practice is to form a relatively small number, say $N=20$, of portfolios of all the assets and apply the above methodology to this smaller number of securities.

The Wald statistic aggregates information across all assets, and so it should deliver a more powerful test of the CAPM than individual t-tests. However, when N is large or even moderately large this approach faces some problems. In the extreme case when $N > T$ the sample covariance matrix is of deficient rank, as already discussed above, which makes the Wald statistic not well defined. Instead, common practice is to present statistics such as: $N^{-1}\sum_{i=1}^{N}\widehat{\alpha}_i$, $N^{-1}\sum_{i=1}^{N}|\widehat{\alpha}_i|$, and $N^{-1}\sum_{i=1}^{N}\widehat{\alpha}_i^2$; see for example Hou, Xue, and Zhang (2015) and Stambaugh and Yuan (2017). The quantity $N^{-1}\sum_{i=1}^{N}|\widehat{\alpha}_i|$ represents the average pricing error and is interpretable in terms of apparently exploitable returns. Under the CAPM all three statistics should be approximately zero. Under the alternative hypothesis, the second and third statistics will tend to infinity with probability tending to one. The first one may fail to do so when some $\alpha_i > 0$ and some $\alpha_i < 0$ leading to $\sum_{i=1}^{N}\alpha_i = 0$. We may calculate the mean and variance of these quantities using the normal limit distribution (as $T \to \infty$) and results of Magnus and Neudecker (1988) for the moments of quadratic forms in normal random variables. For $|\widehat{\alpha}_i|$ we use results about the folded normal distribution (Psarakis and Panaretos (2001)). Under some additional conditions these quantities are approximately normal when both N and T are large provided their null mean is subtracted.

Theorem 32 *Define the mean adjusted statistics*

$$S_1 = \frac{\sqrt{T}\sum_{i=1}^{N}\widehat{\alpha}_i}{\sqrt{\left(1+\frac{\overline{Z}_m^2}{\widehat{\sigma}_m^2}\right)\sum_{i=1}^{N}\sum_{j=1}^{N}\widehat{\sigma}_{ij}}}, \qquad S_3 = \frac{T\sum_{i=1}^{N}\widehat{\alpha}_i^2 - \left(1+\frac{\overline{Z}_m^2}{\widehat{\sigma}_m^2}\right)\sum_{i=1}^{N}\widehat{\sigma}_{ii}}{\sqrt{2\left(1+\frac{\overline{Z}_m^2}{\widehat{\sigma}_m^2}\right)^2 \sum_{i=1}^{N}\sum_{j=1}^{N}\widehat{\sigma}_{ij}^2}}$$

$$S_2 = \frac{\sqrt{T}\sum_{i=1}^{N}|\widehat{\alpha}_i| - \sqrt{\frac{2}{\pi}\left(1+\frac{\overline{Z}_m^2}{\widehat{\sigma}_m^2}\right)}\sum_{i=1}^{N}\sqrt{\widehat{\sigma}_{ii}}}{\sqrt{\left(1+\frac{\overline{Z}_m^2}{\widehat{\sigma}_m^2}\right) \times \left(\left(1-\frac{2}{\pi}\right)\sum_{i=1}^{N}\widehat{\sigma}_{ii} + \sum\sum_{i\neq j}\left(|\widehat{\sigma}_{ij}| - \frac{2}{\pi}\sqrt{\widehat{\sigma}_{ii}\widehat{\sigma}_{jj}}\right)\right)}}.$$

These are approximately standard normal under the null hypothesis, under some regularity conditions that include $N, T \to \infty$ and $N/T \to 0$.

These results can be used to test the CAPM restrictions in the case where N is large. Pesaran and Yamagata (2012) propose instead to combine information from many t statistics to produce a valid test when N is large. A simplified version of their test statistic is $S_{N,T}$, and we give alongside the absolute value version

$$S_{N,T} = \frac{\sum_{i=1}^{N} t_i^2 - N}{\sqrt{2N\left(1 + \frac{1}{N}\sum\sum_{i \neq j} \frac{\hat{\sigma}_{ij}^2}{\hat{\sigma}_{ii}\hat{\sigma}_{jj}}\right)}},$$

$$S_{N,T}^* = \frac{\sum_{i=1}^{N} |t_i| - N\sqrt{\frac{2}{\pi}}}{\sqrt{2N\left(\left(1 - \frac{2}{\pi}\right) + \frac{1}{N}\sum\sum_{i \neq j}\left(\left|\frac{\hat{\sigma}_{ij}^2}{\hat{\sigma}_{ii}\hat{\sigma}_{jj}}\right|^{1/2} - \frac{2}{\pi}\right)\right)}}. \quad (7.21)$$

These are both asymptotically standard normal under the null hypothesis. They also considered the very large N case where $N/T \to \infty$. In this case, they propose sparsifying modifications to $\{\hat{\sigma}_{ij} : i, j = 1, \ldots, N\}$. They show rigorously that $S_{N,T} \to N(0, 1)$ under the null hypothesis.

It is generally found that some of the investment weights of the tangency portfolio are negative. Moreover, as the number of assets increases, it is shown empirically that the percentage of assets corresponding to the tangency portfolio with negative weights approach 50%. These findings apparently contradict the CAPM, since to guarantee equilibrium the investment weights of the tangency portfolio must be all positive. In addition, if most investors in practice choose mainly a portfolio with positive weights, it implies that they do not choose by the MV rule, as selecting an optimal portfolio by the MV rule yields many negative investment weights. Therefore, the existence of negative weights imply that, in practice, investments are not selected by the MV rule; hence the CAPM is not valid. However, Levy and Roll (2010) show that one can rationalize this as due to estimation error, namely within a 95% confidence interval for the estimated weights one can find weights that do satisfy the non-negativity property.

7.3.1 Testing the Black Version of the CAPM

The Black version of the CAPM is slightly more complicated in terms of the restrictions it implies upon the data. There is a cross-equation restriction on α under the null hypothesis. The restricted model to be estimated is nonlinear in the parameters. The unconstrained model is estimated as before using total returns instead of excess returns.

The Black model implies that

$$\alpha = (i_N - \beta)\gamma \quad (7.22)$$

for some unknown scalar parameter γ, where i_N is an N vector of ones. We can rewrite this equation showing the $N - 1$ nonlinear cross-equation restrictions involved

$$\frac{\alpha_1}{1 - \beta_1} = \cdots = \frac{\alpha_N}{1 - \beta_1} = \gamma. \quad (7.23)$$

7.3 Maximum Likelihood Estimation and Testing

Note that one needs at least two assets to test the Black version because otherwise we can always find a γ such that $\gamma = \alpha/(1-\beta)$. When $N \geq 2$, this ratio must be the same across assets, which is a testable restriction.

Maximum likelihood estimation of the unconstrained model is equivalent to equation-by-equation OLS as we have already seen (but with the raw returns instead of the excess returns). For $\theta = (\gamma, \beta_1, \ldots, \beta_N)^\top$ the (constrained) likelihood function of $R_1, \ldots, R_T$ given $R_{m1}, \ldots, R_{mT}$ is

$$\ell(\theta, \Omega_\varepsilon) = c - \frac{T}{2} \log \det(\Omega_\varepsilon)$$
$$- \frac{1}{2} \sum_{t=1}^{T} (R_t - \gamma i_N - \beta(R_{mt} - \gamma))^\top \Omega_\varepsilon^{-1} (R_t - \gamma i_N - \beta(R_{mt} - \gamma)), \quad (7.24)$$

where c is a constant that does not depend on unknown parameters. The MLE of $\theta, \Omega_\varepsilon$ that maximizes $\ell(\theta, \Omega_\varepsilon)$ is denoted $\widehat{\theta}^*, \widehat{\Omega}_\varepsilon^*$. The estimation procedure is nonlinear, but we can simplify the calculation considerably by noticing the particular structure that conditional on γ we have a linear regression. Define for each $\gamma \in \mathbb{R}$

$$\widehat{\beta}_i^*(\gamma) = \frac{\sum_{t=1}^{T}(R_{mt} - \gamma)(R_{it} - \gamma)}{\sum_{t=1}^{T}(R_{mt} - \gamma)^2}$$

$$\widehat{\Omega}_\varepsilon^*(\gamma) = \frac{1}{T} \widehat{\varepsilon}^*(\gamma) \widehat{\varepsilon}^{*\top}(\gamma) = \left(\frac{1}{T} \sum_{t=1}^{T} \widehat{\varepsilon}_{it}^*(\gamma) \widehat{\varepsilon}_{jt}^*(\gamma) \right)_{i,j}.$$

Then search the **profile likelihood** or concentrated likelihood

$$\ell_P(\gamma) = c - \frac{T}{2} \log \det(\widehat{\Omega}_\varepsilon^*(\gamma))$$
$$- \frac{1}{2} \sum_{t=1}^{T} (R_t - \gamma i_N - \widehat{\beta}^*(\gamma)(R_{mt} - \gamma))^\top \Omega_\varepsilon^{-1} (R_t - \gamma i_N - \widehat{\beta}^*(\gamma)(R_{mt} - \gamma)),$$

over the scalar parameter γ. The value of γ that maximizes $\ell_P(\gamma)$ is denoted $\widehat{\gamma}_P$. Then define

$$\widehat{\beta}_P^* = \widehat{\beta}^*(\widehat{\gamma}_P) \quad ; \quad \widehat{\Omega}_{\varepsilon P}^* = \widehat{\Omega}_\varepsilon^*(\widehat{\gamma}_P).$$

In fact, it can be shown that $\widehat{\beta}_P^* = \widehat{\beta}^*$ and $\widehat{\Omega}_{\varepsilon P}^* = \widehat{\Omega}_\varepsilon^*$, so this computational device works.

We may now compute the likelihood ratio test as

$$LR_B = T \left(\log \det(\widehat{\Omega}_\varepsilon^*) - \log \det(\widehat{\Omega}_\varepsilon) \right),$$

where $\widehat{\Omega}_\varepsilon^*$ is the MLE of Ω_ε in the constrained model. This satisfies

$$LR_B \Longrightarrow \chi_{N-1}^2 \quad (7.25)$$

as $T \to \infty$ under the null hypothesis.

The exact distribution of LR_B is not known. CLM made some discussion of this issue and the literature that attempts to address it. Nowadays, simulation methods are popular ways of dealing with this, and we next describe a simple algorithm to obtain the distribution of the test statistic. Let $\varepsilon_t^* \sim N(0, \widehat{\Omega}_\varepsilon^*)$ and compute a new sample (for $t = 1, \ldots, T$)

$$R_t^* = (i_N - \widehat{\beta}^*)\widehat{\gamma}_P + \widehat{\beta}^* R_{mt} + \varepsilon_t^*.$$

For this sample, compute again the test statistic

$$LR_B^* = T\left(\log\det(\widehat{\Omega}_\varepsilon^*) - \log\det(\widehat{\Omega}_\varepsilon)\right)$$

and repeat this B times. Then use the empirical distribution of the B test statistics LR_B^* to define the critical value c_α^* of the test and reject if the original LR_B exceeds c_α^*. Note that one could substitute $\varepsilon_t^* \sim N(0, \widehat{\Omega}_\varepsilon^*)$ by draws from the empirical distribution of the errors (Efron (1979)).

We next consider the Wald statistic. There are two approaches. In the first we represent the null hypothesis as $g = 0$, where g is the $N - 1 \times 1$ vector with typical element

$$g_i = \frac{\alpha_i}{1 - \beta_i} - \frac{\alpha_N}{1 - \beta_N}, \quad i = 1, \ldots, N - 1.$$

Let $\widehat{g} = (\widehat{g}_1, \ldots, \widehat{g}_{N-1})^\top$, where

$$\widehat{g}_i = \frac{\widehat{\alpha}_i}{1 - \widehat{\beta}_i} - \frac{\widehat{\alpha}_N}{1 - \widehat{\beta}_N}, \quad i = 1, \ldots, N - 1.$$

To implement this test we must calculate the large sample variance of the vector $\widehat{g}$. Let $\phi = (\alpha^\top, \beta^\top)^\top$ and $\widehat{\phi} = (\widehat{\alpha}^\top, \widehat{\beta}^\top)^\top$. Then $g = G(\phi)$ and $\widehat{g} = G(\widehat{\phi})$, where $G: \mathbb{R}^{2N} \to \mathbb{R}^{N-1}$ is a known function. We let

$$\Gamma(\phi) = \frac{\partial G}{\partial \phi^\top} = \begin{pmatrix} \frac{\partial G_1}{\partial \alpha_1} & \cdots & \frac{\partial G_1}{\partial \beta_N} \\ & & \\ \frac{\partial G_{N-1}}{\partial \alpha_1} & \cdots & \frac{\partial G_{N-1}}{\partial \beta_N} \end{pmatrix}$$

$$= \begin{bmatrix} \frac{1}{1-\beta_1} & 0 & \cdots & \frac{-1}{1-\beta_N} & -\frac{\alpha_1}{(1-\beta_1)^2} & 0 & \cdots & \frac{\alpha_N}{(1-\beta_N)^2} \\ & & & \vdots & & & & \\ \frac{1}{1-\beta_{N-1}} & 0 & \cdots & \frac{-1}{1-\beta_N} & -\frac{\alpha_{N-1}}{(1-\beta_{N-1})^2} & 0 & \cdots & \frac{\alpha_N}{(1-\beta_N)^2} \end{bmatrix}$$

$$V_g = \Gamma(\phi) \begin{bmatrix} \left(1 + \frac{\mu_m^2}{\sigma_m^2}\right)\Omega_\varepsilon & \frac{-\mu_m}{\sigma_m^2}\Omega_\varepsilon \\ \frac{-\mu_m}{\sigma_m^2}\Omega_\varepsilon & \frac{1}{\sigma_m^2}\Omega_\varepsilon \end{bmatrix} \Gamma(\phi)^\top,$$

7.3 Maximum Likelihood Estimation and Testing

where $\Gamma(\phi)$ is an $N-1 \times 2N$ matrix of partial derivatives, $\mu_m = E(R_m)$, and $\sigma_m^2 = \text{var}(R_m)$. Here, we must require $\beta_i \neq 1$. Then, by the **delta method**, it follows that in large samples $(T \to \infty)$

$$\sqrt{T}(\hat{g} - g) \Longrightarrow N(0, V_g).$$

Let

$$W_B = \hat{g} \hat{V}_g^{-1} \hat{g} \qquad (7.26)$$

$$\hat{V}_g = \hat{\Gamma} \begin{bmatrix} \left(1 + \frac{\bar{Z}_m^2}{\hat{\sigma}_m^2}\right) \hat{\Omega}_\varepsilon & \frac{-\bar{Z}_m}{\hat{\sigma}_m^2} \hat{\Omega}_\varepsilon \\ \frac{-\bar{Z}_m}{\hat{\sigma}_m^2} \hat{\Omega}_\varepsilon & \frac{1}{\hat{\sigma}_m^2} \hat{\Omega}_\varepsilon \end{bmatrix} \hat{\Gamma}^{\mathsf{T}},$$

where $\hat{\Gamma} = \Gamma(\hat{\phi})$. It follows that as $T \to \infty$

$$W_B \Longrightarrow \chi^2_{N-1} \qquad (7.27)$$

under the null hypothesis. This is clearly feasible but hard work. Moreover, it is not possible to find analytic expressions for the exact distribution of this test statistic under normality, although one could use simulation methods to obtain the exact distribution under the normality hypothesis. An alternative way of calculating the Wald statistic is as follows.

Let $\delta = (\gamma, \beta_1, \ldots, \beta_N)^{\mathsf{T}} \in \mathbb{R}^{N+1}$ be the restricted parameters and let $h(\delta) = (\gamma(1-\beta_1), \beta_1, \ldots, \gamma(1-\beta_N), \beta_N)^{\mathsf{T}} \in \mathbb{R}^{2N}$. Then let

$$W = T \min_{\delta \in \mathbb{R}^{N+1}} \left(\hat{\theta} - h(\delta)\right)^{\mathsf{T}} \hat{\Xi}^{-1} \left(\hat{\theta} - h(\delta)\right),$$

where $\hat{\Xi} = \hat{\Omega}_\varepsilon \otimes (X^{\mathsf{T}} X)^{-1}$ is the $2N \times 2N$ estimated covariance matrix of the unrestricted estimates $\hat{\theta}$. Following Chamberlain (1984, Proposition 3), under the null hypothesis $W \Longrightarrow \chi^2_{N-1}$ as $T \to \infty$, while under the alternative hypothesis W grows without bounds.

We can solve (7.22) explicitly for γ as

$$\gamma = \frac{(i-\beta)^{\mathsf{T}} \Omega_\varepsilon^{-1} \alpha}{(i-\beta)^{\mathsf{T}} \Omega_\varepsilon^{-1} (i-\beta)},$$

which shows that we also need $\beta \neq i$ for this restriction to make sense. The estimated γ

$$\hat{\gamma} = \frac{(i-\hat{\beta})^{\mathsf{T}} \hat{\Omega}_\varepsilon^{-1} \hat{\alpha}}{(i-\hat{\beta})^{\mathsf{T}} \hat{\Omega}_\varepsilon^{-1} (i-\hat{\beta})} \qquad (7.28)$$

measures the return on the zero beta portfolio.

7.3.2 Robustness to Normality and Time Series Heteroskedasticity

The CAPM can hold under weaker assumptions than normality (e.g., elliptical symmetry, which includes multivariate t-distributions with heavy tails). The maximum likelihood estimation we have discussed apparently assumes multivariate normal returns (otherwise, it is called the quasi-maximum likelihood estimator (QMLE)). Actually, the QMLEs of α, β are robust to heteroskedasticity, serial correlation, and non-normality, since consistency only requires correct specification of the mean. There is no need to invoke the holy GMM concept to address this robustness issue. However, the exact distribution theory no longer holds when normality does not hold. Furthermore, the large sample theory for the test statistics needs to be adjusted. Specifically, one needs to adjust the standard errors for time series heteroskedasticity, a problem that was solved in regression models and quasi-likelihood framework long ago (White's standard errors).

Suppose that the error term is a martingale difference sequence

$$E(\varepsilon_t | \mathcal{F}_{t-1}, Z_{m1}, \ldots, Z_{mT}) = 0 \tag{7.29}$$

$$E(\varepsilon_t \varepsilon_t^\intercal | \mathcal{F}_{t-1}, Z_{m1}, \ldots, Z_{mT}) = \Omega_{\varepsilon t}, \tag{7.30}$$

where $\Omega_{\varepsilon t}$ is a potentially random (depending on $\mathcal{F}_{t-1}, Z_{mt}$) time varying covariance matrix, and $\mathcal{F}_{t-1}$ is all past information on returns including Z_{t-1}, etc. This is quite a general assumption, but as we shall see in Chapter 11 below it is quite natural to allow for dynamic heteroskedasticity for stock return data. We consider models for $\Omega_{\varepsilon t}$ in Chapter 11, but here we will just allow it to vary freely over time.

Define the following averaged covariance matrices

$$\overline{\Omega}_{\varepsilon T} = \frac{1}{T} \sum_{t=1}^{T} \Omega_{\varepsilon t} \quad ; \quad \Omega_T = \frac{1}{T} \sum_{t=1}^{T} (Z_{mt} - \overline{Z}_m)^2 \Omega_{\varepsilon t},$$

which we assume to have positive definite and finite limits as $T \to \infty$. We can construct robust Wald tests and LM tests based on large sample approximations that in addition dispense with the normality assumptions. Specifically, let

$$\widehat{\Omega}_H = \frac{1}{T} \sum_{t=1}^{T} \left(1 + \frac{\overline{Z}_m^2}{\widehat{\sigma}_m^4} (Z_{mt} - \overline{Z}_m)^2 \right) \widehat{\varepsilon}_t \widehat{\varepsilon}_t^\intercal$$

$$\widetilde{\Omega}_H = \frac{1}{T} \sum_{t=1}^{T} \left(1 - \frac{\overline{Z}_m}{\widehat{\mu}_{m2}} Z_{mt} \right)^2 \widetilde{\varepsilon}_t \widetilde{\varepsilon}_t^\intercal.$$

Then the H-robust Wald statistic and the H-robust LM statistics are given by

$$W_H = T \widehat{\alpha}^\intercal \widehat{\Omega}_H^{-1} \widehat{\alpha} \quad ; \quad LM_H = T \overline{\widetilde{\varepsilon}}^\intercal \widetilde{\Omega}_H^{-1} \overline{\widetilde{\varepsilon}}.$$

7.4 Cross-sectional Regression Tests

Under the null hypothesis (and some further conditions on the form of $\Omega_{\varepsilon t}$, for example, but that do not require normality) we have as $T \to \infty$

$$W_H, LM_H \Longrightarrow \chi_N^2. \tag{7.31}$$

Under the alternative hypothesis, $W_H, LM_H \to \infty$.

7.4 Cross-sectional Regression Tests

Fama and MacBeth (1973), henceforth FM, introduced the two pass regression approach, which is very popular because of its ease of implementation. This approach emphasizes the cross-sectional implications of the CAPM. In the Sharpe–Lintner case the firm's mean excess return $E_i = E(R_i - R_f)$ should be linearly related to the firm's β_i with zero intercept and slope equal to the excess mean return on the market $E_m = E(R_m - R_f)$. In the Black case, the firm's mean return $\mu_i = E(R_i)$ should be linearly related to the firm's β_i with intercept $E(R_0)$ and slope equal to the excess mean return on the market relative to the zero beta portfolio, $E(R_m - R_0)$. Lets proceed to the route. We focus on the Black case. In this case, the null model for returns is

$$R_{it} = \gamma_{0t} + \beta_i \gamma_{1t} + e_{it}, \tag{7.32}$$

where $\gamma_{0t} = R_{0t}$ is the return on the zero-beta portfolio, while $\gamma_{1t} = R_{mt} - R_{0t}$. Let $\Gamma_t = (\gamma_{0t}, \gamma_{1t})^\top$, $t = 1, \ldots, T$. We suppose that e_{it} satisfy $E(e_{it}|\Gamma_1, \ldots \Gamma_T) = 0$ and $E(e_{it} e_{js}|\Gamma_1, \ldots \Gamma_T) = \sigma_{ij}$ if $t = s$ and zero otherwise. It follows that

$$\overline{R}_i = \overline{\gamma}_0 + \beta_i \overline{\gamma}_1 + \overline{e}_i, \tag{7.33}$$

where overbar indicates time series average as usual, e.g., $\overline{R}_i = \sum_{t=1}^T R_{it}/T$. This is now a cross-sectional regression with covariates consisting of the unit vector and the vector of betas. The errors in this regression are small (because of the time series average) but correlated. Let $X = (X_1, \ldots, X_N)^\top$, where $X_i = (1, \beta_i)^\top$, $\overline{R} = (\overline{R}_1, \ldots, \overline{R}_N)^\top$, and $\overline{\Gamma} = (\overline{\gamma}_0, \overline{\gamma}_1)^\top$. Suppose that the betas are observed, then the OLS estimator of $\overline{\Gamma}$ is

$$\widetilde{\Gamma} = (X^\top X)^{-1} X^\top \overline{R}.$$

The parameter $\overline{\Gamma}$ is the time series average of $\Gamma_t = (\gamma_{0t}, \gamma_{1t})^\top$, whereas the parameter of interest is $\Gamma = E(\Gamma_t)$. Therefore,

$$\widetilde{\Gamma} - \Gamma = \widetilde{\Gamma} - \overline{\Gamma} + \overline{\Gamma} - \Gamma = (X^\top X)^{-1} X^\top \overline{e} + \overline{\Gamma} - \Gamma,$$

which has two sources of variation. We suppose that Γ_t is a stationary martingale difference sequence with

$$\Xi = E\left((\Gamma_t - \Gamma)(\Gamma_t - \Gamma)^\top\right)$$

being finite and positive definite. It follows that

$$\operatorname{var}\left(\tilde{\Gamma} - \Gamma\right) = \frac{1}{T}\left((X^\mathsf{T} X)^{-1} X^\mathsf{T} \Omega_\varepsilon X(X^\mathsf{T} X)^{-1} + \Xi\right).$$

Furthermore,

$$\sqrt{T}\left(\tilde{\Gamma} - \Gamma\right) \Longrightarrow N(0, \Psi),$$

where $\Psi = (X^\mathsf{T} X)^{-1} X^\mathsf{T} \Omega_\varepsilon X(X^\mathsf{T} X)^{-1} + \Xi$. The main issue is how to obtain an estimate of Ψ. The FM approach is as follows. Let

$$\tilde{\Gamma}_t = (X^\mathsf{T} X)^{-1} X^\mathsf{T} R_t,$$

where $R_t = (R_{1t}, \ldots, R_{Nt})^\mathsf{T} = X\Gamma_t + e_t$, where $e_t = (e_{1t}, \ldots, e_{Nt})^\mathsf{T}$. It follows that $\tilde{\Gamma}_t - \Gamma = (X^\mathsf{T} X)^{-1} X^\mathsf{T} e_t + \Gamma_t - \Gamma$. They propose

$$\begin{aligned}
\tilde{\Psi} &= \frac{1}{T}\sum_{t=1}^{T}\left(\tilde{\Gamma}_t - \tilde{\Gamma}\right)\left(\tilde{\Gamma}_t - \tilde{\Gamma}\right)^\mathsf{T} \\
&= \frac{1}{T}\sum_{t=1}^{T}\left(X^\mathsf{T} X\right)^{-1} X^\mathsf{T} \left((R_t - \overline{R})(R_t - \overline{R})^\mathsf{T}\right) X\left(X^\mathsf{T} X\right)^{-1} \\
&= \left(X^\mathsf{T} X\right)^{-1} X^\mathsf{T} \left(\frac{1}{T}\sum_{t=1}^{T}(R_t - \overline{R})(R_t - \overline{R})^\mathsf{T}\right) X\left(X^\mathsf{T} X\right)^{-1}.
\end{aligned}$$

We have $R_t - \overline{R} = X(\Gamma_t - \overline{\Gamma}) + e_t - \overline{e}$ so that

$$\begin{aligned}
\frac{1}{T}\sum_{t=1}^{T}(R_t - \overline{R})(R_t - \overline{R})^\mathsf{T} &= \frac{1}{T}\sum_{t=1}^{T}\left(X(\Gamma_t - \overline{\Gamma}) + (e_t - \overline{e})\right)\left(X(\Gamma_t - \overline{\Gamma}) + (e_t - \overline{e})\right)^\mathsf{T} \\
&= X\left(\frac{1}{T}\sum_{t=1}^{T}(\Gamma_t - \overline{\Gamma})(\Gamma_t - \overline{\Gamma})^\mathsf{T}\right)X^\mathsf{T} + \frac{1}{T}\sum_{t=1}^{T}(e_t - \overline{e})(e_t - \overline{e})^\mathsf{T} \\
&\quad + X\frac{1}{T}\sum_{t=1}^{T}(\Gamma_t - \overline{\Gamma})(e_t - \overline{e})^\mathsf{T} + \frac{1}{T}\sum_{t=1}^{T}(e_t - \overline{e})(\Gamma_t - \overline{\Gamma})^\mathsf{T} X^\mathsf{T}.
\end{aligned}$$

Under our assumption that e_t is exogenous with respect to the process Γ_t, the cross product terms converge in probability to zero and it follows that as $T \to \infty$

$$\tilde{\Psi} \xrightarrow{P} \Psi.$$

This means that when the betas are observed we could calculate a confidence interval for γ_0 or γ_1.

We now consider a more general case that allows for testing of the CAPM restrictions. Suppose that S_i is another measure of risk (for example the idiosyncratic

7.4 Cross-sectional Regression Tests

standard deviation $\sigma_{\varepsilon i}$ from the market model). Then for each period t we may embed (7.32) into the more general regression model

$$Z_{it} = \gamma_{0t} + \beta_i \gamma_{1t} + \gamma_{2t}\beta_i^2 + \gamma_{3t} S_i + e_{it}, \tag{7.34}$$

where the coefficients γ_{jt} are the best linear predictor coefficients of Z_{it} (which may be returns or excess returns) at time t by a constant, β_i, β_i^2, and S_i.

Definition 79 *We assume that $\Gamma_t = (\gamma_{0t}, \gamma_{1t}, \gamma_{2t}, \gamma_{3t})^\top$ is a stationary process with finite variance and $\Gamma_t - \Gamma$ is a martingale difference sequence, where $\Gamma = E(\Gamma_t)$. We suppose that e_{it} satisfy $E(e_{it}|\Gamma_1, \ldots \Gamma_T) = 0$ and $E(e_{it}e_{js}|\Gamma_1, \ldots \Gamma_T) = \sigma_{ij}$ if $t = s$ and zero otherwise.*

The Sharpe–Lintner CAPM (with $Z_{it} = R_{it} - R_{ft}$) implies that $E(\gamma_{0t}) = E(\gamma_{2t}) = E(\gamma_{3t}) = 0$ and $\gamma_{1t} = Z_{mt} = R_{mt} - R_{ft}$. The Black CAPM implies that $E(\gamma_{2t}) = E(\gamma_{3t}) = 0$ and $E(\gamma_{0t})$ is the mean of the zero-beta asset, while $E(\gamma_{1t})$ is equal to the excess mean return on the market relative to the zero beta portfolio.

We suppose that β_i and S_i are observed. Let $X = (X_1, \ldots, X_N)^\top$, where $X_i = (1, \beta_i, \beta_i^2, \sqrt{\sigma_{ii}})^\top$, and let $Z_t = (Z_{1t}, \ldots, Z_{Nt})^\top$ and $\overline{Z} = T^{-1} \sum_{t=1}^{T} Z_t$. Then under the null hypothesis

$$Z_t = X\Gamma_t + e_t, \qquad \overline{Z} = X\overline{\Gamma} + \overline{e}, \tag{7.35}$$

where $\overline{\Gamma} = T^{-1} \sum_{t=1}^{T} \Gamma_t$. Then let

$$\widetilde{\Gamma}_t = \left(X^\top W X\right)^{-1} X^\top W Z_t \; ; \; \widetilde{\Gamma} = \left(X^\top W X\right)^{-1} X^\top W \overline{Z} = \frac{1}{T} \sum_{t=1}^{T} \widetilde{\Gamma}_t,$$

where W is an $N \times N$ positive definite symmetric weighting matrix, for example $W = I_N$. We allow for weighting by W because of the cross sectional correlation and heteroskedasticity in the errors. To evaluate the estimator $\widetilde{\Gamma}$ we may first condition on $\Gamma_1, \ldots, \Gamma_T$ and then average over the distribution of these random variables. We may write (7.35) equivalently as

$$Z_t = X\Gamma + u_t, \quad u_t = e_t - X(\Gamma_t - \Gamma), \tag{7.36}$$

and just treat Γ_t as another random variable uncorrelated with e_t. Let

$$\Xi = E\left[(\Gamma_t - \Gamma)(\Gamma_t - \Gamma)^\top\right],$$

which is now a 4×4 covariance matrix.

Theorem 33 *Under the null hypothesis we have: $E(\widetilde{\Gamma}_t|\Gamma_1, \ldots, \Gamma_T) = \Gamma_t$ and $E(\widetilde{\Gamma}|\Gamma_1, \ldots, \Gamma_T) = \frac{1}{T}\sum_{t=1}^{T} \Gamma_t$, and*

$$\Psi = \mathrm{var}(\widetilde{\Gamma}_t) = \left(X^\top W X\right)^{-1} X^\top W \Omega_\varepsilon W X \left(X^\top W X\right)^{-1} + \Xi,$$

and $\text{var}(\widetilde{\Gamma}) = \text{var}(\widetilde{\Gamma}_t)/T$. Furthermore,

$$\sqrt{T}\left(\widetilde{\Gamma} - \Gamma\right) \Longrightarrow N(0, \Psi)$$

as $T \to \infty$. In addition, we have

$$\widetilde{\Psi} = \frac{1}{T}\sum_{t=1}^{T}\left(\widetilde{\Gamma}_t - \widetilde{\Gamma}\right)\left(\widetilde{\Gamma}_t - \widetilde{\Gamma}\right)^{\mathsf{T}} \xrightarrow{P} \Psi$$

as $T \to \infty$. Consider the hypothesis $Q\Gamma = q \in \mathbb{R}^p$, where Q, q are known matrices. The Wald statistic satisfies

$$T\left(Q\widetilde{\Gamma} - q\right)^{\mathsf{T}}\left(Q\widetilde{\Psi}Q^{\mathsf{T}}\right)^{-1}\left(Q\widetilde{\Gamma} - q\right) \Longrightarrow \chi^2_p. \qquad (7.37)$$

In practice, we do not observe β_i and S_i and replace them by consistent estimates. To be specific we suppose that $S_i = \sqrt{\sigma_{ii}}$, then (for the Sharpe–Lintner case) proceed as follows.

First, estimate β_i for each asset ($i = 1, \ldots, N$ with N fixed) from linear regression using the time series data and the mean return or excess return for each asset as well as the idiosyncratic error variance, as defined in (6.6).

Second, estimate the second stage regression using the cross-section of mean returns, betas, and idiosyncratic error variance. Let $\widehat{X}$ be the $N \times 4$ matrix containing a column of ones, a column of $\widehat{\beta}_i$, a column of $\widehat{\beta}_i^2$, and a column of $\widehat{\sigma}_{\varepsilon i}$. Then let

$$\widehat{\Gamma} = (\widehat{\gamma}_0, \widehat{\gamma}_1, \widehat{\gamma}_2, \widehat{\gamma}_3)^{\mathsf{T}} = (\widehat{X}^{\mathsf{T}}W\widehat{X})^{-1}\widehat{X}W\overline{Z} = \frac{1}{T}\sum_{t=1}^{T}\widehat{\Gamma}_t \quad ; \quad \widehat{\Gamma}_t = (\widehat{X}^{\mathsf{T}}W\widehat{X})^{-1}\widehat{X}WZ_t$$

(7.38)

be the weighted least OLS estimate of $\Gamma = (\gamma_0, \gamma_1, \gamma_2, \gamma_3)^{\mathsf{T}}$.

Third, test the hypothesis that $\gamma_j = 0, j = 2, 3$ using t-statistics or Wald statistics like (7.37) but with $\widetilde{\Gamma}$ replaced by $\widehat{\Gamma}$.

The main issue is that in general the FM standard errors are incorrect because they ignore the errors in variables or generated regressor issue caused by using estimated quantities $\widehat{\beta}_i$ and $\widehat{\sigma}_{ii}$. This implies that the estimates of Γ are biased in small samples, but more importantly that the standard errors are inconsistent if applied to individual stocks. The issue of **generated regressors** was studied in Pagan (1984). Shanken (1992) proposes an analytical correction for this case. In the appendix we show the correct asymptotic variance under the large T and small N case. On the other hand, if the cross-section is composed of well-diversified portfolios formed from a large number of base assets, we argue there that the FM standard errors and Wald statistics can

7.4 Cross-sectional Regression Tests

be correct, since in that case the portfolio betas are estimated very precisely, i.e., the measurement error in $\widehat{\beta}_i$ is small.

In fact a key part of the FM methodology is to use a sophisticated portfolio grouping based on different samples to address the measurement error issue and various other statistical issues. We discuss this in more detail in the next section. We next give the full FM algorithm.

1. Estimate the market model to obtain security-level betas for n stocks in period A (**the portfolio formation period**), roughly seven years of monthly data. Sort the securities into N test portfolios (actually, 20 equally weighted portfolios based on the ranked individual betas) containing a roughly equal number of securities.
2. Re-estimate the market model for the individual securities over period B (**the estimation period**), roughly the next five years, to obtain security-level betas and idiosyncratic standard deviations.
3. Finally perform the cross-sectional regressions (7.34) using the portfolio returns from period C (**the testing period**), roughly the next four years, and the portfolio betas etc. calculated as the appropriate linear combination of the individual betas etc. calculated in period B. Test the null hypothesis using individual t-statistics or the Wald test (7.37).

The five year estimation period A is chosen because of possible non-stationary betas and alphas. Portfolio betas are estimated from separate five year periods because of the **regression phenomenon** – high estimated betas are associated with high positive measurement errors and vice versa – which would tend to overestimate (underestimate) the true betas. We explore this issue in Exercise 43. Testing on portfolios rather than individual stocks can mitigate the errors-in-variable problem as estimation errors cancel out each other, as we explore below. Sorting by beta reduces the shrinkage in beta dispersion and statistical power. Subsequent work has used **double-sorted portfolios**, including a secondary characteristic such as size (market capitalization). Performing the tests using data from period C seeks to make the measurement error from period B estimation independent of the error terms in the cross-sectional regression. FM used data from 1926–1968. They considered nine different periods (that is, nine different A-periods, B-periods, and C-periods) with the first testing period being 1935–1938. After various data filters there were between 435 and 845 securities employed per period.

Empirically, FM found that there is a positive and linear relationship between beta risk and return with a high R^2, but $\gamma_0 > 0$ and γ_1 was significantly lower than the market excess return. In Table 3 they find that averaging γ_{it} over the entire period 1934–1962 they can't reject the hypotheses that γ_2, γ_3 are indistinguishable from zero at the 5% level, but the t-statistic for γ_1 is only marginally significant (1.85) at this level when all the control variables are included (Panel D). When $\widehat{\sigma}_{\varepsilon i}$ is dropped the t-statistic for γ_1 becomes significant over the entire period, but the results over the subperiods are not as good. FM suggest this is due to the substantial month to month variability in the parameters. FM also find that behavior through time of $\widehat{\gamma}_{1t}, \widehat{\gamma}_{2t},$

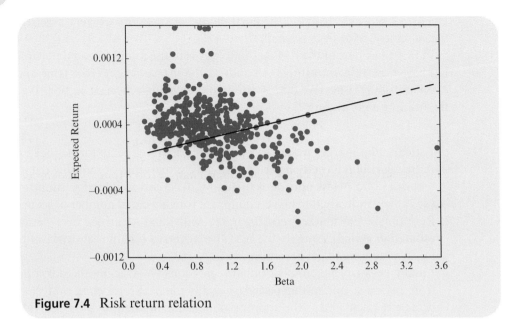

Figure 7.4 Risk return relation

$\widehat{\gamma}_{3t}$ are consistent with the market efficiency hypothesis, i.e. the autocorrelation function of $\widehat{\gamma}_{it}$ is close to zero statistically, although this finding again doesn't seem to be robust to the sample period or to the presence of explanatory variables. FM provide evidence that the portfolio grouping did actually reduce the measurement error problem, since the average standard deviation of the errors in the portfolio regressions were of the order of one-third to one-seventh of the average from the individual stock regressions. Sylvain (2013) replicates the FM study and updates the data to 2010. He also considered double sorted test portfolios with equal weight and value weight and different market proxies. He finds that many of their results are not robust to these minor changes in test portfolios or market proxies. The literature has indeed moved on to multifactor models following the Fama and French papers, and we will consider this in the next chapter.

In Figure 7.4 we show the cross-sectional regression of returns on estimated betas for the S&P500 stocks monthly data. There appears to be a mildly positive relationship, but the explanatory power is quite small.

7.5 Portfolio Grouping

Following FM many researchers used a small number of portfolios drawn from a large universe of assets. There are a number of advantages of this approach. One advantage is that one has to worry less about **survivorship bias**, which arises when using individual stocks because we can only include stocks that are still in business and trading. If we follow a firm through its life cycle, and that includes failure in the end, then the buy and hold return on that firm is minus 100%. In practice one selects a

7.5 Portfolio Grouping

sample that excludes stocks that fail in sample, which is not a representative sample of the universe of stocks. By employing portfolios one can in principle maintain a constant characteristic for the portfolio by rotating the stocks it includes. The usual practice is to sort the universe of stocks according to the observed characteristic at a certain date and then form the portfolios based on the quantiles of the characteristic distribution, i.e., the ranks of the stocks in that distribution. The portfolios are rebalanced periodically to maintain the characteristic value of the portfolio. However, if the distribution of stock characteristics changes over time, then the chosen portfolios are not maintaining constant absolute characteristics but relative characteristics. For example, size or market capitalization has changed over time considerably. In 1921, Coca Cola, a leading Dow Jones index constituent of the time, was valued at roughly $10 million. It is still in the Dow Jones index and hence in any large size portfolio, but is valued as of 2016 at roughly $200 billion. Clearly, size ain't what it used to be. In the case of beta sorted portfolios, there is less of an issue about a strong global trend, but there are some mild trends and variation over time in the level and dispersion of beta, as we see in Chapter 11.

Another econometric issue that portfolio grouping mitigates is the error in variables issue. This is because cross-sectional averaging reduces variance, as we shall see. Suppose that

$$\widehat{\beta}_i = \beta_i + \eta_{Ti}, \qquad (7.39)$$

where η_{Ti} is a mean zero error due to the time series estimation of $\widehat{\beta}_i$, the beta of stock i. Typically, η_{Ti} is small in the sense that $\mathrm{var}(\eta_{Ti}) \leq cT^{-1}$ for some constant c. Consider a portfolio p with (fixed) weights $\{w_{pi}, i=1,\ldots,n\}$ such that $\sum_{i=1}^{n} w_{pi} = 1$, where the number of base assets n is large. The beta of the portfolio is the weighted average of the betas of the constituent assets, i.e., $\widehat{\beta}_p = \sum_{i=1}^{n} w_{pi}\widehat{\beta}_i$. Therefore,

$$\widehat{\beta}_p = \sum_{i=1}^{n} w_{pi}\beta_i + \sum_{i=1}^{n} w_{pi}\eta_{Ti} = \beta_p + \sum_{i=1}^{n} w_{pi}\eta_{Ti} = \beta_p + \eta_{pTi},$$

and the measurement error in the portfolio beta η_{pTi} is mean zero and has variance $\mathrm{var}(\widehat{\beta}_p) = \sum_{i=1}^{n} w_{pi}^2 \mathrm{var}(\eta_{Ti})$ (assuming that the individual errors are cross-sectionally uncorrelated). For example, if $w_{pi} = 1/n$ and $\sigma_{\eta Ti}^2 = \sigma_{\eta}^2/T$ for all i, then the measurement error in the portfolio beta has variance σ_{η}^2/nT, which is of smaller order in magnitude than the measurement error in the individual stock provided n is large. This is a diversification effect, which we will consider in more detail in Chapter 8. Suppose that $p = 1, \ldots, N$ portfolios are considered. By the Cauchy–Schwarz inequality, $\mathrm{cov}(\widehat{\beta}_p, \widehat{\beta}_q)$ is also small in magnitude. This quantity can be exactly zero if we assume that the idiosyncratic errors are cross-sectionally uncorrelated and if the portfolios do not overlap, so that $\{i: w_{pi} \neq 0, w_{qi} \neq 0, p \neq q\}$ is empty. If the portfolios are formed by sorting the individual stocks on some characteristic and then partitioning them into N non-overlapping subsets, then this latter condition is satisfied.

A second econometric issue that arises from portfolio grouping is whether the second stage regression can have good properties. Recall that in linear regression, the necessary condition for the existence of the slope estimates is that the regressors have a positive sample variance. Furthermore, the standard errors decrease with the variability of the regressors. With this in mind it is desirable that the portfolio grouping preserves variation, i.e., if $p = 1, \ldots, P$, then we should seek to maximize

$$\sum_{p=1}^{P} \left(\widehat{\beta}_p - \overline{\widehat{\beta}} \right)^2.$$

One way of achieving this is to form the portfolios on the basis of betas, that is, to sort the estimated betas and then form portfolios of stocks with high betas, not so high betas, and so on. If we had formed portfolios randomly, say alphabetically, then we would find this variation to be very small.

The finance literature has generally used some form of portfolio grouping since FM. As we have seen, it is a very convenient method that has solid foundations. However, there is a perhaps minority view, which argues that portfolio grouping is not the best approach, starting with Roll (1977):

> *Specifically, the widely-used portfolio grouping procedure can support the theory even when it is false. This is because individual asset deviations from exact linearity can cancel out in the formation of portfolios. (Such deviations are not necessarily related to betas.)*

Brennan, Chordia, and Subrahmanyam (1998) argue similarly in favor of using individual stocks. Ang, Liu, and Schwarz (2009) report

> *The literature has argued that creating portfolios reduces idiosyncratic volatility and allows more precise estimates of factor loadings, and consequently risk premia. We show analytically and empirically that smaller standard errors of portfolio beta estimates do not lead to smaller standard errors of cross-sectional coefficient estimates. Factor risk premia standard errors are determined by the cross-sectional distributions of factor loadings and residual risk. Portfolios destroy information by shrinking the dispersion of betas, leading to larger standard errors.*

Connor and Korajczyk (1988) developed the asymptotic principal components (APC) method precisely to handle a large cross section of individual stocks. Their methodology is widely followed and developed in industry. See more recent work by Bai and Ng (2002) and Connor, Hagmann, and Linton (2012). Fan, Liao, and Mincheva (2015) and Pesaran and Yamagata (2017) are two recent studies that work with individual stocks, and we suspect that the big data era will bring many more studies using individual assets.

Portfolio grouping can reduce power. For example, suppose that

$$R_{it} = \gamma_{0t} + \gamma_{1t} \beta_i + \gamma_{2t} S_i + \eta_{it},$$

7.5 Portfolio Grouping

at the individual level, where S_i is some variable that affects expected return and η_{it} is an error term. Then

$$\sum_{i=1}^{n} w_{pi} R_{it} = \gamma_{0t} + \gamma_{1t} \sum_{i=1}^{n} w_{pi}\beta_i + \gamma_{2t} \sum_{i=1}^{n} w_{pi} S_i + \sum_{i=1}^{n} w_{pi}\eta_{it}.$$

It may be that even though $S_1 \neq 0, S_2 \neq 0, \ldots$ the weighted sum $\sum_{i=1}^{n} w_{pi} S_i$ is either zero or close to zero so that the power of the test is reduced. Any accountant knows that aggregating accounts and cumulating them over time helps to reduce the risk of detection for fraudulent transactions.

A final consideration, which is often overlooked, is that the portfolio weights are not deterministic; they are computed from a previous sample of data (from period A as we outlined above). In statistics, this would be called an estimator, and usually one would worry about the consequences of this estimation for subsequent estimation and testing. The segregation of the portfolio formation period from the estimation period is designed to minimize the influence of the portfolio formation on subsequent results, but let us just develop some concerns about this. To be specific, suppose that we calculate market capitalization at time $t=0$ based on observed price and number of shares outstanding, and then sort firms into portfolios on the basis of their market cap so determined. Typically, one might equally weight the top 5% of firms into one portfolio, equally weight the second largest 5% firms into the second portfolio, etc. In this case we would argue that the weights are themselves random variables determined by the specific realization of prices in period zero

$$\widehat{w}_{pi} = w_{pi}(P_{10}, \ldots, P_{n0}),$$

that is, in this case, $w_{pi} = \frac{1}{n_p}$ if $P_{i0} \in [P_{(p)0}, P_{(p+1)0}]$ and $w_{pi}=0$ otherwise, where $P_{(p)0}$ is some cutoff value or quantile of the time zero price empirical distribution (for simplicity we equate size with price here, i.e., number of shares is constant). Let $\widehat{\beta}_i$ be estimated from period B data. Then

$$\widehat{\beta}_p = \sum_{i=1}^{n} \widehat{w}_{pi}\widehat{\beta}_i = \sum_{i=1}^{n} w_{pi}\beta_i + \sum_{i=1}^{n} w_{pi}\eta_{Ti} + \sum_{i=1}^{n}(\widehat{w}_{pi} - w_{pi})\beta_i + \sum_{i=1}^{n}(\widehat{w}_{pi} - w_{pi})\eta_{Ti}$$

$$= \beta_p + \sum_{i=1}^{n} w_{pi}\eta_{Ti} + \sum_{i=1}^{n}(\widehat{w}_{pi} - w_{pi})\beta_i + \sum_{i=1}^{n}(\widehat{w}_{pi} - w_{pi})\eta_{Ti}.$$

The question is, should we take account of the uncertainty introduced by estimating the portfolio weights, and if we do, does it matter? One argument might be that we should just calculate everything conditional on $P_{10}, \ldots, P_{n0}$, as if they were ancillary statistics (Cox and Hinkley (1979)). But they clearly are not ancillary statistics because the marginal distribution of $P_{10}, \ldots, P_{n0}$ surely depends on the parameters $\beta_1, \ldots, \beta_n$. Does it matter? This is more difficult to assess theoretically. Indeed portfolio grouping is an example of **mostly harmless econometrics** because the main thinking and justification does not use a fully specified model.

In conclusion, the portfolio grouping method is a bit of a curate's egg; it has some advantages and some disadvantages. To some extent it was a 1970s solution to certain statistical problems, which can now be addressed by other techniques inside a fully or partially articulated model for individual stock returns. However, it is a practical and convenient method that is widely used and reflects to some extent industry practice.

7.6 Time Varying Model

The starting point of the market model and CAPM testing was that we have a sample of observations independent and identically distributed from a fixed population. This setting was convivial for the development of statistical inference. However, much of the practical implementations acknowledge time variation by working with short, say five year or ten year windows. A number of authors have pointed out the variation of estimated betas over time. Fernandez (2015) calculated the betas of 3,813 companies using 60 monthly returns for each day of December 2001 and January 2002, i.e., for each day the beta is calculated with respect to the S&P500 index using 60 months of past data. He finds that: historical betas change dramatically from one day to the next; only 2,780 companies (out of 3,813) had all estimated betas positive; and historical betas depend very much on which market index is used to calculate them. Indeed, he points out that different industry providers of beta give quite different values for this significant parameter for specific stocks.

We show estimated betas for IBM (against the S&P500) computed from daily stock returns using a five year window over the period 1962–2017. We present the rolling window estimates for each such five year period. It is apparent that there has been substantial variation of the IBM betas over this period with a general upward trend, indicating that it has become a more risky stock. On the other hand the corresponding alphas for IBM are very small and fluctuate above and below zero with no clear trend.

This phenomenon is not limited to individual stocks, In Figure 7.7 we show the time varying beta of the so-called SMB portfolio that we discuss in the next chapter against the market portfolio. The beta is calculated using a ten-year rolling window of monthly data, and one can see clearly the sometimes rapid changes it goes through.

In the presence of time variation in the parameters the usual tests considered above are inconsistent. We next consider a more general framework where time variation is explicitly considered. Suppose that

$$Z_{it} = \alpha_{it} + \beta_{it} Z_{mt} + \varepsilon_{it},$$

where α_{it}, β_{it} vary over time, but otherwise the regression conditions are satisfied, i.e., $E(\varepsilon_{it}|Z_{mt}) = 0$. We may allow $\sigma_{it}^2 = \text{var}(\varepsilon_{it}|Z_{mt})$ to vary over time and asset.

One framework that is implicit in a lot of work is that $\beta_{it} = \beta_{i1}$ in the subperiod 1, $\beta_{it} = \beta_{i2}$ in subperiod 2, etc. In this case, if we use data from subperiod 1 to estimate the parameters and construct, say, the Wald statistic, we can treat the parameters as fixed within our estimation window (one can then aggregate the test statistics across subsamples). This model does not seem very realistic, especially since the choice of

7.6 Time Varying Model

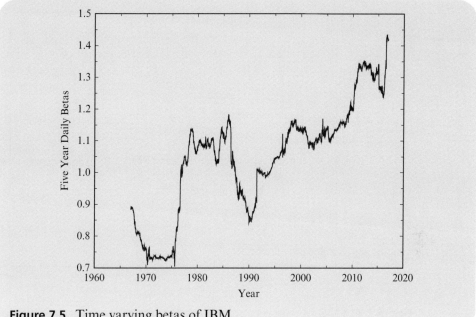

Figure 7.5 Time varying betas of IBM

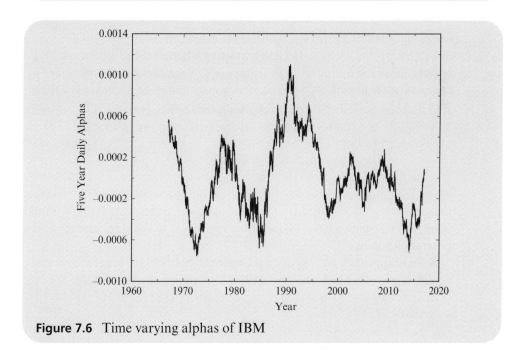

Figure 7.6 Time varying alphas of IBM

starting and ending periods of subsamples are arbitrary. A more realistic model is to assume that the parameters slowly evolve rather than jump at our convenience. Suppose that

$$\alpha_{it} = \alpha_i(t/T), \quad \beta_{it} = \beta_i(t/T), \tag{7.40}$$

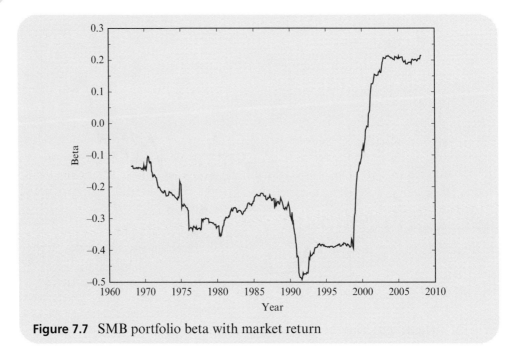

Figure 7.7 SMB portfolio beta with market return

where $\alpha_i(.)$ and $\beta_i(.)$ are left continuous functions (and possibly differentiable). We can reconcile this with the CAPM by testing whether $\alpha_i(u) = 0$ for $u \in [0, 1]$. This can be done using the rolling window framework. Note that it is not necessary to provide a complete specification for σ_{it}^2 as one can construct heteroskedasticity robust inference. Ang and Kristensen (2012) consider long run alphas given by $\int_0^1 \alpha_i(u)du$.

An alternative framework is the unobserved components model

$$\alpha_{it} = \alpha_{i,t-1} + \eta_{it}, \quad \beta_{it} = \beta_{i,t-1} + \xi_{it}, \tag{7.41}$$

where η_{it}, ξ_{it} are i.i.d. shocks with mean zero and variances $\sigma_\eta^2, \sigma_\xi^2$ (Harvey (1998)). One may take $\alpha_{i0} = 0$ and initialize β_{i0} in some other way. In this framework, the issue is to test whether $\sigma_\eta^2 = 0$ (which corresponds to constant and zero α) versus the general alternative.

A final approach is to specify a dynamic multivariate generalized autoregressive conditional heteroskedasticity (GARCH) model or stochastic volatility model, as we discuss in Chapter 11 below.

7.7 Empirical Evidence on the CAPM

There are some practical issues that we will briefly recap. Which assets should be included in the sample? Most studies consider portfolios, but there are a number of studies using individual assets. What sampling frequency should be used? Most studies use monthly data but there are a growing number of studies that employ a

7.7 Empirical Evidence on the CAPM

higher frequency like daily or weekly. How long a time series should be considered? Most studies consider five years or ten years data at a time to mitigate nonstationarity issues. What market portfolio should be used? CRSP indexes are widely used in academic studies, but these are non traded, or rather there are no traded futures contracts on these indexes. What risk free rate should be used? Most US studies take the three month or one month T-bill rate.

There are a number of well established anomalies in the literature. Keim (1983) documented the **size effect**. This is that firms with a low market capitalization seem to earn positive abnormal returns ($\alpha > 0$), while large firms earn negative abnormal returns ($\alpha < 0$). Banz (1981) documented the **value effect**. This is that firms with high value metrics relative to market value seem to earn positive abnormal returns ($\alpha > 0$), while firms with low value metrics relative to market value (also known as growth stocks) have $\alpha < 0$. The typical value metrics are dividend to price ratio (D/P) and book value to market value ratio (B/M). (In accounting, book value is the value of an asset according to its balance sheet account balance. For assets, the value is based on the original cost of the asset less any depreciation, amortization, or impairment costs made against the asset. Traditionally, a company's book value is its total assets minus intangible assets and liabilities. However, in practice, depending on the source of the calculation, book value may sometimes include goodwill or intangible assets, or both. The value inherent in its workforce, part of the intellectual capital of a company, is always ignored. When intangible assets and goodwill are explicitly excluded, the metric is often specified to be "tangible book value.") Finally, Jegadeesh and Titman (1993) documented the **momentum effect**.

Most of the classical studies of the CAPM have been done on the US market, but there are also many from around the world. The Chinese economy and stock market are now very close in size to the US, but they have some special features. The Shanghai Stock Exchange (SSE) and the Shenzhen Stock Exchange (SZSE) were launched in 1990 and 1991, yet price controls are implemented and separate classes of shares are established as a means of capital restrictions. Hong Kong, however, is a special administrative region of China and has a very different institutional setting under the "one country, two systems" principle. The Hong Kong Stock Exchange (SEHK) was established in 1891 as a free open market without restrictions on price fluctuations and capital flows. A-shares and H-shares are two major share classes issued by China-based companies. Primarily, A-shares are limited to domestic investors, listed on the SSE or the SZSE, and denominated in renminbi; and H-shares are limited to foreign investors (investors in Hong Kong and the rest of the world), listed on the SEHK, and denominated in Hong Kong dollars. Even with the recent relaxation of capital restrictions, A- and H-share markets remain segmented to a large extent. A-shares usually enjoy a price premium over their corresponding H-shares despite being issued by the same companies and granting equal ownership rights. The price differential reflects significant institutional impact on stock markets. It is therefore of considerable interest to compare dual-listed AH-shares. Ng (2014) considered: (i) whether there is convergence in AH-share prices; (ii) whether there are long-run cointegrating relationships between AH-share prices; (iii) how the AH-listings differ in terms of risk and return; (iv) whether price disparity implies inefficiencies, i.e., overvaluation of A-shares and/or undervaluation of H-shares; and (v) how risks are priced in the partially segmented markets.

7.7.1 Some Critiques of CAPM Testing

There are a number of issues with the evidence regarding the CAPM. Well known is the **Roll critique**, which says that central to the CAPM predictions is that the market portfolio is mean variance efficient, but one can't actually observe the market portfolio, since it includes many non-traded or infrequently traded assets. It could be that rejections of the CAPM and specific anomalies are therefore not valid. We will see later that asset pricing models based on the multifactor set up avoid this critique by not requiring knowledge of the specific factors. The Roll critique is a bit of a party pooper, but is logically valid. Exercise 47 asks you to show how to test the CAPM when a certain proxy variable for market returns is available. Another line of development has been to distinguish between ex-ante versus ex-post betas and to embed the CAPM in a dynamic framework; we will discuss this later. Taking into account the possible difference between ex-post and ex-ante parameters, in a recent paper, Levy and Roll (2010) show that when only small changes in the sample means and standard deviations are made, the observed market portfolio is mean variance efficient, which according to Roll (1977) implies that the linear CAPM is intact. They employ a novel reverse engineering approach. Levy (1997) shows that the probability of obtaining a positive tangency portfolio based on sample parameters converges to zero exponentially with the number of assets. However, at the same time, very small adjustments in the return parameters, well within the estimation error, yield a positive tangency portfolio. Hence, looking for positive portfolios in the parameter space is somewhat like looking for rational numbers on the number line: if a point in the parameter space is chosen at random it almost surely corresponds to a non-positive portfolio (an irrational number); however, one can find very close points in parameter space corresponding to positive portfolios (rational numbers). Normality or rather elliptical symmetry is crucial to the derivation of the CAPM, but the normal distribution is statistically strongly rejected in the data. Finally, the CAPM only has negligible explanatory power, that is, the cross-sectional $R^{2\prime}s$ are quite small.

7.7.2 Active versus Passive Portfolio Management

Active portfolio management can be defined as attempts to achieve superior returns α through security selection and market timing. Under the CAPM, such superior returns cannot be achieved, so the existence of a large active management industry seems to contradict this theory. **Security selection** involves picking mispriced individual securities, trying to buy low and sell high or short-sell high and buy back low. **Market timing** involves trying to enter the market at troughs and leave at peaks. Grinold and Kahn (1999) is a classic treatment of this subject. **Hedge funds** are often classified by their self-described investment style: event driven, e.g., spin-offs, mergers, bankruptcy reorganisations, etc.; global are funds that invest in any non-US or emerging market equities; global or macro and those investing on the basis of macroeconomic analyses, typically via major interest rate shifts; market neutral are those funds that actively seek to avoid major risk factors, mainly focussing on apparent mispricing. In practice, maintaining market neutrality during crisis periods is hard (Patton (2009)). The asset management industry abounds with conflicts of interest (agency issues) between owners and managers, since the managers are often remunerated according to some defined method of fund valuation that can to some extent be manipulated by them.

So what is to be made of the existence of an active management industry when theoretically it should not deliver what it claims it can?

Example 47 *If an active manager overseeing a $5 billion portfolio could increase the annual return by 0.1%, her services would be worth up to $5 million. Should you invest with her? The role of luck is also important and should be taken account of. Imagine 10,000 managers whose strategy is to park all assets in an index fund but at the end of every year to use a quarter of it to make (independently) a single bet on red or black in a casino. After 10 years, many of them no longer keep their jobs but several survivors have been very successful (the probability of drawing 10 successes in a row is approximately $(1/2)^{10} \simeq 1/1000$). The **infinite monkey theorem** says that if one had an infinite number of monkeys randomly tapping on a keyboard, with probability one, one of them will produce the complete works of Shakespeare. If one has a finite set of characters on the typewriter K and a finite length document n, then the probability that any one monkey would type this is K^{-n}. If there are 47 keys on the standard typewriter, and 884,421 words, so perhaps 5 million characters, in which case the probability is so low that a given monkey will produce the documents. However, the probability that no monkeys would produce it is*

$$\lim_{M \to \infty} \left(1 - K^{-n}\right)^M = 0.$$

Mind you, if the monkey that had typed the complete works of Shakespeare, was then asked to type the complete works of Goethe, and succeeded, then you might draw a different inference.

Passive portfolio management involves tracking a predefined index of securities with no security analysis whatsoever. This amounts to choosing β rather than α. The vehicles of choice are ETFs. There is a big growth in the passive sector. It is much cheaper than active management since there are no costs of information acquisition and analysis, there are lower transaction costs (less frequent trading) and there is also generally greater risk diversification. There are some concerns that this affects price discovery and pins relative prices into a narrow range that does not reflect fundamentals.

7.8 Summary of Chapter

We defined the empirical implementation of portfolio choice. We discussed the two common approaches to testing the CAPM based on the maximum likelihood method and the cross-sectional regression method. We discussed the portfolio grouping method and how it addresses some of the key statistical problems.

7.9 Appendix

7.9.1 Aggregation Results

Theorem 34 *Suppose that returns follow a stationary martingale difference sequence after a constant mean adjustment and possess four moments. Then*

$$\kappa_3(r_A) = \frac{1}{\sqrt{A}}\kappa_3(r) + \frac{3}{\sqrt{A}}\sum_{j=1}^{A-1}\left(1 - \frac{j}{A}\right) E(\widetilde{r}_t^2 \widetilde{r}_{t-j}),$$

where $\widetilde{r}_t = (r_t - E(r_t))/\text{std}(r_t)$. *For the kurtosis we have*

$$\kappa_4(r_A) = \frac{1}{A}\kappa_4(r) + \frac{4}{A}\sum_{j=1}^{A-1}\left(1 - \frac{j}{A}\right) E(\widetilde{r}_t^3 \widetilde{r}_{t-j}) + \frac{6}{A}\sum_{j=1}^{A-1}\left(1 - \frac{j}{A}\right)\left[E(\widetilde{r}_t^2 \widetilde{r}_{t-j}^2) - 1\right]$$

$$+ \frac{6}{A}\sum_{j=1}^{A-1}\sum_{\substack{k=1\\j\neq k}}^{A-1}\left(1 - \frac{j}{A}\right)\left(1 - \frac{k}{A}\right) E(\widetilde{r}_t^2 \widetilde{r}_{t-j} \widetilde{r}_{t-k}).$$

Sketch Proof. We consider weekly returns:

$$\begin{aligned}
(r_1 + r_2 + r_3 + r_4 + r_5)^3 &= r_1^3 + r_2^3 + r_3^3 + r_4^3 + r_5^3 \\
&\quad + 3r_1^2 r_2 + 3r_2^2 r_3 + 3r_3^2 r_4 + 3r_4^2 r_5 \\
&\quad + 3r_1 r_2^2 + 3r_2 r_3^2 + 3r_3 r_4^2 + 3r_4 r_5^2 \\
&\quad + 3r_1^2 r_3 + 3r_2^2 r_4 + 3r_3^2 r_5 + 3r_1 r_3^2 + 3r_2 r_4^2 + 3r_3 r_5^2 \\
&\quad + 3r_1^2 r_4 + 3r_2^2 r_5 + 3r_1 r_4^2 + 3r_2 r_5^2 + 3r_1^2 r_5 + 3r_1 r_5^2 \\
&\quad + 6r_1 r_2 r_3 + 6r_2 r_3 r_4 + 6r_3 r_4 r_5 + 6r_1 r_2 r_4 + 6r_2 r_3 r_5 \\
&\quad + 6r_1 r_2 r_5 + 6r_1 r_3 r_4 + 6r_2 r_4 r_5 + 6r_1 r_3 r_5 + 6r_1 r_4 r_5.
\end{aligned}$$

If we have a martingale difference sequence in expectation this reduces to

$$(5)E(r_1^3) + (12)E(r_2^2 r_1) + (9)E(r_3^2 r_1) + (6)E(r_4^2 r_1) + (3)E(r_5^2 r_1).$$

In general

$$E(r_A^3) = AE(r_t^3) + 3(A-1)E(r_t^2 r_{t-1}) + 3(A-2)E(r_t^2 r_{t-2}) + \cdots + 3E(r_t^2 r_{t+1-A}). \qquad \square$$

This shows that the rate at which aggregation works is the same, but the constants reflect the more complicated dependence structure possible in the martingale difference sequence. In conclusion, the process of aggregation should yield low frequency returns with less skewness and kurtosis, at least when A is large.

7.9 Appendix

7.9.2 Properties of $\sum_{i=1}^{N} |\widehat{\alpha}_i|$

Suppose that X, Y are bivariate standard normal with correlation ρ. We have

$$E(|X|) = E(|Y|) = \sqrt{\frac{2}{\pi}}$$

$$\text{var}(|X|) = 1 - \frac{2}{\pi}.$$

By Psarakis and Panaretos (2001, p123),

$$E(|Y| \,|\, |X|) = |\rho| \times |X|.$$

Therefore

$$E(|Y| \times |X|) = |\rho| \times E(X^2) = |\rho|$$

$$\text{cov}(|Y|, |X|) = |\rho| - \frac{2}{\pi}.$$

Suppose that X, Y are bivariate normal with zero means and variances σ_X^2 and σ_Y^2 and correlation ρ. Then

$$\text{cov}(|Y|, |X|) = \sigma_X \sigma_Y \left(|\rho| - \frac{2}{\pi}\right).$$

We suppose that $\widehat{\alpha}$ is jointly normal or approximately normal with the stated mean and covariance matrix. Then

$$E\left(\sum_{i=1}^{N} |\widehat{\alpha}_i|\right) = \sum_{i=1}^{N} E(|\widehat{\alpha}_i|) = \sum_{i=1}^{N} \sqrt{\text{var}(\widehat{\alpha}_i)} \sqrt{\frac{2}{\pi}} = \sum_{i=1}^{N} \left(1 + \frac{\mu_m^2}{\sigma_m^2}\right) \sigma_{ii} \sqrt{\frac{2}{\pi}}$$

$$\text{var}\left(\sum_{i=1}^{N} |\widehat{\alpha}_i|\right) = \sum_{i=1}^{N} \text{var}(|\widehat{\alpha}_i|) + \sum\sum_{i \neq j} \text{cov}(|\widehat{\alpha}_i|, |\widehat{\alpha}_j|)$$

$$= \left(1 + \frac{\mu_m^2}{\sigma_m^2}\right) \times \left(\left(1 - \frac{2}{\pi}\right) \sum_{i=1}^{N} \sigma_{ii} + \sum\sum_{i \neq j} \left(|\sigma_{ij}| - \frac{2}{\pi} \sqrt{\sigma_{ii} \sigma_{jj}}\right)\right).$$

7.9.3 Two Pass Estimators

We consider here the two pass cross-sectional regression estimators with estimated betas; for simplicity we just consider the case where $X_i = (1, \beta_i)^\mathsf{T}$ and $\Gamma = (\gamma_0, \gamma_1)^\mathsf{T}$. We suppose that

$$\widehat{\beta}_i = \frac{\sum_{t=1}^{T} (R_{mt} - \overline{R}_m) R_{it}}{\sum_{t=1}^{T} (R_{mt} - \overline{R}_m)^2} = \frac{1}{\widehat{\sigma}_m^2} \frac{1}{T} \sum_{t=1}^{T} (R_{mt} - \overline{R}_m) R_{it},$$

where $\hat{\sigma}_m^2 = \sum_{t=1}^T (R_{mt} - \overline{R}_m)^2/T$.
We write
$$Z_t = X\Gamma + u_t = \widehat{X}\Gamma + v_t, \quad \overline{Z} = \widehat{X}\Gamma + \overline{v},$$
where $v_t = (X - \widehat{X})\Gamma + u_t$ and $\overline{v} = (X - \widehat{X})\Gamma + \overline{u}$. We have
$$\widehat{\Gamma}_t = \Gamma + \left(\widehat{X}^\top W\widehat{X}\right)^{-1} \widehat{X}^\top W v_t, \quad \widehat{\Gamma} = \Gamma + \left(\widehat{X}^\top W\widehat{X}\right)^{-1} \widehat{X}^\top W\overline{v}.$$

Since $\widehat{X} \xrightarrow{P} X$ as $T \to \infty$, we have
$$\widehat{X}^\top W\widehat{X} \xrightarrow{P} X^\top WX$$
as $T \to \infty$. Furthermore,
$$\sqrt{T}\left(\widehat{\Gamma} - \Gamma\right) = \left(X^\top WX\right)^{-1} X^\top W\overline{v} + \mathfrak{R}_T, \tag{7.42}$$
where the remainder term $\mathfrak{R}_T \xrightarrow{P} 0$ as $T \to \infty$. The key issue is to obtain var$(\overline{v})$. We have
$$\sqrt{T}\overline{v} = \sqrt{T}\overline{u} - \sqrt{T}(\widehat{X} - X)\Gamma$$
and both terms contribute in general and indeed they might be mutually correlated since they use the same time series. We already calculated var(u_t).
We have
$$\left(X^\top WX\right)^{-1} X^\top W\sqrt{T}\overline{v} = \left(X^\top WX\right)^{-1} X^\top W\sqrt{T}\overline{e} - \sqrt{T}(\overline{\Gamma} - \Gamma)$$
$$- \left(X^\top WX\right)^{-1} X^\top W\sqrt{T}\left(\widehat{\beta} - \beta\right)\gamma_1 \simeq AU_T,$$
where A is a $2 \times (2N+2)$ matrix and U_T is a $(2N+2) \times 1$ vector of stochastic variables:
$$A = \left(\left(X^\top WX\right)^{-1} X^\top W, -1, -1, -\frac{\gamma_1}{\sigma_m^2}\left(X^\top WX\right)^{-1} X^\top W\right);$$
$$U_T = \begin{pmatrix} \frac{1}{\sqrt{T}}\sum_{t=1}^T e_t \\ \frac{1}{\sqrt{T}}\sum_{t=1}^T (\Gamma_t - \Gamma) \\ \frac{1}{\sqrt{T}}\sum_{t=1}^T (R_{mt} - E(R_{mt}))e_t \end{pmatrix},$$
where $\sigma_m^2 = \text{var}(R_{mt})$. Under the null hypothesis, U_T is mean zero and the covariance matrix of U_T is block diagonal with
$$\Lambda = \begin{pmatrix} \Omega_\varepsilon & 0 & 0 \\ 0 & \Xi & 0 \\ 0 & 0 & \sigma_m^2\Omega_\varepsilon \end{pmatrix}.$$

7.9 Appendix

In this case, $\Xi = \text{diag}\{\sigma_0^2, \sigma_m^2 + \sigma_0^2\}$, where $\sigma_0^2 = \text{var}(R_{0t})$ since $\text{var}(\gamma_{1t}) = \text{var}(R_{mt} - R_{0t}) = \sigma_m^2 + \sigma_0^2$ and the zero beta portfolio is uncorrelated with the market return. This follows because we are assuming that $E(e_t e_t^\top | \Gamma_1, \ldots, \Gamma_T) = \Omega_\varepsilon$ does not depend on Γ_t and so $E((R_{mt} - E(R_{mt})) e_t e_t^\top) = E(R_{mt} - E(R_{mt})) \Omega_\varepsilon = 0$. Under time series heteroskedasticity this is not true, and the limiting variance is more complicated. Using a different period B for estimation of β from period C for carrying out the cross-sectional regression can ensure the block diagonal structure even under heteroskedasticity.

Define the following limiting covariance matrix

$$\Phi^* = \lim_{T \to \infty} \text{var}(\sqrt{T} \overline{v}) = \lim_{T \to \infty} \text{var}(\sqrt{T} \overline{u}) + \lim_{T \to \infty} \text{var}\left(\sqrt{T}\left(\widehat{\beta} - \beta\right)\gamma_1\right)$$

$$= A\Lambda A^\top = \Phi + \frac{\gamma_1^2}{\sigma_m^2} \Omega_\varepsilon.$$

Theorem 35 *Under the null hypothesis as $T \to \infty$ with N fixed*

$$\sqrt{T}(\widehat{\Gamma} - \Gamma) \Longrightarrow N(0, \Upsilon)$$

$$\Upsilon = \left(X^\top W X\right)^{-1} X^\top W \Phi^* W X \left(X^\top W X\right)^{-1}.$$

We can also show that the FM standard errors are off target here. We have

$$\widehat{\Gamma}_t - \widehat{\Gamma} = \left(\widehat{X}^\top W \widehat{X}\right)^{-1} \widehat{X}^\top W(v_t - \overline{v}) = \left(\widehat{X}^\top W \widehat{X}\right)^{-1} \widehat{X}^\top W(u_t - \overline{u})$$

and furthermore

$$\widehat{\Psi} = \frac{1}{T} \sum_{t=1}^{T} \left(\widehat{\Gamma}_t - \widehat{\Gamma}\right)\left(\widehat{\Gamma}_t - \widehat{\Gamma}\right)^\top \xrightarrow{P} \Psi$$

so that $\widehat{\Psi}$ ignores the contribution from the preliminary estimation. Instead we can consistently estimate Υ by

$$\widehat{\Upsilon} = \widehat{\Psi} + \frac{\widehat{\gamma}_1^2}{\widehat{\sigma}_m^2} \left(\widehat{X}^\top W \widehat{X}\right)^{-1} \widehat{X}^\top W \widehat{\Omega}_\varepsilon W \widehat{X} \left(\widehat{X}^\top W \widehat{X}\right)^{-1}.$$

The portfolio grouping approach essentially uses a large cross-section n to produce a smaller cross section of assets for testing of size N. Suppose that the market model holds

$$R_{it} = \alpha_i + \beta_i R_{mt} + \varepsilon_{it},$$

where ε_{it} satisfies (6.3). Then (with an abuse of the definition of return as acknowledged earlier) the portfolio $R_{pt} = \sum_{i=1}^{n} w_{pi} R_{it}$, where $\sum_{i=1}^{n} w_{pi} = 1$, satisfies

$$R_{pt} = \alpha_p + \beta_p R_{mt} + \varepsilon_{pt},$$

where $\beta_p = \sum_{i=1}^{n} w_{pi}\beta_i$ and $\varepsilon_{pt} = \sum_{i=1}^{n} w_{pi}\varepsilon_{it}$. We construct $p = 1, 2, \ldots, N$ portfolios that are non overlapping. If the portfolios are well diversified, for example (see the next chapter), say equally weighted on $n_P = n/P$ assets, then the portfolio error variance $\sigma_p^2 = \text{var}(\varepsilon_{pt})$ is small when n is large. In this case $\widehat{\beta}_p$ can be considered to be $\sqrt{nT}$ consistent. It follows that provided $n, T \to \infty$ with N fixed

$$\left(X^\intercal W X\right)^{-1} X^\intercal W \sqrt{T}\widehat{v} = \left(X^\intercal W X\right)^{-1} X^\intercal W \sqrt{T}\overline{e} - \sqrt{T}(\overline{\Gamma} - \Gamma) + \mathfrak{R}_T,$$

where $\mathfrak{R}_T \xrightarrow{P} 0$ and the FM standard errors are consistent.

See Jagannathan, Skoulakis, and Wang (2010) for a more general treatment that allows for serial correlation, time series heteroskedasticity, and a variety of other features.

8 Multifactor Pricing Models

In this chapter we consider multifactor models for returns. From an econometric point of view this is a natural extension of the previous chapter. We discuss some of the key concepts needed in the arbitrage pricing theory, and we discuss the testing of this theory. Multifactor models are now the central econometric tool for a number of purposes, including event studies and portfolio risk management.

8.1 Linear Factor Model

We suppose that one period random returns for firm i are generated by the population model

$$R_i = \alpha_i + b_i^\intercal f + \varepsilon_i, \tag{8.1}$$

where $b_i, f \in \mathbb{R}^K$ with $K \geq 1$ and $i = 1, \ldots, N$. We think of the factors f as being the common component of returns (they are not indexed by i) and the ε_i as being the idiosyncratic component, i.e., firm i specific. Multifactor models are useful because they reduce the effective dimensionality of the covariance matrix of returns, which is helpful in portfolio choice problems. They are used to control for risk in event studies. They are also at the heart of the APT. In the **strict factor model** the idiosyncratic component ε_i is assumed to be uncorrelated with ε_j for $i \neq j$. In the **weak factor model**, we allow ε_i to be correlated with ε_j, although the amount of correlation has to be restricted to guarantee that the APT holds as the number of assets increases. Under some restrictions, the APT implies that the expected returns μ_i of asset i are of the form

$$\mu_i = \lambda_0 + b_i^\intercal \lambda,$$

where $\lambda \in \mathbb{R}^K$ are risk premia associated with the factor vector. If there is a risk free rate R_f, we have $\lambda_0 = R_f$ and $\lambda_k = E(f_k) - R_f$, and it follows that $\mu_i - R_f = b_i^\intercal (E(f) - R_f i_K)$. The CAPM corresponds to the case where $K = 1$ and f_1 is the return on the market portfolio. The APT theory holds more generally provided the linear factor model holds and the size of the economy (N) is large. It doesn't say what the factors are. It does not require normality or any distributional restrictions.

We next address the two key features that are needed for the application of multifactor models and the APT. First, the common factors are pervasive in the sense that they affect a lot of stock returns, i.e., most of the b_i are non zero. Second, if we form a large portfolio with fairly equal weights, we can eliminate the contribution of the idiosyncratic errors to the risk of the portfolio. We will consider these two issues in detail in the next sections. We then consider testing of the APT in the case where the

8.2 Diversification

In this section we examine the notion of diversifiability, which allows for the elimination of idiosyncratic components from the risk of large portfolios. Why and when does the diversification used in (1.33) work? The general principal is that averaging reduces variance provided the correlation between variables is not perfect. Consider a portfolio of two zero mean and finite variance assets X, Y, which has return

$$R_w = wX + (1-w)Y.$$

Each value of $w \in \mathbb{R}$ corresponds to a different portfolio, whose variance is

$$V(w) = w^2 \sigma_X^2 + (1-w)^2 \sigma_Y^2 + 2w(1-w)\rho_{XY}\sigma_X\sigma_Y,$$

where ρ_{XY} is the correlation between X, Y, while σ_X^2 and σ_Y^2 are the variances of the individual assets. Note that if $w \in [0,1]$, then portfolio variance is always at least as good as the worst case individual asset, i.e., $\max_{w\in[0,1]} V(w) \leq \max\{\sigma_X^2, \sigma_Y^2\}$. If short selling is allowed, i.e., $w > 1$ or $w < 0$, then we can have $V(w)$ larger than $\max\{\sigma_X^2, \sigma_Y^2\}$, meaning we can really mess up. On the other hand we can do better than the individual assets, for example, setting $w=0$ or $w=1$, i.e., we have

$$\min_{w \in \mathbb{R}} V(w) \leq \min\{\sigma_X^2, \sigma_Y^2\},$$

with strict inequality if and only if $\rho_{XY} \neq +1$.

Suppose that $\sigma_X^2 = \sigma_Y^2 = 1$, then (solving $dV(w)/dw = 0$) we have

$$w_{opt} = \frac{1}{2}; \quad V(w_{opt}) = \frac{1}{2}(1+\rho_{XY}) \leq 1$$

with strict inequality if and only if $\rho_{XY} \neq +1$. That is, the optimal portfolio improves upon the individual assets unless the correlation between them is exactly plus one. With $\rho_{XY} = 0$, the optimal portfolio achieves half the variance of the individual assets.

We now consider the general case of N assets generated by the linear factor model where the factors have $K \times K$ covariance matrix $\text{var}(f) = \Omega_K$ and the idiosyncratic error terms have $N \times N$ covariance matrix $E(\varepsilon\varepsilon^\top) = \Omega_\varepsilon$. In this case, the variance of a portfolio with weights $w \in \mathbb{R}^N$ is the sum of two terms

$$\text{var}(w^\top R_t) = \underbrace{w^\top B\Omega_K B^\top w}_{\text{common}} + \underbrace{w^\top \Omega_\varepsilon w}_{\text{idiosyncratic}}, \tag{8.2}$$

8.2 Diversification

by virtue of the orthogonality between the factors and the idiosyncratic error terms. For some large portfolios we can eliminate the idiosyncratic contribution to the portfolio risk provided the error covariance matrix satisfies certain restrictions. This needs some clarification.

We give a formal definition of diversifiability.

Definition 80 *We say that the idiosyncratic risk is diversifiable if* $\lim_{N\to\infty} w^\top w = 0$ *implies that*

$$\lim_{N\to\infty} w^\top \Omega_\varepsilon w = 0.$$

This is the sense in which the idiosyncratic component can be made small. We consider the diagonal case first. Suppose that $\Omega_\varepsilon = \text{diag}\{\sigma_1^2, \ldots, \sigma_N^2\}$. Then

$$\text{var}(w^\top \varepsilon) = w^\top \Omega_\varepsilon w = \sum_{i=1}^N w_i^2 \sigma_i^2.$$

We have

$$\sum_{i=1}^N w_i^2 \sigma_i^2 \leq \max_{1 \leq i \leq N} \sigma_i^2 \sum_{i=1}^N w_i^2,$$

and it suffices that $\sigma_i^2 \leq c < \infty$ for all i. So if all the variances are bounded then clearly diversification works for well-balanced portfolios. Equal weighting with $w_i = 1/N$ satisfies the property $\lim_{N\to\infty} w^\top w = 0$ and in fact $w^\top w = 1/N \to 0$ quite fast. For this particular sequence of weights, $\lim_{N\to\infty} w^\top \Omega_\varepsilon w = 0$ can hold under the weaker condition that $\max_{1 \leq i \leq N} \sigma_i^2 / N \to 0$, e.g, $\sigma_i^2 = i^\alpha, 0 \leq \alpha < 1$.

The assumption of uncorrelated errors is considered a bit strong and is stronger than is needed for the diversification property to hold as we next show.

Theorem 36 *Suppose that for all N*

$$\lambda_{\max}(\Omega_\varepsilon) \leq c < \infty. \tag{8.3}$$

Then, $\lim_{N\to\infty} w^\top w = 0$ *implies that*

$$\lim_{N\to\infty} w^\top \Omega_\varepsilon w = 0. \tag{8.4}$$

Proof. The proof of this result is immediate. We have

$$w^\top \Omega_\varepsilon w = w^\top w \frac{w^\top \Omega_\varepsilon w}{w^\top w} \leq \lambda_{\max}(\Omega_\varepsilon) w^\top w \leq c \times w^\top w \to 0 \tag{8.5}$$

as $N \to \infty$. □

Chapter 8 Multifactor Pricing Models

There are several models of cross sectional correlation that are in use. First, one may suppose the block or industry or cluster model, whereby there are, say, G categories of firms with N_g members each, where $g = 1, \ldots, G$. Suppose that within category g, errors can be arbitrarily correlated, but firms from different categories are uncorrelated. Let Ω_g denote the $N_g \times N_g$ category g idiosyncratic error covariance matrix for $g = 1, \ldots, G$. Then, if we reorder the firms according to their category, we have

$$\Omega_\varepsilon = \begin{bmatrix} \Omega_1 & 0 & \cdots & 0 \\ & \ddots & \ddots & \vdots \\ & & & 0 \\ 0 & & & \Omega_G \end{bmatrix}. \tag{8.6}$$

If G is large and $\max_g N_g \leq C$, then (8.3) is easily satisfied. This condition can also be satisfied when both G and N_g are large. In model (8.6) the covariance matrix has many zero elements, in fact $N^2 - \sum_{g=1}^G N_g^2$ of them, and the location of these zeros has for some purposes to be known. A modern variation on this, considered in Fan, Liao, and Mincheva (2013), is to suppose that the error covariance matrix Ω_ε is **sparse**, i.e., has a lot of zeros, but the location of those zeros is unknown.

A second approach is due to Connor and Korajczyk (1993) who assumed that there is some ordering of the cross section such that the process ε_i is strong mixing (Chapter 2).

Example 48 *Suppose that*

$$\text{cov}(\varepsilon_i, \varepsilon_j) = \rho^{|j-i|}$$

for some ρ with $|\rho| < 1$. Then

$$\Omega_\varepsilon = \begin{bmatrix} 1 & \rho & \rho^2 & \cdots & \rho^{N-1} \\ \rho & 1 & \ddots & & \ddots \\ \rho^2 & \ddots & 1 & \rho & \rho^2 \\ \vdots & \ddots & \rho & 1 & \rho \\ \rho^{N-1} & \cdots & \rho^2 & \rho & 1 \end{bmatrix},$$

which has bounded eigenvalues. In fact

$$\frac{1}{N} i^\top \Omega_\varepsilon i = \frac{1}{N} \left[N + 2(N-1)\rho + 2(N-2)\rho^2 + \cdots \right]$$

$$\to 1 + 2 \sum_{j=1}^\infty \rho^j$$

$$= \frac{1+\rho}{1-\rho}.$$

8.2 Diversification

So that the equally weighted portfolio has idiosyncratic variance like $(1+\rho)/N(1-\rho)$ for large N. So diversification eliminates all idiosyncratic risk.

In practice, the raw correlations are high. In Figure 8.1 below we show the empirical distribution (over (i,j)) of the estimated absolute pairwise correlations of S&P500 stocks estimated from daily returns over the period 2000–2010.

By contrast the idiosyncratic errors have smaller correlations but they are still significant. In Figure 8.2 we show the empirical distribution (over (i,j)) of the estimated absolute pairwise correlations of the market model residuals for the same S&P500 stocks in Figure 8.1.

In conclusion, the diversification arguments used in (1.33) can hold more generally even in the presence of correlation between the error terms and large variance terms. If diversification works we obtain that the portfolio variance is dominated by the common components, i.e.,

$$w^\mathsf{T} \Omega_\varepsilon w \simeq 0 \implies \mathrm{var}(w^\mathsf{T} R_t) \simeq \overbrace{w^\mathsf{T} B \Omega_K B^\mathsf{T} w}^{\text{common}}.$$

Consider two well diversified portfolios with weights w_A, w_B. Their covariance is

$$\mathrm{cov}(w_A^\mathsf{T} R_t, w_B^\mathsf{T} R_t) \simeq w_A^\mathsf{T} B \Omega_K B^\mathsf{T} w_B,$$

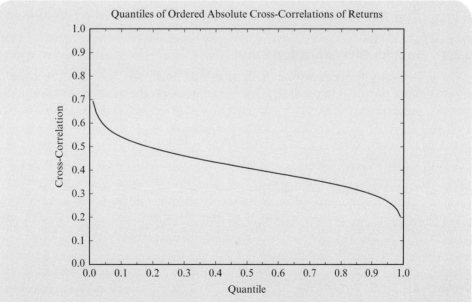

Figure 8.1 Quantiles of ordered (absolute) cross-correlations between S&P500 stocks

Chapter 8 Multifactor Pricing Models

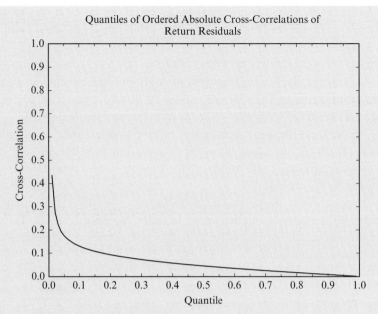

Figure 8.2 Quantiles of ordered (absolute) cross-correlations between S&P500 stocks idiosyncratic errors

because $w_A^\top \Omega_\varepsilon w_B$ is small in magnitude by the Cauchy–Schwarz inequality. That is, in general they may be highly correlated. This is so even if they are mutually orthogonal such that $w_A^\top w_B = 0$ (which could happen if the portfolios are non overlapping).

We next consider some empirical approaches to measuring diversification.

8.2.1 Solnik's Diversification Curve

It is easy to work with equally weighted portfolios and they are a natural benchmark portfolio to consider. Given a set of assets the variance of the equally weighted portfolio is

$$\text{var}\left(\frac{1}{N}\sum_{i=1}^{N} R_i\right) = \overline{\sigma_i^2} + \overline{\sigma}_{ij}$$

$$\overline{\sigma_i^2} = \frac{1}{N}\sum_{i=1}^{N}\text{var}(R_i) \quad ; \quad \overline{\sigma}_{ij} = \frac{2}{N(N-1)}\sum_{j=i+1}^{N}\sum_{i=1}^{N-1}\text{cov}(R_i, R_j).$$

We would like to show how this varies with N by taking a subsample of the assets, but which subsample?

Definition 81 *Solnik (1974) proposed to measure the diversification possibilities of a set on N assets by the sample variance $S(m)$ of a randomly selected equally weighted portfolio of m assets for $m = 1, 2, \ldots, N$.*

8.2 Diversification

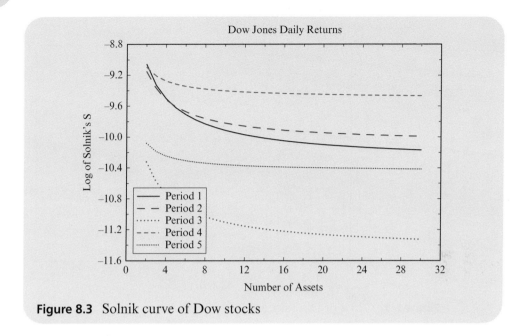

Figure 8.3 Solnik curve of Dow stocks

By considering all possible equally weighted portfolios we remove an obvious bias to do with the particular choice of assets. One can show that Solnik's S satisfies

$$S(m) = \frac{1}{m}\overline{\sigma}_i^2 + \left(1 - \frac{1}{m}\right)\overline{\sigma}_{ij}. \tag{8.7}$$

Furthermore, we see that for m large, $S(m) \to \overline{\sigma}_{ij}$, and in this case the covariance terms dominate the variance terms when there are many assets. For large m, provided $\overline{\sigma}_{ij} > 0$, $\log S(m) \to \log(\overline{\sigma}_{ij})$. Solnik's curve allows one to show how quickly the diversification process occurs.

In Figure 8.3 we show the (log of) Solnik's curve for five subperiods (1970–1980, ...) for the 30 Dow Jones daily returns. The rate at which diversification works is different across subperiods as is the level of variance at which the process settles.

8.2.2 Global Minimum Variance Portfolio

An alternative measure of diversification is the smallest achievable variance from portfolios of the underlying assets. For weights w_{GMV} with $i = (1, \ldots, 1)^\top \in \mathbb{R}^m$

$$w_{GMV} = \frac{\Sigma^{-1} i}{i^\top \Sigma^{-1} i}, \tag{8.8}$$

we achieve the global minimum portfolio variance

$$\sigma_{GMV}^2(m) = \frac{1}{i^\top \Sigma^{-1} i}. \tag{8.9}$$

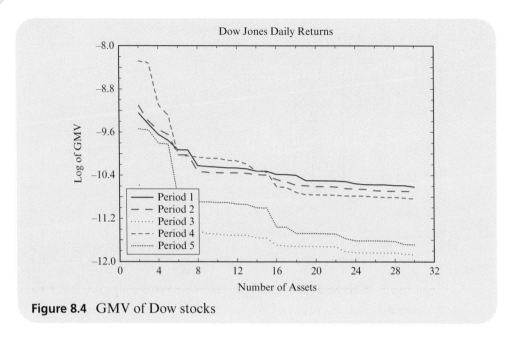

Figure 8.4 GMV of Dow stocks

We compute this for assets $R_1, \ldots, R_m$ with $m = 2, \ldots, N$. In Figure 8.4 we show the log of σ^2_{GMV} for five different subperiods for the 30 Dow Jones daily returns. The rate at which diversification works is different across subperiods as is the level of variance at which the process settles.

This measure is easy to compute with a relatively small number of assets, but when the number of assets increases it becomes unreliable and eventually (when $N > T$) uncomputable.

Some have argued that diversification is harder to achieve nowadays; see for example Connor, Goldberg, and Korajczyk (2010, Figure 2.1), i.e., we now require more stocks to achieve the same risk. Some have even argued for something called a **correlation bubble**, whereby correlation between assets approaches one.

Figure 8.5 below shows one way of measuring this correlation bubble, by rolling window correlations. We have chosen an annual window with $k = 252$ and $j = 0$. This shows the growing correlation between the FTSE100 and FTSE250 over the period since 2006.

8.3 Pervasive Factors

A key assumption in the sequel is that all the included factors are **pervasive**. It is saying that all the factors play an important role in explaining the returns of the assets, essentially, many assets are affected in some way by the factors. It allows the possibility that some sources of risk are not diversified away, otherwise risk would be outlawed and return would always be equal to the risk free rate.

8.3 Pervasive Factors

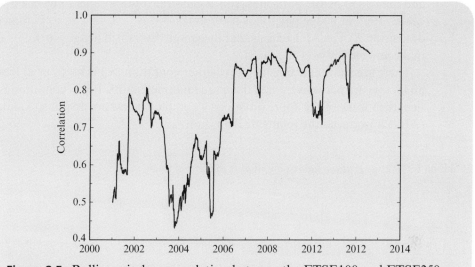

Figure 8.5 Rolling window correlation between the FTSE100 and FTSE250

For example when $K=1$ we might just require that the number of $\beta_i \neq 0$, denoted r, is a large fraction of the sample in the sense that $r(N)/N \geq c > 0$. If this is the case, then there will be some portfolios for which we can't diversify away the common factor. We next give a formal definition. Note that we may normalize the factors to have the identity covariance matrix, without any loss of generality.

Definition 82 *We say that the factors are strongly influential or pervasive when*

$$\frac{1}{N} B^\mathsf{T} B \to M, \tag{8.10}$$

where M is a strictly positive definite matrix.

We have the following result.

Theorem 37 *Suppose that (8.10) holds and that w is a weighting sequence such that $B^\mathsf{T} w \neq 0$, $i^\mathsf{T} w = 1$, and $\lim_{N \to \infty} w^\mathsf{T} w = 0$. Then, we have*

$$\lim_{N \to \infty} w^\mathsf{T} B B^\mathsf{T} w \geq \lambda_{\min}(M) > 0. \tag{8.11}$$

If the diversification condition is satisfied, then this says that for such portfolios, the variance of returns is dominated by the common factor, and this term cannot be eliminated. The common component is not diversifiable. Of course there are also portfolios for which $B^\mathsf{T} w = 0$, and these have already been introduced in Chapter 1, and are called hedge portfolios. It is a classical result in linear algebra that the subspace of $\mathbb{R}^N$ of hedge portfolios is of dimension $N - K$, and so its complement is of dimension K. The APT tells us that those well diversified hedge portfolios should not make any money over their cost.

There are some concerns about whether condition (8.11) holds or at least whether all factors are pervasive, when there are multiple factors. If this condition is not satisfied, then some of the approximations we employ below are no longer valid. Onatskiy (2012) introduced the following definition.

Definition 83 *We say that the factors are weakly influential when*

$$B^\mathsf{T} B \to D, \qquad (8.12)$$

where D is a diagonal matrix.

Under this condition, the contribution of the common components to the variance of the portfolio is of the same magnitude as the idiosyncratic components. This will affect some econometric testing and estimation, discussed below. For example, suppose that B are actually observed (see Section 8.7 below) and f_t are estimated by a cross-sectional regression of returns on B, then the necessary condition for consistency (under i.i.d. errors ref) is that as $N \to \infty$

$$\lambda_{\min}(B^\mathsf{T} B) \to \infty, \qquad (8.13)$$

which is satisfied by (8.10) but not by (8.12). There are intermediate cases between (8.10) and (8.12), where the factors are more influential than the idiosyncratic components.

We may find that some factors are strongly influential, whereas others are not, so that in the multifactor case the matrix B may be rank deficient. One way of modelling this is to say that $B = (B_1, B_2)$, where B_1 are strong factors satisfying (8.11), whereas B_2 are weakish factors that are small in the sense that $\widetilde{B}_2 = \sqrt{T} B_2$ satisfies the strong factor condition but B_2 does not (Bryzgalova (forthcoming)). In that case $B^\mathsf{T} B$ will be of deficient rank when T is large (note that N is fixed in this framework so this is a convenient way of making $B_2^\mathsf{T} B_2$ small).

The conditions for pervasive factors and diversification are related to the conditions for the consistency of least squares regression estimation; see Chapter 2.

8.4 The Econometric Model

We next consider an econometric model for a time series of length T of returns on N assets that obey the same linear factor model. We shall impose some additional restrictions here to facilitate statistical inference.

8.4 The Econometric Model

Definition 84 *The K-factor regression model (for returns or excess returns) can be written*

$$Z_{it} = \alpha_i + \sum_{j=1}^{K} b_{ij} f_{jt} + \varepsilon_{it} \tag{8.14}$$

for $i=1,\ldots,N$ and $t=1,\ldots,T$, or in matrix form

$$Z_t = \alpha + B Z_{Kt} + \varepsilon_t, \tag{8.15}$$

where Z_t is $N \times 1$ returns or excess returns (when the risk free rate R_{ft} is observed), Z_{Kt} is an $K \times 1$ vector of excess returns on the factors, and $\varepsilon_t = (\varepsilon_{1t},\ldots,\varepsilon_{Nt})^\top$ is the vector of idiosyncratic error terms which are independent over time and satisfy

$$E(\varepsilon_t | Z_{K1},\ldots,Z_{KT}) = 0 \quad ; \quad E(\varepsilon_t \varepsilon_t^\top | Z_{K1},\ldots,Z_{KT}) = \Omega_\varepsilon.$$

This is the generalization of the market model (to multiple factors) already discussed. In some cases we assume the following stronger assumption.

Assumption. *The vector ε_t is independent of $Z_{K1},\ldots,Z_{KT}$ and i.i.d. with*

$$\varepsilon_t \sim N(0, \Omega_\varepsilon).$$

There are several different types of factor models that we can distinguish at this point:

1. **Observable factors**

   (a) The factors f are returns to observed traded portfolios (specifically, the returns on portfolios formed on the basis of a security characteristic such as size and value).
   (b) The factors f are observed macro variables such as yield spread etc.

2. **Observable characteristics** models in which the b_i are observable characteristics or depend in a simple way on observable characteristics such as industry or country or even size of the firm.

3. **Statistical factor** models in which both f_{jt} and b_{ij} (and possibly K) are unknown quantities, i.e., parameters.

In each case there are slight differences in cases where there is a risk free asset and in cases where there is not.

In cases 1 and 2, we end up essentially with a panel of linear regression models whose parameters are either the $N \times K$ loadings matrix B or the $T \times K$ factor matrix F, while the other quantities are regressors. Case 1 can be considered a generalization of the market model introduced in Chapter 6. Case 3 poses some genuine differences from the previous chapter in that nothing on the right hand side is directly observed. This is no longer a regression model and needs to be treated in a different way as a model for the covariance matrix of returns. The motivation for considering case 3 is its flexibility with regard to choosing the factors, which sits nicely with the generality of the APT itself.

8.5 Multivariate Tests of the Multibeta Pricing Model with Observable Factors

Tests of the APT are very similar to those for the CAPM, but with vector β replaced by the $N \times K$ matrix B. As in Chapter 7, the mainstream approach assumes that the time series dimension T is large but the cross-section dimension N is fixed, which is consistent with taking the test assets to be portfolios rather than the universe of available stocks. It is also common practice in much of the literature to assume that returns are normally distributed, i.e., (8.4) holds.

8.5.1 Multifactor Pricing Tests with Portfolio Returns as Factors

We first suppose that there is a risk-free asset, that the factor returns F are observable, and that (8.4) holds so that we have a multivariate Gaussian SURE model. The log likelihood function of the data $Z_1, \ldots, Z_T$ (where $Z_{it} = R_{it} - R_{ft}$) conditional on $Z_{K1}, \ldots, Z_{KT}$ is

$$\ell(\alpha, B, \Omega_\varepsilon) = c - \frac{T}{2} \log \det(\Omega_\varepsilon) - \frac{1}{2} \sum_{t=1}^{T} (Z_t - \alpha - BZ_{Kt})^\mathsf{T} \Omega_\varepsilon^{-1} (Z_t - \alpha - BZ_{Kt}).$$

Letting

$$\widehat{\mu}_K = \frac{1}{T} \sum_{t=1}^{T} Z_{Kt} \quad ; \quad \widehat{\mu} = \frac{1}{T} \sum_{t=1}^{T} Z_t$$

$$\widehat{\Omega}_K = \frac{1}{T} \sum_{t=1}^{T} (Z_{Kt} - \widehat{\mu}_K)(Z_{Kt} - \widehat{\mu}_K)^\mathsf{T} \quad ; \quad \widehat{\Omega}_{ZK} = \frac{1}{T} \sum_{t=1}^{T} (Z_t - \widehat{\mu})(Z_{Kt} - \widehat{\mu}_K)^\mathsf{T},$$

the unrestricted MLE of α, B are the equation by equation OLS estimators:

$$\widehat{\alpha} = \widehat{\mu} - \widehat{B}\widehat{\mu}_K \quad ; \quad \widehat{B} = \widehat{\Omega}_{ZK}\widehat{\Omega}_K^{-1}, \quad \widehat{\Omega}_\varepsilon = \frac{1}{T} \sum_{t=1}^{T} \widehat{\varepsilon}_t \widehat{\varepsilon}_t^\mathsf{T}, \qquad (8.16)$$

where $\widehat{\varepsilon}_t = Z_t - \widehat{\alpha} - \widehat{B}Z_{Kt}$, which exists provided $\widehat{\Omega}_K$ is of full rank ($K < T$). Under normality, conditional on the factors we have

$$\widehat{\alpha} \sim N\left(\alpha, \frac{1}{T}(1 + \widehat{\mu}_K^\mathsf{T} \widehat{\Omega}_K^{-1} \widehat{\mu}_K)\Omega_\varepsilon\right). \qquad (8.17)$$

8.5 Multivariate Tests

The multivariate tests of $\alpha = 0$ are similar to those for the CAPM. The F test is

$$\mathcal{F} = \frac{(T - N - K)}{N}(1 + \widehat{\mu}_K^\mathsf{T} \widehat{\Omega}_K^{-1} \widehat{\mu}_K)^{-1} \widehat{\alpha}^\mathsf{T} \widehat{\Omega}_\varepsilon^{-1} \widehat{\alpha}. \tag{8.18}$$

It is exactly distributed as $F_{N,T-N-K}$ under the normality assumption provided that $\widehat{\Omega}_\varepsilon$ is of full rank, which requires that $N < T$. In the absence of normality, under some conditions $N \times \mathcal{F}$ is asymptotically (sample size T gets large) chi-squared with N degrees of freedom, as are the Wald, Lagrange multiplier, and likelihood ratio statistics.

We may alternatively collect parameters equation by equation, that is, let $\Theta = (\theta_1, \ldots, \theta_N)^\mathsf{T}$ be the $N \times (K+1)$ matrix with $\theta_i = (\alpha_i, b_i^\mathsf{T})^\mathsf{T}$, where b_i is the ith row of B, and let X be the $T \times (K+1)$ matrix containing a column of ones and columns of observations on Z_{kt}, $k = 1, \ldots, K$. Then we can write in system notation the estimator of Θ as

$$\widehat{\Theta} = Z^\mathsf{T} X (X^\mathsf{T} X)^{-1};$$

see (7.14). We also let $\theta = \text{vec}(\Theta)$ and $\widehat{\theta} = \text{vec}(\widehat{\Theta})$ be the stacked $N(K+1) \times 1$ vector of parameters and estimators in this ordering. The distribution of $\widehat{\theta}$ is $N(\theta, \Omega_\varepsilon \otimes (X^\mathsf{T} X)^{-1})$ and (8.17) is the marginal distribution of a particular subset of θ.

We may be interested in testing whether $B_2 = 0$, where $B = (B_1, B_2)$ and B_2 is of dimensions $N \times K_2$, that is, we have included some factors whose relevance we may wish to test. The likelihood ratio statistic is

$$LR = T\left(\log \det(\widetilde{\Omega}_\varepsilon) - \log \det(\widehat{\Omega}_\varepsilon)\right),$$

where $\widetilde{\Omega}_\varepsilon$ is the estimated error covariance matrix in the restricted model (with $B_2 = 0$), which is defined as in (8.16) but with only the subset of factors corresponding to B_1. Anderson (1984, Theorem 8.4.1) gives the exact distribution of this test statistic under the null hypothesis (and normality). When the sample size T is large the distribution is approximately $\chi^2_{K_2}$ and this result holds without the condition of normality.

We may also robustify the testing strategy to time series heteroskedasticity as in Chapter 7. Suppose that (7.29) and (7.30) hold, and define

$$W_H = T\widehat{\alpha}^\mathsf{T} \widehat{\Omega}_H^{-1} \widehat{\alpha}$$

$$\widehat{\Omega}_H = \frac{1}{T}\sum_{t=1}^{T} \widehat{\varepsilon}_t \widehat{\varepsilon}_t^\mathsf{T} \left(1 + \left((Z_{Kt} - \widehat{\mu}_K)^\mathsf{T} \widehat{\Omega}_K^{-1} \widehat{\mu}_K\right)^2\right).$$

Under the null hypothesis this statistic is asymptotically χ^2_N.

Suppose now that there is no risk-free asset. Then tests of the APT are completely analogous to those for the zero-beta CAPM. That is, we take $Z_t = R_t$ and $Z_{Kt} = R_{Kt}$ in (8.15). The null restrictions are

$$\alpha = (i_N - Bi_K)\gamma_0,$$

where $\gamma_0 \in \mathbb{R}$ is unknown. To estimate the restricted model one can proceed as in Chapter 7: first estimate B by OLS assuming that γ_0 is known and then search over values of γ_0. The likelihood ratio statistic can be used to test the null hypothesis; it will be approximately chi-squared with $N-1$ degrees of freedom when T is large. We discuss below some of the issues with constructing the Wald statistic.

8.5.2 Multifactor Pricing Tests with Some Macro Factors and Risk Free Rate

We consider the case where there is a risk free rate and some of the factors are traded portfolios $(f_{1t} \in \mathbb{R}^{K_1})$, while others are macro factors $(f_{2t} \in \mathbb{R}^{K_2})$ that are not traded assets. Then the expected returns of the macro factors are not restricted by the null hypothesis.

In this case, the unrestricted model is

$$R_t - R_{ft} = \alpha + B_1(f_{1t} - R_{ft}) + B_2 f_{2t} + \varepsilon_t \qquad (8.19)$$

$$E(f_{2t}) = \mu_{K2}.$$

The null hypothesis consistent with the APT in this case is that

$$\alpha = B_2 \gamma_2, \qquad (8.20)$$

for some unknown $\gamma_2 \in \mathbb{R}^{K_2}$. We can solve for γ_2 explicitly by $\gamma_2 = (B_2^\mathsf{T} B_2)^{-1} B_2 \alpha$ assuming that B_2 is of full rank. It follows that the restrictions can be rewritten as $M_{B_2} \alpha = 0$, where $M_{B_2} = I_{K_2} - B_2(B_2^\mathsf{T} B_2)^{-1} B_2^\mathsf{T}$. The $N \times N$ matrix M_{B_2} is symmetric and idempotent and of rank $N - K_2$. Substituting α into the equation for returns gives

$$R_t - R_{ft} = B_1(f_{1t} - R_{ft}) + B_2(f_{2t} + \gamma_2) + \varepsilon_t,$$

which is linear in B given γ_2 (and linear in γ_2 given B_2). The restricted model can be estimated by first conditioning on γ_2 and solving explicitly for $\widetilde{B}(\gamma_2)$ and then optimizing over the parameters γ_2. Let $\widetilde{B}, \widetilde{\gamma}_2, \widetilde{\Omega}_\varepsilon$ be the restricted MLE. The simplest way to test this hypothesis is using the likelihood ratio statistics, which are asymptotically normal with $N - K_2$ degrees of freedom. We next consider the Wald test or rather a version of the Wald test that does not require explicit use of the restriction $M_{B_2} \alpha = 0$ (which is problematic because of the reduced rank). Let $\widehat{\theta}_i = (\alpha_i, b_{2i}^\mathsf{T})^\mathsf{T}$ and $\widehat{\theta} = (\widehat{\theta}_1^\mathsf{T}, \ldots, \widehat{\theta}_N^\mathsf{T})^\mathsf{T} \in \mathbb{R}^{(K_2+1)N}$, and $\delta = (\gamma_2^\mathsf{T}, b_{21}^\mathsf{T}, \ldots, b_{2N}^\mathsf{T})^\mathsf{T}$. Then let

$$W = T \min_{\delta \in \mathbb{R}^{K_2(1+N)}} \left(\widehat{\theta} - h(\delta)\right)^\mathsf{T} \widehat{\Xi}^{-1} \left(\widehat{\theta} - h(\delta)\right),$$

where $h(\delta) = (b_{21}^\mathsf{T} \gamma_2, b_{21}^\mathsf{T}, \ldots b_{2N}^\mathsf{T} \gamma_2, b_{2N}^\mathsf{T})^\mathsf{T}$ and $\widehat{\Xi} = \widehat{\Omega}_\varepsilon \otimes (X^\mathsf{T} X)^{-1}$. Then by Chamberlain (1984, Theorem 1), we have as $T \to \infty$

$$W \Longrightarrow \chi^2_{N-K_2}.$$

8.5.3 Two Pass Cross-sectional Approach

This is carried out essentially as in the Fama–Macbeth procedure for the CAPM. In this case, including quadratic functions of all the betas is quite exhausting and is usually not attempted. Nevertheless, the cross-sectional method can be useful for testing whether a factor commands a risk premium or not.

1. Estimate b_{ij} by time series regressions for each stock, and denote by $\widehat{B}$ the $N \times K$ matrix of estimates.
2. Then for each t run the cross-sectional regression and estimate the ex-post factor risk premium
$$\widehat{\gamma}_t = (\widehat{X}^\mathsf{T} \widehat{X})^{-1} \widehat{X}^\mathsf{T} Z_t \in \mathbb{R}^{K+1},$$
where Z_t is the n-vector of excess returns and $\widehat{X} = (i, \widehat{B})$.
3. Then let
$$\widehat{\gamma} = (\widehat{\gamma}_0, \ldots, \widehat{\gamma}_K)^\mathsf{T} = \frac{1}{T} \sum_{t=1}^{T} \widehat{\gamma}_t$$

$$\widehat{V} = \frac{1}{T} \sum_{t=1}^{T} (\widehat{\gamma}_t - \widehat{\gamma})(\widehat{\gamma}_t - \widehat{\gamma})^\mathsf{T}.$$

4. Test that γ_j is zero using the FM t-statistic

$$\frac{\widehat{\gamma}_j}{\sqrt{\widehat{V}_{jj}}},$$

which is asymptotically standard normal under the null hypothesis that $\gamma_j = 0$ (under the conditions discussed in Chapter 7).

The cross-sectional regressions are useful for obtaining the factor risk premia in the case where the factors are not traded assets.

8.6 Which Factors to Use?

The market portfolio is a natural candidate, but what else? There are two popular approaches that we consider in turn: the Fama–French sorting method and observable macro factors.

8.6.1 Fama–French Factors

In a series of heavily cited papers, Fama and French (hereafter denoted FF) demonstrate that there have been large return premia associated with size and value. These size and value return premia are evident in US data for the period covered by the CRSP and Compustat database (FF (1992)), in earlier US data (Davis (1994)), and in non-US equity markets (FF (1998); Hodrick, Ng, and Sangmueller (1999)).

FF (1993, 1996, 1998) contend that these return premia can be ascribed to a rational asset pricing paradigm in which the size and value characteristics proxy for assets' sensitivities to pervasive sources of risk in the economy. This approach is now widely used in empirical finance.

We next explain how they constructed their factors. FF (1993) construct six double sorted portfolios formed on two size and three book to market (B/M) portfolios. That is, they first sort the stocks according to their size (at the given date) and divide into two groups: large size and small size: large size and small size. Then they sort each of these groups according to the B/M value and divide each into three further groups. The sorting could equally be done the other way round, and in both ways one obtains six groups of stocks

B/M/Size	1	2
1	high/large	high/small
2	medium/large	medium/small
3	low/large	low/small

The stocks in each bucket are then equally weighted to produce six portfolios, one that measures large size and high B/M (denoted (1,1)), etc. The size factor return is proxied by the difference in return between a portfolio of low-capitalization stocks and a portfolio of high-capitalization stocks, adjusted to have roughly equal book-to-price ratios (SMB). Specifically, this is

$$SMB = \frac{1}{3}\left[\begin{array}{c}\left(\frac{\text{large size}}{\text{high B/M}} - \frac{\text{small size}}{\text{high B/M}}\right) \\ + \left(\frac{\text{large size}}{\text{medium B/M}} - \frac{\text{small size}}{\text{medium B/M}}\right) \\ + \left(\frac{\text{large size}}{\text{low B/M}} - \frac{\text{small size}}{\text{low B/M}}\right)\end{array}\right], \quad (8.21)$$

which you can write using the table as $\{(1,1) - (1,2) + (2,1) - (2,2) + (3,1) - (3,2)\}/3$.

The value factor is proxied by the difference in return between a portfolio of high book-to-price stocks and a portfolio of low book-to-price stocks, adjusted to have roughly equal capitalization (HML), i.e.,

$$HML = \frac{1}{2}\left[\begin{array}{c}\left(\frac{\text{high B/M}}{\text{large size}} - \frac{\text{low B/M}}{\text{large size}}\right) \\ + \left(\frac{\text{high B/M}}{\text{small size}} - \frac{\text{low B/M}}{\text{small size}}\right)\end{array}\right], \quad (8.22)$$

which can be written as $\{(1,1) - (3,2) + (1,2) - (3,2)\}/2$.

These factors are both portfolios of traded assets and can be written $\sum w_{is} R_{it}$ for some weighting sequence w_{is} (the sorting is done at time $s < t$) that satisfies $\sum_{i=1}^{N} w_{is} = 0$, i.e., it is a zero cost portfolio.

The market factor return is proxied by the excess return to a value-weighted market index (MKT). The market return includes all NYSE, AMEX, and NASDAQ firms.

8.6 Which Factors to Use?

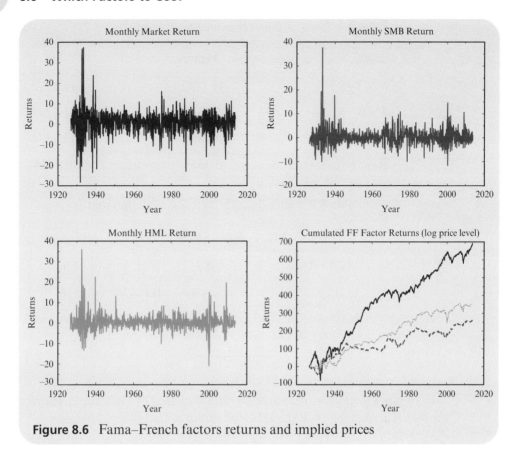

Figure 8.6 Fama–French factors returns and implied prices

SMB and HML for July of year t to June of t+1 include all NYSE, AMEX, and NASDAQ stocks for which there is market equity data for December of t − 1 and June of t, and (positive) book equity data for t − 1. Data on these factors is available from the Ken French web site http://mba.tuck.dartmouth.edu/pages/faculty/ken.french/data_library.html.

In Figure 8.6 below we show the actual percentage returns of the three factors as well as the cumulated returns (which correspond to a price level, although the initial level is fixed arbitrarily here).

In Table 8.1 we give the annualized mean and standard deviation of the monthly return series over the period 1926–2017. To make the return between MKT and SMB HML comparable we added the risk-free rate back into the market return. We find that the mean return on the SMB and HML portfolios is quite small compared with that for the market return, although they are correspondingly less risky. This is not surprising since they are both arbitrage portfolios (positive weights on some securities cancelled by negative weights on others). Indeed, over this period the SMB portfolio appeared to make even less than the riskless asset.

Table 8.2 shows the correlation matrix of the factors. This indicates that there is quite low correlation between the three assets.

Table 8.1 Fama–French factors, mean and std.

Asset	Mean	Std
MKT+Rf	11.1742	18.5558
SMB	2.5374	11.1241
HML	4.7858	12.1354
Rf	3.3313	0.8824

Table 8.2 Fama–French factors correlation matrix

	MKT+Rf	SMB	HML	Rf
MKT+Rf	1	0.3173	0.2421	−0.0176
SMB		1	0.1217	−0.0516
HML			1	0.0183
Rf				1

In Figure 8.7 we show the variance ratio statistics for the three factor returns. This shows the predictability of the HML portfolio at fairly short horizons.

We return to the FF methodology. Having constructed the factors, they then form 25 test portfolios (on size and value characteristics too) and carry out the multivariate regression

$$R_{pt} = \alpha_p + \beta_{p,SMB}SMB_t + \beta_{p,HML}HML_t + \beta_{p,MKT}MKT_t + \varepsilon_{pt}.$$

This is a panel regression model with observed covariates that are the same in each equation p, and so estimation and inference proceeds as described above. They test the restrictions of the APT ($\alpha_p = 0$) using the F statistic. They do not reject the APT restrictions in their sample (although they do reject CAPM when SMB and HML are dropped from the regression). They explain well the size and value anomalies of the CAPM. However, later work has rejected the APT restrictions in the three factor model using different test portfolios, i.e., the model is not consistent with other anomalies (Hou, Xue, and Zhang (2015)). Recent work has extended the three FF factors to four, five, and even more to include in the regression. For example, the momentum factor of Carhart (1997), the own-volatility factor of Goyal and Santa-Clara (2003), the liquidity factor of Amihud and Mendelson (1986) and Pástor and Stambaugh (2003). Hou, Xue, and Zhang (2015) propose an investment factor. Fama and French (2015) themselves propose a five factor model that has a similar motivation. They provide the data for that on French's website.

8.6 Which Factors to Use?

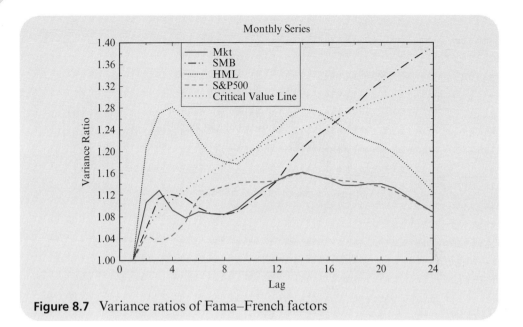

Figure 8.7 Variance ratios of Fama–French factors

8.6.2 Macroeconomic Factors

Another approach is to employ macroeconomic variables as the common sources of risk driving asset prices. One issue is that these series are often nonstationary and so would result in an **unbalanced equation** for returns (which are typically much closer to stationarity). Chan, Chen, and Hsieh (1985) and Chen, Roll, and Ross (1986) employed a number of factors of the **surprise** form

$$f_t = m_t - E_{t-1}(m_t), \tag{8.23}$$

where m_t are observable (potentially nonstationary) macroeconomic variables. By this construction we may hope that f_t is stationary. But more importantly, factors so constructed represent unanticipated information at time $t-1$ and so are not already part of prices. In this way, we are looking for the association of the new information in the factors with the realized return over the same period, as we do in the market model and the FF model. In some cases, they take $E_{t-1}(m_t) = m_{t-1}$ so that f_t is just differenced data. For inflation however, they used the Fama and Gibbons (1982) model to generate $E_{t-1}(m_t)$. This model is

$$I_t = \alpha_{t-1} + \beta TB_{t-1} + \eta_t, \tag{8.24}$$

where I_t is observed monthly inflation in period t, TB_{t-1} is the one month T-bill rate known at time $t-1$, and α_{t-1} is the negative of the unobserved real interest rate, which evolves according to a random walk, i.e., $\alpha_t = \alpha_{t-1} + u_t$, where the processes

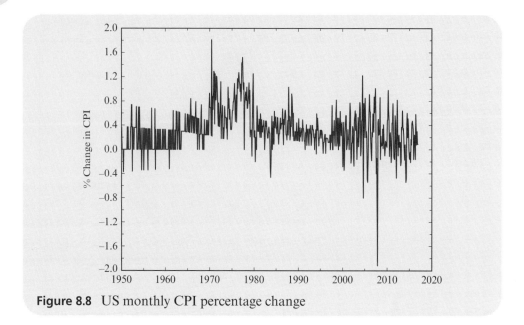

Figure 8.8 US monthly CPI percentage change

$\{u_t\}$ and $\{\eta_t\}$ are mutually independent with mean zero and finite variance. The forecasted value of inflation is $\widehat{\alpha}_{t-1} + \widehat{\beta} TB_{t-1}$, where $\widehat{\beta}$ is estimated from differenced data (the value of β should be equal to one) and $\widehat{\alpha}_{t-1}$ is obtained by the **Kalman filter**. In Figure 8.8 we show the monthly inflation rate since 1950. The series is quite persistent in comparison with stock returns; the first order autocorrelation is around 0.5. The growth of industrial production (monthly series) is similarly persistent relative to stock returns.

Common practice appears to be to ignore the preliminary estimation error in calculating f_t, which potentially biases subsequent results (Pagan (1984)). Chan *et al.* (1985) used monthly data over the period 1958–1984. Specifically:

1. The percentage change in industrial production (lagged by one period).
2. A measure of unexpected inflation.
3. The change in expected inflation.
4. The difference in returns on low-grade (Baa and under) corporate bonds and long-term government bonds (junk spread).
5. The difference in returns on long-term government bonds and short term Tbills (Term spread).

They used 20 test portfolios sorted on the basis of firm size at the beginning of the period. They estimate b_i by time series regressions and then do cross-sectional regressions on $\widehat{b}_i$ to estimate factor risk premia. Some of the macroeconomic factors are not traded assets.

8.7 Observable Characteristic Based Models

They find that average factor risk premia are statistically significant over the entire sample period for industrial production, unexpected inflation, and junk. They also include a market return but find that its associated risk premium is not significant when the risk premia associated with macroeconomic factors are included.

8.7 Observable Characteristic Based Models

These models are commonly used in the asset management industry, i.e., for portfolio risk management, perhaps in combination with other techniques. Rosenberg (1974) considered the multifactor regression model where f are treated as unknown parameters and B is related to observable (time invariant) stock characteristics such as industry or country dummy variables. Other possible characteristics include size and value, which are observed for all stocks. Suppose that (8.14) holds and that for $i = 1, \ldots, N$

$$b_i = Dx_i,$$

where x_i is an observed $J \times 1$ vector of characteristics, and D is a $K \times J$ matrix of unknown parameters. Substituting into the return equation we obtain

$$Z_{it} = \alpha_i + b_i^\mathsf{T} f_t + \varepsilon_{it} = \alpha_i + x_i^\mathsf{T} D^\mathsf{T} f_t + \varepsilon_{it} = x_i^\mathsf{T} f_t^* + \varepsilon_{it},$$

where $f_t^* = D^\mathsf{T} f_t$ is a $J \times 1$ vector of characteristic specific factors for $t = 1, \ldots, T$. We can write this as a cross-sectional regression

$$Z_t = \alpha + X f_t^* + \varepsilon_t$$

for each time period t, where X is the $N \times J$ matrix of observed firm specific characteristics. It follows that

$$E(Z_t | X) = \alpha + X \mu_{f^*} \quad ; \quad \mathrm{var}(Z_t | X) = X \Omega_{f^*} X + \Omega_\varepsilon,$$

where μ_{f^*} and Ω_{f^*} are the mean and variance of f_t^*, which are assumed here to not vary over time. This structure can be used in portfolio choice. That is, we can form portfolios to minimize the conditional variance given a certain conditional mean is achieved.

We suppose (for identification) that the α are orthogonal to the characteristics so that $X^\mathsf{T} \alpha = 0$. In this case

$$P_X Z_t = X f_t^* + P_X \varepsilon_t, \qquad M_X Z_t = \alpha + M_X \varepsilon_t,$$

where $P_X = X(X^\mathsf{T} X)^{-1} X^\mathsf{T}$ and $M_X = I - P_X$ are the projection matrices associated with X. We can estimate the unknown characteristic returns f_t^* using linear cross-sectional OLS regression at each time point and the $\alpha \in \mathbb{R}^N$ by the characteristic purged average return

$$\widehat{f}_t^* = (X^\mathsf{T} X)^{-1} X^\mathsf{T} P_X Z_t = (X^\mathsf{T} X)^{-1} X^\mathsf{T} Z_t, \qquad \widehat{\alpha} = M_X \overline{Z},$$

where $\bar{Z} = \sum_{t=1}^{T} Z_t/T$. For consistency of $\widehat{f_t^*}$ we require that $N \to \infty$, which effectively means using individual assets rather than portfolios. We also require at least (8.13) with B replaced by X. For consistency of α we would also require T to be large.

Connor, Hagmann, and Linton (2012) generalize this idea to allow the beta functions to depend in an unknown fashion on observed characteristics

$$Z_{it} = f_{ut} + \sum_{j=1}^{J} \beta_j(X_{ji}) f_{jt} + \varepsilon_{it}.$$

They propose a nonparametric method for estimating the beta functions and then cross-sectional regression to get the factors. They find that the estimated β_j have nonlinear shapes. In this model, the expected return varies continuously with characteristics, so that

$$\frac{\partial}{\partial x_j} E(Z_{it}|X_i = x) = \beta_j'(x_j) f_{jt}. \qquad (8.25)$$

If X_j is the size characteristic, then (8.25) measures the way expected return changes with size.

8.8 Statistical Factor Models

We now consider statistical factor models for excess returns. We repeat the model in matrix form

$$\mathbf{Z}_{N \times T} = \mu i_T^\mathsf{T} + B_{N \times K} F_{K \times T}^\mathsf{T} + \mathbf{E}_{N \times T} \qquad (8.26)$$

where $B \in \mathbb{R}^{N \times K}$ and $F \in \mathbb{R}^{T \times K}$. We suppose that B, F are full rank matrices and that $T > N > K$. The number of factors K is assumed known and relatively small compared with N, but we will consider later some methods for determining K. In this model, nothing on the right hand side is observed and we may think of B, F as random variables along with $\mathbf{E}$. The model is **bilinear** in B, F, meaning that conditional on B it is linear in F, while conditional on F it is linear in B. We will exploit this bilinearity below. In particular, we may think of one or other of B, F as parameters of interest (in which case we will condition on their sample values) and the other variable as a random variable to be averaged over. This corresponds to what happens in observed factor models and observed characteristics models. For simplicity, we assume that the the random variables B, F, and $\mathbf{E}$ are all mutually independent of each other. We may therefore compute conditional expectations given B (which corresponds to treating B as unknown parameters) or conditional expectations given F (which corresponds to treating F as unknown parameters).

We suppose that the idiosyncratic error is uncorrelated over time, i.e., it satisfies

$$E(\varepsilon_{it}\varepsilon_{js}) = \begin{cases} \sigma_{ij}, & \text{if } t = s \\ 0, & \text{else.} \end{cases} \qquad (8.27)$$

We may rewrite (8.26) in different ways according to our purpose.

8.8 Statistical Factor Models

Definition 85 *For each time period write*

$$Z_t = \mu + Bf_t + \varepsilon_t, \qquad (8.28)$$

where $Z_t, \varepsilon_t \in \mathbb{R}^N$ and $E(\varepsilon_t) = 0$ and $E(\varepsilon_t \varepsilon_t^\top) = \Omega_\varepsilon$, where $\Omega_\varepsilon = (\sigma_{\varepsilon ij})_{i,j}$ is an $N \times N$ covariance matrix. We assume that f_t are random variables from a common or stationary distribution and B are fixed quantities. Without loss of generality we suppose that $E(f_t) = 0$, which is consistent with Z_t being centered random variables.

We consider a special case of this model that restricts the error covariance matrix further.

Definition 86 *Strict factor structure – suppose that the idiosyncratic error is uncorrelated in the cross section, i.e.,*

$$E(\varepsilon_t \varepsilon_t^\top) = D, \qquad (8.29)$$

where D is a diagonal matrix that does not vary with time.

The second version of the general model (8.26) involves rewriting the population factor model for the excess return vector Z_i.

Definition 87 *For each asset i, write*

$$Z_i = \mu_i i_T + Fb_i + \varepsilon_i, \qquad (8.30)$$

where $Z_i, \varepsilon_i \in \mathbb{R}^T$, and $E(\varepsilon_i \varepsilon_i^\top) = \Xi_\varepsilon$, where Ξ_ε is a $T \times T$ covariance matrix. Consistent with (8.27) we assume that $\Xi_\varepsilon = \sigma_{\varepsilon i}^2 I_T$. We assume that the b_i are $K \times 1$ random variables from a common distribution and F are fixed quantities. Without loss of generality we suppose that $E(b_i) = 0$, which is consistent with Z_i being centered random variables.

This captures the idea that the cross section of returns are driven by a smaller number of quantities.

8.8.1 Identifying the Factors in Asset Returns

We first suppose that (8.28) holds. We have $E(Z_t) = \mu$ so this moment tells us nothing about B, F. Furthermore, we can't use the conditional expectation like a regression model since none of the right hand side is observable. Consider the variance covariance matrix of returns

$$\Sigma_{N \times N} = E\left[(Z_t - E(Z_t))(Z_t - E(Z_t))^\top\right].$$

This symmetric matrix has $N(N+1)/2$ free parameters that can be consistently estimated by the time series sample average when T is large. Under the strict factor structure we have (treating B as fixed and F as random)

$$\Sigma = B\Omega_K B^\mathsf{T} + D, \qquad (8.31)$$

where $\Omega_K = E(f_t f_t^\mathsf{T})$. The right hand side has $NK + K(K+1)/2 + N$ parameters, which is less than $N(N+1)/2$ provided $K < N$.

However, when the factors are unknown there is an identification issue. One can write for any nonsingular L

$$B^* = BL^{-1} \; ; \; f_t^* = Lf_t$$

and (B^*, f_t^*) are observationally equivalent to (B, f_t) since

$$Bf_t = BL^{-1}Lf_t = B^* f_t^*.$$

To resolve this problem, we shall suppose that $E(f_t f_t^\mathsf{T}) = \Omega_K = I_K$, that is, the factors are orthogonal with unit scale. This amounts to imposing $K(K+1)/2$ equality restrictions. It follows that

$$\Sigma = BB^\mathsf{T} + D. \qquad (8.32)$$

Then B, D are essentially unique. Actually, in this case B is only unique up to an orthonormal matrix transformation Q with $QQ^\mathsf{T} = I_K$. One can say that the subspace of $\mathbb{R}^N$ generated by B, called the **factor space**, is uniquely determined, but the basis with which it is defined is not unique. A further restriction is to require that the matrix $B^\mathsf{T} D^{-1} B$ be diagonal, which imposes an additional set of restrictions on B, D. This uniquely pins the parameters down, at least up to the sign of the elements of the matrix B.

We now turn to estimation of the model parameters. Define the sample mean and sample covariance matrix of returns as follows

$$\widehat{\mu} = \frac{1}{T}\sum_{t=1}^T Z_t \quad ; \quad \widehat{\Sigma} = \frac{1}{T}\sum_{t=1}^T (Z_t - \widehat{\mu})(Z_t - \widehat{\mu})^\mathsf{T}. \qquad (8.33)$$

We can estimate the parameters B, D, μ by maximum likelihood factor analysis provided $N < T$ (*small N and large T*). Supposing that Z_t is normally distributed with mean vector μ and covariance matrix Σ, we obtain the log likelihood function for $Z_1, \ldots, Z_T$

$$\ell(B, D, \mu) = c - \frac{T}{2}\log\det(BB^\mathsf{T} + D) - \frac{1}{2}\sum_{t=1}^T (Z_t - \mu)^\mathsf{T}(BB^\mathsf{T} + D)^{-1}(Z_t - \mu).$$

There are a total of $NK + 2N$ free parameters. The first order conditions for μ are linear and the MLE is just the sample mean given in (8.33). However, the first order conditions for B, D are nonlinear and can't be solved explicitly. They need to be solved

8.8 Statistical Factor Models

by iterative numerical procedures. Let $\widehat{B}, \widehat{D}$ be the maximum likelihood estimators of B, D, and let $\widehat{\theta}$ be the vector of dimensions $N(K+1) \times 1$ that contains all these parameter estimates and θ be the vector of corresponding true values. Provided T is large and N is fixed, these will be consistent and approximately normally distributed, that is

$$\sqrt{T}\left(\widehat{\theta} - \theta\right) \Longrightarrow N(0, \Xi)$$

for some matrix Ξ. The limiting variance Ξ is quite complicated.

The factor model imposes some restrictions on the covariance matrix of returns. One can test the restrictions of the model by the likelihood ratio method, that is

$$LR = T\left(\log\det(\widetilde{\Sigma}) - \log\det(\widehat{\Sigma})\right),$$

where $\widetilde{\Sigma} = \widehat{B}\widehat{B}^{\mathsf{T}} + \widehat{D}$ is the restricted covariance matrix estimator. Under the null hypothesis that the restrictions are valid, this statistic is asymptotically χ_p^2 with $p = ((N-K)^2 - N - K)/2$ (see Anderson (1984, p563)).

For some purposes the factors f_t themselves are of interest. We can then estimate the factors by cross-sectional regression, i.e., OLS or GLS, for each time period t, for example

$$\widehat{f}_t = (\widehat{B}^{\mathsf{T}}\widehat{D}^{-1}\widehat{B})^{-1}\widehat{B}^{\mathsf{T}}\widehat{D}^{-1}(Z_t - \widehat{\mu}). \tag{8.34}$$

These can be seen as returns on portfolios with portfolio weights; see below.

These estimates are hard to analyze statistically because: (1) replacing B with $\widehat{B}$ creates an errors in variables problem; and (2) the consistency of the maximum likelihood procedure for estimating B was established under the assumption that N is fixed, but for estimation of f_t we should like N to be large. A resolution of this issue is discussed later. Nevertheless, as $T \to \infty$

$$\widehat{f}_t \xrightarrow{P} f_t^* = (B^{\mathsf{T}}D^{-1}B)^{-1}B^{\mathsf{T}}D^{-1}(Z_t - \mu),$$

and f_t^* is the best linear unbiased estimator of f_t based on the given cross-sectional data of size N.

8.8.2 Factor Mimicking Portfolios

Portfolios that hedge or "mimic" the factors are the basic components of various portfolio strategies. The mimicking portfolio for a given factor is the portfolio with the maximum correlation with the factor. If all assets are correctly priced, then each investor's portfolio should be some combination of cash and the mimicking portfolios. Other portfolios have the same level of expected return and sensitivities to the factor but greater variance.

We next show how to interpret the cross-sectional GLS and OLS estimates of factor returns

$$\widehat{f}_{jt} = \widehat{w}_j^{\mathsf{T}}(Z_t - \widehat{\mu}), \quad j = 1, \ldots, K,$$

where $\widehat{W} = \widehat{D}^{-1}\widehat{B}(\widehat{B}^{\mathsf{T}}\widehat{D}^{-1}\widehat{B})^{-1} = (\widehat{w}_1, \ldots, \widehat{w}_K)$, as the returns to factor-mimicking portfolios. The factors are (conditionally) linear functions of excess returns with weights $\widehat{w}_j, j = 1, \ldots, K$. We may renormalize the weights $\widehat{w}_j$ so that they sum to one and are proper unit cost portfolio weights. Note that $\widehat{W}^{\mathsf{T}}\widehat{B} = I_K$ so that the jth factor mimicking portfolio has unit exposure to factor j and zero exposure to the other factors. These weights solve the following problem.

Definition 88 *Let $w_j \in \mathbb{R}^N, j = 1, \ldots, K$ minimize the objective functions*

$$w_j^{\mathsf{T}} \widehat{D} w_j \tag{8.35}$$

subject to the equality restrictions:

$$w_j^{\mathsf{T}} \widehat{b}_h = 0 \quad \text{for all } h \neq j \; ; \quad w_j^{\mathsf{T}} \widehat{b}_j = 1.$$

The set of portfolios that are hedged against factors h, $h \neq j$, is of dimension $N - K$. We are finding the one with the smallest idiosyncratic variance. We have $\widehat{W}^{\mathsf{T}}\widehat{W} = (\widehat{B}^{\mathsf{T}}\widehat{D}^{-1}\widehat{B})^{-1}$, which goes to zero under (8.10). This says that the portfolio weights are well spread out.

One value of the factor model is in dimensionality reduction. It is important in portfolio choice and asset allocation, which usually involves an inverse covariance matrix that has to be estimated. Note that when

$$\Sigma = BB^{\mathsf{T}} + D,$$

we have

$$\Sigma^{-1} = D^{-1} - D^{-1}B(I_K + B^{\mathsf{T}}D^{-1}B)^{-1}B^{\mathsf{T}}D^{-1},$$

which only involves inverting the $K \times K$ matrix $I_K + B^{\mathsf{T}}D^{-1}B$, which is of full rank.

In the observed factor model the identification restrictions are not needed and the unconditional covariance matrix of returns (averaging over the factors) is $\Sigma = B\Omega_K B^{\mathsf{T}} + D = B^* B^{*\mathsf{T}} + D$, where $B^* = B\Omega_K^{1/2}$.

8.8.3 Asymptotic Principal Components
Principal Components

An alternative to maximum likelihood factor analysis is asymptotic principal components introduced by Connor and Korajczyk (1988). This procedure is effective in the case where T is *small and n is large*, which is the opposite of the case where the maximum likelihood works.

8.8 Statistical Factor Models

Gregory Connor is Professor of Finance at Maynooth University. Connor's research interests are in portfolio risk analysis and related financial econometrics topics, particularly factor modelling. He has a number of widely cited publications in academic journals including the *Journal of Economic Theory*, *Journal of Finance*, *Journal of Financial Economics*, *Review of Financial Studies*, and *Journal of Econometrics and Econometrica*. Prior to moving to Maynooth in 2008, Connor was Professor of Finance at the London School of Economics, and previous to that he served as Assistant Professor of Finance at the Haas School of Business, University of California, Berkeley and the Kellogg School of Management, Northwestern University. He earned his MA and PhD (Economics) from Yale University and his BA (Economics) from Georgetown University. He has made a number of seminal contributions in the literature on multifactor pricing. He is best known for introducing asymptotic principal component analysis to empirical asset pricing.

We recall the definition of principal components. Let a mean zero random vector X have covariance matrix Σ. The first principal component of X is the normalized combination $w^\mathsf{T} X$ that has maximum variance. This is the solution to the optimization problem $\max_{w^\mathsf{T} w=1} w^\mathsf{T} \Sigma w$. This can be shown to be the eigenvector corresponding to the largest eigenvalue of Σ. The second principal component maximizes variance amongst all normalized linear combinations that are orthogonal to the first one; it is the second eigenvector. We will apply this notion more generally.

We work with model (8.30) and define the $K \times K$ covariance matrix of the loadings $E(b_i b_i^\mathsf{T}) = \Omega_b$. Define the $T \times T$ covariance matrix Ψ_N

$$\Psi_N = \frac{1}{N} \sum_{i=1}^{N} E\left[(Z_i - E(Z_i))(Z_i - E(Z_i))^\mathsf{T}\right]. \tag{8.36}$$

Under the factor structure (treating F as fixed and B as random), it follows that

$$\Psi_N = F\Omega_b F^\mathsf{T} + \overline{\sigma}_\varepsilon^2 I_T, \tag{8.37}$$

$$\overline{\sigma}_\varepsilon^2 = \frac{1}{N} \sum_{i=1}^{N} \sigma_{\varepsilon i}^2.$$

The quantity $\overline{\sigma}_\varepsilon^2$ depends on N and so does Ψ_N but for notational simplicity we dispense with the subscript on Ψ_N. As $N \to \infty$, we assume that $\Psi_N \to \Psi_\infty$ for some nonsingular matrix Ψ_∞, so that the idiosyncratic variances have well defined average values.

The $T \times T$ covariance matrix Ψ in general has $T(T+1)/2$ free parameters. We suppose that $N > T > K$. Under the factor model these parameters are restricted in the sense that the right hand side of (8.37) has only $TK + K(K+1)/2 + 1$ free parameters. This is the value of the factor model expressed in how it simplifies this covariance matrix. The issue, though, is how to estimate the unknown parameters of the covariance matrix implied by the factor structure.

Before we proceed, we note that there is an identification problem in the sense that there are many observationally equivalent versions of F, b_i. For any nonsingular matrix L we can define $F^* = FL$ and $b_i^* = L^{-1} b_i$, and then notice that $Fb_i = FLL^{-1} b_i = F^* b_i^*$. Therefore we must impose some normalization, which amounts to a choice of L. For example, we may take $L = (F^\mathsf{T} F)^{-1/2}$ and then $F^{*\mathsf{T}} F^* = (F^\mathsf{T} F)^{-1/2} F^\mathsf{T} F (F^\mathsf{T} F)^{-1/2} = I_K$.

Definition 89 *Suppose that the loading covariance matrix is diagonal and that the factors are orthonormal*

$$\Omega_b = \mathrm{diag}\{\gamma_1, \ldots, \gamma_K\} = \Gamma \quad ; \quad F^\mathsf{T} F = I_K,$$

where without loss of generality $\gamma_1 \geq \ldots \geq \gamma_K$.

It follows that

$$\Psi = F \Gamma F^\mathsf{T} + \overline{\sigma}_\varepsilon^2 I_T. \tag{8.38}$$

We next show that $F, \Gamma, \overline{\sigma}_\varepsilon^2$ are essentially unique (F is unique up to sign).

Theorem 38 *F can be identified as the eigenvectors corresponding to the K largest eigenvalues of the $T \times T$ covariance matrix Ψ, denoted*

$$F = \mathrm{eigvec}_K(\Psi),$$

where $\mathrm{eigvec}_K(\Psi)$ denotes the eigenvectors corresponding to the K largest eigenvalues of Ψ.

We give the proof here because it is constructive. Recall that for any symmetric $T \times T$ matrix Ψ we have the unique eigendecomposition

$$\Psi = Q \Lambda Q^\mathsf{T} = \sum_{t=1}^{T} \lambda_t q_t q_t^\mathsf{T}, \tag{8.39}$$

where eigenvectors Q satisfy $QQ^\mathsf{T} = Q^\mathsf{T} Q = I_T$ and eigenvalues $\Lambda = \mathrm{diag}\{\lambda_1, \ldots, \lambda_T\}$ ordered from largest to smallest. Let G be a $T \times (T - K)$ matrix such that $Q = (F_{T \times K}, G_{T \times T-K})$ satisfies $QQ^\mathsf{T} = Q^\mathsf{T} Q = I_T$. Provided F has full rank K, this Q exists

8.8 Statistical Factor Models

and is unique. We have

$$\Psi = F\Gamma F^\mathsf{T} + \bar{\sigma}_\varepsilon^2 I_T$$

$$= (F, G) \begin{bmatrix} \Gamma & 0 \\ 0 & 0 \end{bmatrix} (F, G)^\mathsf{T} + \bar{\sigma}_\varepsilon^2 I_T$$

$$= Q \begin{bmatrix} \Gamma + \sigma^2 I_K & 0 \\ 0 & \bar{\sigma}_\varepsilon^2 I_{T-K} \end{bmatrix} Q^\mathsf{T} = Q\Lambda Q^\mathsf{T}.$$

The $\Gamma, \bar{\sigma}_\varepsilon^2$ are also uniquely identified by this. The result follows. □

We next define the estimation procedure. The APC estimates (in the first pass) factor returns rather than factor betas using the above identification argument. The procedure is as follows.

1. Compute the $T \times T$ sample covariance matrix of excess returns

$$\widehat{\Psi} = \frac{1}{N} \sum_{i=1}^{N} (Z_i - \overline{Z})(Z_i - \overline{Z})^\mathsf{T}, \quad \overline{Z} = \frac{1}{N} \sum_{i=1}^{N} Z_i.$$

2. Do the empirical eigendecomposition of the covariance matrix and take

$$\widehat{F} = \text{eigvec}_K\left(\widehat{\Psi}\right); \quad \widehat{\sigma}_\varepsilon^2 = \text{eigval}_{K+1}\left(\widehat{\Psi}\right); \quad \widehat{\gamma}_j = \text{eigval}_j\left(\widehat{\Psi}\right) - \text{eigval}_{K+1}\left(\widehat{\Psi}\right),$$
$$j = 1, \ldots, K.$$

These estimates can be shown to be consistent and asymptotically normal when N is large. Let $\widehat{\theta}$ be the vector of dimensions $K(T+1) + 1 \times 1$ that contains all these parameter estimates and let θ be the vector of corresponding true values. Provided N is large and T is fixed, we have

$$\sqrt{N}\left(\widehat{\theta} - \theta\right) \Longrightarrow N(0, \Upsilon)$$

for some matrix Υ (Connor and Korajczyk (1988)). Given the estimates of F, the factor betas can be estimated by time-series OLS regression

$$\widehat{b}_i = (\widehat{F}^\mathsf{T} \widehat{F})^{-1} \widehat{F}^\mathsf{T} (Z_i - \widehat{\mu}_i i_T), \quad (8.40)$$

which approximate $\widehat{b}_i^* = (F^\mathsf{T} F)^{-1} F^\mathsf{T} (Z_i - \mu_i i_T)$, which is the best linear unbiased estimator of b_i. One can test the specification of this model using the likelihood ratio

$$LR = N\left(\log \det\left(\widetilde{\Psi}\right) - \log \det\left(\widehat{\Psi}\right)\right),$$

where $\widetilde{\Psi} = \widehat{F}\widehat{\Gamma}\widehat{F}^\intercal + \widehat{\sigma}_\varepsilon^2 I_T$. Under the null hypothesis that the model is correctly specified this statistic is approximately χ_p^2 with $p = ((T-K)^2 - T - K)/2$.

One problem with this approach is that it assumes time series homoskedasticity for the idiosyncratic error, which is not a good assumption; see Chapter 11. Jones (2001) extended the APC procedure analysis to allow for time varying average idiosyncratic variances. In this case the $T \times T$ covariance matrix is

$$\Psi = F\Gamma F^\intercal + D, \tag{8.41}$$

$$D = \mathrm{diag}\{\overline{\sigma}_1^2, \ldots, \overline{\sigma}_T^2\}, \quad \overline{\sigma}_t^2 = \frac{1}{N}\sum_{i=1}^N E(\varepsilon_{it}^2).$$

Jones' method involves an iterative application of APC. That is, given first round estimates defined above, calculate the time series heteroskedasticity of the idiosyncratic errors

$$\widehat{D} = \mathrm{diag}\{\widehat{\sigma}_1^2, \ldots, \widehat{\sigma}_T^2\}, \quad \widehat{\sigma}_t^2 = \frac{1}{N}\sum_{i=1}^N \left(Z_{it} - \widehat{b}_i^\intercal \widehat{f}_t\right)^2.$$

Then compute the factors from the centered estimate

$$\widehat{F} = \mathrm{eigvec}_K\left(\widetilde{\Psi} - \widehat{D}\right)$$

and the loadings as in (8.40) and iterate.

Time series heteroskedasticity is harder to accommodate in the maximum likelihood factor model approach because this relies on large T and small N. If we were to allow unrestricted time series heteroskedasticity there we would not have enough information to estimate them consistently because N is assumed fixed and is small relative to T.

Example 49 *In Figure 8.9 we show the eigenvalues of $\widehat{\Sigma}$ for daily S&P500 returns ($N = 441$, $T = 2732$).*

Example 50 *In Figure 8.10 we show the eigenvalues of $\widehat{\Psi}_N$ for monthly S&P500 returns ($N = 441$, $T = 124$). In Figure 8.11 we show the dominant principal component.*

8.8.4 Bai and Ng

Bai and Ng (2002) consider a general setting for the estimation of a statistical factor model where both N and T can be large. They suppose that

$$Z_{it} = b_i^\intercal f_t + \varepsilon_{it},$$

8.8 Statistical Factor Models

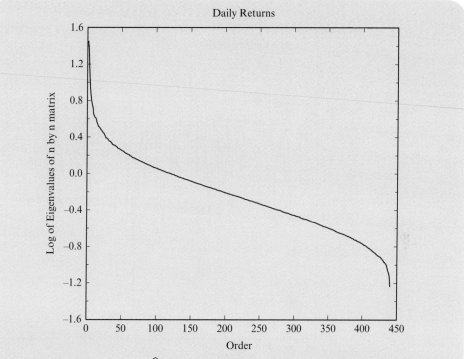

Figure 8.9 Eigenvalues of $\widehat{\Sigma}$ for daily S&P500 stocks, $N = 441$ and $T = 2732$

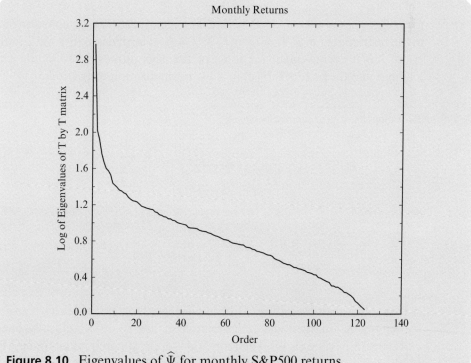

Figure 8.10 Eigenvalues of $\widehat{\Psi}$ for monthly S&P500 returns

310 Chapter 8 Multifactor Pricing Models

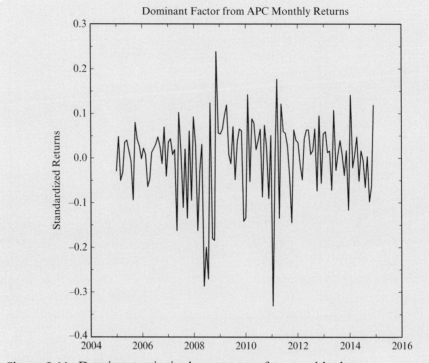

Figure 8.11 Dominant principal component for monthly data

for $i=1,\ldots,N$, $t=1,\ldots,T$ as before. For identification they consider one or other of the restrictions: $B^\mathsf{T} B/N = I_K$ or $F^\mathsf{T} F/T = I_K$. The division by N or T in this is irrelevant, but it is convenient for some of the asymptotics. They jointly estimate the loadings and the factors subject to these normalization restrictions.

Definition 90 *Define the sum of squared residuals*

$$Q_K(B,F) = \frac{1}{NT}\sum_{t=1}^{T}\sum_{i=1}^{N}\left(Z_{it} - b_i^\mathsf{T} f_t\right)^2 = \frac{1}{NT}\sum_{t=1}^{T}(Z_t - Bf_t)^\mathsf{T}(Z_t - Bf_t)$$

and let given $B \in \mathbb{R}^{N \times K}$, $F \in \mathbb{R}^{K \times T}$. They define estimators based on the constrained least squares

$$\widehat{B}, \widehat{F} = \arg\min_{B^\mathsf{T} B/N = I_K \text{ or } F^\mathsf{T} F/T = I_K} Q_K(B,F). \tag{8.42}$$

In practice, their procedure can be implemented as iterative least squares: first cross-section regression, then time series regression, and so on. Specifically, one iteratively calculates (8.34) and (8.40). Their method is approximately equal to the APC method (under the same normalization) when T is fixed but N is large.

8.8 Statistical Factor Models

They derive some properties of their estimation procedure when N and T are large under quite general conditions on the error terms. In particular, they allow for both time series and cross sectional heteroskedasticity and weak dependence. It is worth discussing some of their assumptions. The random variables ε_{it} satisfy

1. $E(\varepsilon_{it}) = 0$, $E(|\varepsilon_{it}|^8) \leq C$.
2. $N^{-1} \sum_{i=1}^{N} E(\varepsilon_{it}\varepsilon_{is}) = \gamma_N(s,t)$ with $|\gamma_N(s,t)| \leq C$ and $T^{-1} \sum_{s=1}^{T} \sum_{t=1}^{T} |\gamma_N(s,t)| \leq C$.
3. $E(\varepsilon_{it}\varepsilon_{jt}) = \tau_{ij,t}$ with $|\tau_{ij,t}| \leq \tau_{ij}$ with $N^{-1} \sum_{i=1}^{N} \sum_{j=1}^{N} |\tau_{ij}| \leq C$.
4. $E(\varepsilon_{it}\varepsilon_{js}) = \tau_{ij,ts}$ with $N^{-1}T^{-1} \sum_{i=1}^{N} \sum_{j=1}^{N} \sum_{s=1}^{T} \sum_{t=1}^{T} |\tau_{ij,ts}| \leq C$.
5. For every t,s

$$E\left(\left|\frac{1}{\sqrt{N}} \sum_{i=1}^{N} (\varepsilon_{is}\varepsilon_{it} - E(\varepsilon_{is}\varepsilon_{it}))\right|^4\right) \leq C.$$

The assumptions are quite weak in some respects. The dependence restrictions are specified in terms of the second moments rather than in terms of mixing coefficients. Sufficient conditions for the bounds to hold are that the process ε_{it} is stationary and mixing across both i,t (Connor and Korajczyk (1993)) with algebraic decay of a certain order. The eight moment assumption is perhaps the strongest assumption, and for daily stock return data this may be hard to justify.

Under their conditions the estimated factor space $\{F_t^0 A : A \text{ is } K \times K \text{ nonsingular}\}$ is consistent. Specifically, they show that

$$\frac{1}{T} \sum_{t=1}^{T} \left\|\widehat{F}_t - F_t^0 A\right\|^2 \xrightarrow{P} 0$$

for some nonsingular $K \times K$ matrix A. Under an additional assumption that $\sum_{s=1}^{T} |\gamma_N(s,t)| \leq C < \infty$ for all t, T they show that $\|\widehat{F}_t - F_t^0 A\|^2 \xrightarrow{P} 0$ for each t.

They also propose a model selection method to determine the number of factors K. This involves estimating a number of factor models with different dimensions and then comparing the models according to goodness of fit, adjusting for the complexity of the model that is being estimated. They propose a penalty of the form

$$M_{NT}(K) = \frac{1}{NT} \sum_{t=1}^{T} \sum_{i=1}^{N} \widehat{\varepsilon}_{it}^2(K) + Kg(N,T),$$

where $\widehat{\varepsilon}_{it}(K)$ are the residuals for factor model of dimension K, and $g(N,T)$ is a penalty factor, for example

$$g(N,T) = \frac{N+T}{NT}.$$

They show that choosing K to minimize $M_{NT}(K)$ consistently estimates the order of the factor model, that is $\widehat{K} \to K_0$ with probability one, where $\widehat{K} = \arg\min_{1 \leq K \leq \overline{K}} M_{NT}(K)$ and $K_0 \leq \overline{K}$ is the true value of K. Here, $\overline{K}$ is a fixed upper bound that has to be chosen in advance; $\overline{K} = 1000$ usually works.

8.9 Testing the APT with Estimated Factors

Common practice is to proceed as if the estimation error in obtaining the factors was ignorable for the purposes of testing the APT. That is, with the estimated factors we may construct $\widehat{Z}_{Kt}$ and the MLE of B, α conditional on F, and construct the F test (8.18) using the estimated factors, and compare with the F critical values. Of course, this is hard to justify, and in practice may not work well.

8.10 The MacKinlay Critique

CLM and others note the pre-test bias of multifactor models, based on size and value-related factors, that improve upon the pricing performance of the CAPM. Fama and French claim that there are pervasive factors associated with size and value characteristics and that is why they carry risk premia. Other researchers claim that it is irrationality, not risk premia, which best explain the return premia associated with these characteristics. MacKinlay (1995) uses the notion of approximate arbitrage to differentiate between risk-factor based and non-risk-factor based explanations. Suppose that the CAPM is rejected but an additional factor (say associated with size) eliminates the rejection. There are two explanations. Either size explains the pricing anomaly because there is a pervasive factor associated with size, or size explains the pricing anomaly because investors are irrationally opposed to holding small stocks, and there is no pervasive factor associated with size. Under the alternative hypothesis the GRS statistic is distributed as a noncentral F with noncentrality parameter δ that depends upon the Sharpe ratio of the size factor (assuming normality). In the first case, this is bounded by the Sharpe ratio of the tangency portfolio $\delta < Ts_q^2$. In the second case, the Sharpe ratio associated with size can be made unboundedly large as the number of assets grows to infinity; the variability associated with size is diversifiable. These are quite different predictions potentially. MacKinlay (1995) finds that his test statistics are too big to rationalize under the first scenario, and thereby provides a critique of the Fama–French approach. Despite this, the Fama–French approach is the central methodology to address a variety of issues, although the number of factors one should include seems to be increasing over time, and indeed which factors (beyond the first three) is subject to passionate debate.

8.11 Summary of Chapter

We discussed the multifactor pricing models of widespread current use, and the key concepts of diversification and pervasiveness. We discussed the Fama–French approach, the macro factor approach, and the statistical factor approach.

8.12 Appendix

Proof of Theorem 37. Note that BB^T is a positive semidefinite rank K symmetric matrix, meaning it has K positive eigenvalues $\lambda_1, \ldots, \lambda_K$ and $N - K$ zero ones.

8.12 Appendix

Furthermore, the non zero eigenvalues of BB^T are exactly the eigenvalues of the $K \times K$ matrix $B^T B$ (although the eigenvectors are different). This is because

$$BB^T q_j = \lambda_j q_j$$

$$B^T BB^T q_j = B^T B \widetilde{q}_j = \lambda_j B^T q_j = \lambda_j \widetilde{q}_j,$$

where $\widetilde{q}_j = B^T q_j$. Therefore,

$$w^T BB^T w = w^T w \frac{w^T BB^T w}{w^T w} \leq w^T w \lambda_{\max}(BB^T) = w^T w \lambda_{\max}(B^T B).$$

For the lower bound we note that for w such that $B^T w \neq 0$

$$w^T BB^T w = w^T w \frac{w^T BB^T w}{w^T w} \geq w^T w \lambda_{\min}^+(BB^T) = w^T w \lambda_{\min}(B^T B) = Nw^T w \lambda_{\min}(B^T B/N),$$

where $\lambda_{\min}^+(BB^T)$ means the smallest positive eigenvalue. Then, since $w^T w \geq 1/N$, the result follows. The latter follows by Cauchy–Schwarz inequality since

$$1 = (w^T i)^2 \leq w^T w \times i^T i.$$

□

9 Present Value Relations

In this chapter we discuss how prices should be determined by fundamentals and we discuss methods that are designed to establish whether in fact prices are so determined. We discuss the volatility tests developed by Shiller and the predictive regressions methodology.

9.1 Fundamentals versus Bubbles

There are two competing views of financial markets. Are stock prices driven by fundamental values or rational expectations of what those fundamental values will be in the future? This is the view maintained in the efficient markets hypothesis, the CAPM, etc. Or are stock prices driven by animal spirits, fads, bubbles, and irrational exuberance? This latter literature is informed by **behavioral finance**; see Thaler (2016). Perhaps both these views have something useful to say, and sometimes one and sometimes the other prevails.

One typically thinks of bubbles as a pervasive market wide phenomenon with rapid increases in prices ultimately followed by a crash. There are many well-known examples of bubbles in financial history: tulips in Amsterdam (1634–37), the Mississippi bubble (1719–20), and the South Sea Bubble (1720). In the twentieth century, the roaring '20s and the turn of the century tech bubble are well known. More recently the words Bitcoin and bubble have become synonymous. Many academic studies have been devoted to explaining why these events were not well explained by fundamental factors. On the other hand, there are authors who dispute the bubble explanations. Garber (1990) proposes market fundamental explanations for the three famous historical bubbles. Pástor and Veronesi (2006) argue that the turn of the twentieth century tech bubble was at least partly explained by an increase in the uncertainty about average future profitability in the late 1990s, and the rational response of investors to that uncertainty.

Market crashes are also well studied. Famous examples of stock market crashes include: the US markets on October 24, 1929 and October 19, 1987. Each of these market events have been the subject of a lot of research and newspaper coverage as well as government reports. Typically these reports do not identify a single causal explanation but several factors that played a role. So even in these well-known extreme cases it is hard to pin specific blame. Shin (2010) argues that a major factor in the 1987 crash was the **endogenous risk** induced by the dynamic hedging strategies of portfolio insurance companies.

9.1 Fundamentals versus Bubbles

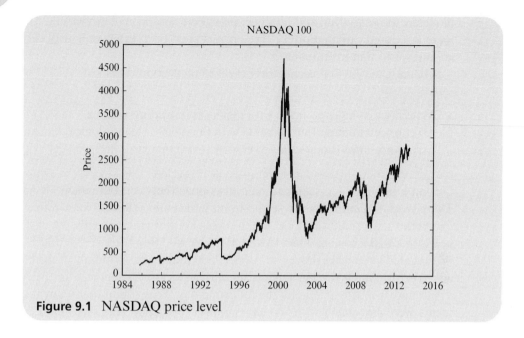

Figure 9.1 NASDAQ price level

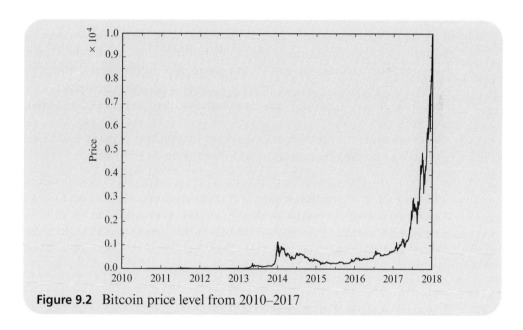

Figure 9.2 Bitcoin price level from 2010–2017

Shiller (2000) provides some evidence on extremely optimistic stock market appreciations around the world. In his Table 6.1 he lists those international stock markets with some extremely large one-year real appreciations such as the Philippines during Dec 1985–Dec 1986 when prices increased 683.4%, and Denmark during April 1971–April 1972 when real prices rose 122.9%. In some cases prices kept on rising more modestly the year after, and in other cases there was a smaller reversal of fortune. Is there a rational explanation for these prices rises, or are they only explicable due to animal spirits and irrationality?

Scheinkman (2014) presents three stylized facts about bubbles:

1. Asset price bubbles coincide with increases in trading volume.
2. Asset price bubbles often coincide with financial or technological innovation.
3. Asset price implosions seems to coincide with increases in the assets's supply.

During a bubble there seems to be a disconnect between fundamentals and prices. There is a lot of empirical work questioning the connection between prices and fundamentals in normal times, i.e., when prices are not rising rapidly. We just mention here some well known studies. Cutler, Poterba, and Summers (1989) and Fair (2002) investigated large price moves on the S&P500 (daily and intradaily in the latter case) and tried to match the movements up with news stories reported in the *New York Times* and *Wall Street Journal* the following day. They found that many large movements were associated with monetary policy, but there remained many significant movements that they could not find explanations for. Fundamentals for stock prices are difficult to measure and potentially include all macroeconomic news that can drive the market, so it is difficult to make strong conclusions. Roll (1984b) is a famous study of the market for orange juice futures for which the fundamentals are pretty simple to measure – weather, in particular the number of frost days in Florida. He found some disconnect between the movements in orange juice futures and the weather. Online betting markets such as Betfair provide a closer connection between prices and fundamentals, since the payoff is purely determined by the result of a sports match or election.

In other contexts, fundamentals are hard to measure as many businesses seem to have little or no tangible assets, as the following quote indicates

In 2015 Uber, the world's largest taxi company, owns no vehicles; Facebook, the world's most popular media owner, creates no content; Alibaba, the most valuable retailer, has no inventory; and Airbnb, the world's largest accommodation provider, owns no real estate.

9.2 Present Value Relations

We present the dividend discount model which determines prices from future cash-flows and discount rates in a rational way under risk neutrality. Let R denote the required one-period return (discount rate), which does not vary with time t. Then

$$R = E_t(R_{t+1}) = E_t\left(\frac{P_{t+1} + D_{t+1}}{P_t} - 1\right), \tag{9.1}$$

9.2 Present Value Relations

where E_t denotes expectation at time t. Here, D denotes future dividends or some other cash flow such as earnings, depending on the application. We can write

$$P_t = E_t\left(\frac{P_{t+1} + D_{t+1}}{1+R}\right) = \frac{E_t(P_{t+1})}{1+R} + \frac{E_t(D_{t+1})}{1+R},$$

which is an equation or set of equations (for each t) in the price sequence. Applying the same logic to future prices we have

$$P_{t+1} = \frac{E_{t+1}(P_{t+2})}{1+R} + \frac{E_{t+1}(D_{t+2})}{1+R},$$

so that (using the law of iterated expectation) and continuing

$$P_t = \frac{E_t\left(E_{t+1}(P_{t+2}) + E_{t+1}(D_{t+2})\right)}{(1+R)^2} + \frac{E_t(D_{t+1})}{1+R}$$

$$= \frac{E_t(D_{t+1})}{1+R} + \frac{E_t(D_{t+2})}{(1+R)^2} + \frac{E_t(P_{t+2})}{(1+R)^2}$$

$$= \frac{E_t(D_{t+1})}{1+R} + \frac{E_t(D_{t+2})}{(1+R)^2} + \cdots + \frac{E_t(D_{t+k})}{(1+R)^k} + \frac{E_t(P_{t+k})}{(1+R)^k}.$$

We assume the absence of price bubbles at infinity, i.e., the discounted terminal price goes to zero with horizon k,

$$\lim_{k\to\infty} \frac{E_t(P_{t+k})}{(1+R)^k} = 0. \tag{9.2}$$

Then, the sum converges to

$$P_t = E_t\left(\left[\sum_{i=1}^{\infty}\left(\frac{1}{1+R}\right)^i D_{t+i}\right]\right) = \sum_{i=1}^{\infty}\left(\frac{1}{1+R}\right)^i E_t(D_{t+i}), \tag{9.3}$$

which gives the unique solution to (9.1). Therefore, prices are set equal to the present discounted value of expected future cash flows or dividends. In this sense prices are driven by fundamentals. In this model, stock prices themselves are not necessarily martingales because dividends are part of the profit of the investor, and $E_t(P_{t+1}) = (1+R)P_t - E_t(D_{t+1})$. However, the implied price where dividends are reinvested is a martingale.

Example 51 *The Gordon Growth Model.* Suppose that

$$D_{t+1} = (1+G)D_t\xi_{t+1}, \tag{9.4}$$

where ξ_{t+1} is i.i.d. with mean one (the deterministic case has $\xi_{t+1} = 1$ for all t). We have

$$E_t(D_{t+i}) = (1+G)^i D_t E_t(\xi_{t+1} \cdots \xi_{t+i}) = (1+G)^i D_t.$$

In this case

$$P_t = \sum_{i=1}^{\infty} \left(\frac{1+G}{1+R}\right)^i D_t = \frac{(1+G)}{R-G} D_t.$$

Hence with a constant dividend growth rate, stock prices depend only upon innovations to expected dividends. The dividend price ratio is constant $D_t/P_t = (R-G)/(1+G)$, and

$$\frac{\partial \log P_t}{\partial \log G} = \frac{G(1+R)}{(R-G)(1+G)} > 0 \quad ; \quad \frac{\partial \log P_t}{\partial \log R} = \frac{-R}{R-G} < 0.$$

Note that prices can be very sensitive to changes in G or R when $R \simeq G$.

In the Gordon Growth Model, dividends are non stationary with growing mean; moreover the process is not difference stationary. An alternative model is to suppose that $\triangle D_t$ is a stationary process. In this case we can argue that prices are also difference stationary, taking differences of (9.3). Furthermore, we can argue that P_t and D_t are cointegrated of order one with cointegrating vector $(1, \frac{1}{R})$, since

$$P_t - \frac{D_t}{R} = \left(\frac{1}{R}\right) E_t \left(\sum_{i=0}^{\infty} \left(\frac{1}{1+R}\right)^i \triangle D_{t+1+i}\right) \tag{9.5}$$

is a stationary process, where $\triangle D_{t+1+i} = D_{t+1+i} - D_{t+i}$. We can test this hypothesis directly given a time series on prices and dividends.

We examine the result (9.5) further by considering some specific models for the dividend process.

Example 52 *Suppose that dividends follow a random walk plus drift*

$$D_{t+1} = \mu + D_t + \eta_{t+1} \tag{9.6}$$

with η_t i.i.d. with mean zero. Then we have $E_t[\triangle D_{t+1+i}] = E_t[\mu + \eta_{t+1+i}] = \mu$, in which case

$$P_t = \frac{D_t}{R} + \mu \frac{1+R}{R^2}$$

and the process $P_t - \frac{D_t}{R}$ is actually degenerate, i.e., it is nonstochastic. In the case where η_t is an AR(1) process with parameter $\rho \in (-1, 1)$ we obtain

$$P_t - \frac{D_t}{R} = \mu \frac{1+R}{R^2} + \left(\sum_{i=1}^{\infty} \left(\frac{1}{1+R}\right)^i \sum_{j=1}^{i} \rho^j\right) \eta_t = a(\mu, R) + b(R, \rho) \eta_t,$$

and the right hand side is a non-degenerate stationary process.

9.3 Rational Bubbles

Example 53 *An alternative model for dividends is that they are trend stationary*

$$D_{t+1} = \mu + \beta t + \eta_{t+1}, \tag{9.7}$$

with η_t i.i.d. with mean zero. In this case, we may show that

$$P_t - \frac{D_t}{R} = \frac{1}{R}\sum_{i=0}^{\infty} \left(\frac{1}{1+R}\right)^i E_t \left(D_{t+i+i} - D_{t+i}\right) = \beta \frac{1+R}{R^2} - \frac{1}{R}\eta_t,$$

which is stochastic and stationary.

In both these examples, dividends and prices can generally be negative, which is a motivation for modelling the log of dividends and the log of prices instead. We will take up this approach below, but first we consider the concept of rational bubbles.

9.3 Rational Bubbles

There are models of rational bubbles that start from the same pricing equation (9.1) but do not assume the condition

$$\lim_{k \to \infty} \frac{E_t \left[P_{t+k}\right]}{(1+R)^k} = 0. \tag{9.8}$$

Specifically, write

$$P_t = P_t^* + B_t,$$

where P_t^* is the fundamental price determined from (9.3) and B_t is the bubble process that satisfies

$$B_t = \frac{E_t(B_{t+1})}{1+R}. \tag{9.9}$$

This means that the bubble component is growing in expectation. This is a rational bubble because it fits in precisely with the rational pricing framework above. The sequence $\{P_t\}$ solves the equation (9.1).

Blanchard and Watson (1982) proposed the following simple bubble model, a switching AR(1) process, that is consistent with (9.9).

Definition 91 *Suppose that*

$$B_{t+1} = \begin{cases} \frac{1+R}{\pi} B_t + \eta_{t+1} & \text{with probability } \pi \\ \eta_{t+1} & \text{with probability } 1 - \pi \end{cases} \tag{9.10}$$

where η_{t+1} is an i.i.d. shock variable.

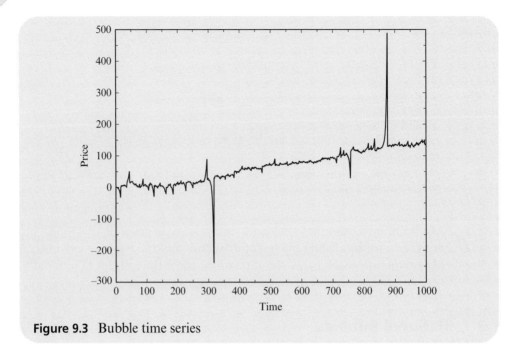

Figure 9.3 Bubble time series

During a bubble period the process grows explosively, but each period there is the possibility that the bubble collapses; in fact it does so with constant probability $1 - \pi$. It follows that the expected duration of a given bubble is $1/(1 - \pi)$. In Figure 9.3 we show a typical trajectory from the case where $\pi = 0.75$, $R = 0.001$, and $\eta \sim N(0, 1)$ superimposed on a rational price process. It is apparent from the series that the bubbles in this case are quite extreme but also short lived; there are both positive and negative bubbles. Furthermore, the bubble series goes substantially negative for a period of time.

Some authors have emphasized logical issues with the concept of rational bubbles. First, the whole framework requires an infinite horizon so that rational bubbles cannot exist on finite-lived assets. Furthermore, because asset prices are bounded below by zero (for assets with limited liability), then negative bubbles cannot exist. Likewise, if there is an upper bound on the asset price (e.g. there exists a high-priced substitute in perfectly elastic supply), then positive bubbles cannot exist. If positive bubbles can exist, but negative bubbles are ruled out, then a bubble can never start off, it must have existed from the beginning of trading. Similar critiques can be addressed to many financial theories, such as the Froot and Stein (1998) model; see Högh, Linton, and Nielsen (2006).

CLM argue that even if these models are logically flawed, they do point us to outcomes we do observe, such as a small amount of persistent return predictability having large effects on prices. Even if one does not believe that rational bubbles exist in practice, the rational bubble literature is informative because it tells us what phenomena we may observe in a world of near-rational bubbles (a small amount of persistent return predictability having large effects on prices).

In any case, popular discussion of bubbles seems often to be referring to irrational bubbles rather than rational ones, and there is a wide literature proposing different bubble models. Brunnermeier (2008) surveys the literature and provides a useful classification. He argues there are four general classes of coherent bubble models:

1. *Rational bubbles*. All participants are rational and equally informed. This implies that bubbles have to follow an explosive path, which is testable. In a rational bubble setting an investor only holds a bubble asset if the bubble grows in expectations ad infinitum.
2. In contrast, in the other bubble models a rational investor might hold an overpriced asset if he thinks he can resell it in the future to a less informed trader or someone who holds biased beliefs or behaves irrationally (the greater fool theory):
    (a) *Information asymmetry*. Bubbles can emerge under general conditions and not everyone knows there is a bubble going on.
    (b) *Limited arbitrage*. Interaction between rational and behavioural traders. Limits to arbitrage prevent rational investors from eradicating the price impact of the irrational guys.
    (c) *Heterogeneous beliefs*. Individual beliefs perhaps contain biases, and they disagree about the fundamental value.

9.4 Econometric Bubble Detection

We discuss briefly some methodology for identifying bubbles in data.

Peter Charles Bonest Phillips (born 1948) is Sterling Professor of Economics and Professor of Statistics at Yale University, where he has taught since 1979. He holds part-time positions at the University of Auckland, Singapore Management University, and the University of Southampton. He was co-director of the Center for Financial Econometrics in the Sim Kee Boon Institute for Financial Economics at Singapore Management University between 2008 and 2013. He is the founding editor of the journal *Econometric Theory*. He has supervised numerous PhD students and mentored many young academics. According to the February 2018 ranking of economists by Research Papers in Economics, he is the 7th most influential economist in the world. He has over 300 published journal articles and advanced diverse areas of research in econometrics, including most recently climate econometrics. His work on continuous time, finite-sample theory, trending time series, unit roots and cointegration, long-range dependent time series, and panel data econometrics have influenced applied research in economics and finance, and more widely in the social sciences.

> He introduced the use of the functional central limit theorem to derive asymptotic distributions in regressions with nonstationary time series and explore the nature of spurious regression phenomena. His most widely cited articles deal with general methods of unit root testing in time series. He has recently developed powerful methods for testing for the presence of bubbles in asset prices and real time bubble detection methods which are widely used by central banks.

Phillips and Yu (2010) and Phillips, Shi, and Yu (2012) have developed a lot of machinery to detect (after the fact) bubbles in financial time series. An explosive price process is $p_t = \delta p_{t-1} + \varepsilon_t$ with $\delta > 1$. In this case, prices increase very rapidly. This model is perhaps too extreme to be consistent with empirical reality. For a realistic model we need to moderate δ and/or mix with nonexplosive periods like the Blanchard and Watson model. A simple version of the Phillips and Yu model is as follows: the time series model for log prices is composed of three regimes:

$$p_t = \begin{cases} p_{t-1} + \varepsilon_t & \text{if } t < \tau_e \\ \delta_T p_{t-1} + \varepsilon_t & \text{if } \tau_e \leq t \leq \tau_f \\ p_{t-1} + \varepsilon_t & \text{if } t > \tau_f. \end{cases}$$

The process is a martingale during $[1, \tau_e]$ and $[\tau_f, T]$ but has an explosive incident in the middle of the sample. The parameter $\delta_T > 1$ controls the degree of explosiveness. If $\delta_T = 1 + c$ for some positive c, then the price process p_t rapidly explodes to extreme values. This may be appropriate in some hyperinflation environments, but in more modest cases, it is a bit too strong. Phillips and Yu suppose that

$$\delta_T = 1 + cT^{-\alpha} \tag{9.11}$$

for $c > 0$ and $0 < \alpha < 1$, which is a sophisticated way of saying that the explosiveness is relatively mild. They show how to estimate τ_e and τ_f from sample data, that is, they can detect the beginning and end of the bubble. They extend to multiple bubbles. Their methodology is based on rolling window unit root tests. An alternative approach is to work with returns. Their model implies that during a bubble epoch, returns are positively autocorrelated, and in fact returns should have small positive autocorrelations at each lag. This situation is well suited to the variance ratio method. For the univariate bubble process with nontrivial bubble epoch (i.e., $c > 0$ and $(\tau_f - \tau_e)/T \to \tau_0 > 0$), we have, as $T \to \infty$

$$\widetilde{VR}(p) \xrightarrow{P} p \tag{9.12}$$

for all p, so that the variance ratio statistic is greater than one for all p and gets larger with horizon. Essentially, the bubble period dominates all the sample statistics, and all return autocorrelations converge to one inside the bubble period, thereby making the variance ratio equal to the maximum it can achieve (because $1 + 2\sum_{j=1}^{p-1}(1 - \frac{j}{p}) = p$). Empirically, this prediction is hard to find in the data; see Tables 3.5, 3.6, and 3.7.

Sornette (2003) considers the following class of models for logarithmic prices

$$E(\log(p_t)) = A + B(t_c - t)^m + C(t_c - t)^m \cos(\omega \log(t_c - t) - \phi), \quad (9.13)$$

where $m \in (0, 1)$ and t_c is the crash point. By the way, the model is only valid for $t < t_c$. This is motivated by some physical systems such as are found in geophysics. He also provides prediction of future bubbles on his website.

9.5 Shiller Excess Volatility Tests

Shiller (1981) argues that the stock market is inefficient because it is too variable in comparison with fundamentals. The stock price should equal the expected discounted value of future dividends. However dividends are very stable; they fluctuate very little about an upward trend. Expected dividends should therefore also fluctuate little, and consequently stock prices should be stable. In fact, stock prices fluctuate wildly. Shiller (1981) shows how the variability of the dividend sets an upper bound to the variability of the stock price.

Suppose that P_t^* is a rational unbiased forecast of P_t, i.e.,

$$E(P_t^* | \mathcal{F}_t) = P_t, \quad (9.14)$$

where $\mathcal{F}_t$ is a set of available information that contains current price. Then we can write

$$P_t^* = P_t + e_t,$$

where $E(e_t | P_t) = 0$ by the law of iterated expectation. It follows that $\text{cov}(P_t, e_t) = 0$ and so

$$\text{var}(P_t^*) = \text{var}(P_t) + \text{var}(P_t^* - P_t) \geq \text{var}(P_t).$$

In this case, P_t^* has to be constructed from the dividend stream, and we discuss how this is implemented next. Shiller (1981) provides a test of whether shock prices move too much to be justified by subsequent changes in dividends. The basis of the test is the valuation model (9.3). Consider the **perfect foresight price**, i.e., when investors know the actual realization of the future dividend stream

$$P_t^* = \sum_{s=0}^{\infty} \frac{D_{t+s}}{(1 + R)^s}.$$

Then we have $\text{var}(P_t^*) \geq \text{var}(P_t)$, where P_t is the actually observed market price, which is based on a smaller information set. However, we haven't observed the entire dividend stream yet (except for failed firms), so how can we calculate P_t^*? First note that for any $G > 0$ with

$$P_t^R = \frac{P_t}{(1+G)^t}, \quad D_t^R = \frac{D_t}{(1+G)^t},$$

then from (9.3) it follows that

$$P_t^R = \sum_{s=0}^{\infty} \left(\frac{1+G}{1+R}\right)^{s+1} D_{t+s}^R = \sum_{s=0}^{\infty} \left(\frac{1}{1+\overline{R}}\right)^{s+1} D_{t+s}^R$$

$$= \sum_{s=0}^{M} \left(\frac{1}{1+\overline{R}}\right)^{s+1} D_{t+s}^R + \frac{E_t\left[P_{t+M}^R\right]}{(1+\overline{R})^M},$$

provided $G < R$. Shiller assumed that aggregate real dividends D_t^R on the stock follow a finite variance stationary stochastic process and that G is the deterministic growth rate of dividends, for example, the Gordon Growth Model (9.4). It follows that P_t^R is also a stationary process. The discount factor $\overline{R} = (R - G)/(1 + G)$ for the detrended series satisfies $\overline{R} = E(D_t^R)/E(P_t^R)$. Shiller estimates the perfect foresight price by

$$\widehat{P}_t^{R*} = \sum_{s=0}^{T-t} \frac{\widehat{D}_{t+s}^R}{(1+\widehat{\overline{R}})^s} + \frac{\widehat{P}_T^R}{(1+\widehat{\overline{R}})^{T-t}}, \qquad (9.15)$$

where the terminal log price $\widehat{P}_T^R$ is estimated as the average price over his hundred year sample. He estimates the trend rate G by regressing $\log(P_t)$ on a constant and time trend and lets $\widehat{D}_t^R = D_t/(1+\widehat{G})^t$. He estimates the discount rate $\widehat{\overline{R}}$ as the average of detrended real dividend divided by the average detrended real price, around 0.045 in his sample. He uses the real S&P500 and Dow Jones index and dividend series from 1871 to 1979.

Shiller found that the variance bound is violated empirically. The variance of the price is much larger than the variance of the ex-post rational price. One may conclude that the stock market is inefficient. This inefficiency means that one can forecast the rate-of-return on stocks from the dividend yield (the dividend/price ratio). A profitable trading rule is to buy when the dividend yield is high (because the price is then too low) and to sell when the dividend yield is low (because the price is then too high).

He finds that the variance inequality is substantially violated – the variance is 5–13 times too high relative to the perfect foresight price variance. This paper was very influential, but there was a lively debate for many years as to the methods and results.

In Figure (9.4) below we show the estimated trend line for the S&P500 series to illustrate how this method works.

One line of disagreement is represented by Marsh and Merton (1984) who questioned whether real dividends on the market can be described by a finite variance stationary stochastic process with a deterministic exponential trend. They examine the implications of an alternative dividend process for the variance inequality. They argue that (9.3) is a constraint on future dividends and not on the current rational

9.5 Shiller Excess Volatility Tests

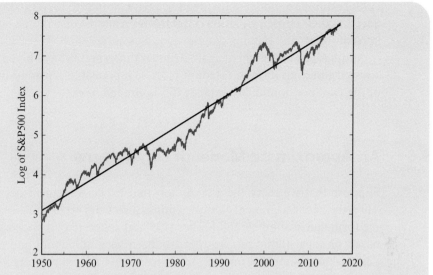

Figure 9.4 The logarithm of S&P500 index with trend line fitted from 1950 to 2017

stock price – an unanticipated change in the stock price must necessarily cause a change in expected future dividends. They then suggest that dividend policies adopted by managers relate to the firm's intrinsic value using Lintner's (1956) stylized facts based on a set of interviews with managers. These are:

1. Managers believe firms should have some long term target payout ratio.
2. In setting dividends, they focus on the change in existing payouts, and not on the level.
3. A major unanticipated and nontransitory change in earnings would be an important reason to change dividends.
4. Most managers try to avoid making changes in dividends that stand a good chance of having to be reversed within the near future.

They suggest an alternative dividend process. Let E_t be the real permanent earnings at time t and V_t be the firm's intrinsic value, so that $E_t = RV_t$. Then suppose that

$$D_{t+1} - D_t = GD_t + \sum_{k=0}^{n} \gamma_k \left(E_{t+1-k} - E_{t-k} - GE_{t-k} \right), \qquad (9.16)$$

where n is the horizon and $\gamma_k \geq 0$. Managers set dividends to grow at rate G but deviate from this long run growth path in response to changes in permanent earnings that deviate from their long-run growth path. This process meets all the stylized facts above. They then show that under some conditions we can expect a reversal of the variance inequality, i.e., $\text{var}(p^*) \leq \text{var}(p)$. In another paper, Marsh and Merton

(1996) estimate a model of dividend behavior similar to their theoretical model. They find that their chosen equation outperforms a simple distributed lag with time trend, in terms of R^2 and passing diagnostic tests.

Another line of argument against Shiller's interpretation of his empirical results has stressed time variation in discount rates as a plausible explanation for the violation of the variance bound. We explore this approach next.

9.6 An Approximate Model of Log Returns

Suppose that we relax the assumption that expected returns are constant in (9.1). Then the above arguments are very difficult to work with and we do not obtain simple formulae. Campbell and Shiller (1988ab) suggested an approximation argument based on logarithmic returns that has been very influential and that allows one to derive similar simple formulae. A further reason for considering logarithmic returns is that this seems to provide a better econometric fit.

Calculating log return from the formula for one-period arithmetic return gives

$$r_{t+1} \equiv \log(P_{t+1} + D_{t+1}) - \log(P_t)$$
$$= \log(P_{t+1}(1 + D_{t+1}/P_{t+1})) - \log(P_t)$$
$$= p_{t+1} - p_t + \log(1 + \exp(d_{t+1} - p_{t+1})).$$

They linearized this equation using the first-order Taylor approximation around average values $f(x_{t+1}) \simeq f(\bar{x}) + f'(\bar{x})(x_{t+1} - \bar{x})$.

Letting $x_{t+1} = d_{t+1} - p_{t+1}$, $\bar{x} = \bar{d} - \bar{p}$ and $f(x) = \log(1 + \exp(x))$, we find

$$r_{t+1} \simeq k + \rho \, p_{t+1} + (1 - \rho) \, d_{t+1} - p_t, \qquad (9.17)$$

where $k = -\log(\rho) - (1 - \rho)\log(1/\rho - 1)$ with

$$\rho = \frac{1}{1 + \exp(\bar{d} - \bar{p})} \simeq \frac{1}{1 + \bar{D}/\bar{P}}.$$

You should be asking yourself, what does $\simeq$ mean here? It means that I want to ignore the remaining terms. Technically, the approximation has a remainder term that is proportional to $(x_{t+1} - \bar{x})^2$. The general idea is to linearize some nonlinear dynamic system around its equilibrium value. Theoretically, this approximation is valid if the variance of the log dividend price ratio is small (compared with the other stuff) or if the system is approximately linear; see below for discussion.

9.6 An Approximate Model of Log Returns

We can write equation (9.17) as an equation for p_t, and solving the equation forward and imposing the no-bubbles condition gives

$$p_t = k + \rho\, p_{t+1} + (1-\rho)\, d_{t+1} - r_{t+1}$$

$$= \frac{k}{1-\rho} + E_t\left(\sum_{j=0}^{\infty} \rho^j \left[(1-\rho)\, d_{t+1+j} - r_{t+1+j}\right]\right)$$

$$= \underbrace{\frac{k}{1-\rho} + (1-\rho)\sum_{j=0}^{\infty} \rho^j E_t(d_{t+1+j})}_{p_{dt}} - \underbrace{\sum_{j=0}^{\infty} \rho^j E_t(r_{t+1+j})}_{p_{rt}}.$$

This says that prices reflect both expected future dividend flow and expected future returns. An unexpectedly good stock return occurs because: (a) the current dividend went up; or (b) expectations of future dividends go up; or (c) because expectations of future returns go down. The first two terms are a standard cash flow effect and the third is an expected return or risk premium effect: the price goes up if the risk premium or risk-free interest rate go down. This is important since one large anomaly in observed behavior is the large size of realized return innovations relative to realized dividend innovations.

We may rewrite this equation in several ways. First, we can relate the long horizon returns to fundamentals as follows

$$\underbrace{\sum_{j=0}^{\infty} \rho^j E_t(r_{t+1+j})}_{\text{long horizon returns}} = \underbrace{\widehat{d_t - p_t}}_{\text{fundamentals}} + \frac{k}{1-\rho} + \underbrace{\sum_{j=0}^{\infty} \rho^j E_t(\Delta d_{t+1+j})}_{\text{small?}}. \qquad (9.18)$$

This is one starting point for predictive regression modelling of long horizon returns on fundamentals; see below.

We may also write (9.17) as an expression for the dividend price ratio

$$d_t - p_t = \rho\,(d_{t+1} - p_{t+1}) + k + \Delta d_{t+1} - r_{t+1}.$$

Solving forward and imposing the no-bubbles condition gives

$$d_t - p_t = \frac{k}{1-\rho} + \sum_{j=0}^{\infty} \rho^j \left(E_t(\Delta d_{t+j+1}) - E_t(r_{t+j+1})\right).$$

Variation in the price–dividend ratio occurs because of variation in dividend growth or discount factors.

Similarly returns and return innovations can be written in terms of two separate components, i.e., the return innovation is a linear combination of the innovation to discounted expected dividends and the innovation to discounted expected returns. Applying $E_{t+1} - E_t$ to both sides of (9.17) we obtain

$$r_{t+1} - E_t(r_{t+1}) = \rho\left(p_{t+1} - E_t(p_{t+1})\right) + (1-\rho)\left(d_{t+1} - E_t(d_{t+1})\right).$$

Solving forward and imposing the no-bubbles condition gives

$$r_{t+1} - E_t(r_{t+1}) = \underbrace{(E_{t+1} - E_t)\sum_{j=0}^{\infty}\rho^j \Delta d_{t+1+j}}_{\eta_{d,t+1}} - \underbrace{(E_{t+1} - E_t)\sum_{j=1}^{\infty}\rho^j r_{t+1+j}}_{\eta_{r,t+1}},$$

where the unanticipated shocks $\eta_{d,t+1}$ and $\eta_{r,t+1}$ are related to dividend flow and expected returns, respectively. This is important since one large anomaly in observed behavior is the large size of realized return innovations relative to realized dividend innovations.

9.6.1 The Impact of Cash Flow and Discount Rate Innovations

We next consider the simple case where expected returns are a constant plus an AR(1) process

$$E_t(r_{t+1}) = r + \mu_t$$

$$\mu_{t+1} = \phi\mu_t + \xi_{t+1}, \quad -1 < \phi < 1.$$

The process μ_t captures the idea that expected returns may vary slowly over time (it is not necessarily observed by the econometrician). Inserting into the formula for realized return innovation as a combination of dividend and expected return innovations gives

$$p_{rt} \equiv \sum_{j=0}^{\infty} \rho^j E_t(r_{t+1+j}) = \frac{r}{1-\rho} + \frac{\mu_t}{1-\rho\phi}.$$

If the innovations are very persistent (ϕ is close to 1), then the stock price is sensitive to innovations in expected returns. If the average dividend yield is high, then ρ is close to one, increasing the sensitivity of stock prices to an innovation to expected returns. If $\rho\phi \simeq 1$ this might help to explain the findings of excess return volatility. Restating in terms of realized returns we obtain

$$r_{t+1} = r + \mu_t + \eta_{d,t+1} - \frac{\rho\xi_{t+1}}{1-\rho\phi}.$$

We next make the simplifying assumption that the innovations to dividends and the innovations to expected returns are uncorrelated, and we obtain

$$\text{var}(r_{t+1}) = \sigma_d^2 + \sigma_\mu^2 \left(\frac{1+\rho^2 - 2\rho\phi}{(1-\phi\rho)^2}\right) \simeq \sigma_d^2 + \frac{2\sigma_\mu^2}{1-\phi}, \tag{9.19}$$

where the approximation uses $\rho \simeq 1$, which implies that

$$\frac{1+\rho^2 - 2\rho\phi}{(1-\phi\rho)^2} \simeq \frac{2}{1-\phi}.$$

Therefore, the big volatility in returns could be due to a time varying expected returns process, in which shocks to expected returns persist for a long time.

9.6.2 A Word of Caution about Linearization: Nonlinear Dynamics and Flat Earthers

The linearization arguments seem plausible, and CLM give some numerical support to this in terms of the correlation between the exact and approximate returns. In the same way, if I walk from my home to my office (15 minutes) I can ignore the curvature of the Earth. On the other hand, flying from New York to Beijing assuming that the Earth was flat would not be a good idea. If one is concerned with fundamentally long term issues, relying on the linearization arguments advocated by Campbell and Shiller (1988ab) seems a little presumptuous.

Example 54 *Suppose that $x_0 \in [0, 1]$ and*

$$x_t = 4x_{t-1}(1 - x_{t-1}). \tag{9.20}$$

*This is a nonlinear difference equation called the logistic equation. It is a poster child for **chaos theory**, immortalized in the movie **Jurassic Park**. Even though there is no stochastic error at all in this process ($\text{var}(x) = 0$), it behaves like a stochastic process in many ways. In particular, it possesses sensitive dependence on initial conditions (the **butterfly effect**) whereby even if $x_0 - x_0' = 10^{-42}$, say, we have eventually that the trajectories of the processes x_t, x_t' diverge. Small differences in starting points can matter over long horizons.*

9.7 Predictive Regressions

A common econometric approach to modelling time varying expected returns is based on the so-called predictive regressions. A simple version of this is the linear predictive regression specification

$$r_{t+1} = \beta^\top x_t + \varepsilon_{t+1}, \tag{9.21}$$

where $E(\varepsilon_{t+1}|\mathcal{F}_t) = 0$, where $\mathcal{F}_t$ is information available at time t including x_t. Typically, x_t is a vector of predictors including dividend/price ratio, earnings/price ratio, or term structure variables lagged by one period. Under the conditional moment restriction, we can estimate β by the OLS estimator. We then have an estimate of the time varying expected return (or risk premium if r is excess of the risk free rate) $E(r_{t+1}|\mathcal{F}_t)$. The stock return process as we have seen is itself not very autocorrelated whereas the covariate process often is strongly autocorrelated. This raises some econometric issues.

Chapter 9   Present Value Relations

Suppose that the covariate process is univariate and a stationary AR(1) process

$$x_{t+1} = \rho x_t + \eta_{t+1}, \tag{9.22}$$

where ρ describes the persistence of x_t. Suppose also that the shocks are i.i.d. but contemporaneously correlated so that

$$\begin{pmatrix} \varepsilon_t \\ \eta_t \end{pmatrix} \sim 0, \begin{pmatrix} \sigma_{\varepsilon\varepsilon} & \sigma_{\varepsilon\eta} \\ \sigma_{\varepsilon\eta} & \sigma_{\eta\eta} \end{pmatrix}. \tag{9.23}$$

This says that the regressors x_t are **predetermined** but are not **strictly exogenous**. One can estimate β consistently by OLS of r_{t+1} on x_t, but the estimator is biased in small samples. Stambaugh (1999) argues that in relevant cases the estimator will be badly biased: the larger ρ is, the larger is the bias of $\widehat{\beta}$. This is sometimes called the **Stambaugh bias**. We have

$$E(x_t \varepsilon_{t+j}) = \begin{cases} 0 & \text{if } j \geq 1 \\ \neq 0 & \text{if } j \leq 0, \end{cases}$$

i.e., the regressors are predetermined but not strictly exogenous. One may also worry about omitted variable bias here, because there may be many such predictors of future returns and leaving anyone out of the equation (9.21) would violate the conditional moment restriction.

Empirically we may find $\widehat{\beta}$ quite large and statistically significant even though returns themselves have very weak autocorrelation as we saw in Chapter 3. These two findings are not necessarily incompatible.

Example 55 *Suppose that (9.21), (9.22), and (9.23) hold. Then*

$$\text{cov}(r_t, r_{t-k}) = 0 \tag{9.24}$$

for all k, if and only if

$$\frac{\sigma_{\varepsilon\eta}}{\sigma_{\eta\eta}} = -\frac{\beta\rho}{1-\rho^2}.$$

That is, if $\beta, \rho > 0$ we may find (9.24) provided $\sigma_{\varepsilon\eta}$ is sufficiently negative. This corresponds to the case where shocks to the dividend/price ratio are contemporaneously negatively correlated with innovations to returns, which is quite plausible.

Typically, dividends are measured at a monthly frequency over a trailing annual horizon, that is, they are really an overlapping aggregated series, which is essentially why they are so autocorrelated. A lot of work has looked at long horizon returns

9.7 Predictive Regressions

using overlapping returns data. In the above model (9.21), aggregation produces the following linear regression

$$r_{t+1} + \cdots + r_{t+K} = \beta(1 + \rho + \cdots + \rho^K)x_t + \varepsilon_{t+1} + \cdots + \varepsilon_{t+K} + \beta\eta_{t+1}$$
$$+ \cdots + \beta\rho^{K-1}\eta_{t+K-1}.$$

We can write this model as

$$r_{t+1:t+K} = r_{t+1} + \cdots + r_{t+K} = \beta(K)x_t + u_{t,t+K}, \qquad (9.25)$$

where $u_{t,t+K}$ is an innovation that satisfies $E(u_{t,t+K}|x_t) = 0$. This model is defined for all t, the highest frequency of the data, so one appears to have a large sample of data. Suppose that returns r_{t+1} obey **rw1**, then the aggregated returns $r_{t+1:t+K}$, for $K > 1$, will generally not be independent, in fact $\text{cov}(r_{t+1:t+K}, r_{t+j+1:t+j+K}) \neq 0$ for $j = 1, \ldots, K$. Likewise the error terms in (9.25) satisfy $\text{cov}(u_{t,t+K}, u_{t+j,t+j+K}) \neq 0$, and in fact form an $MA(K)$ process. The OLS estimator of $\beta(K)$ from (9.25) is consistent, but the standard errors need to be adjusted to take account of the serial correlation induced by the overlapping observations in the standard errors (Hansen and Hodrick (1980), henceforth HH). Let

$$\widehat{\beta}(K) = \left(\sum_{t=1}^{T-K} x_t x_t^\mathsf{T}\right)^{-1} \sum_{t=1}^{T-K} x_t r_{t+1:t+K}$$

be the OLS estimator of $\beta(K)$ for $K = 1, 2, \ldots$, where $x_t \in \mathbb{R}^K$ are the regressors, and let $\widehat{u}_t(K) = r_{t+1:t+K} - \widehat{\beta}^\mathsf{T}(K)x_t$ be the least squares residuals. Hansen and Hodrick (1980) proposed standard errors that involved estimating the autocovariance functions of $u_{t,t+K}$ and x_t out to lag $\pm K - 1$. They are not guaranteed to be positive definite, and in practice are not so when K is quite large.

Under the null hypothesis that $\beta = 0$, $r_{t+1:t+K} = \varepsilon_{t+1} + \cdots + \varepsilon_{t+K}$. The Hodrick (1992), henceforth H, standard errors for $\widehat{\beta}(K)$ are derived from the matrix

$$\widehat{V}(K) = \left(\sum_{t=1}^{T-K} x_t x_t^\mathsf{T}\right)^{-1} \sum_{t=K}^{T} \widehat{u}_t^2(1) \left(\sum_{k=0}^{K-1} x_{t-k}\right) \left(\sum_{k=0}^{K-1} x_{t-k}\right)^\mathsf{T} \left(\sum_{t=1}^{T-K} x_t x_t^\mathsf{T}\right)^{-1}, \qquad (9.26)$$

and these are guaranteed to be positive definite. The Hansen and Hodrick (1980) and Hodrick (1992) standard errors are both biased and noisy, especially when ρ is large and/or K is large. Reliability of finite-sample inference depends upon the persistence of the dynamic system. Empirically, the dividend yield ratio is very persistent (i.e. close to unit root or long memory series).

CLM report results for 1927–1994 and two subperiods 1927–1951 and 1951–1994 in their Table 7.1. They use real variables, i.e., adjusted for inflation. They find that the coefficients on the dividend price ratio are always positive, which says that low prices relative to dividends leads one to anticipate higher returns and hence prices in the future. The coefficients are strongly statistically significant for the long horizon regression:

> *At a horizon of 1-month, the regression results are rather unimpressive: The R^2 statistics never exceed 2%, and the t-statistics exceed 2 only in the post-World War II subsample. The striking fact about the table is how much stronger the results become when one increases the horizon. At a 2 year horizon the R^2 statistic is 14% for the full sample ... at a 4-year horizon the R^2 statistic is 26% for the full sample.*

The regressions on the term premium variable show similar predictability (although the sign of the effect is not consistent across the subperiods 1927–1951 and 1952–1994). Cochrane (1999, p37) describes this as one of the three most important facts in finance in his survey *New Facts in Finance*

> *Now, we know that ...*
> *[Fact] 2. Returns are predictable. In particular: Variables including the dividend/price (d/p) ratio and term premium can predict substantial amounts of stock return variation. This phenomenon occurs over business cycle and longer horizons. Daily, weekly, and monthly stock returns are still close to unpredictable ...*

Clearly, the R^2 of the regressions increases with the horizon, but you may recall from your first econometrics course that one can't really compare the R^2 of regressions with different dependent variables. Subsequent work has cast doubt on the robustness of these findings. Boudoukh, Richardson, and Whitelaw (2008) argue that:

> *The prevailing view in finance is that the evidence for long-horizon stock return predictability is significantly stronger than that for short horizons. We show that for persistent regressors, a characteristic of most of the predictive variables used in the literature, the estimators are almost perfectly correlated across horizons under the null hypothesis of no predictability. For the persistence levels of dividend yields, the analytical correlation is 99% between the 1 and 2 year horizon estimators and 94% between the 1 and 5 year horizons. Common sampling error across equations leads to ordinary least squares coefficient estimates and R^2s that are roughly proportional to the horizon under the null hypothesis. This is the precise pattern found in the data.*

We investigate this issue below. In Figure 9.5 below we show P_{t+1}/P_t at the annual frequency for the S&P500 index. It has a mean value of 1.057 and standard deviation 17.8%.

In Figure 9.6 we show the S&P500 dividend yield, which is the 12-month dividend per share divided by the price (D_{t+1}/P_t). This series is quite persistent and different from the stock return series.

We computed the predictive regressions (9.25) using the Shiller real dividend/price monthly data. We consider $K \in \{1, 3, 12, 24, 36, 48, 60\}$, and we work with estimation windows defined by $t = i - 240, \ldots, i + 240$, where the window center varies from $i = 1890, \ldots, 1990$.

9.7 Predictive Regressions

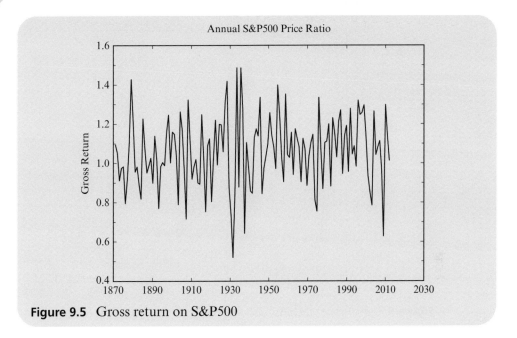

Figure 9.5 Gross return on S&P500

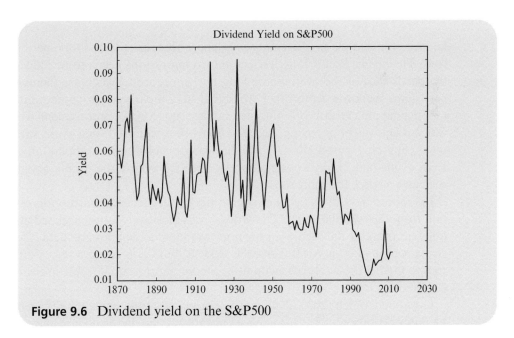

Figure 9.6 Dividend yield on the S&P500

In Figure 9.7 below we show the rolling window R^2 of these regressions for each K plotted against i. We find $R^2(i,1) \leq \cdots \leq R^2(i,60)$, i.e., predictability increases with horizon, consistent with the CLM results. However, we find a lot of variability over i: the 40 year periods centered on 1970 and 1940 both have high R^2 at long horizon, but the intermediate period and subsequent periods do not. It appears that the

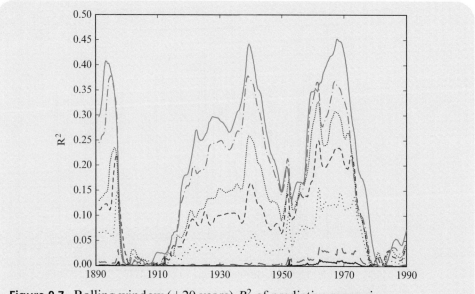

Figure 9.7 Rolling window (±20 years) R^2 of predictive regression

finding of in sample predictability is fragile with respect to the time period considered. The higher R^2 for long horizon return regressions needs to be judged against the persistence of the overlapping long horizon returns, which are themselves very persistent. We know from other work that in comparing overlapping returns with non-overlapping returns one obtains a modest (up to 50%) improvement in efficiency, which is equivalent to having, say, 50% more observations. If one works with $K = 60$, then using only post-1950 data one has roughly 14 non-overlapping observations. Even if one had 21 such observations would one believe the results obtained from such an exercise? Probably not.

In Figure 9.9 we show the daily overlapping return series for 60 months of aggregation from the S&P500 series. The series is very positively autocorrelated and totally different from the daily series or the non-overlapping long horizon data. Clearly, one step ahead prediction for this series is evident, but the predictive regression method tries to predict these returns 60 months into the future, and it is not so clear that one can do this.

Goyal and Welch (2003, 2008) consider out of sample predictability of the equity premium (equity returns) using classical ratios as predictors. They find that return forecasts based on dividend yields and a number of other variables do not work out of sample. They compared forecasts of returns at time $t + 1$ formed by estimating the regression using data up to time t, with forecasts that use the sample mean in the same period. They find that the unconditional sample mean produces a better out-of-sample prediction than do the conditional return-forecasting regressions. Their influential results confirm the economic and statistical lack of predictability of the predictive regression out of sample.

9.8 Summary of Chapter

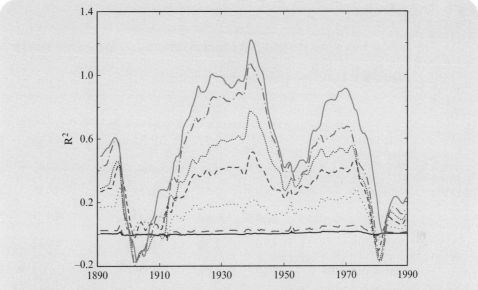

Figure 9.8 Rolling window (±20 years) slope coefficent of predictive regression

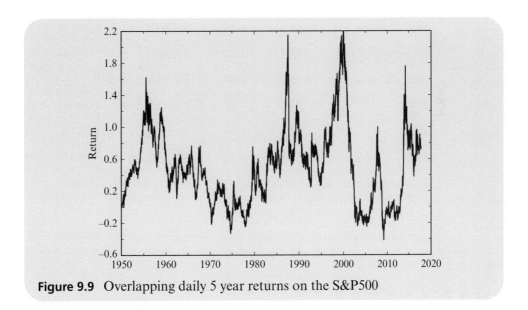

Figure 9.9 Overlapping daily 5 year returns on the S&P500

9.8 Summary of Chapter

We described a simple relation that asset prices should have with fundamentals in the absence of rational bubbles. We then discussed some of the econometric tests of these models predictions. We considered also the model with time varying expected return that can explain the high variability of stock prices relative to dividends and other fundamentals.

9.9 Appendix

In Table 9.1 we show the price and annual dividend yield for Dow stocks in 2013.

Table 9.1 Dividend yield on the Dow stocks

	P	%D/P		P	%D/P
Alcoa Inc.	9.32	1.39	JP Morgan	48.88	2.86
AmEx	61.69	1.3	Coke	37.42	2.94
Boeing	75.03	2.65	McD	93.9	3.35
Bank of America	12.03	0.67	MMM	103.23	2.50
Caterpillar	95.61	2.36	Merck	41.42	4.18
Cisco Systems	20.99	2.86	MSFT	28.01	3.57
Chevron	114.96	3.38	Pfizer	27.29	3.59
du Pont	46.94	3.66	P&Gamble	76.54	3.08
Walt Disney	55.61	1.71	AT&T	35.36	5.12
General Electric	23.29	3.39	Travelers	80.39	2.44
Home Depot	67.52	1.98	United Health	57.32	1.73
HP	16.79	3.14	United Tech	90.78	2.46
IBM	200.98	1.89	Verizon	44.4	4.71
Intel	21.12	4.47	Wall Mart	69.3	2.51
Johnson&Johnson	76.16	3.36%	Exxon Mobil	88.36	2.76

10 Intertemporal Equilibrium Pricing

In this chapter we discuss dynamic equilibrium models and their predictions. We then discuss the empirical methodology used to test these models. Finally, we look at some of the more recent macro-finance models that attempt to reconcile the empirical findings.

10.1 Dynamic Representative Agent Models

A common paradigm in economics is the **representative agent** framework. In this setting one supposes that there is a hypothetical being who is called the representative agent, and that this agent makes decisions regarding investment and consumption to maximize his or her utility. These decisions then determine relationships between observed outcomes. We are invited to believe that this framework represents the working of the entire economy of eight billion persons worldwide, and that differences of opinion, differences in wealth, and other divergences can be ignored by the aggregation into the representative agent. There are some results that justify aggregation from an economy with many individual decision-makers to a representative agent, which justify this monocultural approach. Like all models it is a simplification, and the proof of concept lies in whether it is useful. We will review the theory, predictions, and empirical performance of this approach.

We suppose that preferences are represented by a concave, positively sloped, time-invariant per period utility function U for consumption, and a constant rate of time preference δ. The agent invests in risky assets and consumes the proceeds over time. The representative agent maximizes the discounted expected utility of lifetime consumption at time t

$$E_t\left(\sum_{j=0}^{\infty} \delta^j U(C_{t+j})\right)$$

subject to a budget constraint, which we do not specify here, that reflects his initial endowment or wealth and the investment outcomes he achieves. The risky assets have a stochastic gross return $1 + R_{i,t+1}$ whose distributions are known to the representative agent. He takes expectations with respect to his past information, which we denote by E_t. This optimization problem can be characterized by the associated first order conditions, which are often called the **Euler equations**, whereby

$$U'(C_t) = \delta E_t\left((1 + R_{i,t+1})\, U'(C_{t+1})\right) \tag{10.1}$$

for each asset i.

10.2 The Stochastic Discount Factor

Defining $M_{t+1} = \delta U'(C_{t+1})/U'(C_t)$ and rearranging the first-order conditions, we obtain

$$1 = E_t\left((1 + R_{i,t+1})\, M_{t+1}\right) \tag{10.2}$$

for each asset i. The random variable M_t is called the **stochastic discount factor** (SDF) or **pricing kernel**. It is the ratio of marginal utilities between each "investment" date-state and "realized return" date-state, weighted by pure time preference; it is random but satisfies $M_t > 0$ with probability one. We can obtain a pricing formula for any asset as follows

$$P_t = E_t\left(M_{t+1} X_{t+1}\right), \tag{10.3}$$

where X_{t+1} is the cash flow in period $t+1$ (e.g., $P_{t+1} + D_{t+1}$). The SDF is of central importance in asset pricing.

This relationship can be derived more generally from a no-arbitrage assumption in a complete markets setting: there does not exist a negative-cost portfolio with a uniformly non-negative payoff. Given no-arbitrage, by the separating hyperplane theorem there exists a strictly positive random variable M_{t+1} such that (10.3) holds for all i. Note that the stochastic discount factor of any investor is an allowable non-arbitrage pricing kernel.

Suppose that we replace the true probability weights in the basic pricing expectation by "hypothetical" probability weights that are transformed by the SDF, that is, we let $\mathbb{P}^*$ be the probability measure with $\mathbb{P}^* \propto M_t \mathbb{P}$. Then, we can write

$$E_t(X_{t+1} M_{t+1}) = E_t^*(X_{t+1}), \tag{10.4}$$

which says that the value of random future cash flows is determined by their expected magnitude under this transformed probability. This new hypothetical probability measure $\mathbb{P}^*$ is called the equivalent martingale measure or the **risk neutral probability** and E_t^* denotes expectation with respect to this probability measure.

We may relate this model to the discounted cash flow model. Define the multiperiod stochastic discount function in terms of the one period SDFs, i.e.,

$$M_{t,t+k} = M_{t+k-1,t+k} \times \cdots \times M_{t,t+1}.$$

Then, letting

$$R_{t+1} = \frac{P_{t+1} + D_{t+1}}{P_t},$$

10.3 The Consumption Capital Asset Pricing Model

we have

$$P_t = E_t(M_{t+1}(P_{t+1} + D_{t+1}))$$
$$= E_t(M_{t+1}(E_{t+1}(M_{t+1,t+2}(P_{t+2} + D_{t+2})))) + E_t[M_{t+1}D_{t+1}]$$
$$= E_t(M_{t+1}D_{t+1}) + E_t(M_{t,t+2}(P_{t+2} + D_{t+2})),$$

and repeating this we obtain

$$P_t = E_t \left(\sum_{j=1}^{\infty} M_{t,t+j} D_{t+j} \right), \qquad (10.5)$$

under the no-bubbles condition. This shows that prices are the risk adjusted present discounted value of future cash flows. The discount factor here is stochastic and time varying.

10.3 The Consumption Capital Asset Pricing Model

We next rewrite the first order condition (10.2) by adding and subtracting to obtain

$$1 = E_t\left((1 + R_{i,t+1}) M_{t+1}\right) = E_t\left(1 + R_{i,t+1}\right) E_t\left(M_{t+1}\right) + \text{cov}_t\left(R_{i,t+1}, M_{t+1}\right).$$

Let R_{0t} denote an asset such that $\text{cov}_t(R_{0,t+1}, M_{t+1}) = 0$, i.e., a zero beta or risk free asset. Then

$$E_t(M_{t+1}) = \frac{1}{E_t(1 + R_{0,t+1})}.$$

Substituting and rearranging we obtain for any asset i

$$E_t(R_{i,t+1} - R_{0,t+1}) = -\text{cov}_t(R_{i,t+1}, M_{t+1}) \times E_t(1 + R_{0,t+1})$$
$$= -\text{cov}_t\left(R_{i,t+1}, \delta \frac{U'(C_{t+1})}{U'(C_t)}\right) \times E_t(1 + R_{0,t+1})$$
$$= \frac{\text{cov}_t(R_{i,t+1}, M_{t+1})}{\text{var}_t(M_{t+1})} \times -\text{var}_t(M_{t+1}) E_t(1 + R_{0,t+1}).$$

This says that an asset whose covariance with M_t is negative tends to have low returns when the investor's marginal utility of consumption is high i.e., when consumption is low. The larger is the negative covariance, the larger the risk premium one requires to hold such an asset.

Suppose there is an asset R_{mt} that pays off exactly M_t. Then

$$E_t(R_{m,t+1} - R_{0,t+1}) = -\text{var}_t(M_{t+1}) \times E_t(1 + R_{0,t+1}),$$

and we obtain the following result.

Theorem 39 *We have*

$$E_t(R_{i,t+1} - R_{0,t+1}) = \beta_{im,t} E_t(R_{m,t+1} - R_{0,t+1}) \qquad (10.6)$$

$$\beta_{im,t} = \frac{\mathrm{cov}_t(R_{i,t+1}, R_{m,t+1})}{\mathrm{var}_t(R_{m,t+1})} = \frac{\mathrm{cov}_t\left(R_{i,t+1}, \delta \frac{U'(C_{t+1})}{U'(C_t)}\right)}{\mathrm{var}_t\left(\delta \frac{U'(C_{t+1})}{U'(C_t)}\right)}.$$

This pricing model is called the consumption CAPM, and $\beta_{im,t}$ is called the consumption beta. It depends on the conditional moments taken at time t, and one has to specify a model to make this a practically relevant relationship, unlike in the unconditional case. In fact, one has only to specify a model for $\beta_{im,t}$ (and suppose that $R_{0,t+1} = R_{f,t+1}$), since then one has

$$R_{i,t+1} - R_{f,t+1} = \beta_{im,t}(R_{m,t+1} - R_{f,t+1}) + \varepsilon_{i,t+1},$$

where $E_t(\varepsilon_{i,t+1}) = 0$. We discussed in Chapter 7 two possible models for $\beta_{im,t}$, given by (7.40) and (7.41). Instead, one can work directly with the conditional moment restrictions (10.2), which is what we do below.

We can also derive an unconditional version of (10.6). Let R_{0t} denote an asset such that $\mathrm{cov}(R_{0t}, M_t) = 0$, then

$$E(R_{it} - R_{0t}) = \beta_{im} E(R_{mt} - R_{0t}), \qquad \beta_{im} = \frac{\mathrm{cov}(R_{it}, R_{mt})}{\mathrm{var}(R_{mt})},$$

but note that $\beta_{im} \neq E(\beta_{im,t})$. This can typically be implemented without further specification. Furthermore, we have

$$E^2(R_{it} - R_{0t}) = \frac{\mathrm{cov}(R_{it}, M_t)^2}{E^2(M_t)} \leq \frac{\mathrm{var}(R_{it})\mathrm{var}(M_t)}{E^2(M_t)},$$

which is equivalent to

$$\frac{\mathrm{var}(M_t)}{E^2(M_t)} \geq \frac{E^2(R_{it} - R_{0t})}{\mathrm{var}(R_{it})} \qquad (10.7)$$

for all assets i. This is the so-called **Hansen–Jagannathan** bound that the stochastic discount factor should obey. It states that the stochastic discount factor should be quite variable to be consistent with the pricing model. This is a useful restriction that one can check for a given specification of M_t.

In practice, we don't observe R_{mt} or R_{0t} and we don't observe the agent's information set. Suppose we work with the original formulation, where $M_{t+1} = \delta U'(C_{t+1})/U'(C_t)$. Then we need to specify $U(.)$ in order to estimate β_{ic} from

10.3 The Consumption Capital Asset Pricing Model

consumption data. Consider the CRRA class of utility functions with parameter γ that measures risk aversion

$$U(C_t) = \frac{C_t^{1-\gamma} - 1}{1 - \gamma}.$$

In this case the stochastic discount factor is

$$M_{t+1} = \delta \left(\frac{C_{t+1}}{C_t}\right)^{-\gamma}.$$

The risk free interest rate satisfies

$$1 + R_{ft} = \frac{1}{\delta} \left(\frac{E_t(C_{t+1})}{C_t}\right)^{\gamma},$$

which varies over time, but is known at time t. If we specify a process for C_{t+1} then we can obtain explicit solutions for the risk free rate etc.

Example 56 *Suppose that $C_{t+1} = C_t \eta_{t+1}$, where $\eta_t \geq 0$ is i.i.d. with $E(\eta_t) = 1 + \mu$. Then the risk free rate is constant and satisfies*

$$1 + R_{ft} = \frac{1}{\delta}(1 + \mu)^{\gamma}.$$

10.3.1 Econometric Testing

The model has cross-sectional predictions (relative risk premia are proportional to consumption betas), time-series predictions (expected returns vary with expected consumption growth rates etc.), and joint time-series and cross-sectional predictions. The standard CAPM only has cross-sectional predictions. We next consider how to empirically test the consumption CAPM.

Hansen and Singleton (1982) developed the GMM methodology for this purpose. The first order condition is again

$$E\left((1 + R_{it+1})\delta \left(\frac{C_{t+1}}{C_t}\right)^{-\gamma} - 1 \bigg| \mathcal{F}_t\right) = 0, \tag{10.8}$$

where $\mathcal{F}_t$ is the agent's information set. If we replace $\mathcal{F}_t$ by any coarser information set (of observed data) $\mathcal{F}_t^*$, then this conditional moment restriction continues to hold. We will suppose that returns and consumption are observable along with some other data. Let $\theta = (\delta, \gamma)^\mathsf{T}$ and define the vector

$$g(X_{t+1}, \theta) = \left(\begin{pmatrix} 1 + R_{1,t+1} \\ \vdots \\ 1 + R_{n,t+1} \end{pmatrix} \delta \left(\frac{C_{t+1}}{C_t}\right)^{-\gamma} - 1\right) \otimes Z_t \in \mathbb{R}^q$$

for each parameter value θ. Here, Z_t are observed **instruments** included in $\mathcal{F}_t^*$, and X_t denotes the observed data including returns, consumption, and instruments. By the law of iterated expectation, the unconditional moment restrictions must hold

$$E(g(X_{t+1}, \theta)) = 0, \tag{10.9}$$

where E denotes the unconditional expectation. We can approximate the unconditional expectation by the sample average, so long as the data are stationary and this is the basis for estimation. We estimate the parameters θ by the GMM method using q sample moments. In practice, there are many many instruments $Z_t \in \mathcal{F}_t^*$ that could be used so that the parameters θ are overidentified, i.e., $q > 2$.

Definition 92 *Let $\widehat{\theta}_{GMM}$ solve the following optimization problem*

$$\min_{\theta \in \mathbb{R}^p} G_T(\theta)^\top W_T G_T(\theta), \quad \text{where } G_T(\theta) = \frac{1}{T} \sum_{t=1}^{T-1} g(X_{t+1}, \theta) \tag{10.10}$$

and W_T is a $q \times q$ weighting matrix. For example, $W_T = I_q$.

This objective function is nonlinear in θ. Under some conditions this method is consistent and the estimators are normally distributed in large samples. Suppose that as $T \to \infty$

$$\text{var}\left[T^{1/2} G_T(\theta_0)\right] \to \Omega, \quad \frac{\partial G_T(\theta_0)}{\partial \theta} \xrightarrow{P} \Gamma,$$

where Γ, Ω are of full rank.

Theorem 40 *Under some regularity conditions, $\widehat{\theta}_{GMM}$ is consistent and asymptotically normal*

$$T^{1/2}(\widehat{\theta}_{GMM} - \theta_0) \Longrightarrow N(0, V(W)),$$

where with W the probability limit of the sequence W_T, we have

$$V(W) = (\Gamma^\top W \Gamma)^{-1} \Gamma^\top W \Omega W \Gamma (\Gamma^\top W \Gamma)^{-1}.$$

If we take $W_{opt} = \Omega^{-1}$, then the asymptotic variance of $\widehat{\theta}_{GMM}$ is $V(W_{opt}) = (\Gamma^\top \Omega^{-1} \Gamma)^{-1}$, and satisfies

$$(\Gamma^\top \Omega^{-1} \Gamma)^{-1} \leq (\Gamma^\top W \Gamma)^{-1} \Gamma^\top W \Omega W \Gamma (\Gamma^\top W \Gamma)^{-1}$$

for all weighting matrices W. In practice, we take W_T to be a consistent estimator of Ω^{-1}. The resulting estimator is consistent and asymptotically normal, and optimal within this class.

10.3 The Consumption Capital Asset Pricing Model

One can construct confidence intervals for θ using t-statistics or Wald statistics based on a consistent estimator $\widehat{V}$ of V. Specifically, let

$$\widehat{\gamma}_{GMM} \pm z_{\alpha/2} \sqrt{\frac{\widehat{V}_{\gamma\gamma}}{T}},$$

where $\widehat{V}_{\gamma\gamma}$ is a consistent estimator of $V_{\gamma\gamma}$, the component of V corresponding to γ. This is a symmetric, coverage $1 - \alpha$ confidence interval. We take

$$\widehat{\Gamma} = \frac{\partial G_T(\widehat{\theta}_{GMM})}{\partial \theta} \quad ; \quad \widehat{\Omega} = \frac{1}{T}\sum_{t=1}^{T-1} g(X_{t+1}, \widehat{\theta}_{GMM}) g(X_{t+1}, \widehat{\theta}_{GMM})^{\mathsf{T}},$$

which are consistent estimates under (10.8).

One can also test whether overidentifying restrictions hold using the J-test (Hansen (1982)) based on the optimal GMM, i.e.,

$$J = \|G_T(\widehat{\theta})\|_{W_{opt}}. \tag{10.11}$$

This test statistic is asymptotically χ^2_{q-p} under the null hypothesis that all of the q moments are correctly specified and $J \to \infty$ when some of the moment conditions are invalid. However, the validity of the overidentifying restrictions is neither sufficient nor necessary for the validity of the moment conditions implied by the underlying economic model, and therefore provides little information on the possibility of identifying the parameters of interest.

A. Ronald "Ron" Gallant is a leading American econometrician. Gallant is a Professor of Economics and a Liberal Arts Research Professor at Pennsylvania State University. Prior to joining the Penn State faculty he was the Hanes Corporation Foundation Professor of Business Administration, Fuqua School of Business, Duke University, with a secondary appointment in the Department of Economics, Duke University, as well as a Distinguished Scientist in Residence, Department of Economics, New York University, both in the United States. He is also Adjunct Professor, Department of Statistics, University of North Carolina at Chapel Hill. Before joining the Duke faculty, he was Henry A. Latane Distinguished Professor of Economics at the University of North Carolina at Chapel Hill. He retains emeritus status there. Previously he was, successively, Assistant, Associate, Full, and Drexel Professor of Statistics and Economics at North Carolina State University. He received his AB in mathematics from San Diego State University, his MBA in marketing from UCLA, and his PhD in statistics from Iowa State University. He is a Fellow of both the Econometrics Society and the American Statistical Association. He serves on the Board of Directors of the National Bureau of Economic Research and has served on the Board of

Chapter 10 Intertemporal Equilibrium Pricing

Directors of the American Statistical Association and on the Board of Trustees of the National Institute of Statistical Sciences. He is co-editor of the *Journal of Econometrics* and past editor of *The Journal of Business and Economic Statistics*. In collaboration with George Tauchen of Duke University, Gallant developed the Efficient Method of Moments (EMM). He is one of the leaders in nonlinear time series analysis, nonlinear econometrics, nonlinear dynamic systems, and econometrics theory. I believe he is responsible for introducing the term seminonparametrics.

A commonly adopted simplification is to work with the log linear version, which follows from a lognormal distributional specification. Suppose that $W = (1 + R_{i,t+1})\delta(C_{t+1}/C_t)^{-\gamma}$ is conditionally lognormally distributed so that $\log(W) = r_{i,t+1} + \log(\delta) - \gamma g_{t+1}$ is normally distributed, where $g_{t+1} = \log(C_{t+1}/C_t)$. Then using $\log E(W) = E(\log(W)) + \frac{1}{2}\mathrm{var}(\log(W))$ for lognormal W (see (16.1)) we obtain

$$0 = E_t(r_{i,t+1}) + \log(\delta) - \gamma E_t(g_{t+1}) + \frac{1}{2}(\mathrm{var}_t(r_{i,t+1}) + \gamma^2 \mathrm{var}_t(g_{t+1})$$
$$- 2\gamma \mathrm{cov}_t(r_{i,t+1}, g_{t+1})). \tag{10.12}$$

The presence of the conditional variance and covariance terms means that we can't directly apply GMM to deliver estimation of the parameters γ, δ. Suppose that $X_{t+1} = (r_{1,t+1}, \ldots, r_{n,t+1}, g_{t+1})^\mathsf{T}$ obeys a stationary VAR(P) with

$$X_t = \mu + \sum_{j=1}^{P} A_j X_{t-j} + V_t, \tag{10.13}$$

where V_t is i.i.d. normal with mean zero and variance matrix Σ (Hansen and Singleton (1982, 1983)). In this case, the conditional moments are determined by the dynamic parameter matrices A_j and the error variance matrix: $E_t(r_{i,t+1})$ and $E_t(g_{t+1})$ are generally time varying, while $\sigma_i^2 = \mathrm{var}_t(r_{i,t+1})$, $\sigma_c^2 = \mathrm{var}_t(g_{t+1})$, and $\sigma_{ic} = \mathrm{cov}_t(r_{i,t+1}, g_{t+1})$ are constant over time. Define the linearized error process

$$u_{it+1} = r_{i,t+1} - \gamma g_{t+1} + \log\delta + \frac{1}{2}\left(\sigma_i^2 + \gamma^2 \sigma_c^2 - 2\gamma \sigma_{ic}\right). \tag{10.14}$$

It follows that for $i = 1, \ldots, n$ and $Z_t \in \mathcal{F}_t$, $E(u_{it+1}) = 0$, $E(u_{it+1}Z_t) = 0$. Hansen and Singleton (1982) show how to estimate this model by maximum likelihood, which is to estimate the VAR (10.13) subject to some restrictions on $\mu, A_1, \ldots, A_P, \Sigma$ implied by (10.12); see below for an example. One can also estimate this by GMM using the moment conditions $E(V_t|X_{t-1},\ldots) = 0$ and $E(V_t V_t^\mathsf{T} - \Sigma|X_{t-1},\ldots) = 0$ from the VAR. Either way is more complicated and requires more assumptions than (10.8).

Example 57 Suppose that $A_j = 0, j = 1, \ldots, P$. Then the conditional moments are constant and we have the n restrictions

$$\mu_i + \log(\delta) - \gamma\mu_c + \frac{1}{2}\left(\sigma_{ii} + \gamma^2 \sigma_c - 2\gamma\sigma_{ic}\right) = \frac{1}{2}\sigma_c\gamma^2 - (\mu_c + 2\sigma_{ic})\gamma + \frac{1}{2}\sigma_{ii} + \mu_i + \tau = 0$$

on the mean and variance matrix of X_t that involve the unknowns δ, γ or $\gamma, \tau = \log(\delta)$. One could first estimate the unconditional moments μ, Σ without restriction and then impose these restrictions on them to obtain estimates of γ, δ. With two assets we can exactly solve for γ, τ.

If we are only interested in γ then we can write (10.14) as $r_{i,t+1} = \pi_i + \gamma g_{t+1} + u_{it+1}$, where $\pi_i = \log \delta + [\sigma_i^2 + \gamma^2 \sigma_c^2 - 2\gamma \sigma_{ic}]/2$ is an unknown intercept parameter, and $E(u_{it+1}|\mathcal{F}_t) = 0$. Note that if $\gamma = 0$ (risk neutrality) this implies that returns are serially uncorrelated. We can estimate (π, γ) by instrumental variables using past values of returns and consumption growth or indeed any variable known at time t, which is a special case of GMM, that is, we take

$$g(X_t, \theta) = \begin{pmatrix} r_{1,t+1} - \pi_1 - \gamma g_{t+1} \\ \vdots \\ r_{n,t+1} - \pi_n - \gamma g_{t+1} \end{pmatrix} \otimes Z_t \in \mathbb{R}^q$$

in (10.10), where $\theta = (\pi_1, \ldots, \pi_n, \gamma)^\top \in \mathbb{R}^p$ with $p = n+1$ and $X_t = (r_{1,t+1}, \ldots, r_{n,t+1}, g_{t+1}, Z_t)^\top$. The linear in parameters form of the moment function allows closed form (two-stage least squares) computation of the estimates. Unfortunately, the empirical results are not good; see Table 8.2 in CLM. We update some of these results to include data up to 2017 and find estimates of γ around -2 for the CRSP market index using lagged consumption growth and lagged returns as instruments. The estimated γ has the wrong sign, but is not statistically significant.

Empirically this model performs very poorly, at least in its standard formulation. The empirical failure of the consumption CAPM is among the most important anomalies of asset pricing theory (Cochrane (2001)).

Consumption is not very variable. The growth of real annual per capita expenditure variable (rPCEa) is shown below in Figure 10.1. Its mean is $\bar{g} = 0.0134$ and standard deviation $s_g = 0.0127$, which is much less than the variation of stock returns.

In the next chapter we discuss models for conditional variance. It is possible to extend the VAR model (10.13) to include a dynamic model for the conditional covariance matrix, which implies looser restrictions.

10.4 The Equity Premium Puzzle and the Risk Free Rate Puzzle

In the lognormal framework (10.13) (with $r_{f,t+1}$ included as a state variable), one obtains

$$r_{f,t+1} = \gamma E_t(g_{t+1}) - \log(\delta) - \frac{\gamma^2}{2}\sigma_c^2, \tag{10.15}$$

and subtracting one obtains an unconditional pricing equation for the market risk premium in terms of covariation with consumption growth

$$E(r_{m,t+1} - r_{f,t+1}) = \gamma \sigma_{mc} - \frac{\sigma_m^2}{2} \leq \gamma \sigma_{mc}. \tag{10.16}$$

Chapter 10 Intertemporal Equilibrium Pricing

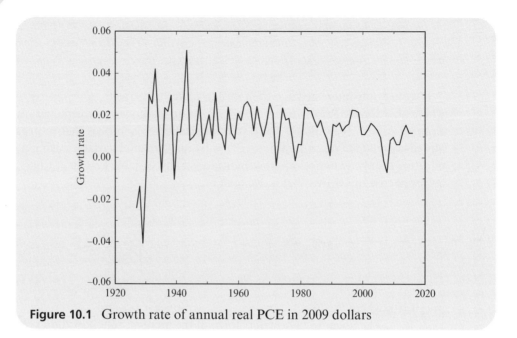

Figure 10.1 Growth rate of annual real PCE in 2009 dollars

Empirically, σ_{mc} is very small relative to the observed return premium of equities over fixed income securities; hence this implies a very high coefficient of risk aversion γ. Typically, one requires γ around 20 to be consistent with this equation. If γ is set high enough to explain the observed equity risk premia, it is too high (given average consumption growth) to explain observed risk-free returns! The rate of pure time preference is driven below zero.

Mehra and Prescott (1985):

> *Historically the average return on equity has far exceeded the average return on short-term virtually default-free debt. Over the ninety-year period 1889–1978 the average real annual yield on the Standard and Poor 500 Index was seven percent, while the average yield on short-term debt was less than one percent. The question addressed in this paper is whether this large differential in average yields can be accounted for by models that abstract from transactions costs, liquidity constraints and other frictions absent in the Arrow–Debreu set-up. Our finding is that it cannot be, at least not for the class of economies considered.*

10.5 Explanations for the Puzzles

A large number of explanations for or resolutions to the puzzle have been proposed. These include:

1. A contention that the equity premium does not exist: that the puzzle is a statistical illusion.

10.5 Explanations for the Puzzles

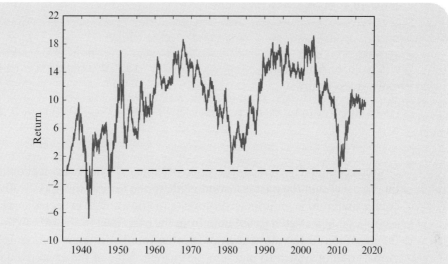

Figure 10.2 Rolling window trailing 10 year gross nominal returns on the CRSP value weighted index

2. Modifications to the assumed preferences of investors.
3. Imperfections in the model of risk aversion.
4. Does the representative agent model make any sense?

10.5.1 Statistical Illusion

The most basic explanation is that there is no puzzle to explain: that there is no equity premium, or that the equity premium is not well defined. Fernandez, Ortiz Pizarro and Fernández Acín (2015) distinguish between four versions of the premium: the historical equity premium, the expected equity premium, the required equity premium, and the implied equity premium. Most empirical work involves measuring the historical equity premium under the assumption that this is a sound statistical estimate of the expected equity premium. Fernandez, Ortiz Pizarro and Fernández Acín (2015) argue that the variations in the value of the equity premium that different authors report are very substantial and depend on the period considered, the market index considered, the risk free rate considered, as well as the method of computing returns. If one works with annual data over the period 1900–2016 there are only 116 years of data, which is not a large number of years statistically to measure the value of the mean return even assuming that these data are independent when the variance is as large as it is (the rolling window of mean returns in Figure 10.2 shows how much variation there is in the historical equity premium over different ten year periods). The returns also vary greatly depending on which points are included and the starting point. Using annual data starting from the top of the market in 1929 or starting from the bottom of the market in 1932 (leading to estimates of equity premium of 1% lower per year), or ending at the top in 2000 (versus bottom in 2002) or top in 2007 (versus bottom in 2009 or

Table 10.1 Market risk premium

	μ	med	σ	$IQR/1.349$	$\rho(1)$
(1926–2016) Annualized daily excess returns	7.320	15.120	16.906	10.473	0.0679
(1926–2016) Annual excess returns	8.48	10.735	20.29	20.167	0.0214

beyond) completely changes the value obtained. To use historical data one must rely on the belief that the past is a good guide to the future, and specifically, the historical equity premium is an unbiased estimate of the average expected equity premium. It seems a bit of a stretch to use data from the early part of the twentieth century before the invention of the motor car and with a totally different demographic profile and global trading system to measure the current value of the expected equity premium. In addition, one could argue that focussing on the US market is flawed due to sample selection bias. The US equity market is the most intensively studied in equity market research. Not coincidentally, it had the best equity market performance in the twentieth century; others (e.g. Russia, Germany, and China) produced a gross return of zero (for some period) due to bankruptcy events.

In Table 10.1 we report the estimated market risk premium using the FF market factors, the annualized daily return series and the annual return series. For the daily return series there are $n = 24034$ observations for which $1/\sqrt{n} = 0.00645$ and for the annual return series $n = 90$ for which $1/\sqrt{n} = 0.1054$.

In Figure 10.2 we show the ten year rolling window annualized returns for the FF market return time series, and in Figure 10.3 we show its histogram. This shows two things. First, there is considerable variation in long horizon returns around the very long run average of around 10% per year. Second, the series itself is quite predictable, a predictability that has been manufactured out of the rolling window construction.

For comparison, in Figure 10.4 we show the quarterly US differenced log of PCE (seasonally adjusted). This series has a positive mean and is positively autocorrelated. In Figure 10.5 we show the autocorrelation out to 24 lags along with the 95% Bartlett confidence intervals.

10.5.2 Market Frictions

The perfect markets formulation of the investment problem is somewhat unrealistic. If some or all assets cannot be sold short by the agent, then the stochastic discount factor equation is much weaker. In this case we obtain the weaker inequality restrictions

$$E((1 + R_{it}) M_t) \leq 1.$$

The inequality version of the stochastic discount factor does not aggregate across investors. Hence aggregate consumption is not directly relevant.

10.5 Explanations for the Puzzles

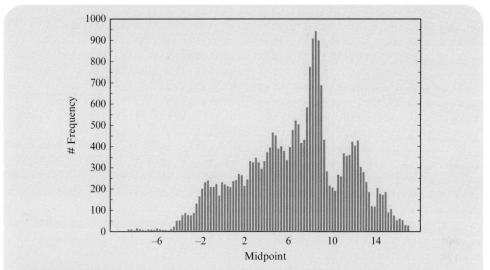

Figure 10.3 Distribution of the annual risk premium on the FF market factor from ten years of daily data

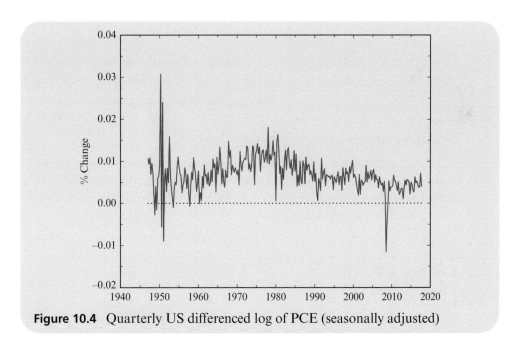

Figure 10.4 Quarterly US differenced log of PCE (seasonally adjusted)

10.5.3 Separating Risk Aversion and Intertemporal Substitution

The standard multiperiod VNM utility function is elegant but may not provide an accurate representation of investor decision-making. The multiperiod VNM utility has a single parameter, which governs both the elasticity of intertemporal substitution and Arrow–Pratt risk aversion. Elasticity of intertemporal substitution (EIS) reflects

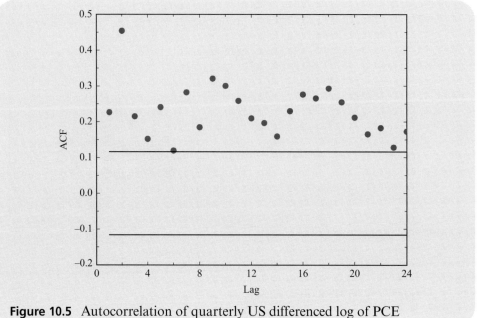

Figure 10.5 Autocorrelation of quarterly US differenced log of PCE

the extent to which consumers are willing to smooth certain consumption through time, while risk aversion relates to the extent to which consumers are willing to smooth consumption across uncertain states of nature.

Elasticity of Intertemporal Substitution

Consider an investor who consumes or saves in period zero and consumes in period one with no risk. Suppose that the risk-free interest rate is R_f. The elasticity of intertemporal substitution is defined as the percentage change in optimal consumption growth for a percentage change in the risk-free interest rate:

$$EIS = \frac{\%\partial(C_1/C_0)}{\%\partial(R_f)}.$$

In the CRRA case it is easy to show $EIS = \frac{1}{\gamma}$. Note that EIS is an intertemporal concept with no connection to risk.

Arrow–Pratt Risk Aversion

The Arrow–Pratt coefficients of absolute and relative risk aversion are defined in (1.8). They are both pure risk concepts with no intertemporal component whereas EIS is a pure intertemporal concept with no risk component. In the multiperiod VNM framework they are inextricably linked together.

10.5 Explanations for the Puzzles

The Epstein–Zin–Weil "Utility" Function

We now consider an approach to preference specification that separates intertemporal substitution from risk aversion. Defining the value function $V_t = E_t(\sum_{j=0}^{\infty} \delta^j U(C_{t+j}))$ of the standard optimization problem, we may write the recursive equation $V_t = U(C_t) + \delta E_t(V_{t+1})$. Epstein and Zin (1989) generalize this equation to $V_t = F(C_t, R_t(V_{t+1}))$, where $R_t(V_{t+1}) = G^{-1}(E_t(G(V_{t+1})))$ for some F, G, and in particular $G(x) = \frac{x^{1-\gamma}}{1-\gamma}$ and $F(x,y) = ((1-\delta)x^{1-\rho} + \delta y^{1-\rho})^{\frac{1}{1-\rho}}$, in which case

$$V_t = \left\{ (1-\delta) C_t^{1-\rho} + \delta \left((1-\alpha)^{\frac{1}{1-\gamma}} \left(E_t \left[\frac{V_{t+1}^{1-\gamma}}{1-\gamma} \right] \right)^{\frac{1}{1-\gamma}} \right)^{1-\rho} \right\}^{\frac{1}{1-\rho}}$$

$$= \left\{ (1-\delta) C_t^{1-\rho} + \delta \left(\left(E_t \left[V_{t+1}^{1-\gamma} \right] \right)^{\frac{1-\rho}{1-\gamma}} \right) \right\}^{\frac{1}{1-\rho}},$$

where δ is the discount factor, γ is the coefficient of relative risk aversion, and $1/\rho$ is the elasticity of intertemporal substitution. Suppose now that the representative agent maximizes V_t subject to the same budget constraint as before. It may be shown that

$$M_{t+1} = \delta \left(\frac{C_{t+1}}{C_t} \right)^{-\rho \frac{\gamma-1}{\rho-1}} R_{mt+1}^{\frac{\rho-\gamma}{1-\rho}}, \tag{10.17}$$

where R_{mt+1} is the return on the wealth portfolio $R_{mt+1} = W_{t+1}/(W_t - C_t)$, and therefore, we have the conditional moment restrictions

$$E_t \left(\left\{ \delta \left(\frac{C_{t+1}}{C_t} \right)^{-\frac{1}{\psi}} \right\}^\theta \left\{ \frac{1}{(1+R_{m,t+1})} \right\}^{1-\theta} (1+R_{i,t+1}) - 1 \right) = 0,$$

where $\theta = (1-\gamma)/(1-\rho)$ and $\psi = 1/\rho$. We may apply the GMM methodology to estimate the parameters δ, γ, and θ based on consumption, stock return, and other macro data. The issue is how to measure consumption and wealth accurately and obtain a long enough time series of such measurements to carry out the estimation precisely.

Suppose that consumption and the return on the market portfolio are jointly lognormal. Then as before one obtains the following linearized equations:

$$r_{f,t+1} = -\log \delta + \frac{\theta-1}{2} \sigma_m^2 - \frac{\theta}{2\psi^2} \sigma_c^2 + \frac{1}{\psi} E_t(g_{t+1})$$

$$E_t(r_{i,t+1}) - r_{f,t+1} = \theta \frac{\sigma_{ic}}{\psi} + (1-\theta) \sigma_{im} - \frac{\sigma_i^2}{2}, \tag{10.18}$$

which may be compared with (10.15) and (10.16). This says that consumption betas and market portfolio betas both affect asset risk premia. For the market return, $E_t(r_{m,t+1}) - r_{f,t+1} = (1/2 - \theta)\sigma_m^2 + (\theta/\psi)\sigma_{mc}$.

Eliminating Consumption Campbell (1993) points out that there is information in the intertemporal budget constraint and shows how to use this to eliminate consumption from (10.18). Linearizing this equation and substituting, he obtained

$$E_t(r_{i,t+1}) - r_{f,t+1} = \gamma \sigma_{im} + (\gamma - 1)\sigma_{ih} - \frac{\sigma_i^2}{2} \qquad (10.19)$$

$$\sigma_{ih} = \text{cov}_t\left(r_{i,t+1}, \sum_{j=1}^{\infty} \rho^j \{E_{t+1}(r_{m,t+1+j}) - E_t(r_{m,t+1+j})\}\right). \qquad (10.20)$$

In this representation, risk premia depend on market betas and on **changing opportunity set betas**. This says that covariation with news about future returns to the market affects risk premia. Note that the coefficients are γ and $1-\gamma$ rather than θ and $1-\theta$ so the EIS parameter is not present in this equation. For the market return, $E_t[r_{m,t+1}] - r_{f,t+1} = (\gamma - 1/2)\sigma_m^2 + (\gamma - 1)\sigma_{mh}$. If $\sigma_{mh} = 0$ (because stock returns are unforecastable), then $\gamma = (E[r_{m,t+1}] - r_{f,t+1})/\sigma_m^2 + 1/2$, then according to Table 10.1, $\gamma \simeq 2.6$, which is a much more modest degree of risk aversion.

We need one more step to bring this to data, that is we need to model all the state variables to deliver forecasts of future market returns. Let $X_t \in \mathbb{R}^k$ be the observed state variables, where r_{mt} is the first element of this vector, and suppose that the state variables are generated by a k-dimensional $VAR(1)$ process, i.e.,

$$X_{t+1} = \mu + AX_t + \varepsilon_{t+1}, \qquad (10.21)$$

where $A = (a_{ij})$ is an unknown parameter matrix and ε_{t+1} is an error vector with $E(\varepsilon_{t+1}|X_t, X_{t-1}, \ldots) = 0$. Let $e_1^\top = (1, 0, \ldots, 0)$, then

$$r_{m,t+1} = e_1^\top X_{t+1} = e_1^\top \mu + e_1^\top A X_t + e_1^\top \varepsilon_{t+1}. \qquad (10.22)$$

In this model we can forecast the future by $E_t(X_{t+1}) = \mu + AX_t$, which implies that $E_t(r_{m,t+1}) = e_1^\top \mu + e_1^\top A X_t$, and $E_t(X_{t+j}) = \sum_{k=0}^{j-1} A^k \mu + A^j X_t$, which implies that $E_t(r_{m,t+j}) = e_1^\top \sum_{k=0}^{j-1} A^k \mu + e_1^\top A^j X_t$. Therefore,

$$E_{t+1}\left(\sum_{j=1}^{\infty} \rho^j r_{m,t+1+j}\right) - E_t\left(\sum_{j=1}^{\infty} \rho^j r_{m,t+1+j}\right)$$

$$= \sum_{j=1}^{\infty} \rho^j e_1^\top \left(\sum_{k=0}^{j-1} A^k \mu + A^j X_{t+1}\right) - \sum_{j=1}^{\infty} \rho^j e_1^\top \left(\sum_{k=0}^{j} A^k \mu + A^{j+1} X_t\right)$$

$$= -e_1^\top \sum_{j=1}^{\infty} \rho^j A^j \mu + e_1^\top \sum_{j=1}^{\infty} \rho^j A^j \varepsilon_{t+1}$$

$$= e_1^\top \rho A (1 - \rho A)^{-1} (\mu + \varepsilon_{t+1}) \equiv \varphi^\top (\mu + \varepsilon_{t+1}).$$

10.6 Other Asset Pricing Approaches

The factor betas φ are nonlinear combinations of the VAR coefficients, and the extra-market factors are the VAR innovations. Let $\sigma_{ik} = \text{cov}(r_{i,t+1}, \varepsilon_{k,t+1})$ and inserting this into the above equation (10.19) gives

$$E_t\left(r_{i,t+1} - r_{f,t+1}\right) = -\frac{\sigma_i^2}{2} + \gamma\sigma_{i1} + (\gamma - 1)\sum_{k=1}^{K}\varphi_k\sigma_{ik},$$

which (except for the log expectation adjustment) is identical to the multi-factor pricing models tested in Chapter 8.

10.6 Other Asset Pricing Approaches

10.6.1 Habits Models

An alternative refinement of the standard model is to suppose that the agent has consumption habits that he doesn't like to change; we all know people like that. Constantinides (1990) specified this in difference form

$$U = E_t\left(\sum_{j=0}^{\infty}\delta^j\frac{(C_{t+j} - X_{t+j})^{1-\gamma} - 1}{1-\gamma}\right),$$

where habit X_t is, for example, some level of previous consumption. This model gives additional flexibility and fits the data better. Campbell and Cochrane (1999) develop this model further. Letting $S_t = (C_t - X_t)/C_t$, they assume the following dynamic process

$$g_{t+1} = g + u_{t+1}$$

$$s_{t+1} = \log S_{t+1} = (1-\phi)\bar{s} + \phi s_t + \lambda(s_t)u_{t+1},$$

where u_{t+1} is a normally distributed shock, $\bar{s}$ is a steady state level, and $\lambda(.)$ is some (specific) function of the state. The habit is assumed to depend only on aggregate consumption. A key feature of the model is that the local risk aversion is time varying, i.e.,

$$\frac{-CU_{CC}}{U_C} = \frac{\gamma}{S_t}. \tag{10.23}$$

Solving the model is somewhat complicated. They find that the model matches the level of the riskless rate and the equity premium as well as the volatility of the price–dividend ratio.

10.6.2 Hyperbolic Discounting

Hyperbolic discounting replaces δ^j by the more slowly declining $(1 + \alpha j)^{-\beta/\alpha}$ function with parameters α, β. Laibson (1997) introduces a simple way of bringing in this discounting mechanism into dynamic choice models. In his model, the representative agent maximizes

$$U(C_t) + \beta E_t \left(\sum_{j=1}^{\infty} \delta^j U(C_{t+j}) \right)$$

subject to the usual budget constraint. When $\beta = 1$, we are back to the standard formulation, but when $\beta < 1$ this represents different preferences with regard to the future. In particular, it implies a sharp drop in valuation of next periods benefits compared with subsequent periods.

10.6.3 Lettau and Ludvigson (2001)

Denote log consumption by c, log wealth by w, and let r_w be the (log) net return on aggregate wealth. By linearization and solving forward they obtain the equation

$$c_t - w_t = \sum_{i=1}^{\infty} \rho_w^i E_t (r_{w,t+i} - g_{t+i}).$$

Wealth is composed of asset holdings A and human capital H, and its rate of return can be written as

$$r_{w,t} = \omega r_{a,t} + (1 - \omega) r_{h,t},$$

where $\omega \in [0, 1]$. Approximating the nonstationary component of human capital by aggregate labour income, they obtain

$$\overbrace{c_t - \omega a_t - (1 - \omega) y_t}^{cay_t} = (1 - \omega) z_t + \sum_{i=1}^{\infty} \rho_w^i E_t \left(\omega r_{a,t+i} + (1 - \omega) r_{h,t+i} - \Delta c_{t+i} \right).$$

The assumption is that consumption, asset values and income are cointegrated. Their residual summarizes expectations of future returns on the market portfolio. Using US quarterly stock market data, they find that fluctuations in the consumption–wealth ratio (cay) are strong predictors of both real stock returns and excess returns over a Treasury bill rate. They find that this variable is a better forecaster of future returns at short and intermediate horizons than is the dividend yield, the dividend payout ratio, and several other popular forecasting variables. We show this updated series in Figure 10.6.

10.6.4 Long Run Risks Model

Bansal and Yaron (2004) propose a model that combines several features of the earlier models. First, they use Epstein–Zin preferences, which allow EIS and risk aversion to differ. The first order condition they work with is

$$E_t \left(\left\{ \delta \left(\frac{C_{t+1}}{C_t} \right)^{-\frac{1}{\psi}} \right\}^{\theta} \left\{ \frac{1}{(1 + R_{a,t+1})} \right\}^{1-\theta} (1 + R_{i,t+1}) - 1 \right) = 0,$$

10.6 Other Asset Pricing Approaches

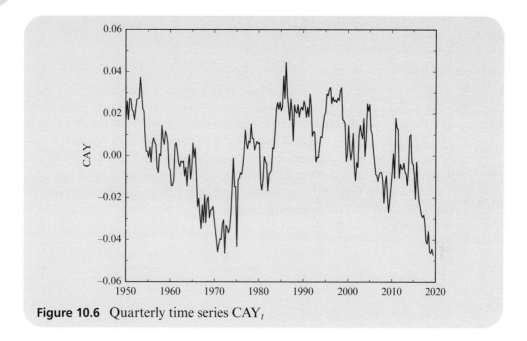

Figure 10.6 Quarterly time series CAY_t

where R_a is the gross return on an asset that delivers aggregate consumption as its dividend each period (like but not equal to the market portfolio). They work with a log linear approximation to this, in which case the stochastic discount factor is linear in the growth rate of consumption and also the return on the asset

$$m_{t+1} = \theta \log \delta - \frac{\theta}{\psi} g_{t+1} + (\theta - 1) r_{a,t+1} \tag{10.24}$$

$$r_{a,t+1} = \kappa_0 + \kappa_1 z_{t+1} - z_t + g_t,$$

where $r_{a,t+1} = \log(1 + R_{a,t+1})$, $g_{t+1} = \log(C_{t+1}/C_t)$, and $z_t = \log(P_t/C_t)$.

The second ingredient is to specify consumption growth as having a small predictable component (the long-run risk) so that consumption news in the present affects expectations of future consumption growth. This adds an additional risk factor to the Bansal–Yaron model not present in models that assume that consumption growth is i.i.d., and helps increase the correlation between consumption growth and stock returns. By making consumption growth persistent, the long-run risk can increase the impact of present consumption growth on the difference between present discounted values of dividend streams. It is the difference between these values that ultimately drives returns in consumption-based models. The third ingredient is allowing for time-varying volatility in consumption growth. This reflects time-varying economic uncertainty and is a further source of investor uncertainty and risk. They specify dynamics for consumption and dividend growth rates as follows:

$$g_{t+1} = \mu + x_t + \sigma_t \eta_{t+1}$$

$$g_{d,t+1} = \mu_d + \phi x_t + \varphi_d \sigma_t u_{t+1},$$

where the unobserved state variables satisfy:

$$x_{t+1} = \rho x_t + \varphi_e \sigma_t e_{t+1}$$

$$\sigma_{t+1}^2 = \sigma^2 + \nu_1(\sigma_t^2 - \sigma^2) + \sigma_w w_{t+1}.$$

The variable x is the long run risk. They assume that the innovations $e_{t+1}, w_{t+1}, \eta_{t+1}, u_{t+1}$ are standard normal and i.i.d. The stochastic volatility process σ_t^2 is the conditional variance of consumption growth and also drives the volatility of the dividend growth rate. We will discuss the properties of these processes in Chapter 11.

They assume that both risk aversion and the IES are greater than one. They solve the model and find the following results:

Innovation to SDF

$$m_{t+1} - E_t(m_{t+1}) = \lambda_{m,\eta} \overbrace{\sigma_t \eta_{t+1}}^{\text{consumption shock}} - \lambda_{m,e} \overbrace{\sigma_t e_{t+1}}^{\text{LRR shock}} - \lambda_{m,w} \sigma_w \overbrace{w_{t+1}}^{\text{shock to vol}}.$$

There are three shocks driving the SDF:

Equity premium

$$E_t(r_{m,t+1} - r_{f,t+1}) = \beta_{m,e}\lambda_{m,e}\sigma_t^2 + \beta_{m,w}\lambda_{m,w}\sigma_w^2 - \frac{1}{2}\text{var}_t(r_{m,t+1})$$

$$\text{var}_t(r_{m,t+1}) = (\beta_{m,e}^2 + \varphi_d^2)\sigma_t^2 + \beta_{m,w}^2\sigma_w^2.$$

Risk return relationship

$$E_t(r_{m,t+1} - r_{f,t+1}) = \tau_0 + \tau_1 \text{var}_t(r_{m,t+1}).$$

This captures the idea that news about growth rates and economic uncertainty (i.e., consumption volatility) alters perceptions regarding long-term expected growth rates and economic uncertainty and that asset prices will be fairly sensitive to small growth rate and consumption volatility news.

They calibrated the model with a very high persistence for x, $\rho = 0.98$, risk aversion of 10, and IES of 1.5. At these values, a reduction in economic uncertainty or better long-run growth prospects leads to a plausible rise in the wealth–consumption and the price–dividend ratios. The model is capable of justifying the observed magnitudes of the equity premium, the risk-free rate, and the volatility of the market return, dividend-yield, and the risk-free rate. Further, it captures the volatility feedback effect, that is, the negative correlation between return news and return volatility news. As in the data, dividend yields predict future returns and the volatility of returns is time-varying. Bansal and Yaron show that there is a significant negative correlation between price–dividend ratios and consumption volatility. About half of the variability in equity prices is due to fluctuations in expected growth rates, and the remainder is due to fluctuations in the cost of capital.

10.7 Summary of Chapter

Constantinides and Ghosh (2011) derive an estimation methodology for the log linear model based on inverting the linear relationship between observable log price–consumption ratio z_t and the log price/dividend ratio z_{mt} and the unobserved state variables x_t and σ_t^2. In particular, they show that

$$m_{t+1} = c_1 + c_2 \Delta c_{t+1} + c_3 \left(r_{f,t+1} - \frac{1}{\kappa} r_{f,t} \right) + c_4 \left(z_{m,t+1} - \frac{1}{\kappa} z_{m,t} \right),$$

where c_j are nonlinear functions of the parameters θ, ψ, δ, etc. This can be used to estimate these parameters from the data by GMM. They find mixed results. The model matches the unconditional moments of aggregate dividend and consumption growth rates. As predicted, the long-run risk forecasts consumption and dividend growth. However, the implied risk-free rate is too high and insufficiently variable. The market price–dividend ratio is not sufficiently variable and the model requires greater persistence in consumption and dividend growth than is observed. At long horizons, the conditional variance does not forecast the equity premium but does forecast consumption and dividend growth, contrary to the model's predictions. Moreover, the J-statistic p-value is less than 0.03 in all specifications considered.

10.7 Summary of Chapter

We consider the dynamic representative agent model of consumption and investment and its associated first order conditions. We discussed the econometric methodology for testing this theory and its lack of success empirically. We described the equity premium puzzle and the various proposed solutions to this.

11 Volatility

In this chapter we consider three different approaches to measuring volatility: methods based on derivative prices, methods based on high frequency data, and methods based on discrete time models for medium frequency data.

11.1 Why is Volatility Important?

Volatility measurement and estimation are central to finance for several reasons. First, **asset pricing**. For example, the conditional CAPM says that for some $\gamma_t > 0$

$$E_{t-1}(r_{i,t}) - r_f = \beta_{i,t}\gamma_t, \quad \beta_{i,t} = \frac{\text{cov}_{t-1}(r_{i,t}, r_{m,t})}{\text{var}_{t-1}(r_{m,t})},$$

which involves the conditional second moments of individual asset returns and market returns. Second, **risk management or Value at Risk**. A well known measure of risk called the Value at Risk (which is treated below in Chapter 14) can, for a large class of models, be written as follows

$$VaR_t(\alpha) = \mu_t + \sigma_t \times q_\alpha, \tag{11.1}$$

where μ_t is the conditional mean and σ_t is the conditional standard deviation of the returns on the asset in question, while q_α is the α-quantile of the innovation distribution. Third, **portfolio allocation** is a major practical problem that institutional and retail investors have to solve. The most common version of this is the optimization problem

$$\min_{w \in S} w^\top \text{var}_{t-1}(r_t) w \quad \text{s.t.} = w^\top E_{t-1}(r_t) = m,$$

where S is some set of portfolio weights that satisfy at least $\sum_{i=1}^n w_i = 1$, and the conditional mean vector $E_{t-1}(r_t)$ and variance matrix $\text{var}_{t-1}(r_t)$ are key inputs that have to be estimated from data. Fourth, volatility can be interpreted as a measure of the stability of the underlying system, and as such captures some notion of **market quality**, whereby greater volatility indicates poorer quality. This argument has to be mitigated by the fact that volatility as usually measured contains both upside and downside variation. High volatility does not necessarily connote with poor market quality (otherwise the North Korean stock market would be the best on earth). To put this in perspective, consider airline safety, which can be measured by the frequency of plane crashes. Figure 11.1 shows the accidental fatalities per mile during

11.2 Implied Volatility from Option Prices

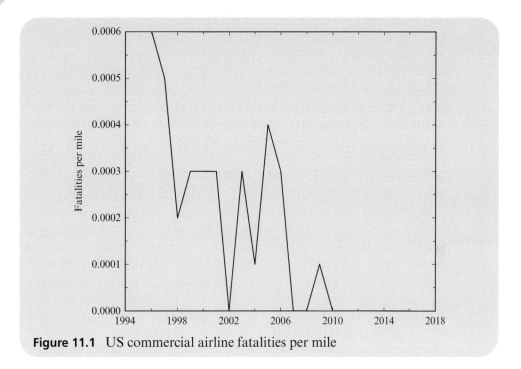

Figure 11.1 US commercial airline fatalities per mile

US commercial flights since 1996 (available from the National Transportation Safety Board at www.ntsb.gov/Pages/default.aspx; this excludes terrorist incidents, which affects the figures for 2001). This shows that the risk of flying in commercial aircraft has reduced considerably over this 20 year horizon. But as we shall see, the volatility of the US stock market has not shown such a clear downward trend, and it is not clear that it ever would. Finally, we may care about the consequences of residual volatility in carrying out various statistical inference procedures while testing the CAPM or the APT, for example.

We consider three different approaches to measuring and defining volatility. These approaches differ according to whether they are measuring **ex-ante** or **ex-post** volatility, and of course the data requirements.

11.2 Implied Volatility from Option Prices

Suppose that stock prices P follow the process

$$d \log P(t) = \mu dt + \sigma dB(t), \tag{11.2}$$

where B is Brownian motion, i.e., for all t, s, $B(t+s) - B(t)$ is normally distributed with mean zero and variance s with independent increments. This is a continuous time model, which we study in greater detail in Chapter 12. In this particular model, prices are lognormally distributed at all time periods and volatility is constant per unit time and measured by the parameter σ or σ^2. The question we consider is how to estimate this quantity given some data on prices and option prices.

Suppose you have a European (exercisable only at maturity) call option on the stock with strike price K and time to maturity τ. Black and Scholes (1973) showed that the value of this option should be (5.40). The value of the option increases with the volatility. In practice we observe call option prices C along with the price of the underlying instrument P. Suppose that we observe C, P, K, r_f, and τ. Then we can invert the relation to obtain σ^2, that is, we define the value of σ^2, denoted σ^2_{BS}, as the value that solves the equation

$$C = C_{BS}(P, K, r_f, \tau, \sigma^2_{BS}). \tag{11.3}$$

In general, this does not yield a closed form solution, but it is a simple problem to compute $\sigma^2_{BS}(P, K, r_f, \tau)$ numerically, and indeed this function is performed by many commercially available portable calculators. The quantity σ^2_{BS} is called the implied volatility. One can do this at every time period where we have these observations thereby generating a time series of volatility $\sigma^2_{BS}(t)$, $t = 1, \ldots, T$. In fact, we generally have for each day multiple values of K, τ, i.e., there are many different option contracts with the same underlying asset price P, and so one gets multiple implied volatilities for each time period.

In practice, some adjustments are made to this simple principle in computing the VIX. This is the ticker symbol for the Chicago Board Options Exchange (CBOE) Market Volatility Index, a popular measure of the implied volatility of S&P500 index options. This is often referred to as the **fear index**. It represents one measure of the market's expectation of stock market volatility over the next 30 day period. It is quoted in percentage points and translates, roughly, to the expected movement in the S&P500 index over the next 30 day period (which is then annualized). The VIX is calculated and disseminated in real-time by the CBOE. CBOE (2010) gives a full description of the current methodology. The old methodology was based on averaging implied volatilities from (11.3), but this changed in 2000 to the current approach, which is based on averaging option prices on the S&P500. The population quantity it estimates can be shown to be (Martin (2017))

$$VIX_t^2 = \frac{2r_{ft}}{T-t}\left(\int_0^{F_{t,T}} \frac{1}{K^2} put_{t,T}(K)dK + \int_{F_{t,T}}^{\infty} \frac{1}{K^2} call_{t,T}(K)dK\right),$$

where T is the time of expiration, F is the forward index level derived from index option prices, and K is the strike price of the put or call option, while $put_{t,T}$ and $call_{t,T}$ are the put and call option prices. Carr and Wu (2006) present the theoretical underpinning of the *VIX* volatility measure, and argue that since the new method is based on market prices and not on the Black–Scholes formula, it is a valid volatility measure under more general conditions than (11.2).

In Figure 11.2 we show the daily time series of closing VIX prices since 1990.

In Figure 11.3 we report the histogram of the VIX time series, which shows how positively skewed its distribution is. Volatility may well be better measured on a

11.2 Implied Volatility from Option Prices

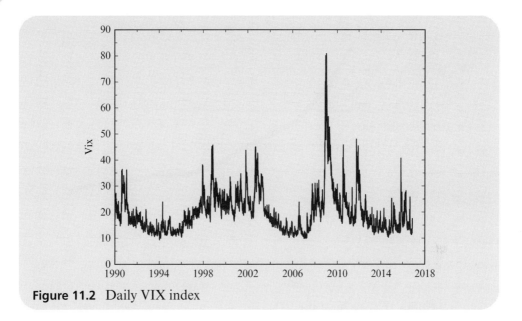

Figure 11.2 Daily VIX index

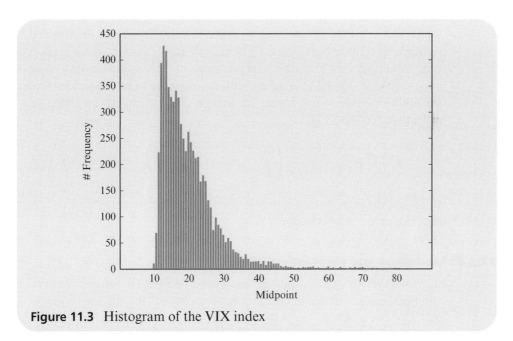

Figure 11.3 Histogram of the VIX index

log scale as some have argued, just like the Richter scale for measuring earthquake intensity, since the log of VIX has a much more symmetric distribution.

This time series is very persistent in comparison with daily returns as we show in Figure 11.4. Part, but not all, of the reason for this is that it is measuring volatility over a future 30 day period on a daily basis, so that consecutive daily values of the VIX have considerable overlap.

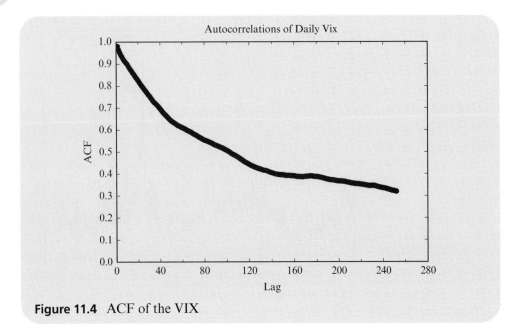

Figure 11.4 ACF of the VIX

The VIX is a forward looking measure of volatility that is determined by market participants trading based on their beliefs about the future, which gives this measure some advantages. In fact, the interpretation of VIX as a volatility measure can be made precise in that VIX_t^2 is the conditional variance of returns under the risk neutral probability measure; see Chapter 10. Martin (2017) proposes an alternative volatility measure called the SVIX

$$SVIX_t^2 = \frac{2r_{ft}}{(T-t)F_{t,T}} \left(\int_0^{F_{t,T}} put_{t,T}(K)dK + \int_{F_{t,T}}^{\infty} call_{t,T}(K)dK \right), \quad (11.4)$$

which has the interpretation that it is the conditional variance of returns under the objective probability measure.

11.2.1 Modelling and Forecasting VIX

For a number of applications we may need to forecast future values of VIX. It is common practice to use a linear time series model for the logarithm of VIX, $y_t = \log VIX_t$. For example, the AR(∞) class of models (2.22), which yields the forecast (2.23), whereby the VIX is forecast by the exponential of this forecast. The EWMA method (2.24) is also often used for forecasting.

11.3 Intra Period Volatility

Suppose that we are interested in the volatility of low frequency (e.g., monthly) returns r_t, but we also have higher frequency (e.g., daily) data r_{t_j}, $j = 1, \ldots, m_t$, where m is

11.3 Intra Period Volatility

the total number of observations inside each period, assumed constant for simplicity. We may estimate the volatility of stock at time t, which we denote by $\widehat{\sigma}^2_{t+1}$, as the intraperiod sample variance

$$\widehat{\sigma}^2_{t+1} = \frac{1}{m_t} \sum_{j=1}^{m_t} \left(r_{t_j} - \frac{1}{m_t} \sum_{j=1}^{m_t} r_{t_j} \right)^2. \tag{11.5}$$

In some cases people use $\widehat{\sigma}^2_{t+1} = \sum_{j=1}^{m_t} r_{t_j}^2 / m_t$, because mean daily returns are small and so their square is even smaller and do not contribute much in total so can be ignored. In other cases, for example Schwert (1989), authors use an adjustment that allows for serial correlation based on the Roll argument, i.e., to purge the short term noise component. That is, they estimate the underlying volatility by

$$\widehat{\sigma}^2_{t+1} = \frac{1}{m} \sum_{j=1}^{m} \left(r_{t_j}^2 + 2 r_{t_j} r_{t_{j-1}} \right). \tag{11.6}$$

This seems a natural way of capturing intra period variability. It can be considered an **ex-post** measure of volatility, meaning that it is a measure of the volatility that happened in the period $t, t+1$, which is why we give it the $t+1$ subscript. It is not what was anticipated to happen at time t, i.e., it is not the conditional variance of returns given past information. This method can be applied to raw returns or to the residuals from a factor model, in which case (11.5) and (11.6) are measuring ex-post idiosyncratic volatility.

The issue with this approach is that it relies on mixed frequency data (daily data to estimate a monthly volatility), and it is not clear how to interpret σ^2_{t+1} and $\widehat{\sigma}^2_{t+1}$ in terms of plausible discrete time models of r_{t_j}, although we can try. Suppose that

$$r_{t_j} = \mu_t + \sigma_t \varepsilon_{t_j}, \tag{11.7}$$

where low frequency period $t=1,\ldots,T$ and high frequency period $t_j, j=1,\ldots,m_t$. The mean μ_t and variance σ_t^2 of returns vary over the long period but are constant within each period. We may suppose that ε_{t_j} and $\varepsilon_{t_j}^2 - 1$ both satisfy **rw3** but may be dependent and heteroskedastic. In this case the estimates of σ_t^2 can be given an interpretation. For consistency, we would require that the number of intraperiod observations m_t be large to obtain a consistent estimate.

In Figure 11.5 we present estimates of the annualized standard deviation of the S&P500 based on computing the intra year variance of the daily return series. That is, we compute (11.5), multiply by 252, and take the square root, for $t=1950,\ldots,2017$. Apart from 2017, we have around 252 observations per year, so the estimates should be fairly accurate. The figure shows the substantial variation from year to year of this volatility measure and the relatively low values it took throughout the 1950s and 1960s. The financial crisis period 2009 and 2010 saw very high levels of volatility.

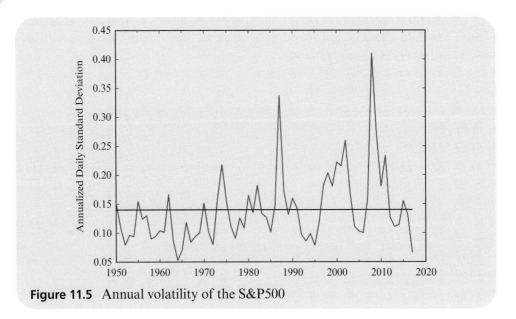

Figure 11.5 Annual volatility of the S&P500

11.3.1 Continuous Record or Infill Framework

The model (11.7) is a bit contrived, and it is not easy to justify why volatility should be constant within a year and then change to a new level for the next year. This approach has recently been given a proper interpretation inside the framework of continuous time processes or "continuous record asymptotics" (Foster and Nelson (1996)), when the number of within period observations is large. This is consistent with the current use of intraday data on prices. Typically, the trading day is from around 9am to 4pm, and one observes prices continuously throughout this day. The following table gives the number of observations per day associated with specific frequencies, and illustrates how we may be justified in using an asymptotic approximation.

Frequency	n (returns)
Hourly	7
10 mins	42
5 mins	84
1 min	420
10 secs	2520
1 sec	25200
1 millisecond	25200000

We now present a model that is consistent with the use of infill asymptotics and an estimator of volatility. Suppose that stock prices P follow (11.2). In this case, returns are normally distributed with mean and variance that depends on the sampling interval.

11.3 Intra Period Volatility

Definition 93 *Suppose that we observe prices at times $t_j = j/n \in [0,1], j = 0, 1, \ldots, n$, whence*

$$r_{t_j} = \log P(t_j) - \log P(t_{j-1}).$$

Define the **realized volatility** *(RV)*

$$\widehat{\sigma}^2_{[0,1]} = \sum_{j=1}^{n} r_{t_j}^2. \tag{11.8}$$

This is like (11.5) except that we do not subtract the mean and we do not normalize by the number of periods considered – in this framework where the variance shrinks with the length of the interval the extra normalization is not needed. We can write

$$r_{t_j} \sim N(\mu/n, \sigma^2/n) = \frac{\mu}{n} + \frac{\sigma}{\sqrt{n}} z_j,$$

where z_j are standard normal. This corresponds to the model (11.7) with the additional assumption of normality.

Theorem 41 *As $n \to \infty$, $\widehat{\sigma}^2_{[0,1]} \xrightarrow{P} \sigma^2$. Furthermore,*

$$\sqrt{n}\left(\widehat{\sigma}^2_{[0,1]} - \sigma^2\right) \Longrightarrow N(0, 2\sigma^4).$$

Proof. We have

$$\widehat{\sigma}^2_{[0,1]} = \sum_{j=1}^{n} r_{t_j}^2 = \sum_{j=1}^{n} \left(\frac{\mu}{n} + \frac{\sigma}{\sqrt{n}} z_j\right)^2 = \sigma^2 \frac{1}{n} \sum_{j=1}^{n} z_j^2 + \frac{\mu^2}{n} + \frac{1}{n}\frac{2\mu}{\sqrt{n}} \sum_{j=1}^{n} z_j,$$

Therefore, the first result follows by the LLN and the second result by the CLT. □

This result continues to hold under much more general conditions, specifically for diffusion processes, stochastic volatility processes, and so on; see Chapter 12.

Torben G. Andersen is the Nathan S. and Mary P. Sharp Professor of Finance at Northwestern University. He joined the faculty in 1991 and is a Faculty Research Associate of the National Bureau of Economic Research and an International Fellow of the Center for Research in Econometric Analysis of Economic Time Series (CREATES) in Aarhus, Denmark. He has published widely in asset pricing, empirical finance, and empirical market microstructure. His work centers on modelling of volatility fluctuations in financial returns with applications to asset and derivatives pricing, portfolio selection, and the term structure of interest rates. His current work explores the use of large data sets of high-frequency data for volatility forecasting, portfolio choice, and risk management. He received his PhD in Economics from Yale University.

Realized volatility requires one to observe high frequency intraday stock returns. This type of data is not always available for a long period of time or for a broad cross section of stocks, and so one might seek alternative methods to estimate volatility. Another common method of measuring volatility within a period is to use the so-called **realized range**, which does not require such large amounts of intraday data. Yahoo, Bloomberg, etc all report the daily opening price, closing price, the intraday high price, and the intraday low price: P_O, P_C, P_H, P_L. Most authors' work with the daily closing price and returns as we have described them have been computed this way. Measures of volatility can be computed from these prices as follows.

Definition 94 *The realized range relative on day t is defined as*

$$v_t = \frac{P_{Ht} - P_{Lt}}{P_{Lt}}. \tag{11.9}$$

This has the interpretation as the return that an omniscient trader could achieve by buying once at the low and selling at the high price within the day. Some authors compute this measure with P_{Ot} or P_{Ct} in the denominator.

Definition 95 *The Parkinson (1980) volatility estimator is*

$$\tilde{\sigma}^2 = \frac{(\log P_{Ht} - \log P_{Lt})^2}{4 \log 2} = \frac{(\log(1 + v_t))^2}{4 \log 2}. \tag{11.10}$$

These estimators can be given an interpretation inside continuous time models. Specifically, under the Brownian motion model assumption (11.2), the Parkinson estimator is equal to σ^2 provided the high price and the low price are measured perfectly accurately, i.e., true prices are recorded continuously and

$$P_{Ht} = \sup_{s \text{ in day } t} P_s \quad ; \quad P_{Lt} = \inf_{s \text{ in day } t} P_s.$$

The estimator $\tilde{\sigma}^2$ is equal to σ^2 in this case. In practice, the estimators are computed on a finite time grid (because only a finite number of transactions occur during the day), i.e.,

$$P_{Ht} = \max_{1 \leq j \leq n} P_{t_j} \quad ; \quad P_{Lt} = \min_{1 \leq j \leq n} P_{t_j},$$

11.3 Intra Period Volatility

where $t_1, \ldots, t_n$ are the grid points (observed by Yahoo but not by the econometrician). In this framework, specifically, $\tilde{\sigma}^2$ converges to σ^2 under the Brownian motion model assumption (11.2) as the grid size $n \to \infty$. However, it turns out to be an inefficient estimator of σ^2 under the model assumption in comparison with realized volatility that is based on observing prices at all grid points.

Garman and Klass (1980) introduced an estimator that is more efficient than the Parkinson estimator under the no drift Brownian motion case

$$V_t^{GK} = 0.5 \left(\log P_t^H - \log P_t^L\right)^2 - (2\log 2 - 1)\left(\log P_t^C - \log P_t^O\right)^2. \quad (11.11)$$

Rogers and Satchell (1991) introduce an estimator

$$V_t^{RS} = (\log P_t^H - \log P_t^C)(\log P_t^H - \log P_t^O) + (\log P_t^L - \log P_t^C)(\log P_t^L - \log P_t^O) \quad (11.12)$$

that is exactly unbiased in the case where the process has a non zero drift. These are also consistent estimators of σ^2 inside the model (11.2). For this special case one can achieve a similar objective to that achieved by realized volatility without observing the entire transaction record. See Chou, Chou, and Liu (2009) for a discussion of range based volatility estimators. These estimators are easy to compute for a wide variety of financial instruments including individual stocks and indexes. These estimators all have interpretations as measures of **scale** (a more general statistical notion that includes variance or standard deviation as special cases) under more general settings than Brownian motion, although they may not be directly comparable to realized volatility measures outside the simple case.

How does $(\log P_C - \log P_O)^2$ compare with $(\log P_H - \log P_L)^2$? For the S&P500 daily return series since 1983, this correlation is around 0.90. This suggests that absent a measurement on the high and low price, one may get a not totally useless measure of daily volatility by looking at the squared daily return or the squared intraday return. However, Anderson and Bollerslev (1998) argue that using squared daily return is a poor proxy for volatility because although unbiased they are very noisy.

In Figure 11.6 we show the Parkinson estimator for the S&P500 over the period 1990–2017, which may be compared with the VIX.

This series is also very persistent as is shown by the autocorrelation function in Figure 11.7, although it is not as persistent as the VIX, since each day the series is constructed from different prices. Its distribution is also positively skewed (Figure 11.8), i.e., there is a long right tail.

In Tables 11.1 and 11.2 we present the most volatile days on US and UK stock markets. We measure volatility by (11.9); − means $P_C < P_O$, + means $P_C > P_O$.

The US market appears to be slightly more volatile than the UK market, which is a bit surprising since the S&P500 represents a market value of around $22.6 trillion against the $1.9 trillion value of the FTSE100. Some possible explanations of this are given below. Both US and UK markets were dominated by events in 2008 and 1987. Circuit breakers now limit the worst case (or perhaps spread it out over several days), so that it is not possible to have a stock market crash of more than 10% in one day any more.

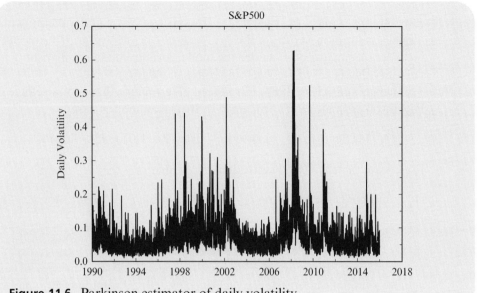

Figure 11.6 Parkinson estimator of daily volatility

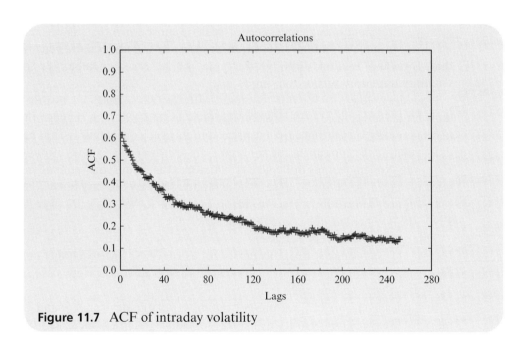

Figure 11.7 ACF of intraday volatility

11.3.2 Modelling and Forecasting RV

These measures are all ex-post, and for many applications we need ex-ante measures. To forecast future values of RV or range based volatility it is common practice to use a linear time series model for the logarithm of RV, $y_t = \log RV_t$. For example,

11.3 Intra Period Volatility

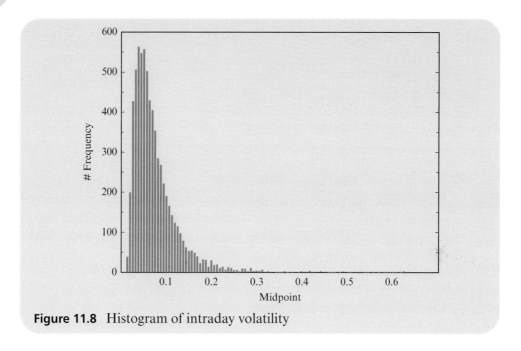

Figure 11.8 Histogram of intraday volatility

Table 11.1 The FTSE100 top 20 most volatile days since 1984

Date	Volatility
19871020	0.131−
19871022	0.115−
20081010	0.112−
19971028	0.096−
20081024	0.096−
20081006	0.094−
20081008	0.094−
20080919	0.093+
20081124	0.090+
20081015	0.084−
19871019	0.081−
20020920	0.080+
20081013	0.076+
20010921	0.076−
20081029	0.075+
20110809	0.074+
20090114	0.074−
20080122	0.074+
20020715	0.071−
20081016	0.070−

Table 11.2 The S&P500 top 20 most volatile days since 1960

Date	Volatility
19871019	0.257−
19871020	0.123+
20081010	0.107−
20081009	0.106−
20081113	0.104+
20081028	0.101+
20081015	0.100−
20081120	0.097−
20081013	0.094+
20080929	0.093−
19871026	0.092−
20100506	0.090−
20081201	0.089+
19620529	0.089+
19871021	0.087+
20081016	0.087+
20081006	0.085−
20081022	0.085−
20020724	0.081+
19980831	0.080−

the AR(∞) class of models (2.22). This yields the forecast (2.23), whereby the RV is forecast by the exponential of 2.23. The EWMA method (2.24) is also often used.

11.4 Cross-sectional Volatility

Another measure of market wide volatility is called **cross-sectional volatility**.

Definition 96 *The annualized daily cross-sectional volatility is*

$$\hat{\sigma}_t = \sqrt{\frac{252}{n-1} \sum_{i=1}^{n} (r_{it} - \bar{r}_t)^2}.$$

This measures the cross-sectional dispersion in returns. In Figure 11.9 we show the evolution of this measure for the Dow stocks.

Suppose that stock returns obey a single factor model as in Chapter 8

$$r_{it} = \alpha_i + \beta_i r_{mt} + \varepsilon_{it},$$

11.5 Empirical Studies

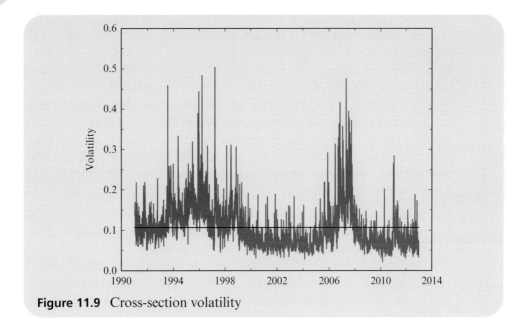

Figure 11.9 Cross-section volatility

where the idiosyncratic error term satisfies $\text{var}(\varepsilon_{it}) = \sigma^2_{\varepsilon it}$, which may vary over time. Then for large n we can obtain

$$\frac{1}{n}\sum_{i=1}^{n}(r_{it}-\bar{r}_t)^2 = \frac{1}{n}\sum_{i=1}^{n}(\alpha_i-\bar{\alpha})^2 + \frac{1}{n}\sum_{i=1}^{n}(\beta_i-\bar{\beta})^2 r_{mt}^2 + \frac{1}{n}\sum_{i=1}^{n}(\varepsilon_{it}-\bar{\varepsilon}_t)^2,$$
$$\simeq \sigma^2_\alpha + \sigma^2_\beta r_{mt}^2 + \sigma^2_{\varepsilon t}.$$

In the second line we assume a **random coefficient** interpretation for the market model where α_i, β_i are random variables with $\sigma^2_\alpha = \text{var}(\alpha)$ and $\sigma^2_\beta = \text{var}(\beta)$. The cross-sectional variance of stock returns could vary over time due to variation of the common factor r_{mt} or the idiosyncratic variance $\sigma^2_{\varepsilon t}$ or both. If the idiosyncratic variance was constant over time, then this measure is perfectly correlated with the squared common factor (market volatility), but otherwise time varying idiosyncratic volatility could contribute to the cross-sectional volatility. Campbell, Lettau, Malkiel, and Xu (2001) argued that over the period 1962–1997 there has been a noticeable increase in firm-level volatility relative to market volatility. Connor, Korajczyk, and Linton (2006) extended the time period and found a subsequent decline in the firm-level volatility.

11.5 Empirical Studies

What are the properties of volatility and what causes volatility, or rather which factors affect volatility and lead it to change over time and to vary across firms and countries? We discuss some empirical studies that have tried to address these questions.

Schwert (1989) (updated in Schwert (2011)) examines monthly US market volatility over a long history of the US market. To measure volatility he computes the average of squared daily returns on the market index within each month for 1885–1987. A full description of this data and its construction is given on his website, which is listed in the data appendix. He actually corrects for serial correlation, which could be caused by non-synchronous trading, within the month, so that he uses (11.6). He finds that this estimated volatility is positively serially correlated and the percentage change in volatility is negatively serially correlated. Therefore, volatility is not a random walk. He considers a number of model specifications for volatility including ARMA models and time series regression models with macroeconomic predictors. He finds that:

1. The average level of volatility is higher during (NBER dated) recessions.
2. The level of volatility during the Great Depression was very high.
3. The effect of financial leverage on volatility is small.
4. There is weak evidence that macroeconomic volatility can help to predict financial asset volatility and stronger evidence for the reverse prediction.
5. The number of trading days in the month is positively related to stock volatility (trading days per year on the NYSE are around 252).
6. Share trading volume growth is positively related to stock volatility.

We next consider the comparison of volatility within trading days, overnight, and during the weekend. The calendar time hypothesis and trading time hypothesis make quite different predictions about these quantities. Fama (1965) compared daily variances with weekend variances. Weekend variances should be three times daily. He finds that the actual variance discrepancy is much smaller, around 22%. French and Roll (1986) extend this work. They classify days into a number of different categories depending on the number of elapsed days since a closing price. This could be 1 for a normal trading day, 3 for a Monday following a weekend, 4 for a Tuesday following a Monday holiday period, etc. They measure volatility by the sample variance for days within each category. A typical trading day may be 8 hours long out of 24 hours (say 8–4). The weekend, i.e., Friday close to Monday open, contains 66 hours. They find that the per hour return variance is 70 times larger during a trading hour than during a weekend hour. They propose several explanations:

1. Volatility is caused by public information which is more likely to arrive during normal business hours.
2. Volatility is caused by private information which affects prices when informed investors trade.
3. Volatility is caused by pricing errors that occur during trading.

They find that although there is some evidence of mispricing, information explanations 1 and 2 contribute more to total volatility. To distinguish between explanations 1 and 2 they use the fact that in 1968, NYSE was closed every Wednesday because of the **paperwork crisis**, but otherwise was a regular business day producing the usual amount of public information. They find in favor of the second hypothesis that

11.5 Empirical Studies

volatility is caused by private information that affects prices when informed investors trade.

Many authors have argued that the introduction of computerized trading and the increased prevalence of High Frequency Trading strategies in the period post 2005 has led to an increase in volatility (see Linton, O'Hara, and Zigrand (2013)). A direct comparison of volatility before and after would be problematic here because of the Global Financial Crisis, which raised volatility during the same period that HFT was becoming more prevalent. There are a number of studies that have investigated this question with natural experiments methodology (Hendershott, Brogaard, and Riordan (2014) and Brogaard, Hendershott, and Riordan (2014)), but the conclusions one can draw from such work are event specific; see Angrist and Pischke (2009). One implication of this hypothesis is that, *ceteris paribus*, the ratio of intraday to overnight volatility should have increased during this period because trading is not taking place during the market close period. One could just compare the daily return volatility from the intraday segment with the daily return volatility from the overnight segment, as in French and Roll (1986). However, this would ignore both fast and slow variation in volatility through business cycle and other causal factors. Also, overnight raw returns are very heavy tailed and so sample (unconditional) variances are not very reliable. Linton and Wu (2016) develop a dynamic model for overnight and intraday returns. They show that the ratio of hourly intraday volatility to hourly overnight volatility over the more recent period 2000–2015 is more like 9, which is a big reduction. They also found that the ratio of overnight to intraday volatility seems to have increased over the period 2000–2015 for large stocks.

Bartram, Brown, and Stulz (2012) study volatility across countries. They compare the annual volatility of US firms with the annual volatility (standard deviation of weekly, Friday to Friday US dollar returns) of firms from the rest of the world (50 countries) over the period 1990–2006. They used a matching methodology. They find American exceptionalism: US firms are more volatile than rest of the world firms after controlling for firm size, longevity, market to book value, and other characteristics. They find that this volatility difference is mostly attributable to foreign firms having lower idiosyncratic risk than comparable US firms, where idiosyncratic risk is computed form the residuals of a model with observed factors; see Chapter 8. The difference in idiosyncratic risk between foreign and comparable US firms is related to both country and firm characteristics using Fama–Macbeth regressions. High idiosyncratic risk can result from factors that decrease welfare as well as from factors that increase welfare. Put differently, there is good idiosyncratic volatility and bad idiosyncratic volatility. Idiosyncratic volatility that results from instability or from noise trading worsens welfare. Idiosyncratic volatility that is the product of greater risk taking and more entrepreneurship can improve welfare and increase economic growth. They find that the higher idiosyncratic volatility of the US is associated with factors that we would expect to be associated with greater economic welfare. In particular, they find that idiosyncratic volatility increases with investor protection, with stock market development, and with innovation. They also find that firm-level variables that are associated with innovation and growth opportunities are associated with greater idiosyncratic volatility. US firms have a significantly higher share of R&D in the sum of capital expenditures and R&D than comparable firms in foreign countries.

Table 11.3 Idiosyncratic volatility of Dow stocks

	σ	σ_ε		σ	σ_ε
Alcoa Inc.	0.2151	0.1676	JP Morgan	0.0991	0.0806
AmEx	0.1851	0.1160	Coke	0.2234	0.1591
Boeing	0.2350	0.1717	McD	0.1491	0.1367
Bank of America	0.1540	0.1191	MMM	0.1458	0.1217
Caterpillar	0.1595	0.1168	Merck	0.1972	0.1811
Cisco Systems	0.1103	0.0923	MSFT	0.2092	0.1818
Chevron	0.2151	0.1695	Pfizer	0.2057	0.1901
du Pont	0.1504	0.1251	P & Gamble	0.1045	0.0872
Walt Disney	0.1408	0.0980	AT&T	0.1327	0.1058
General Electric	0.2267	0.2007	Travelers	0.1703	0.1402
Home Depot	0.2200	0.1862	United Health	0.2132	0.1968
HP	0.1937	0.1614	United Tech	0.1852	0.1578
IBM	0.2014	0.1686	Verizon	0.1280	0.1034
Intel	0.1318	0.0978	Wall Mart	0.1268	0.1022
Johnson&Johnson	0.2268	0.1880	Exxon Mobil	0.1470	0.1218

In Table 11.3 we give the annualized standard deviation and the idiosyncratic standard deviation (residuals from the market model regression) for the Dow Jones 30 stocks over the period 1990–2012. There is quite a range of volatility across firms.

11.6 Discrete Time Series Models

We next consider a class of discrete time series models for capturing the notion of volatility and how it changes over time. The key properties of stock returns and asset returns generally are that:

1. Marginal distributions of stock returns are leptokurtic. See Mandelbrot (1963) and Fama (1965).
2. The scale of stock returns does not appear to be constant over time.
3. Stock returns are almost uncorrelated but dependent. Highly volatile periods tend to cluster together.

11.6 Discrete Time Series Models

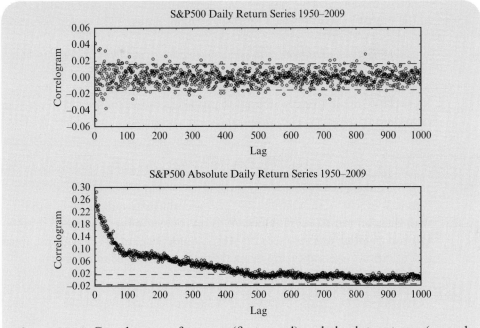

Figure 11.10 Correlogram of returns (first panel) and absolute returns (second panel)

In Figure 11.10 we show the correlogram of the S&P500 stock index daily return series from 1950–2017 along with the correlogram of the absolute value of this return. This shows the strong positive dependence in the absolute values of returns.

Linear models cannot well capture all these phenomena. Specifically suppose for summable weights ($\sum_{i=0}^{\infty} |a_i| < \infty$)

$$y_t = \sum_{i=0}^{\infty} a_i \varepsilon_{t-i}, \tag{11.13}$$

where ε_s is i.i.d. with mean zero and finite variance σ_ε^2. This is called a linear process. It has conditional mean (given $\mathcal{F}_{t-1} = \{y_{t-1}, y_{t-2}, \ldots\}$)

$$E(y_t|\mathcal{F}_{t-1}) = \sum_{i=1}^{\infty} b_i y_{t-i},$$

for some coefficients b_i (under an invertibility condition; see Chapter 2). However, the model implies constant conditional variance

$$\text{var}(y_t|\mathcal{F}_{t-1}) = \sigma_\varepsilon^2 a_0^2,$$

which is unrealistic. This suggests we need to examine **nonlinear models**. As the winner of the Nobel Prize in Physics, Stanislaw Ulam, notes:

Using the term nonlinear to describe a time series model is like saying that zoology is the study of nonelephant animals.

Press (1967) made an early attempt to propose a model for security returns that allows time varying volatility. The compound events model says that prices are determined by

$$\log P_t = \log P_0 + \sum_{k=1}^{N_t} Y_k + X_t,$$

where $X_t \sim N(0, \sigma_1^2 t)$, while $Y_1, Y_2, \ldots$, are mutually independent random variables with distribution $N(\theta, \sigma_2^2)$. Here, N_t is an integer valued random variable, in fact a **Poisson counting process** (see Chapter 12) independent of Y_k with parameter λt. We can interpret N_t as capturing news arrival intensity. This implies that

$$r_t = \log P_t - \log P_{t-1} = \sum_{k=N_{t-1}+1}^{N_t} Y_k + \varepsilon_t,$$

where $\varepsilon_t \sim N(0, \sigma_1^2)$, and $E(\varepsilon_t \varepsilon_{t-\tau}) = 0$ for all $\tau \neq 0$. Basically, returns are a sum of a random number of i.i.d. normal random variables rather like in the non-trading model of Chapter 5. Returns are independent over time because ε_t and Y_k are independent, but

$$\text{var}(r_t | N_1, \ldots, N_t) = \sigma_1^2 + \Delta N_t \sigma_2^2,$$

where $\Delta N_t = N_t - N_{t-1} - 1$, which changes over time. This implies that the unconditional moments satisfy $E(r_t) = \theta \lambda$; $\text{var}(r_t) = \sigma_1^2 + \lambda(\theta^2 + \sigma_2^2)$. The distribution of r_t is in the class of stable distributions and is leptokurtic. It is more peaked in the vicinity of the mean than a comparable normal variable. One can derive expressions for the higher cumulants also in terms of the underlying parameters. Press proposed to estimate the model by matching cumulants.

Rosenberg (1972) provided another innovative model for security returns that predates a lot of subsequent work. He specified

$$r_t = m_t + V_t^{1/2} \eta_t \quad t = 1, \ldots, T,$$

where η_t are serially independent random variables with mean zero, variance one, and kurtosis γ, while V_t, the variances of the price changes, obey a stochastic process that can be forecasted. He shows that when the variance of a population of random variables is time varying, the population kurtosis is greater than the kurtosis of the probability distribution of the individual random variables. Therefore, the high kurtosis observed in the distribution of security prices can be explained by high kurtosis in the individual price changes, non-stationarity in the variance of price changes, or any combination of these two causes. He finds evidence that r_t obeys a normal distribution with predictably fluctuating variance.

11.7 Engle's ARCH Model

The breakthrough model for financial volatility was Engle's (1982) autoregressive conditional heteroskedasticity (ARCH) model. Let $\mathcal{F}_t = \{y_t, y_{t-1}, \ldots\}$ be currently available information and let

$$\mu_t = E(y_t | \mathcal{F}_{t-1}) \quad \text{and} \quad \sigma_t^2 = \text{var}(y_t | \mathcal{F}_{t-1}). \tag{11.14}$$

We first assume for simplicity that $\mu_t = 0$, so the question is only about modelling the conditional variance σ_t^2. The ARCH model specifies this as a dynamic equation.

Definition 97 *ARCH(1). For all t, let*

$$\sigma_t^2 = \omega + \gamma y_{t-1}^2. \tag{11.15}$$

The conditional variance depends linearly on past squared returns. The parameters ω, γ determine the properties of the process. Provided $\omega > 0$ and $\gamma \geq 0$, then $\sigma_t^2 > 0$ with probability one. This seems like a minimal requirement for a model of volatility. If past squared returns are large, then the conditional variance will be large.

Robert Fry Engle III (born 1942) is the winner of the 2003 Nobel Memorial Prize in Economic Sciences, sharing the award with Clive Granger, "for methods of analyzing economic time series with time-varying volatility (ARCH)." He was born in Syracuse, New York into a Quaker family and went on to graduate from Williams College with a BS in physics. He earned an MS in physics and a PhD in economics, both from Cornell University in 1966 and 1969 respectively. After completing his PhD, Engle became Professor of Economics at the Massachusetts Institute of Technology from 1969 to 1977. He joined the faculty of the University of California, San Diego (UCSD) in 1975, wherefrom he retired in 2003. He now holds positions of Professor Emeritus and Research Professor at UCSD. He currently teaches at New York University, Stern School of Business where he is the Michael Armellino Professor in Management of Financial Services. At New York University, Engle teaches for the Master of Science in Risk Management Program for Executives, which is offered in partnership with the Amsterdam Institute of Finance. Engle's most important contribution was his path-breaking discovery of a method for analyzing unpredictable movements in financial market prices and interest rates. Accurate characterization and prediction of these volatile movements are essential for quantifying and effectively managing risk. For example, risk measurement plays a key role in pricing options and financial derivatives. Previous researchers had either assumed constant volatility or had used simple devices to approximate it. Engle developed new statistical models of volatility that captured the tendency of stock prices and other financial variables to move between high volatility and low volatility periods ("Autoregressive Conditional Heteroskedasticity: ARCH"). These statistical models have become essential tools of modern arbitrage pricing theory and practice. Engle was the central founder and director of NYU-Stern's Volatility Institute which publishes weekly data on systemic risk across countries on its V-LAB site.

As usual with dynamic processes, there is a question about the **initial condition**. There are two general approaches to this. A convenient mathematical approach is to assume that the process started in the infinite past, i.e., $t = 0, \pm 1, \ldots$ in which case we don't need to specify the initial condition. This approach requires stationarity. With this assumption many calculations are easy and simple to state. An alternative approach is to specify some initial condition $\sigma_1^2 > 0$, and then to define the process from this starting point. This approach is well suited to the case where the process is nonstationary.

Suppose also that
$$y_t = \varepsilon_t \sigma_t \tag{11.16}$$
with ε_t i.i.d. mean zero and variance one. Then (11.15) and (11.16) along with the initial condition provide a complete specification of the return process.

11.7.1 Weak Stationarity

We next consider the question of **stationarity**, i.e., whether this process implies that y_t is a stationary process and if so, what the stationary variance is. We first suppose that the process is weakly stationary and see what parameter values are consistent with this assumption. Suppose that $\text{var}(y_t) = \sigma^2 < \infty$ and this does not change with time. Then it must satisfy
$$\sigma^2 = E(\sigma_t^2) = \omega + \gamma E(y_{t-1}^2) = \omega + \gamma \text{var}(y_{t-1}) = \omega + \gamma \sigma^2,$$

Then, provided $|\gamma| < 1$, we obtain
$$\sigma^2 = \frac{\omega}{1 - \gamma}. \tag{11.17}$$

The equation has this unique solution under the stated condition. On the other hand if $\gamma > 1$, then (11.17) would be negative, which is not permissible. Another way of looking at this question is to suppose that the process starts at some given initial value σ_1^2, and to ask what happens to volatility far in the future. Again, provided $|\gamma| < 1$
$$\begin{aligned} E(\sigma_t^2) &= \omega + \gamma E(\sigma_{t-1}^2) \\ &= \omega + \gamma \omega + \gamma^2 E(\sigma_{t-2}^2) \\ &= \omega \sum_{j=1}^{t} \gamma^{j-1} + \gamma^t E(\sigma_1^2) \\ &\to \frac{\omega}{1 - \gamma} \end{aligned}$$

as $t \to \infty$ regardless of $E(\sigma_1^2)$ provided $E(\sigma_1^2) < \infty$. In this case, we say that the process y_t is **asymptotically weakly stationary**. If the initial random variable satisfies $E(\sigma_1^2) = \frac{\omega}{1-\gamma}$, then $E(\sigma_t^2) = E(\sigma_1^2)$ exactly for all t.

We will consider below the distinction between weak and strong stationarity, which is a key issue for this class of models.

11.7.2 Marginal Distribution of Returns

We next ask about the distribution of y_t, or more specifically, we ask about higher cumulants of the process using the same type of logic we used to calculate the variance of y_t. The general finding is that y will be **heavy tailed** even if the innovation ε_t in (11.16) is standard normal.

Theorem 42 *Suppose that ε_t is standard normal, in which case $E(\varepsilon_t^4) = 3E^2(\varepsilon_t^2) = 3$. Suppose that the process y_t is weakly stationary and possesses finite (time invariant) fourth moments. Then, provided $\gamma < 1/\sqrt{3}$, the excess kurtosis is*

$$\kappa_4 = \frac{6\gamma^2}{(1-3\gamma^2)} \geq 0. \qquad (11.18)$$

If $\gamma \geq 1/3^{1/2}$, then $E(y_t^4) = \infty$. The existence of moments is important for the interpretation of the sample correlogram of y_t and y_t^2, and inference about these quantities. Some authors have argued that this restriction on γ is too strong, since we expect $E(y_t^4) < \infty$ more or less and we might expect γ to be large; see Chapter 14 for more discussion. The distribution of y_t is a normal variance mixture with the mixing distribution given by the stationary distribution of the process σ_t^2. Exercise 55 asks you to calculate the condition for weak stationarity and the kurtosis of y_t when ε_t is not Gaussian.

11.7.3 Dependence Property

The next issue is to understand the dependence embodied in the ARCH process. We write

$$y_t^2 = \sigma_t^2 \varepsilon_t^2 = \sigma_t^2 + \sigma_t^2(\varepsilon_t^2 - 1) = \omega + \gamma y_{t-1}^2 + \eta_t,$$

where $\eta_t = y_t^2 - \sigma_t^2 = \sigma_t^2(\varepsilon_t^2 - 1)$ is a mean zero innovation uncorrelated with its past, albeit heteroskedastic, i.e., an MDS. This is an AR(1) process for squared returns. Likewise, we can write

$$\sigma_t^2 = \omega + \gamma \sigma_{t-1}^2 + \gamma \eta_{t-1}$$

with an MDS shock process. It follows that (assuming the process y_t^2 is weakly stationary)

$$\operatorname{corr}(y_t^2, y_{t-j}^2) = \frac{\operatorname{cov}(y_t^2, y_{t-j}^2)}{\operatorname{var}(y_t^2)} = \operatorname{corr}(\sigma_t^2, \sigma_{t-j}^2) = \gamma^j > 0.$$

Squared returns and volatility are both positively dependent. This model generates **volatility clustering**, whereby volatile periods tend to be followed by volatile periods and quiet periods tend to be followed by quiet periods.

11.8 The GARCH Model

Although the ARCH(1) model implies heavy tails and volatility clustering, it does not in practice generate enough of either. The ARCH(p) model for p big does a bit better but at a price in terms of parsimony, which badly effects estimation. There are also many inequality restrictions that should be imposed in order to guarantee positivity of the conditional variance process, but which are hard to impose in practice and which if not imposed can be violated in estimation. Bollerslev (1986) introduced the GARCH(p,q) process.

Definition 98 *GARCH(p,q) for all t*

$$\sigma_t^2 = \omega + \sum_{k=1}^{p} \beta_k \sigma_{t-k}^2 + \sum_{j=1}^{q} \gamma_j y_{t-j}^2. \tag{11.19}$$

This allows for a more parsimonious representation of long ARCH(p); empirically, GARCH(1,1) often does better than ARCH(12), say, and with only three unknown parameters.

Tim Peter Bollerslev (born 1958) is currently the Juanita and Clifton Kreps Professor of Economics at Duke University. Bollerslev is known for his ideas for measuring and forecasting financial market volatility and for the GARCH model. He received his MSc in economics and mathematics in 1983 from the Aarhus University in Denmark. He continued his studies in the US, earning his PhD in 1986 from the University of California at San Diego with a thesis titled "Generalized Autoregressive Conditional Heteroskedasticity with Applications in Finance" written under the supervision of Robert F. Engle (Nobel Prize in Economics winner in 2003). After his graduate studies, Bollerslev taught at the Northwestern University between 1986 and 1995 and at the University of Virginia between 1996 and 1998. Since 1998 he has been the Juanita and Clifton Kreps Professor of Economics at Duke University.

We first discuss the restrictions needed on the parameters to make sure that σ_t^2 is positive with probability one. For the GARCH(1,1) process we require that $\gamma, \beta \geq 0$ and $\omega > 0$, but this is not necessary in higher order models. Nelson and Cao (1992)

11.8 The GARCH Model

work out the necessary and sufficient conditions for the general GARCH(p,q) model. For example, in the GARCH(1,2) model they show that the sufficient conditions are that:

$$\beta, \gamma_1 \geq 0 \quad ; \quad \beta\gamma_1 + \gamma_2 \geq 0. \tag{11.20}$$

This allows $\gamma_2 < 0$ provided the other parameters compensate. To some extent this just makes life more complicated because writing down these inequalities that characterize the positivity region is very tedious; to impose this in estimation is challenging to say the least.

We next consider the issue of weak stationarity in the GARCH(p,q) model.

Theorem 43 *The process y_t is weakly stationary and has finite unconditional variance*

$$\sigma^2 = \mathrm{var}(y_t) = E(\sigma_t^2) = \frac{\omega}{1 - \sum_{k=1}^{p} \beta_k - \sum_{j=1}^{q} \gamma_j},$$

provided

$$\sum_{k=1}^{p} \beta_k + \sum_{j=1}^{q} \gamma_j < 1. \tag{11.21}$$

Under some conditions we can represent GARCH(p,q) processes as an infinite discounted sum.

Definition 99 *ARCH(∞) model. Suppose that*

$$\sigma_t^2 = \psi_0 + \sum_{j=1}^{\infty} \psi_j y_{t-j}^2, \tag{11.22}$$

where ψ_j satisfy $\sum_{j=1}^{\infty} |\psi_j| < \infty$.

The GARCH(p,q) model is the special case of (11.22) where ψ_j depend only on the β, γ, ω parameters. In the GARCH(1,1) process it is easy to see that (assuming an infinite past) (11.22) holds with: $\psi_0 = \omega/(1 - \beta)$, $\psi_j = \gamma\beta^{j-1}$, $j = 1, 2, \ldots$, where the sum is well defined provided that $0 \leq \beta < 1$.

In practice, estimated parameters often lie close to the boundary of the region defined by (11.21). In Table 11.4 we show estimates of the GARCH(1,1) model for daily S&P500 stock index return series from 1950–2018. The computations were carried out in `Eviews`.

We show in Figure (11.11) the conditional standard deviation for the daily series, which shows the peak value in October 1987. In Figure (11.12) we show the standardized residuals $\hat{\varepsilon}_t$ from the estimated model. This shows that the model has reduced the maximum achieved negative value from -21 to around -11. Although this is a big reduction, it still is hard to rationalize within a normal distribution, since such values would be extremely unlikely to occur.

Chapter 11 Volatility

Table 11.4 GARCH(1,1) parameter estimates

	Daily	Weekly	Monthly
ρ_1	0.0893 (0.00807)	0.002871 (0.021709)	0.020311 (0.049513)
ρ_2	−0.0253 (0.00784)	0.032247 (0.022462)	−0.058747 (0.045839)
ω	8.55E−07 (6.57E−08)	1.14E−05 (2.50E−06)	0.000103 (4.63E−05)
β	0.9094 (0.00237)	0.845708 (0.014920)	0.870139 (0.040540)
γ	0.08293 (0.00186)	0.131007 (0.012950)	0.074298 (0.027528)

Note: Standard errors in parentheses. These estimates are for the raw data series and refer to the AR(2)-GARCH(1,1) model: $y_t = c + \rho_1 y_{t-1} + \rho_2 y_{t-2} + \overbrace{\varepsilon_t \sigma_t}^{u_t}$ and $\sigma_t^2 = \omega + \beta \sigma_{t-1}^2 + \gamma u_{t-1}^2$

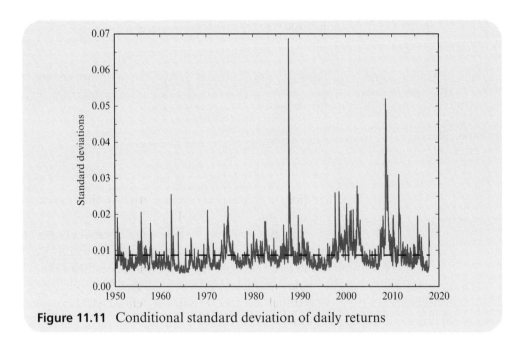

Figure 11.11 Conditional standard deviation of daily returns

Bampinas, Ladopoulos, and Panagiotidis (2018) estimated the GARCH(1,1) (with a constant mean) using daily data from January 2008 to December 2011 for all the constituents of the S&P1500. They found mean values for $\beta = 0.878$ and $\gamma = 0.114$ with cross-sectional standard deviations of 0.079 and 0.056 respectively, suggesting that there is quite a variation across smaller stocks in the values taken by these two key parameters, but nevertheless $\beta + \gamma$ is often quite close to one.

We next discuss a special case of the GARCH process where (11.21) fails.

11.8 The GARCH Model

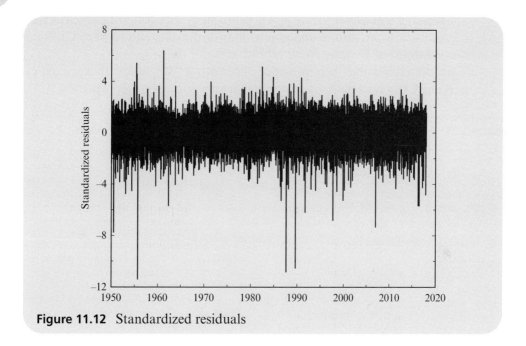

Figure 11.12 Standardized residuals

Definition 100 *The IGARCH(1,1) model satisfies $\beta + \gamma = 1$, i.e., for $\beta \in (0,1)$*

$$\sigma_t^2 = \omega + \beta \sigma_{t-1}^2 + (1-\beta) y_{t-1}^2.$$

This violates the conditions for covariance stationarity. In this case the unconditional variance of returns is infinite and the conditional variance converges to infinity with time horizon. However, returns can still be a strongly stationary process, as we discuss more formally below. We can also show that returns are weakly dependent in this case (in contrast to the unit root case for linear time series). Note the difference between conditional variance and unconditional variance. The process y_t has infinite variance but the conditional variance σ_t^2 is a well-defined, stationary, and weakly dependent process. This means that

$$\frac{1}{\sqrt{T}} \sum_{t=1}^{T} y_t - E(y_t) \tag{11.23}$$

does not obey a CLT, and so many standard tests based on the sample mean or correlogram are not valid. However, for any function λ such that $E(\lambda(y_t)^2) < \infty$, we have the CLT

$$\frac{1}{\sqrt{T}} \sum_{t=1}^{T} (\lambda(y_t) - E(\lambda(y_t))) \Longrightarrow N(0, V), \tag{11.24}$$

where V is some finite variance. There are many robust statistics such as the quantilogram that work with bounded functions of the data and so obey CLTs under these conditions. In Exercise 56 we explore some of the properties of IGARCH processes including how they behave under differencing. We close with a special case of the IGARCH model (with no intercept) that has some special properties.

Example 58 *A special case of the IGARCH model is called* **the Riskmetrics model** *(also known as the EWMA). Suppose that*

$$y_t = \sigma_t \varepsilon_t$$
$$\sigma_t^2 = \beta \sigma_{t-1}^2 + (1-\beta) y_{t-1}^2,$$

which is an IGARCH process but with no intercept. We can also write this as

$$\sigma_t^2 = \sigma_{t-1}^2 \left[\beta + (1-\beta)\varepsilon_{t-1}^2\right],$$

so that $\log \sigma_t^2$ *is a random walk with i.i.d. innovation, i.e.,*

$$\log \sigma_t^2 = \log \sigma_{t-1}^2 + \eta_{t-1}, \qquad \text{where } \eta_{t-1} = \log(\beta + (1-\beta)\varepsilon_{t-1}^2).$$

The properties of this process depend on the expectation of η_{t-1}. *Specifically:*

(1) *If* $E(\eta_{t-1}) > 0$, *then* $\log \sigma_t^2$ *is a random walk with positive drift and so* $\sigma_t^2 \to \infty$ *with probability one;*

(2) *If* $E(\eta_{t-1}) < 0$, *then* $\log \sigma_t^2$ *is a random walk with negative drift and so* $\sigma_t^2 \to 0$ *with probability one;*

(3) *If* $E(\eta_{t-1}) = 0$, *then* $\log \sigma_t^2$ *is a driftless random walk (and so recurrent; see Chapter 12). In any of cases,* $\log \sigma_t^2$ *and hence* σ_t^2 *is nonstationary.*

If we assume that $E(\varepsilon_t^2) = 1$, then by Jensen's inequality

$$E(\eta_{t-1}) = E\left[\log(\beta + (1-\beta)\varepsilon_{t-1}^2)\right]$$
$$< \log(\beta + (1-\beta)E(\varepsilon_{t-1}^2))$$
$$= \log(\beta + (1-\beta)) = 0,$$

so that $\log \sigma_t^2$ is a random walk with negative drift and so $\sigma_t^2 \to 0$ with probability one. However, the process y_t is weakly stationary since for all t

$$E\left[\sigma_t^2\right] = \prod_{s=1}^{t} E\left[\beta + (1-\beta)\varepsilon_s^2\right] E\left[\sigma_0^2\right] = E\left[\sigma_0^2\right].$$

This shows some of the strange things that can happen with nonlinear processes.

11.8 The GARCH Model

11.8.1 Strong Stationarity

We now formally consider the issue of strong stationarity. Consider the GARCH(1,1) process

$$y_t = \sigma_t \varepsilon_t, \quad \sigma_t^2 = \omega + \beta \sigma_{t-1}^2 + \gamma y_{t-1}^2 \tag{11.25}$$

with ε_t i.i.d. nondegenerate, and $\beta \geq 0$ and $\omega, \gamma > 0$. We usually assume that $E(\varepsilon_t) = 0$ and $\text{var}(\varepsilon_t) = 1$ so that σ_t^2 is interpreted as a conditional variance. In that case, as we have seen, necessary and sufficient conditions for weak stationarity are that $\beta + \gamma < 1$. We next provide the conditions for strong stationarity. These just assume that ε_t is i.i.d. and do not specify its moments.

Theorem 44 *(Nelson (1990a)) Let σ_t^2 be initialized at $t=0$ with some random value $\sigma_0^2 > 0$, and suppose that*

$$E\left[\log(\beta + \gamma \varepsilon_t^2)\right] < 0.$$

Then:

(1) The process

$$_u\sigma_t^2 = \omega \left(1 + \sum_{j=1}^{\infty} \prod_{i=1}^{j} \left(\gamma \varepsilon_{t-i}^2 + \beta\right)\right)$$

is strictly stationary;

(2) $_u\sigma_t^2 \in [\frac{\omega}{1-\beta}, \infty)$;

(3) as $t \to \infty$, $\sigma_t^2 - {}_u\sigma_t^2 \to 0$ with probability one; and

(4) the distribution of σ_t^2 converges, as $t \to \infty$, to the non-degenerate and well-defined distribution of $_u\sigma_t^2$.

When $E(\varepsilon_t^2) = 1$, we have by Jensen's inequality

$$E\left(\log(\beta + \gamma \varepsilon_t^2)\right) < \log\left(E(\beta + \gamma \varepsilon_t^2)\right) = \log(\beta + \gamma).$$

Therefore, it can be that $E(\log(\beta + \gamma \varepsilon_t^2)) < 0$ even when $\beta + \gamma \geq 1$. In the ARCH(1) Gaussian case, we have strong stationarity provided only that $\gamma < 3.5$. That is, in the region $1 \leq \gamma < 3.5$, the process y_t is strongly stationarity but not weakly stationarity.

Example 59 *Suppose that (11.25) holds and that ε_t has the Cauchy distribution (in which case $E(\varepsilon_t^2) = \infty$). Then, we have*

$$E\left(\log(\beta + \gamma \varepsilon_t^2)\right) = 2\log(\beta^{1/2} + \gamma^{1/2}).$$

In this case, the necessary and sufficient conditions for stationarity are $\beta^{1/2} + \gamma^{1/2} < 1$, which restrict the parameter space more than the conditions $\beta + \gamma < 1$ that are required for weak stationarity. In this case, not even the conditional variance is defined, but the process σ_t^2 is well defined and is strongly stationary under these conditions. In that case we might interpret σ_t^2 as a **dynamic scale process** *and ignore the variance terminology altogether.*

11.8.2 Dependence Properties and Forecasting

GARCH processes are weakly dependent under some conditions, meaning that their dependence on the past dies out relatively rapidly. We write

$$\sigma_t^2 = \omega + (\beta + \gamma)\sigma_{t-1}^2 + \gamma\eta_{t-1} \quad (11.26)$$

$$y_t^2 = \sigma_t^2 \varepsilon_t^2 = \sigma_t^2 + \sigma_t^2(\varepsilon_t^2 - 1) = \omega + (\beta + \gamma)y_{t-1}^2 + \eta_t - \beta\eta_{t-1},$$

where $\eta_t = y_t^2 - \sigma_t^2 = \sigma_t^2(\varepsilon_t^2 - 1)$ is an MDS. Therefore, σ_t^2 follows an AR(1) process while squared returns follow an ARMA(1,1) process, both with heteroskedastic innovations. Provided the fourth order moments of y_t are uniformly bounded, it follows that for some constant c

$$\text{cov}(y_t^2, y_{t-k}^2) \leq c(\beta + \gamma)^{k-1}.$$

Therefore, provided $\beta + \gamma < 1$ (plus the restrictions needed for the fourth moments to be bounded), the covariance function of y_t^2 decays exponentially fast. To forecast future volatility, one can use the AR representation (11.26). Specifically,

$$\sigma_{T+j+1|T}^2 = \omega^* + (\beta + \gamma)^j \sigma_{T+1}^2,$$

where $\omega^* = \omega(1 + (\beta + \gamma) + \cdots + (\beta + \gamma)^{j-1})$. The value σ_{T+1}^2 can be expressed in terms of $y_T^2, y_{T-1}^2, \ldots$.

Mikosch and Starica (2005) argue that the covariance function is not so useful for the GARCH type of processes, because the theory for sample autocovariance functions typically requires moments of fourth order to exist. Instead one may consider mixing measures of dependence such as strong mixing or beta mixing.

Theorem 45 *Suppose that ε_t is a sequence of i.i.d. real-valued random variables with mean zero and variance one. Suppose that ε_t has a continuous density (with respect to Lebesgue measure on real line), and its density is positive on $\mathbb{R}$. Then, a sufficient condition for the GARCH(1,1) process y_t be β-mixing with exponential decay, i.e., for some $\rho < 1$*

$$\beta(k) \leq c\rho^k$$

for all k, is that the process is weakly stationary, i.e., $\beta + \gamma < 1$ (Carrasco and Chen (2002, Corollary 6)).

Actually it has been shown that the IGARCH process is also mixing with exponential decay under some mild additional conditions (Meitz and Saikkonnen (2008)) so that the dependence property is not coupled with the moment existence (i.e., the weak stationarity condition).

11.8 The GARCH Model

11.8.3 Interpretation of the Model

For the purposes of understanding and for the derivation of properties like stationarity it is convenient to assume that ε_t is i.i.d. But for other purposes, such as for estimation, one often only requires conditional moment specifications, i.e., we might not require the innovation to be i.i.d. Drost and Nijman (1993) propose a classification of GARCH models according to the properties of the innovation ε_t.

Definition 101 *Strong GARCH.* ε_t *is i.i.d. with*

$$E(\varepsilon_t) = 0 \text{ and } E(\varepsilon_t^2) = 1.$$

This is a full model specification and is useful for deriving properties like stationarity and mixing. As we have seen, this class of processes generates heavy tailed distributions for the unconditional distribution of y_t. However, it restricts all conditional cumulants of y_t given past information $\mathcal{F}_{t-1}$ to be constant and indeed to be the cumulants of ε_t. That is

$$\kappa_j(y_t|\mathcal{F}_{t-1}) = \kappa_j(\varepsilon_t)$$

for all j for which this is well defined.

Definition 102 *Semi-strong GARCH.* ε_t *satisfies*

$$E(\varepsilon_t|\mathcal{F}_{t-1}) = 0 \text{ and } E(\varepsilon_t^2|\mathcal{F}_{t-1}) = 1.$$

This is not a full model specification. Under this condition it follows that

$$E(r_t|\mathcal{F}_{t-1}) = 0 \text{ and } E(r_t^2|\mathcal{F}_{t-1}) = \sigma_t^2,$$

that is, the quantity σ_t^2 is truly the conditional variance. However, this specification does not restrict, for example, the third and fourth conditional cumulants of the return process; the return process could have an asymmetric and heavy tailed distribution, which could vary over time. For example $E\left(y_t^4|\mathcal{F}_{t-1}\right) = E\left(\varepsilon_t^4|\mathcal{F}_{t-1}\right)\sigma_t^4$, and so the excess conditional kurtosis satisfies

$$\kappa_4(y_t|\mathcal{F}_{t-1}) = E\left(\varepsilon_t^4|\mathcal{F}_{t-1}\right) - 3 = \kappa_4(\varepsilon_t|\mathcal{F}_{t-1}),$$

which may vary over time. In practice, this can be important. The class of semi-strong GARCH processes is much larger than the class of strong GARCH processes, which allows it more flexibility in fitting the data.

Finally, we may consider the following.

Definition 103 *Weak GARCH. Suppose that y_t is mean zero and uncorrelated and*

$$\sigma_t^2 = E_L(y_t^2|I_{t-1}),$$

where $I_{t-1} = \{1, y_{t-1}, y_{t-2}, \ldots, y_{t-1}^2, y_{t-2}^2, \ldots\}$, and E_L denotes best linear projection on this set.

This class of processes is even larger than the class of semi-strong processes. In this case there are no restrictions on $\kappa_j(y_t|\mathcal{F}_{t-1})$ whatsoever.

Drost and Nijman (1993) show that conventional strong and semi-strong GARCH processes are not closed under temporal aggregation. Specifically, suppose that one has a GARCH(1,1) model for daily data, does this imply that the monthly data follows some GARCH(p,q) process? They showed that the answer is no, even when the definition of the process is semi-strong. The weak GARCH class is closed under temporal aggregation. This is in contrast with the corresponding results for linear time series models, where aggregation preserves model structure. This says that this model is essentially frequency specific. Continuous time models, which we will consider in the next chapter, have the feature that one has a prediction for the model structure for observed time series whatever the frequency of the data.

11.8.4 News Impact Curve

The GARCH class of models has imposed a very specific functional form in terms of how past values of y_t affect σ_t^2. We consider in this section whether this is appropriate and what alternative specifications have been proposed.

Definition 104 *The news impact curve is the relationship between σ_t^2 and $y_{t-1} = y$ holding past values σ_{t-1}^2 constant at some level σ^2.*

Example 60 *For the GARCH process, the news impact curve is*

$$m(y, \sigma^2) = \omega + \gamma y^2 + \beta \sigma^2.$$

This embodies three properties; it is:
(1) a separable function of σ^2, meaning $\partial^2 m(y, \sigma^2)/\partial y \partial \sigma^2 = 0$;
(2) an even function of news y, $m(y, \sigma^2) = m(-y, \sigma^2)$; and
(3) a quadratic function of y.

The standard GARCH process does not allow asymmetric news impact curves: good news and bad news have the same effect on volatility. The martingale hypothesis says that $\text{cov}(y_t, y_{t-j}^2) = 0$ for $j = 1, 2, \ldots$ The standard GARCH model with symmetrically distributed errors (such as standard normal) implies that $\text{cov}(\sigma_t^2, y_{t-j}) = 0$ and

11.9 Asymmetric Volatility Models and Other Specifications

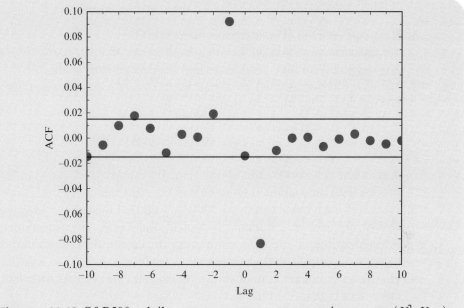

Figure 11.13 S&P500 daily return cross autocovariance $\text{cov}(Y_t^2, Y_{t-j})$, $j = -10, \ldots, 10$

$\text{cov}(y_t^2, y_{t-j}) = 0$ for $j = 1, 2, \ldots$. However, there is strong empirical evidence for a negative correlation between y_t^2 and y_{t-j} and a positive correlation between y_t^2 and y_{t+j}. In Figure 11.13 we show this for the S&P500 daily return series

There are two alternative explanations for this association:

1. **Leverage hypothesis**: negative returns lower equity price thereby increasing corporate leverage, thereby increasing equity return volatility.
2. **Volatility-feedback hypothesis**: an anticipated increase in volatility would raise the required rate of return, in turn necessitating an immediate stock-price decline to allow for higher future returns.

In the next section we discuss models that capture the leverage effect.

11.9 Asymmetric Volatility Models and Other Specifications

Definition 105 *Nelson (1991) introduced the EGARCH(p,q) process. Let $h_t = \log \sigma_t^2$ and*

$$h_t = \omega + \sum_{j=1}^{p} \gamma_j \left(\delta \varepsilon_{t-j} + \theta \left| \varepsilon_{t-j} \right| \right) + \sum_{k=1}^{q} \beta_k h_{t-k},$$

where $\varepsilon_t = \frac{y_t - \mu_t}{\sigma_t}$ is i.i.d.

Chapter 11 Volatility

This paper contained four innovations:

1. He models the log of volatility not the level. There are therefore no problems with parameter restrictions. On the other hand **inliers**, that is, observations for which y_t are very small (so $\log(y_t)$ are large and negative) are influential.
2. He allows asymmetric effect of past shocks ε_{t-j} on current volatility when $\delta \neq 0$. His model has the property that

$$\operatorname{cov}(y_t^2, y_{t-j}) \neq 0$$

even when ε_t is symmetric about zero. By comparison, the standard GARCH process with innovation ε_t symmetric about zero, has $\operatorname{cov}(y_t^2, y_{t-j}) = 0$ for all j.
3. He makes the innovations to volatility ε_{t-j} i.i.d. This ensures that h_t is a linear process. Strong and weak stationarity coincide for h_t. The downside of this assumption is that distribution theory is quite tricky because of the repeated exponential and logarithmic transformations involved.
4. He allows leptokurtic innovations. Specifically, he worked with the generalized error distribution (GED)

$$f(\varepsilon) = \frac{\nu \exp\left(-\frac{1}{2}|\varepsilon/\lambda|^\nu\right)}{\lambda 2^{(1+1/\nu)} \Gamma(1/\nu)}, \quad \lambda = \left(2^{(-2/\nu)} \Gamma(1/\nu)/\Gamma(3/\nu)\right)^{1/2},$$

where Γ is the **Gamma function**, that is $\Gamma(x) = \int_0^\infty s^{x-1} e^{-s} ds$. The GED family of errors includes the normal ($\nu = 2$), uniform ($\nu = \infty$), and Laplace ($\nu = 1$) as special cases. The distribution is symmetric about zero for all ν, and has finite second moments for $\nu > 1$. For this density

$$E(|\varepsilon_t|) = \frac{\lambda 2^{1/\nu} \Gamma(2/\nu)}{\Gamma(1/\nu)}.$$

In Table 11.5 we show estimates of the EGARCH model obtained via Eviews. For all three frequencies the parameter δ is significant.

An alternative approach with only some of the above features is the Glosten, Jagannathan, and Runkle (1993) (GJR) model.

Definition 106 *GJR GARCH(1,1). For all t*

$$\sigma_t^2 = \omega + \beta_1 \sigma_{t-1}^2 + \gamma y_{t-1}^2 + \delta y_{t-1}^2 1(y_{t-1} < 0). \tag{11.27}$$

This model is similar to the GARCH process, except that the news impact curve is allowed to be asymmetric when $\delta \neq 0$. In Table 11.6 we show estimates of this model.

In Figure 11.14 we compare the estimated news impact curves of the GARCH model and the GJR asymmetric GARCH model.

11.9 Asymmetric Volatility Models and Other Specifications

Table 11.5 Estimated EGARCH model

	Daily	Weekly	Monthly
ρ_1	0.119808 (0.008811)	0.017711 (0.021473)	−0.018609 (0.046477)
ρ_2	−0.034664 (0.008938)	0.035808 (0.021383)	−0.034550 (0.043348)
ω	−0.236662 (0.018375)	−0.520846 (0.081954)	−0.847544 (0.486045)
β	0.986377 (0.001680)	0.956470 (0.008808)	0.889015 (0.070006)
γ	0.136142 (0.006927)	0.215399 (0.027070)	0.172956 (0.093206)
δ	−0.064461 (0.004093)	−0.101905 (0.015014)	−0.091206 (0.047421)
ν	1.377342 (0.013240)	1.700196 (0.060307)	1.469910 (0.092301)

Note: Standard errors in parentheses. These estimates are for the S&P500 data series and refer to the AR(2)-EGARCH(1) model $y_t = c + \rho_1 y_{t-1} + \rho_2 y_{t-2} + \varepsilon_t \sigma_t$ and $\log \sigma_t^2 = \omega + \beta \sigma_{t-1}^2 + \gamma |\varepsilon_{t-1}| + \delta \varepsilon_{t-1}$.

Table 11.6 Estimation of asymmetric GJR GARCH model

	Daily	Weekly	Monthly
ρ_1	0.138788 (0.009524)	0.007065 (0.022000)	0.014661 (0.045131)
ρ_2	−0.01906 (0.009449)	0.051815 (0.022044)	−0.018694 (0.045083)
$\omega(\times 1000)$	0.0000721 (0.0000064)	0.00130 (0.000242)	0.862000 (0.249000)
β	0.920489 (0.002243)	0.850348 (0.015580)	0.442481 (0.176365)
γ	0.034018 (0.002613)	0.047885 (0.013504)	−0.076662 (0.042047)
δ	0.078782 (0.003302)	0.140013 (0.020349)	0.266916 (0.094669)

Note: Standard errors in parentheses. These estimates are for the S&P500 data series and refer to the AR(2)-GJR GARCH(1,1) model $y_t = c + \rho_1 y_{t-1} + \rho_2 y_{t-2} + \varepsilon_t \sigma_t$ and $\sigma_t^2 = \omega + \beta \sigma_{t-1}^2 + \gamma u_{t-1}^2 + \delta u_{t-1}^2 1(u_{t-1} < 0)$.

11.9.1 Other Variations on the GARCH Model

There have been a proliferation of different models since the original papers, too many to list here. Hansen and Lunde (2005) compare the GARCH(1,1) model with some of these other models according to forecast performance. The comparison is made using realized volatility as the target. They looked at daily IBM stock returns and exchange rates. They find that for exchange rates the GARCH(1,1) model is the best,

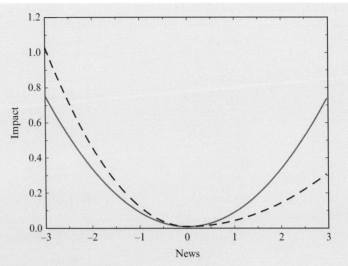

Figure 11.14 Comparison of the estimated news impact curves from GARCH(1,1) and GJR(1,1) for daily S&P500 returns

whereas for IBM stock returns, models that account for leverage such as the GJR do better.

11.10 Mean and Variance Dynamics

We next consider the joint specification of the mean and the variance, i.e., we allow μ_t in (11.14) to be dynamically specified. A common specification is to assume that the mean is an ARMA process with heteroskedastic innovations. Suppose that

$$A(L)y_t = u_t, \quad u_t = \sigma_t \varepsilon_t, \tag{11.28}$$

where ε_t are i.i.d. with $E(\varepsilon_t) = 0$ and $E(\varepsilon_t^2) = 1$, while $A(L) = 1 + \sum_{j=1}^{\infty} a_j L^j$ are lag polynomials with $Ly_t = y_{t-1}$. The AR(∞) class of processes includes both finite order AR and ARMA processes. These models can arise under efficient markets hypothesis combined with some market microstructure issues. We suppose that σ_t^2 depends only on past values in which case

$$E(y_t | \mathcal{F}_{t-1}) = \sum_{j=1}^{\infty} a_j y_{t-j}$$

$$\text{var}(y_t | \mathcal{F}_{t-1}) = \sigma_t^2.$$

The question is, how to specify σ_t^2? There are two options:

1. In Approach 1 we specify a GARCH process where the innovation is the square of the mean zero process u_t, i.e., we write

$$\sigma_t^2 = \omega + \beta \sigma_{t-1}^2 + \gamma u_{t-1}^2. \tag{11.29}$$

11.10 Mean and Variance Dynamics

This is the usual approach implemented in Eviews.
2. In Approach 2 we specify a GARCH process where the innovation is the return process y_t, i.e., we write

$$\sigma_t^2 = \omega + \beta \sigma_{t-1}^2 + \gamma y_{t-1}^2. \tag{11.30}$$

In Approach 1, it is easy to establish weak stationarity for the process u_t by directly applying the results obtained above. We may write

$$u_{t-1}^2 = \left(\sum_{j=1}^{\infty} a_j y_{t-1-j} \right)^2,$$

which shows that in terms of y the volatility process in Approach 1 will be complicated. This is the approach implemented in most softwares. In Approach 2, the innovation is quite complicated, since

$$y_{t-1}^2 = (B(L) u_{t-1})^2,$$

where $B(L) = A(L)^{-1}$, so stationarity and so forth is not so obvious. On the other hand it is very easy to implement estimation in this model. Does it matter? The mean effect is usually quite small but not zero (Lumsdaine and Ng (1999)).

11.10.1 GARCH in Mean

We may expect that the mean and the variance process are related. This is captured in the GARCH in mean model, Engle, Lilien and Robbins (1987).

Definition 107 *Suppose that*

$$y_t = g(\sigma_t^2; \alpha) + \varepsilon_t \sigma_t, \tag{11.31}$$

where g is a measurable function, and σ_t^2 is a GARCH process depending on lagged squared y_t or lagged squared u_t as discussed above. Here, α are unknown parameters (to be estimated along with the parameters of σ_t^2). Specifically: $g(\sigma^2; \alpha) = \alpha_0 + \alpha_1 \sigma^2$, $g(\sigma^2; \alpha) = \alpha_0 + \alpha_1 \sigma$, and $g(\sigma^2; \alpha) = \alpha_0 + \alpha_1 \log \sigma^2$ are common choices available in EVIEWS.

This model captures the idea that risk and return should be related. Merton (1973) in continuous time partial equilibrium model shows that

$$E\left((r_{mt} - r_{ft}) | \mathcal{F}_{t-1} \right) = \kappa \mathrm{var}\left((r_{mt} - r_{ft}) | \mathcal{F}_{t-1} \right), \tag{11.32}$$

where r_{mt}, r_{ft} are the returns on the market portfolio and risk-free asset respectively. In this case, κ is the Arrow–Pratt measure of relative risk aversion. The Bansal and Yaron asset pricing model delivers a similar formula for the market risk premium; see Chapter 10.

Chapter 11 Volatility

Table 11.7 Estimates of GARCH in mean model

	Daily	Weekly	Monthly
α	0.081504 (0.029699)	0.121757 (0.076905)	0.415873 (0.327167)
ω	6.49E−07 (7.48E−08)	1.13E−05 (2.53E−06)	0.000125 (0.072803)
β	0.916160 (0.002356)	0.846601 (0.014707)	0.858988 (0.044015)
γ	0.079801 (0.001737)	0.130387 (0.012697)	0.072803 (0.027614)

Note: Standard errors in parentheses. These estimates are for the S&P500 data series and refer to the GARCH(1,1) in mean model $y_t = c + \alpha \sigma_t + \varepsilon_t \sigma_t$ and $\sigma_t^2 = \omega + \beta \sigma_{t-1}^2 + \gamma u_{t-1}^2$.

The conditional mean $E(y_t|\mathcal{F}_{t-1})$ is now a complicated nonlinear function of $y_{t-1}, y_{t-2}, \ldots$ although in the case $g(\sigma^2; \alpha) = \alpha_0 + \alpha_1 \sigma^2$, where σ_t^2 is GARCH(1,1), it follows that

$$\text{cov}(y_t, y_{t-j}) = \kappa \text{cov}(\sigma_t^2, \sigma_{t-j}^2) = \kappa'(\beta + \gamma)^j$$

for some positive constant κ'.

The GARCH in mean model can be estimated in Eviews. Some authors find small but significant effects. In Table 11.7 we report results of estimating some GARCH in mean specifications on the S&P500 daily return data over the period 1950–2002.

11.10.2 Covariate Effects

Suppose that there are covariates x_t that effect the volatility of returns such as macroeconomic conditions. We consider **component models** of the form

$$y_t = g(x_t)^{1/2} v_t \varepsilon_t, \tag{11.33}$$

where ε_t is the usual error (i.i.d. or MDS), $g(x_t)$ is the covariate effect, and v_t^2 is the purely dynamic volatility component such as

$$v_t^2 = \omega + \beta v_{t-1}^2 + \gamma u_{t-1}^2,$$

where $u_t = v_t \varepsilon_t = y_t / g(x_t)$. For example

$$g(x) = \exp\left(\pi^\top x\right),$$

where π are unknown parameters. In this case, $\text{var}(y_t|\mathcal{F}_{t-1}, x_t) = \exp(\pi^\top x_t) v_t^2$, and the covariates affect the level of volatility. It seems reasonable to assume for identification purposes that $E(v_t^2|x_t)$ does not vary with x_t, in which case

$$E(y_t^2 - \exp\left(\pi^\top x_t\right) c | x_t) = 0$$

for some constant c, which is a conditional moment restriction that can be used to estimate the parameters π, c. Once g is estimated, one can work with the rescaled returns $y_t/g(x_t)^{1/2}$, which follow a pure GARCH(1,1) process, and estimate the dynamic parameters $\theta = (\omega, \beta, \gamma)^\mathsf{T}$ using the methods described below.

11.11 Estimation of Parameters

11.11.1 Gaussian Likelihood

The usual method of parameter estimation is based on the Gaussian Likelihood. Suppose that

$$y_t = \mu_t(\theta) + \sigma_t(\theta)\varepsilon_t,$$

where $\varepsilon_t \sim N(0,1)$ and $\mu_t(\theta), \sigma_t(\theta)$ depend only on past observations $\mathcal{F}_{t-1} = \{y_{t-1}, y_{t-2}, \ldots\}$, and $\theta \in \Theta \subset \mathbb{R}^p$ is a vector of unknown parameters. Then the conditional distribution of y_t is normal

$$y_t | \mathcal{F}_{t-1} \sim N(\mu_t, \sigma_t^2).$$

Therefore, by the **prediction error decomposition** the sample log likelihood (conditional on the first observation) is (up to a constant that does not depend on parameter values)

$$\ell(\theta) = \sum_{t=2}^{T} \ell_t(\theta), \quad \ell_t(\theta) = -\frac{1}{2}\log \sigma_t^2(\theta) - \frac{1}{2}\left(\frac{y_t - \mu_t(\theta)}{\sigma_t(\theta)}\right)^2, \tag{11.34}$$

where $\sigma_t^2(\theta)$ and perhaps $\mu_t(\theta)$ are built up by recursions from some starting values, which we discuss below. We define $\hat{\theta}$ as the value of θ that maximizes $\ell(\theta)$ over the parameter space Θ.

We next discuss the recursions needed to compute $\sigma_t^2(\theta)$. Specifically, in the pure GARCH(1,1) case we have $\theta = (\omega, \beta, \gamma)^\mathsf{T}$ and

$$\sigma_t^2(\theta) = \omega + \beta \sigma_{t-1}^2(\theta) + \gamma y_{t-1}^2, \quad t = 2, \ldots, T.$$

We can compute these recursions for all parameter values θ, given some starting values. There are several approaches to this for the GARCH(1,1) model:

1. $\sigma_1^2(\theta) = \frac{\omega}{1-\beta-\gamma}$;
2. $\sigma_1^2 = \frac{1}{T}\sum_{t=1}^{T} y_t^2$;
3. $\sigma_1^2 = y_1^2$; and
4. Treat σ_1^2 as an unknown parameter to estimate along with θ.

The first approach imposes weak stationarity in the sense that the initial value is taken from the implied mean of the volatility. The second approach does not impose anything, except that in the absence of weak stationarity the random variable σ_1^2 may diverge with sample size. The third approach is arbitrary but does not impose anything with regard to stationarity or otherwise. Likewise the final approach does not impose anything with regard to stationarity, but on the other hand requires finding another parameter in the optimization.

We have seen that there are some restrictions on value of the parameters in order that $\sigma_t^2(\theta) \geq 0$, in which case the parameter space Θ is restricted by the inequality restrictions. Also, in some cases one might want to impose weak or strong stationarity, which yields additional inequality restrictions. The usual algorithms for doing the optimization ignore the inequality restrictions. Some use analytic expression for the first and/or second derivatives of the log likelihood, and some use numerical derivatives in some places. Software exists on the web to compute the estimates in a number of languages. EVIEWS uses the **Marquadt algorithm** as default. Brooks, Burke, and Persand (2001) discuss different software and how they implement the maximization with some surprising conclusions!

In the pure GARCH case the likelihood derivatives are defined below in a recursive fashion

$$\frac{\partial \ell_t}{\partial \theta}(\theta) = -\frac{1}{2}\left(\varepsilon_t^2(\theta) - 1\right)\frac{\partial \log \sigma_t^2(\theta)}{\partial \theta}, \quad \varepsilon_t^2(\theta) = \frac{y_t^2}{\sigma_t^2(\theta)}$$

$$\frac{\partial^2 \ell_t}{\partial \theta \partial \theta^\top}(\theta) = -\frac{1}{2}\left(\varepsilon_t^2(\theta) - 1\right)\frac{\partial^2 \log \sigma_t^2(\theta)}{\partial \theta \partial \theta^\top} + \frac{1}{2}\varepsilon_t^2(\theta)\frac{\partial \log \sigma_t^2(\theta)}{\partial \theta}\frac{\partial \log \sigma_t^2(\theta)}{\partial \theta^\top}$$

$$\frac{\partial \log \sigma_t^2(\theta)}{\partial \theta} = \frac{1}{\sigma_t^2(\theta)}\frac{\partial \sigma_t^2(\theta)}{\partial \theta},$$

where for $t = 2, \ldots, T$ we define recursively (from some starting values):

$$\frac{\partial \sigma_t^2(\theta)}{\partial \omega} = 1 + \beta\frac{\partial \sigma_{t-1}^2(\theta)}{\partial \omega}; \quad \frac{\partial \sigma_t^2(\theta)}{\partial \beta} = \sigma_{t-1}^2(\theta) + \beta\frac{\partial \sigma_{t-1}^2(\theta)}{\partial \beta}; \quad \frac{\partial \sigma_t^2(\theta)}{\partial \gamma} = y_{t-1}^2 + \beta\frac{\partial \sigma_{t-1}^2(\theta)}{\partial \gamma}.$$

One can expect the MLE to be consistent, asymptotically normal, and even efficient under the full specification. In fact, this estimator is also consistent and asymptotically normal under weaker conditions, specifically that the conditional mean and the conditional variance are correctly specified (i.e., semi-strong GARCH), but the conditional normality of the error distribution is not required. In this more general setting the estimation procedure is sometimes referred to as the **Quasi Maximum Likelihood Estimator** or QMLE. The large sample results for the QMLE are stated in quite general terms in Bollerslev and Wooldridge (1992).

Theorem 46 *Under the conditions of Bollerslev and Wooldridge (1992),*

$$T^{1/2}\left(\widehat{\theta} - \theta_0\right) \Longrightarrow N(0, \mathcal{J}^{-1}\mathcal{I}\mathcal{J}^{-1})$$

$$\mathcal{J} = E\left(\frac{1}{T}\frac{\partial \ell_T^2(\theta_0)}{\partial \theta \partial \theta^\top}\right) \quad ; \quad \mathcal{I} = E\left(\frac{\partial \ell_t}{\partial \theta}\frac{\partial \ell_t}{\partial \theta^\top}(\theta_0)\right).$$

11.11 Estimation of Parameters

If ε_t is i.i.d., $\mathcal{I} \propto \mathcal{J}$. In particular

$$\mathcal{I} = \frac{1}{4} E\left((\varepsilon_t^2 - 1)^2\right) E\left(\frac{\partial \log \sigma_t^2(\theta_0)}{\partial \theta} \frac{\partial \log \sigma_t^2(\theta_0)}{\partial \theta^\top}\right)$$

$$\mathcal{J} = \frac{1}{2} E\left(\frac{\partial \log \sigma_t^2(\theta_0)}{\partial \theta} \frac{\partial \log \sigma_t^2(\theta_0)}{\partial \theta^\top}\right)$$

$$\mathcal{J}^{-1} \mathcal{I} \mathcal{J}^{-1} = E\left((\varepsilon_t^2 - 1)^2\right) \left[E\left(\frac{\partial \log \sigma_t^2(\theta_0)}{\partial \theta} \frac{\partial \log \sigma_t^2(\theta_0)}{\partial \theta^\top}\right)\right]^{-1}.$$

When $\varepsilon_t \sim N(0, 1)$, $E\left((\varepsilon_t^2 - 1)^2\right) = E\left(\varepsilon_t^4\right) - 1 = 2$.

The theory is based on Taylor series expansion and the CLT for martingale difference sequences to the standardized score function and the LLN for the standardized Hessian matrix. The particular form of the asymptotic variance is valid because the score function is a martingale difference sequence, i.e., with probability one

$$E\left(\frac{\partial \ell_t}{\partial \theta}(\theta_0) \middle| \mathcal{F}_{t-1}\right) = 0 \tag{11.35}$$

under the correct conditional mean and conditional variance (semi-strong specification). In this case, the correlation between $\frac{\partial \ell_t}{\partial \theta}(\theta_0)$ and $\frac{\partial \ell_s}{\partial \theta}(\theta_0)$ is zero for any $t \neq s$.

The large sample theory is difficult to work out from primitive conditions even for simple models. The following are some notable contributions. Lumsdaine (1996) included the IGARCH case but she assumed strong stationarity and symmetric unimodal i.i.d. ε_t with $E(\varepsilon_t^{32}) < \infty$. Lee and Hansen (1994) prove consistency and asymptotic normality for the GARCH(1,1) model under weaker conditional moment conditions and allow for semi-strong processes with some higher level assumptions. Hall and Yao (2003) assume weak stationarity and show that if $E(\varepsilon_t^4) < \infty$ the asymptotic normality holds, but also establish limiting behavior (no-normal) under weaker moment conditions. Jensen and Rahbek (2004) established consistency and asymptotic normality of the QMLE in a strong GARCH model without strict stationarity (actually for only a subset of the parameters when the process is non-stationary). There are some recent results for the consistency and asymptotical normality of the EGARCH under primitive conditions, although the theory is less complete. Straumann (2004) comes close.

We next turn to the question of standard errors and inference. There are three different sorts of estimates of the asymptotic covariance matrix of $\widehat{\theta}$. Define

$$\widehat{\mathcal{J}} = \frac{1}{T} \frac{\partial^2 \ell(\widehat{\theta})}{\partial \theta \partial \theta^\top} \quad ; \quad \widehat{\mathcal{I}} = \frac{1}{T} \sum_{t=1}^{T} \frac{\partial \ell_t}{\partial \theta} \frac{\partial \ell_t}{\partial \theta^\top}(\widehat{\theta}). \tag{11.36}$$

Gaussian Errors

The simplest estimate of the asymptotic variance matrix takes the i.i.d. Gaussian structure used in the estimation seriously and uses either $\widehat{\mathcal{J}}^{-1}$ or $\widehat{\mathcal{I}}^{-1}$, which are only consistent under the strong Gaussian GARCH model, i.e., Gaussian i.i.d. errors. These are how the default standard errors are obtained in Eviews, for example.

i.i.d. Errors

In the second level we may assume that the errors ε_t are i.i.d. but not necessarily Gaussian. In that case, the asymptotic variance of $\widehat{\theta}$ is

$$E\left((\varepsilon_t^2 - 1)^2\right) \mathcal{J}^{-1},$$

where $E\left((\varepsilon_t^2 - 1)^2\right)$ may not be equal to two. In this case, we may estimate the asymptotic variance consistently by

$$\frac{1}{T}\sum_{t=1}^{T}(\widehat{\varepsilon}_t^2 - 1)^2 \widehat{\mathcal{J}}^{-1}, \tag{11.37}$$

where $\widehat{\varepsilon}_t = y_t/\sigma_t(\widehat{\theta})$.

Martingale Difference Sequence Errors

Finally, if we are only willing to assume semi-strong GARCH model such as ε_t and $\varepsilon_t^2 - 1$ are martingale difference sequences, then we should take the full sandwich estimator

$$\widehat{\mathcal{J}}^{-1}\widehat{\mathcal{I}}\widehat{\mathcal{J}}^{-1}, \tag{11.38}$$

which will be consistent under general conditions, since $\widehat{\mathcal{J}} \xrightarrow{P} \mathcal{J}, \widehat{\mathcal{I}} \xrightarrow{P} \mathcal{I}$. This can be used to provide confidence intervals about the parameters or to test hypotheses about them, for example the absence of heteroskedasticity would correspond to the null hypothesis that $\beta = \gamma = 0$. The general restriction $R\theta = r$ with R a $q \times p$ matrix with $q < p$ and r a $q \times 1$ vector may be tested using the Wald statistic

$$W = T\left(R\widehat{\theta} - r\right)^{\mathsf{T}} \left(R\widehat{\mathcal{J}}^{-1}\widehat{\mathcal{I}}\widehat{\mathcal{J}}^{-1}R^{\mathsf{T}}\right)^{-1} \left(R\widehat{\theta} - r\right), \tag{11.39}$$

which is distributed as a chi-squared with q degrees of freedom in large samples.

Target Variance

Estimation of GARCH models is big business and there are many competing softwares for implementing the procedures. Typically, for financial returns one finds small intercepts and a large parameter on lagged dependent volatility. The two parameter estimates are highly correlated (colinear), and this can lead to difficulties. Table 11.8 shows a typical correlation matrix between estimated parameters.

The **target variance** approach proposed by Engle and Sheppard (2001) is an attempt to improve the estimation of GARCH models. For a weakly stationary GARCH(1,1) process we have

$$E(y_t^2) = \frac{\omega}{1 - \beta - \gamma},$$

11.11 Estimation of Parameters

Table 11.8 Correlation matrix of estimated parameters from GARCH model

	c	ω	β	γ
c	1.0000	0.1214	0.1596	−0.2322
ω	0.1214	1.0000	−0.7936	0.2895
β	0.1596	−0.7936	1.0000	−0.7600
γ	−0.2322	0.2895	−0.7600	1.0000

Note: These estimates refer to the GARCH(1,1) model computed on the daily S&P500 series $y_t = c + \varepsilon_t \sigma_t$ and $\sigma_t^2 = \omega + \beta \sigma_{t-1}^2 + \gamma u_{t-1}^2$.

so that $\omega = E(y_t^2)(1 - \beta - \gamma)$. They suggest replacing $E(y_t^2)$ by $\sum_{t=1}^{T} y_t^2 / T$ in the variance recursion of the likelihood, so that

$$\sigma_t^2(\beta, \gamma) = \frac{1}{T} \sum_{t=1}^{T} y_t^2 (1 - \beta - \gamma) + \beta \sigma_{t-1}^2(\theta) + \gamma y_{t-1}^2. \tag{11.40}$$

This profiling method means that we only have two parameters to optimize over.

In practice the method appears to improve the computation of estimates in GARCH models. The downside is that distribution theory is much more complicated due to the lack of martingale property in the score function for β, γ. If the error is i.i.d. then one can get a closed form expression for the asymptotic variance, but under only the semi-strong form specification one has to use some sort of Newey–West standard errors.

11.11.2 Alternative Estimation Methods

We consider here some alternatives to the Gaussian QMLE.

Second Order Properties

One approach is to exploit the known second order properties (the correlogram) of the process. As we have shown for a GARCH(1,1) process $x_t \equiv y_t^2$ can be written as an ARMA(1,1) process

$$x_t = \omega + \phi x_{t-1} + \eta_t + \theta \eta_{t-1},$$

where $\eta_t = x_t - \sigma_t^2$ is a martingale difference sequence with respect to $\mathcal{F}_{t-1}$, while $\phi = \gamma + \beta > 0$ and $\theta = -\beta < 0$.

The autocorrelation function, $\rho(k)$ of x_t solves the following Yule–Walker equations,

$$\rho(k) = \phi \rho(k-1), k = 2, 3, \ldots, \quad \rho(1) = \frac{(1 + \phi\theta)(\phi + \theta)}{1 + \theta^2 + 2\phi\theta}.$$

Chapter 11 Volatility

Table 11.9 Daily GARCH in mean t-error

	α	ρ_1	ρ_2	ν	ω	β	γ
parameter estimate	2.503	0.0845	−0.0366	6.663	6.33e−7	0.9170	0.0781
standard error	1.040	0.0079	0.0078	0.2951	8.69e−8	0.0042	0.0042

$y_t = c + \alpha\sigma_t^2 + \rho_1 y_{t-1} + \rho_2 y_{t-2} + \varepsilon_t \sigma_t, \varepsilon_t \sim t_\nu$ and $\sigma_t^2 = \omega + \beta\sigma_{t-1}^2 + \gamma u_{t-1}^2$

This yields a quadratic equation for θ given ϕ and $\rho(.)$. One can solve these equations to express the parameters of interest in closed form

$$\phi = \frac{\rho(k)}{\rho(k-1)}, \quad k = 2, 3, \ldots,$$

$$\theta = \frac{-b - (b^2 - 4)^{1/2}}{2}, \quad b \equiv \frac{\phi^2 + 1 - 2\rho(1)\phi}{\phi - \rho(1)}, \quad \omega = \sigma^2(1-\phi), \quad \sigma^2 \equiv E(y_t^2).$$

One can use these equations to estimate ω, β, γ by plugging in the sample autocorrelations and sample variance. This gives an explicit closed form estimator (Kristensen and Linton (2006)). The Whittle estimator of Giraitis and Robinson (2001) uses the second order properties of y_t more systematically, but at the cost of requiring numerical procedures as in the likelihood method.

Alternative Likelihood

In practice one often finds that the rescaled residuals $\hat{\varepsilon}_t$ are leptokurtic and incompatible with the normal distribution, and one often rejects formal tests of this hypothesis. Some authors advocate using alternative likelihood criteria. The most popular alternative to the normal distribution is the t-distribution with known or unknown degrees of freedom ν. In this case, the likelihood function is (apart from constants) $\ell(\theta, \nu) = \sum_{t=1}^{T} \ell_t(\theta, \nu)$, where

$$\ell_t(\theta, \nu) = -\sigma_t(\theta) - \frac{\nu+1}{2} \log\left(1 + \frac{y_t^2}{(\nu-2)\sigma_t^2(\theta)}\right) + \log\Gamma\left(\frac{\nu+1}{2}\right)$$
$$- \frac{1}{2}\log(\nu-2) - \log\Gamma\left(\frac{\nu}{2}\right). \tag{11.41}$$

This is implemented in EVIEWS. In Table (11.9) we show parameter estimates for daily data obtained by this method.

Note that the estimated degrees of freedom are around 6.66 with quite a tight standard error. For this distribution, the value −11 or less obtained in Figure 11.12 is still unlikely but occurs with probability in the region of 7.8×10^{-6} rather than 2×10^{-28} as for the normal distribution.

The GARCH in mean effect is positive and significant although not strongly so. The unconditional standard deviation is around 0.011. The standard errors in this case are

11.11 Estimation of Parameters

calculated as if the t-distribution were correctly specified, i.e., from the information matrix.

The estimators $\widehat{\theta}, \widehat{\nu}$ are consistent and asymptotically normal provided the model is correct. In fact, $\widehat{\theta}$ remains consistent under some departures from the t-distribution. Newey and Steigerwald (1997) investigated the consistency of estimated GARCH parameters using non-Gaussian QMLE objective functions. They showed that if the assumed error and true error are symmetric then the QMLE is still consistent but otherwise it may not be. This is in contrast with the Gaussian QMLE which is consistent regardless of the shape of the error distribution (provided only that $E(\varepsilon_t^4) < \infty$).

Robust Estimation

There is evidence that for some medium and high frequency data the rescaled errors ε_t are heavy tailed; see Chapter 14. Suppose that the GARCH model holds but that $E(\varepsilon_t^4) = \infty$, what should be done? One approach is to use a parametric model for the error distribution that allows for heavy tails, such as the t-distribution. But this particular choice will yield inconsistent estimates when the error distribution is asymmetric. We may consider asymmetric error distributions, but the corresponding MLEs are not robust to misspecification of the error distribution. Another line of work has proposed estimators that are robust to heavy tailed error distributions. A leading example of this are the log based estimators of Peng and Yao (2003). They propose to estimate the parameters θ by minimizing (either of) the objective functions:

$$Q_T(\theta) = \sum_{t=1}^{T} \left|\log y_t^2 - \log \sigma_t^2(\theta)\right|, \quad Q_T(\theta) = \sum_{t=1}^{T} \left(\log y_t^2 - \log \sigma_t^2(\theta)\right)^2.$$

Provided that ε_t is i.i.d. (strong GARCH) the parameters β, γ are consistently estimated in both cases provided only $E(\log \varepsilon_t^2) < \infty$. The estimated intercept is affected by the error distribution in both cases. However, if one replaces the condition that $E(\varepsilon_t^2|\mathcal{F}_{t-1}) = 1$ by a slightly different normalizing condition, then we can obtain consistency of all parameters. In the least absolute deviation (LAD) criterion one obtains consistency of the estimated intercept parameter ω when ε_t^2 has median one or in fact if the conditional median is one. This is because

$$\text{med}\left(\log y_t^2|\mathcal{F}_{t-1}\right) = \text{med}\left(\log \varepsilon_t^2|\mathcal{F}_{t-1}\right) + \log \sigma_t^2 = \log \text{med}\left(\varepsilon_t^2|\mathcal{F}_{t-1}\right) + \log \sigma_t^2.$$

If $\text{med}\left(\varepsilon_t^2|\mathcal{F}_{t-1}\right) = 1$, then $\text{med}\left(\log y_t^2|\mathcal{F}_{t-1}\right) = \log \sigma_t^2$, and the objective function represents a correctly specified median regression. Therefore, with a slight change of normalizing assumption (which is required anyway because the conditional variance may not exist) this estimator consistently estimates all parameters of the GARCH process. For the least squares criterion we have

$$E\left(\log y_t^2|\mathcal{F}_{t-1}\right) = E\left(\log \varepsilon_t^2|\mathcal{F}_{t-1}\right) + \log \sigma_t^2(\theta_0)$$

and in this case, it suffices that $E\left(\log \varepsilon_t^2|\mathcal{F}_{t-1}\right) = 0$. Their estimators are asymptotically normal under weak moment and dependence conditions.

11.12 Stochastic Volatility Models

The class of stochastic volatility models are popular and capture many of the same data features as the GARCH class of models. Ghysels, Harvey, and Renault (1996) provide a review.

Definition 108 *Taylor (1982). Suppose that*

$$h_t = \log \sigma_t^2 = \alpha_0 + \alpha_1 \log \sigma_{t-1}^2 + \sigma_\eta \eta_t \tag{11.42}$$

$$y_t = \varepsilon_t \sigma_t,$$

where (ε_t, η_t) are i.i.d. mean zero and variance one and $E(\sigma_t^2) < \infty$. This is called the SV(1) model.

The main difference from the GARCH class of processes is that there are two shocks ε_t, η_t rather than one. The process is specified in log form like the EGARCH model, which implies that $\sigma_t \geq 0$ no matter what parameter values we have. The properties of stochastic volatility models are similar to those of GARCH models. The stationarity question is easy to address in this class of models. Provided $|\alpha_1| < 1$, the process h_t is strongly stationary, and so is y_t. The process y_t is also weakly stationary provided $E(\sigma_t^2) = E(\exp(h_t)) < \infty$, which is the case when η_t is standard normal.

We distinguish between the case where ε_t is independent of η_t (called the **no-leverage** case) and the case where this does not hold.

Example 61 *A leading case is where*

$$\begin{pmatrix} \varepsilon_t \\ \eta_t \end{pmatrix} \sim N\left(\begin{pmatrix} 0 \\ 0 \end{pmatrix}, \begin{pmatrix} 1 & \rho \\ \rho & 1 \end{pmatrix} \right), \tag{11.43}$$

where the parameter ρ measures the correlation between the shocks and hence the degree of leverage allowed for.

The random variable y_t is leptokurtic even though it is driven by normal shocks. The process $\log \sigma_t^2$ is linear, i.e.,

$$h_t = \psi_0 + \sum_{j=1}^{\infty} \psi_j \eta_{t-j}, \tag{11.44}$$

where ψ_j depend on $\alpha_0, \alpha_1, \sigma_\eta$. Assuming that ε_t, η_t are mutually independent and standard normal we have:

$$h_t \sim N\left(\frac{\alpha_0}{1-\alpha_1}, \frac{\sigma_\eta^2}{1-\alpha_1^2} \right) = N\left(\mu_h, \sigma_h^2 \right)$$

11.12 Stochastic Volatility Models

$$y_t|\sigma_t \sim N(0, \sigma_t^2).$$

It follows that y_t is a normal mixture with lognormal mixing distribution. This distribution is heavy tailed but not very heavy tailed, i.e., all moments exist.

Second, we may show that the series y_t^2 and $\log y_t^2$ are autocorrelated. Assume for simplicity that (ε_t, η_t) are mutually independent. The autocorrelation function of y_t^2 is (Jacquier, Polson, and Rossi (1994)),

$$\text{corr}(y_t^2, y_{t-j}^2) = \frac{\exp(\sigma_h^2 \alpha_1^j) - 1}{3\exp(\sigma_h^2) - 1}, \quad j \neq 0.$$

The correlation function of $\log y_t^2$ is easier to derive, since $\log y_t^2 = \log \varepsilon_t^2 + h_t$, where $\log \sigma_t^2$ is an AR(1) process and $\log \varepsilon_t^2$ is i.i.d. This process behaves like an ARMA(1,1) in terms of its second order properties (autocovariance function). However, we may show that $\text{cov}(y_t^2, y_{t-j}) = 0$ at all leads and lags, unlike the EGARCH model, say. If ε_t, η_t are mutually dependent, then the autocorrelation function $\text{corr}(y_t^2, y_{t-j}^2)$ is more complicated but qualitatively similar. In this case, one finds that $\text{cov}(y_t^2, y_{t-j}) \neq 0$, which justifies calling this the leverage case.

We next consider how to estimate the parameters. We first consider the likelihood method. To construct the likelihood we want the distribution of $y_t|y_{t-1}, \ldots, y_1$, whereas we are given the distribution of $y_t|\sigma_t, y_{t-1}, \ldots, \sigma_t, y_1$, where σ_t is a latent unobserved variable, or equivalently $y_t|\eta_t, y_{t-1}, \ldots, \eta_1, y_1$. We have

$$f(y_t|y_{t-1}, \ldots, y_1) = \int f(y_t|y_{t-1}, \ldots, y_1, \eta_t, \ldots, \eta_1) dP_\eta(\eta_t, \ldots, \eta_1), \quad (11.45)$$

where P_η is the distribution of $(\eta_t, \ldots, \eta_1)$. To compute this integral is quite difficult in practice even for the canonical case where η_t is i.i.d. Gaussian, since the t-dimensional integral is not given in closed form. The literature has pursued simulation methods in order to calculate this (Danielsson (1994)). Other popular methods include indirect estimation (Gourieroux, Monfort, and Renault (1993) and Gallant and Tauchen (1996)).

An alternative estimation method is to use the method of moments applied to the autocovariance function $\gamma_{\log y^2}(j) = \text{cov}(\log y_t^2, \log y_{t-j}^2)$, $j = 1, \ldots, J$. Under the full specification, $\gamma_{\log y^2}(j) = f(\theta; j)$, where f is a known function. We estimate the covariance function of $\log y^2$ using the sample data and then find the value of θ that minimizes the discrepancy between the population implied function $f(\theta; j)$ and the sample estimates, i.e., by GMM.

We note one significant difference between SV models and the GARCH type of models. This is that σ_t^2 is not the conditional variance of returns given past returns. In fact, one can derive the quantity $\text{var}(y_t|\mathcal{F}_{t-1})$ numerically from the full specification by use of the **Kalman filter**, but analytical expressions for this quantity are not available. This makes comparative statics, e.g., investigation of news impact curves, complicated. And it makes computation of conditional variances for portfolio analysis more time consuming than for the GARCH case.

On the other hand this class of models has good links with continuous time models. Actually, both GARCH models and SV models can be embedded in a sequence of "infill" approximations such that they converge in distribution to (different) diffusion processes (Nelson (1990b)).

11.13 Long Memory

The GARCH(1,1) process

$$\sigma_t^2 = \omega + \beta \sigma_{t-1}^2 + \gamma y_{t-1}^2$$

is of the ARCH(∞) form (provided the process is weakly stationary, which requires $\gamma + \beta < 1$), i.e., we can write

$$\sigma_t^2 = \psi_0(\theta) + \sum_{j=1}^{\infty} \psi_j(\theta) y_{t-j}^2 \qquad (11.46)$$

for constants $\psi_j(\theta)$ satisfying $\psi_j(\theta) = \gamma \beta^{j-1}$. These coefficients decay very rapidly to zero, which implies that the actual amount of memory in the process is quite limited. It implies the same rate of decay in the autocorrelation function of y_t^2. The empirical evidence on the autocorrelation function of y_t^2 suggests rather slower decay. Therefore, we consider time series processes that are consistent with such slow decay.

Definition 109 *A stationary short memory process X with finite variance satisfies*

$$\sum_{j=-\infty}^{\infty} |\text{cov}(X_t, X_{t-j})| < \infty. \qquad (11.47)$$

For example, a stationary AR(p) process $X_t = \mu + \rho_1 X_{t-1} + \cdots + \rho_p X_{t-p} + \varepsilon_t$, where ε_t is i.i.d. with mean zero and finite variance has $|\text{cov}(X_t, X_{t-j})| \leq c\rho^j$ for some ρ with $0 < \rho < 1$, which is clearly summable.

Definition 110 *A long memory process X_t has*

$$\sum_{j=-\infty}^{\infty} |\text{cov}(X_t, X_{t-j})| = \infty. \qquad (11.48)$$

We next give a general model of the autocovariance function that may satisfy this condition. Suppose that

$$\text{cov}(X_t, X_{t-j}) \simeq Cj^{2d-1} \qquad (11.49)$$

11.13 Long Memory

for some $C>0$ and $d \in \mathbb{R}$. There are several cases of interest:

1. If $d \geq 0$, the process is long memory. If $d < 0$, the process is short memory.
2. If $d \in (-1/2, 1/2)$, then X_t is stationary and invertible.
3. If $d > 1/2$, then X_t is nonstationary. If $d = 1$, this corresponds to unit root process.

This class of models nests the random walk or unit root case, which corresponds to the $d=1$ case, but it does so in a more subtle way than the AR(1) process, because it allows for a whole range of nonstationary behavior when $d \in [1/2, 1]$. Define the fractional differencing operator $(1-L)^d$. For $d < 1$, one can define the binomial expansion of $(1-L)^d$ (think of L as a small number) and its inverse

$$(1-L)^d = \sum_{j=0}^{\infty} \frac{\Gamma(j-d)}{\Gamma(j+1)\Gamma(-d)} L^j \quad ; \quad (1-L)^{-d} = \sum_{j=0}^{\infty} \frac{\Gamma(j+d)}{\Gamma(j+1)\Gamma(d)} L^j.$$

Definition 111 *Define for any d,*

$$(1-L)^d X_t = \varepsilon_t,$$

where ε_t is an i.i.d. mean zero series. We can write $X_t = (1-L)^{-d}\varepsilon_t = \sum_{j=0}^{\infty} \psi_j \varepsilon_{t-j}$, where $\psi_j = \Gamma(j+d)/\Gamma(j+1)\Gamma(d)$. It follows that X_t satisfies (11.49) and is a special case of the class of fractional autoregressive integrated moving average (FARIMA) processes.

We now turn to long memory GARCH type volatility models. Suppose that

$$\sigma_t^2 = \psi_0(\theta) + \sum_{j=1}^{\infty} \psi_j(\theta) y_{t-j}^2 \tag{11.50}$$

for some parameter vector θ. Specifically, suppose that $\psi_j = Cj^{-\theta}$ for some $\theta > 0$, as in (11.49). The coefficients satisfy $\sum_{j=1}^{\infty} \psi_j^2(\theta) < \infty$ provided $\theta > 1/2$, and satisfy $\sum_{j=1}^{\infty} |\psi_j(\theta)| < \infty$ provided $\theta > 1$.

Definition 112 *Baillie, Bollerslev, and Mikkelsen (1996) define the fractional integrated GARCH process (FIGARCH) where*

$$(1-L)^d \sigma_t^2 = \omega + \gamma \eta_{t-1} = \omega + \gamma \sigma_{t-1}^2 (\varepsilon_{t-1}^2 - 1),$$

where ε_t is an i.i.d. mean zero and variance one series, and $\eta_{t-1} = \sigma_{t-1}^2(\varepsilon_{t-1}^2 - 1)$ is therefore MDS.

When $d=1$ we have the IGARCH process. When $d<1$ we apply the binomial expansion to express σ_t^2 in the form (11.50) with slowly decaying coefficients. By substitution we also obtain the expression

$$(1-L)^d y_t^2 = \omega + \eta_t - (1-\gamma)\eta_{t-1},$$

Table 11.10 Estimated d by frequency

	Daily	Weekly	Monthly
y_t	−0.0181	−0.0026	0.0079
y_t^2	0.1484	0.2862	0.0986
$\log y_t^2$	0.2638	0.2453	0.1782

Note: Estimation of d is by Whittle likelihood for the returns on the S&P500 index for the period 1955–2002 for three different data frequencies. Replace $y_t = 0$ by ε in $\log y_t^2$.

which can be written in the form given above. One can extend this class of processes to allow also short memory filters $A(L)$, $B(L)$, so that, for example, $A(L)(1-L)^d \sigma_t^2 = B(L)(\omega + \gamma \eta_{t-1})$. Estimation of the parameter d can be carried out by a variety of methods (see Robinson (1994)).

Evidence for Long Memory

Breidt, Crato, and de Lima (1998) considered the daily CRSP value weighted index. They find that some series have d larger than 1/2 in squared returns and log squared returns. Lo (1991) finds little evidence of long memory in stock returns y_t using the rescaled range statistic. In Table 11.10 we give estimates of d for daily stock return data.

The estimated values are quite low in this case, and the standard errors are small. The estimated values of d reduce further if a short range model is fitted first. Evidence for long memory in daily stock returns based on correlogram may be questionable due to the issues raised by the paper of Mikosch and Starica (2000).

In practice the following class of processes are widely used in place of long memory processes.

Definition 113 *The heterogeneous autoregressive model (HAR) satisfies*

$$\sigma_t^2 = \omega + \theta_1 \left(\frac{1}{K_1} \sum_{j=N_1+1}^{N_1+K_1} y_{t-j}^2 \right) + \ldots + \theta_r \left(\frac{1}{K_r} \sum_{j=N_r+1}^{N_r+K_r} y_{t-j}^2 \right),$$

where $K_1, \ldots, K_r$ and $N_1, \ldots, N_r$ are specified, such as $K_1 = 5$, $K_2 = 22$, etc.

It has been fashionable recently to work with this process, which is a kind of poor man's long memory process. It is a special case of the ARCH(∞) class of processes with free parameters $\omega, \theta_1, \ldots, \theta_r$, so there is little more to say about this process in terms of its properties or estimation. In practice, however, these processes can well approximate long memory processes.

11.14 Multivariate Models

We now consider models for a vector of time series. We focus on zero mean series, which could be the residuals from a VAR equation.

Definition 114 *The conditional covariance matrix of some $n \times 1$ vector of the mean zero series y_t*

$$\Sigma_t = E(y_t y_t^\top | \mathcal{F}_{t-1}), \tag{11.51}$$

where $\mathcal{F}_{t-1}$ is the information set consisting of the past return history of all the assets.

The conditional covariance matrix contains on the diagonals the conditional variance of each asset but on the off diagonals it contains the conditional covariance between the two assets. This matrix is important for asset pricing and portfolio choice considerations alongside the conditional mean (which we have assumed is zero here). In particular, we may define the conditional $\beta_{im,t}$ of asset i with respect to the market portfolio as the ratio of the conditional covariance to the conditional variance of the market return

$$\beta_{it} = \frac{\text{cov}(r_{it}, r_{mt} | \mathcal{F}_{t-1})}{\text{var}(r_{mt} | \mathcal{F}_{t-1})},$$

and the conditional CAPM says that $E(r_{it} - r_{ft} | \mathcal{F}_{t-1}) = \beta_{it} \gamma_t$, where γ_t is the time varying risk premium.

Definition 115 *Let $h_t = \text{vech}(\Sigma_t)$ denote the unique elements. Here, vech denotes the unique upper (or lower) triangle of the square matrix (including diagonal elements). Bollerslev, Engle, and Wooldridge (1988) defined a general dynamic model for the conditional covariance matrix of the form*

$$h_t = A + B h_{t-1} + C \text{vech}(y_{t-1} y_{t-1}^\top), \tag{11.52}$$

where A is an $n(n+1)/2 \times 1$ parameter vector, while B, C are $n(n+1)/2 \times n(n+1)/2$ matrices.

The cross section is naturally large with asset returns data. This naive GARCH extension has $\frac{n^2(n+1)^2}{2} + \frac{n(n+1)}{2}$ parameters, so with modest $n = 1000$ this requires estimating five hundred billion ($5 * 10^{11}$) parameters! There are simply too many parameters for estimation and too many for interpretation. In particular, the conditional variance of asset i depends on all the lagged values $y_{j,t-s} y_{k,t-s}$ for $j, k = 1, \ldots, n$ and $s = 1, 2, \ldots$, each one of these terms has a parameter associated with it, and sorting out their meaning is an impossible task. In addition, one usually wants to ensure that the conditional covariance matrix is positive definite, but this model makes it very difficult to do so in estimation as this involves a complicated system of inequality restrictions across the many parameter values. In practice therefore this model is only feasible for at most $n = 3$.

The main properties of the univariate GARCH process can be extended to this general multivariate model. Specifically, we can write the outer product of returns as a vector autoregression moving average (VARMA) process in observable data $x_t = \text{vech}(y_t y_t^T)$. Specifically,

$$x_t = h_t + \text{vech}\left(\Sigma_t^{1/2}\left(\varepsilon_t \varepsilon_t^T - I_n\right)\Sigma_t^{1/2}\right)$$
$$= A + B h_{t-1} + C x_{t-1} + \eta_t$$
$$= A + (B + C) x_{t-1} + \eta_t - B \eta_{t-1},$$

where $\eta_t = \text{vech}(\Sigma_t^{1/2}(\varepsilon_t \varepsilon_t^T - I_n)\Sigma_t^{1/2})$ satisfies $E(\eta_t | \mathcal{F}_{t-1}) = 0$. This shows that the generalization of squared returns follows a VARMA process. Note however that the matrices $y_t y_t^T$ and $\varepsilon_t \varepsilon_t^T$ are of rank one so that the process x_t has a lot of cross-sectional dependence in its shock process. We can obtain conditions for strong and weak stationarity and mixing from the parameters and error distribution. For example, provided $I - B - C$ is invertible we have weak stationarity of returns and the unconditional variance matrix has unique elements $E(x_t) = [I - (B + C)]^{-1} A$.

Estimation can in principle be done through the (quasi) likelihood function of $y_T, \ldots, y_2$ given y_1

$$\ell(\theta; y_T, \ldots, y_2, y_1) = \text{const} - \frac{1}{2} \sum_{t=1}^{T} \log \det(\Sigma_t(\theta)) - \frac{1}{2} \sum_{t=1}^{T} y_t^T \Sigma_t(\theta)^{-1} y_t, \quad (11.53)$$

where θ is the full set of parameters in A, B, C. However, in practice the log likelihood function can be relatively flat unless the sample size is very big because the number of parameters is large. Indeed, computing $\Sigma_t(\theta)^{-1}$ can be challenging without further restrictions.

There are many approaches to reducing the number of parameters needed to describe the conditional covariance matrix. An early extension is the so-called BEKK model (Baba, Engle, Kraft, and Kroner (2000)).

Definition 116 *Suppose that*

$$\Sigma_t = A A^T + B \Sigma_{t-1} B^T + C y_{t-1} y_{t-1}^T C^T, \quad (11.54)$$

where A, B, C are $n \times n$ unrestricted parameter matrices, in which case

$$h_t = D_+(A \otimes A) D_- \text{vech}(I) + D_+(B \otimes B) D_- h_{t-1} + D_+(C \otimes C) D_- \text{vech}(y_{t-1} y_{t-1}^T),$$

where D_+, D_- are matrices of zeros and ones such that $\text{vech}(A) = D_+ \text{vec}(A)$ and $D_- \text{vech}(A) = \text{vec}(A)$ for any matrix A.

This gives a big reduction in the number of parameters and imposes symmetry and positive definiteness on Σ_t automatically. The discussion around stationarity and

11.14 Multivariate Models

dependence is the same for this process, namely provided $I - D_+(B \otimes B)D_- h_{t-1} - D_+(C \otimes C)D_+$ is invertible we have weak stationarity of returns. However, there are still a lot of parameters to estimate for n large, and the meaning of the individual parameters is not so clear. In practice these first generation models are not widely used outside the bivariate or trivariate case. We next consider a number of approaches to reducing the dimensionality of the parameter space.

11.14.1 Factor GARCH Models

Suppose that returns are generated by a factor model with observable factors, that is,

$$y_t = \alpha + Bf_t + \varepsilon_t, \qquad (11.55)$$

where y_t, α, and ε_t are $n \times 1$ vectors, while f_t is a $K \times 1$ vector of portfolio returns such as the Fama–French factor portfolios, where $K < n$. Suppose that $E(\varepsilon_t \varepsilon_t^\top) = \Sigma_0$ does not vary over time and that the errors and factors are mutually uncorrelated conditional on past information. Suppose that each factor follows a univariate GARCH(1,1) process with conditional variance σ_{kt}^2 that satisfies

$$\sigma_{kt}^2 = \omega_k + \beta_k \sigma_{k,t-1}^2 + \gamma_k f_{k,t-1}^2$$

for some parameters $(\omega_k, \beta_k, \gamma_k)$ and that the factors are mutually uncorrelated conditional on past information. Then the conditional covariance matrix of y_t satisfies

$$\Sigma_t = \Sigma_0 + \sum_{k=1}^K b_k b_k^\top \sigma_{kt}^2 = \Sigma_0 + B\Lambda_t B^\top, \qquad (11.56)$$

which is symmetric and positive definite. Here, Λ_t is the diagonal matrix with entries σ_{kt}^2. The factor loading matrix B can be estimated by OLS; see Chapter 8. The matrix Σ_0 is often taken to be diagonal, but in any case it can be estimated from the residuals, as we saw in Chapter 7. The parameters of σ_{kt}^2 can be estimated by univariate GARCH fit on the observed factors; see Engle, Ng, and Rothschild (1990). The Fama–French factors, for example, are not mutually orthogonal, and in that case one can (at least contemporaneously) orthogonalize them by premultiplying by the matrix $\widehat{\Omega}_f^{-1/2}$, i.e., fit univariate GARCH process to each element of the orthogonalized Fama–French factors $f_t^* = \widehat{\Omega}_f^{-1/2} f_t$ (and don't forget to correct the expression for Σ_t).

Sentana (1998) provides a discussion of factor GARCH models when the factors are not observed. Specifically, suppose that

$$y_t = \alpha + Bf_t + u_t$$

$$\begin{pmatrix} f_t \\ u_t \end{pmatrix} \Big| I_{t-1} \sim \left(\begin{pmatrix} 0 \\ 0 \end{pmatrix}, \begin{pmatrix} \Lambda_t & 0 \\ 0 & \Gamma \end{pmatrix} \right),$$

where $I_t = \{y_t, f_t, y_{t-1}, f_{t-1}, \ldots\}$, where $y_t \in \mathbb{R}^n, f_t \in \mathbb{R}^K$. In this case the factors f_t are latent, i.e., they are not observed. We suppose that B is of full column rank K and Γ

is a covariance matrix. Here, Λ_t is a $K \times K$ positive definite time varying conditional variance matrix with a specification like (11.52). It follows that

$$y_t | I_{t-1} \sim 0, B\Lambda_t B^\top + \Gamma.$$

Diebold and Nerlove (1989) assume that the matrix Λ_t is diagonal with

$$\lambda_{jjt} = \text{var}(f_{jt} | I_{t-1}) = \omega_j + \beta_j \lambda_{jj,t-1} + \gamma_j f_{j,t-1}^2.$$

To obtain the distribution of $y_t | \mathbb{Y}_{t-1}$, where $\mathbb{Y}_{t-1} = \{y_{t-1}, \ldots\}$, one has to integrate out over $\mathbb{F}_{t-1} = \{f_t, f_{t-1}, \ldots\}$. In these models, estimation can be tricky because of the latent variables. Diebold and Nerlove (1989) used an "approximate" Kalman filter.

11.14.2 CCC and DCC

The **constant conditional covariance** model is due to Bollerslev (1990).

Definition 117 *Suppose that*

$$E(y_t y_t^\top | \mathcal{F}_{t-1}) = \Sigma_t = D_t R D_t, \qquad D_t = \text{diag}\{\sigma_{1t}, \ldots, \sigma_{nt}\}$$

$$\sigma_{it}^2 = \omega_i + \beta_i \sigma_{i,t-1}^2 + \gamma_i y_{i,t-1}^2,$$

where R is a time invariant correlation matrix

$$R_{ij} = \frac{E(\varepsilon_{it} \varepsilon_{jt})}{\left(E(\varepsilon_{it}^2) E(\varepsilon_{jt}^2)\right)^{1/2}} = E(\varepsilon_{it} \varepsilon_{jt}).$$

Provided $\beta_i, \gamma_i \geq 0$ and $\omega_i > 0$, the matrix Σ_t is automatically symmetric and positive definite.

Estimation of the parameters of this model is simple. First, estimate univariate GARCH processes by, for example, Gaussian QMLE. Then estimate R by the sample correlation matrix of the standardized residuals ($\widehat{\varepsilon}_{it} = y_{it}/\widehat{\sigma}_{it}$)

$$\widehat{R}_{ij} = \frac{\frac{1}{T} \sum_{t=1}^T \widehat{\varepsilon}_{it} \widehat{\varepsilon}_{jt}}{\left(\frac{1}{T} \sum_{t=1}^T \widehat{\varepsilon}_{it}^2 \frac{1}{T} \sum_{t=1}^T \widehat{\varepsilon}_{jt}^2\right)^{1/2}}.$$

The estimated conditional covariance matrix is then given by

$$\widehat{\Sigma}_t = \widehat{D}_t \widehat{R} \widehat{D}_t, \qquad \widehat{D}_t = \text{diag}\{\widehat{\sigma}_{it}\}.$$

This allows for time variation in the conditional covariance matrix, but of a rather restrictive sort.

The **dynamic conditional covariance** model of Engle and Sheppard (2001) generalizes this model to allow for time variation in the conditional correlation.

11.14 Multivariate Models

Definition 118 *Suppose that*
$$\Sigma_t = D_t R_t D_t, \quad D_t = \text{diag}\{\sigma_{it}\},$$
where for each $i = 1, \ldots, n$
$$\sigma_{it}^2 = \omega_i + \beta_i \sigma_{it}^2 + \gamma_i y_{i,t-1}^2.$$
The matrix R_t has typical element
$$r_{ij,t} = \frac{q_{ij,t}}{(q_{ii,t} q_{jj,t})^{1/2}}$$
$$q_{ij,t} = c_{ij} + b_{ij} q_{ij,t-1} + a_{ij} \varepsilon_{i,t-1} \varepsilon_{j,t-1}$$
for coefficients a_{ij}, b_{ij}, c_{ij}. The model innovation is the dynamic equation for
$$q_{ij,t} = \text{cov}(\varepsilon_{i,t}, \varepsilon_{jt} | \mathcal{F}_{t-1}),$$
which is not bounded, whereas $r_{ij,t} \in [-1, 1]$. If $c_{ij} = c$, $b_{ij} = b$, $a_{ij} = a$, then Σ_t is guaranteed to be positive semidefinite (provided also the individual GARCH processes are).

Estimation of this model is a little more involved than for the CCC model, but the first step is the same – estimate the univariate GARCH processes and standardize returns by the fitted value $\hat{\sigma}_{it}$. In the second step one maximizes the derived likelihood $\ell(a, b, c)$ with respect to parameters a, b, c.

11.14.3 Multivariate Stochastic Volatility Models

Harvey, Ruiz, and Shephard (1994) introduced the multivariate SV model.

Definition 119 *Suppose that*
$$y_{it} = \exp(h_{it}/2) \varepsilon_{it},$$
where $\varepsilon_t \sim N(0, \Omega)$, where Ω is a correlation matrix (has diagonals one) while
$$h_{it} = \alpha_{0i} + \alpha_{1i} h_{it-1} + \eta_{it}$$
$$\eta_t \sim N(0, \Sigma), \quad \eta_t = (\eta_{1t}, \ldots, \eta_{nt})^\top.$$

They show how to estimate the parameters using some ad hoc methods in the special case where $\Omega = I$ and ε_t and η_t are mutually independent vectors. The technology to estimate these models has now improved, and Bayesian Markov chain Monte Carlo (MCMC) methods can compute the parameters of these models effectively in the case where ε_t and η_t are correlated multivariate normals, but sadly this is not available in EVIEWS yet.

11.15 Nonparametric and Semiparametric Models

Parametric models arise frequently in economics and are of central importance. However, such models only arise when one has imposed specific functional forms on utility or production functions, like CARA. Without these ad hoc assumptions one only gets much milder restrictions on functional form like concavity, symmetry, homogeneity, etc. The nonparametric approach is based on the belief that parametric models are usually misspecified and may result in incorrect inferences. By not restricting the functional form one obtains valid inferences for a much larger range of circumstances. In practice, the applicability depends on the sample size and quality of data available.

Smoothing techniques have a long history starting at least in 1857 when the German economist Engel found the law named after him. He analyzed Belgian data on household expenditure, using what we would now call the regressogram. Whittaker (1923) used a graduation method for regression curve estimation which one would now call spline smoothing, or the Hodrick–Prescott filter in macroeconomics. Nadaraya (1964) and Watson (1964) provided an extension for general random design based on kernel methods. These methods have developed considerably in the last 25 years, and are now frequently used by applied statisticians. The massive increase in computing power as well as the increased availability of large cross-sectional and high-frequency financial time-series datasets are partly responsible for the popularity of these methods. We consider three general areas for the use of nonparametric methods in volatility modelling. Härdle and Linton (1994) provide a review of nonparametric methods.

11.15.1 Error Density

We suppose that returns follow a standard strong GARCH model but that the error distribution is be of unknown functional form, i.e.,

$$y_t = \varepsilon_t \sigma_t, \quad \sigma_t^2 = \omega + \beta \sigma_{t-1}^2 + \gamma y_{t-1}^2, \tag{11.57}$$

where ε_t is i.i.d. with density f. The evidence is that the density of the standardized residuals $\varepsilon_t = y_t/\sigma_t$ is non-Gaussian for whatever class of parametric models for σ_t are considered. In Figure 11.15 we show the estimated density of the standardized residuals for the S&P500 data in comparison with a normal density.

One may use a t-distribution or some other heavy-tailed parametric distribution, but the likelihood functions associated with this choice do not deliver robust estimates of the parameters of interest (Newey and Steigerwald (1997)). The semiparametric approach allows one to jointly estimate the distribution of the shock along with the parameters of the GARCH process (Engle and Gonzalez-Rivera (1991), Linton (1993), and Drost, Klaassen, and Werker (1997)).

The conditional Log Likelihood of $y_T, \ldots, y_2$ given y_1 with known f is

$$\ell(\theta) = -\frac{1}{2} \sum_{t=1}^{T} \log \sigma_t^2(\theta) - \sum_{t=1}^{T} \log f\left(\frac{y_t}{\sigma_t(\theta)}\right).$$

11.15 Nonparametric and Semiparametric Models

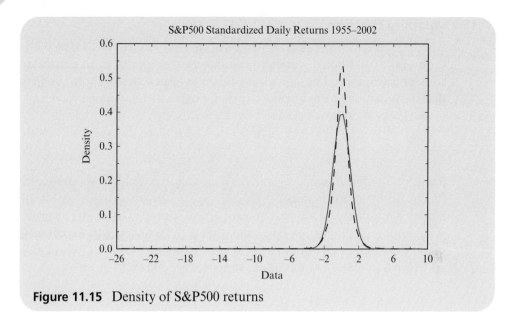

Figure 11.15 Density of S&P500 returns

Score functions are of the form

$$\frac{\partial \ell}{\partial \theta}(\theta) = -\frac{1}{2} \sum_{t=1}^{T} \left(\varepsilon_t(\theta) \frac{f'}{f}(\varepsilon(\theta)) + 1 \right) \frac{\partial \log \sigma_t^2(\theta)}{\partial \theta}.$$

The estimation procedure is as follows.

1. First, estimate parameters θ consistently by some method such as QMLE and obtain standardized residuals.
2. Second, estimate the error density f nonparametrically based on the standardized residuals.
3. Third, use the estimated error density to re-estimate parameters doing two step estimated MLE.

In the semiparametric model where f is unknown, there is an identification issue. This is because one parameter (overall scale) in σ_t^2 is not identified jointly with f. In that case one can set $\omega = 1$ without loss of generality, or one can assume that $E(\varepsilon_t^2) = 1$.

Hafner and Rombouts (2007) consider the multivariate case. The unrestricted nonparametric estimation of the error vector density suffers from the **curse of dimensionality** (Stone (1980)). The elliptically symmetric class of multivariate densities (Definition 29) is essentially one dimensional and obviates this issue. This class of distributions is important in finance theory as the CAPM holds if returns are elliptically distributed. The normal distribution is included as a special case as are many others (Vorkink (2003)).

11.15.2 Volatility Function

Another issue with parametric models such as GARCH(1,1) is that the functional form of the conditional variance itself may be too restrictive. In particular, it is not so clear why it should be a quadratic function of past observations. Suppose instead that the functional form of $\sigma_t^2(\mathcal{F}_{t-1})$ is allowed to be nonparametric, i.e.,

$$\sigma_t^2 = g(y_{t-1}, \ldots, y_{t-p}) \tag{11.58}$$

for some unknown function g and fixed lag length p. This allows completely general functional form, for example $\partial^2 g(y_1, \ldots, y_p)/\partial y_i \partial y_j \neq 0$. However, this allows only limited dependence on the past in comparison with the GARCH(1,1) process, which is a function of all past $y's$. One can estimate the function g by kernel or series estimation techniques (Whistler (1990), Pagan and Hong (1991), Chen (2007)). Härdle and Tsybakov (1997). For example, suppose $p=1$. Then let

$$\widehat{g}(y) = \frac{\sum_{t=2}^{T} K\left(\frac{y-y_{t-1}}{h}\right) y_t^2}{\sum_{t=2}^{T} K\left(\frac{y-y_{t-1}}{h}\right)}, \tag{11.59}$$

where h is the bandwidth and K is the kernel. This estimator is a local weighted average of squared returns. See Härdle and Linton (1994) for more discussion. In Figure 11.16 we show nonparametric estimates of the first four conditional cumulants of the daily return series (which are estimated similarly to (11.59)) for the daily S&P500 return data. The conditional variances show a consistency of pattern not shared by the other cumulants. In particular, the conditional skewness and kurtosis are quite noisy.

We may also consider adjusting for the conditional mean in estimating the volatility. This can be done in two ways. Define

$$\widehat{g}_m(y) = \frac{\sum_{t=2}^{T} K\left(\frac{y-y_{t-1}}{h}\right) y_t^2}{\sum_{t=2}^{T} K\left(\frac{y-y_{t-1}}{h}\right)} - \left(\frac{\sum_{t=2}^{T} K\left(\frac{y-y_{t-1}}{h}\right) y_t}{\sum_{t=2}^{T} K\left(\frac{y-y_{t-1}}{h}\right)}\right)^2,$$

$$\widehat{g}_r(y) = \frac{\sum_{t=2}^{T} K\left(\frac{y-y_{t-1}}{h}\right) \widehat{u}_t^2}{\sum_{t=2}^{T} K\left(\frac{y-y_{t-1}}{h}\right)}, \quad \widehat{u}_t = y_t - \frac{\sum_{t=2}^{T} K\left(\frac{y-y_{t-1}}{h}\right) y_t}{\sum_{t=2}^{T} K\left(\frac{y-y_{t-1}}{h}\right)}.$$

Fan and Yao (1998) argue that the residual approach, $\widehat{g}_r(y)$, is better in terms of estimator performance: it has the same variance but its bias does not depend on the curvature of the conditional mean $E(y_t|y_{t-1}=y)$.

11.15 Nonparametric and Semiparametric Models

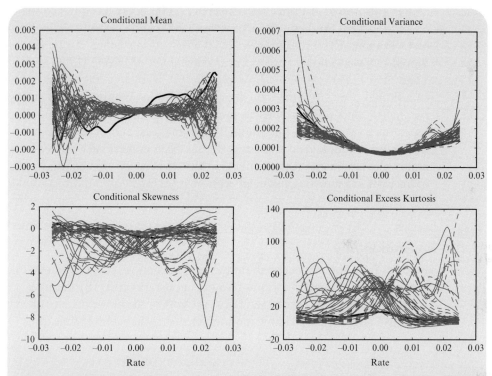

Figure 11.16 Shows the nonparametrically estimated conditional comulants $cum_j(y_t|y_{t-k})$, for $j = 1, 2, 3, 4$ and $k = 1, \ldots, 50$. The black curve is the case $k = 1$; all the other curves are shown in grey

Jianqing Fan (born 1962) is a statistician and financial econometrician. He is currently the Frederick L. Moore '18 Professor of Finance, a Professor of Statistics, and a former Chairman of Department of Operations Research and Financial Engineering (2012–2015) at Princeton University. He is also the Dean of School of Data Science at Fudan University since 2015. Fan is interested in statistical theory and methods in data science, finance, economics, risk management, machine learning, computational biology, and biostatistics, with a particular focus on high-dimensional statistics, nonparametric modelling, longitudinal and functional data analysis, nonlinear time series, and wavelets, among other areas. After receiving his PhD in Statistics from the University of California at Berkeley in 1989, he joined the mathematics faculty at the University of North Carolina at Chapel Hill (1989–2003) and the University of California at Los Angeles (1997–2000). He was then appointed Professor of Statistics and Chairman at the Chinese University of Hong Kong (2000–2003), and as a Professor at Princeton University (2003–). He has directed the Committee of Statistical Studies at Princeton since 2006 and currently chairs Department of Operations Research and Financial Engineering (since 2012). He has coauthored two well-known books (*Local Polynomial Modeling* (1996) and *Nonlinear Time Series: Parametric and Nonparametric Methods* (2003)) and authored or coauthored over 170 articles on finance, economics, computational biology, semiparametric and non-parametric modelling, statistical learning, nonlinear time series, survival analysis, longitudinal data analysis, and other aspects of theoretical and methodological statistics. He has been consistently ranked as a top 10 highly-cited mathematical scientist. He has received various awards in recognition of his work on statistics, financial econometrics, and computational biology.

11.15.3 Nonstationarity

There has been some concern about the issue of stationarity, i.e., whether it is appropriate to assume it, and if not how to model nonstationarity. By taking $\beta + \gamma > 1$ one can have nonstationary processes, but at the cost of the non-existence of variance. An alternative approach is to allow the coefficients to change over time, thus

$$\sigma_t^2 = \omega_t + \beta_t \sigma_{t-1}^2 + \gamma_t y_{t-1}^2. \tag{11.60}$$

In Table 11.11 we estimate the GARCH model for each decade, which shows parameter non-constancy.

This corresponds to taking the parameters (11.60) to be constant within a decade and then to change across decade in an arbitrary fashion. An alternative approach is to allow the parameters to vary smoothly over time and to exploit this local constancy by using smoothing methods.

Starica (2003) compares GARCH(1,1) forecasts with forecasts from the process

$$y_t = \sigma_t \varepsilon_t = \sigma(t/T) \varepsilon_t, \tag{11.61}$$

where $\sigma(u)$ is an unknown function of (rescaled) time t/T, and ε_t is i.i.d. mean zero and variance one. In this model the conditional and unconditional variance of y_t are both equal to $\sigma^2(t/T)$. The process y_t^2 is uncorrelated in the population and so is quite a different process from the GARCH process, that is, $\text{cov}(y_t^2, y_{t-j}^2) = 0$ for all j. However, note that the sample autocovariance satisfies

$$\frac{1}{T} \sum_{t=j+1}^{T} y_t^2 y_{t-j}^2 - \left(\frac{1}{T} \sum_{t=j+1}^{T} y_t^2 \right)^2 \xrightarrow{P} \int_0^1 \sigma^2(u) \sigma^2(u-s) du - \left(\int_0^1 \sigma^2(u) du \right)^2 \neq 0$$

for any j with $j = j_0 + sT$, where $s \in [0, 1)$ and j_0 is any fixed integer. The process y_t^2 behaves as if it is an autocorrelated series in this respect.

11.15 Nonparametric and Semiparametric Models

Table 11.11 Estimated GARCH model by decade

	50–60	60–70	70–80	80–90	90–00	00–09
c	0.0213	0.0019	-0.0106	0.0111	0.0043	-0.0234
ρ_1	0.1584	0.2229	0.2429	0.0635	0.0603	-0.0814
ρ_2	-0.0977	-0.0288	-0.0563	-0.0033	0.0120	-0.0445
ω	0.0425	0.0166	0.0039	0.0605	0.0132	0.0121
β	0.8330	0.8086	0.9543	0.8620	0.9230	0.9416
γ	0.0584	0.0574	0.0073	0.0362	0.0016	-0.0179
δ	0.0692	0.2031	0.0691	0.0980	0.1264	0.1278
R^2	0.0165	0.0320	0.0515	0.0025	0.0000	0.0171
mper	0.0607	0.1941	0.1866	0.0602	0.0723	-0.1259
vper	0.9260	0.9676	0.9962	0.9472	0.9878	0.9876
μ_{year}	0.1199	0.0714	0.0346	0.0939	0.0768	0.0161
σ_{year}	0.1144	0.1080	0.1520	0.1616	0.1571	0.1492

$y_t = c + \rho_1 y_{t-1} + \rho_2 y_{t-2} + \varepsilon_t \sigma_t$ and $\sigma_t^2 = \omega + \beta \sigma_{t-1}^2 + \gamma u_{t-1}^2 + \delta u_{t-1}^2 1(u_{t-1} < 0)$.

We can estimate $\sigma^2(u)$ by

$$\widehat{\sigma}^2(u) = \frac{1}{Th} \sum_{t=1}^{T} K_h(u - t/T) y_t^2,$$

where $K_h(.) = K(./h)/h$ and K is a kernel. One can use one-sided or two-sided kernels here depending on the objectives. For one-sided kernels K with support $[-1, 0]$ we can see that

$$\widehat{\sigma}_t^2 = \widehat{\sigma}^2(t/T) = \sum_{j=1}^{t} w_j y_{t-j}^2,$$

and so it is a weighted average of past squared returns. In this respect it looks quite similar to the GARCH process itself. He argues that this simple model does better than GARCH in terms of out of sample performance, although his paper was never published. In Figure 11.17 we show the estimated time varying mean, variance, skewness, and kurtosis of the S&P500 daily return series, which are all computed using kernel methods.

The locally varying model is not dynamic in any way. It predicts, for example, that $\text{cov}(y_t^2, y_s^2) = E(\sigma^2(t/T)\sigma^2(s/T)(\varepsilon_t^2 - 1)(\varepsilon_s^2 - 1)) = 0$ for any $t \neq s$, which seems a little unrealistic. Consider the process

$$\sigma_t^2 = \omega(t/T) + \beta(t/T)\sigma_{t-1}^2 + \gamma(t/T)y_{t-1}^2,$$

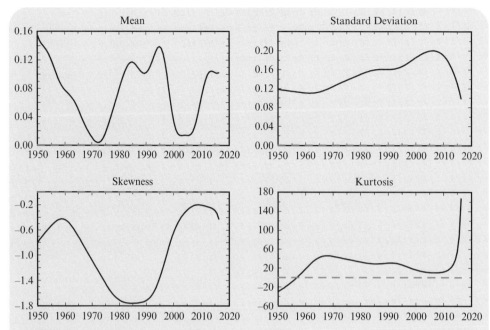

Figure 11.17 Time varying cumulants of S&P500 returns. The mean and standardard deviations are annualized

where ω, β, and γ are smooth but unknown functions of rescaled time. This captures the slow evolutionary change idea but also allows for the autogenerative idea of AR and GARCH models. This class of processes is nonstationary but can be viewed as locally stationary along the lines of Dahlhaus (1997), provided that $\beta(u) + \gamma(u) \leq c < 1$. In this way the unconditional variance of returns exists for each t, T, i.e., $E(\sigma_t^2) = \omega(t/T)/(1 - \beta(t/T) - \gamma(t/T)) < \infty$, but it can change slowly over time as can the persistence of the process. The model can be estimated by local QMLE, that is, we maximize

$$\ell^{loc}(\theta; u) = -\sum_{t=1}^{T} K_h(u - t/T) \left[\log \sigma_t^2(\theta) + \frac{y_t^2}{\sigma_t^2(\theta)},\right]$$

with respect to the local parameters $\theta = (\omega, \beta, \gamma)$ for each $u \in [0, 1]$. Dahlhaus and Subba Rao (2006) provided a comprehensive theory of such processes. This method can be adapted to allow for discontinuous changes by considering only one-sided kernels.

Engle and Rangel (2008) impose some restrictions that makes the unconditional variance $\sigma^2(t/T) = \omega(t/T)/(1 - \beta(t/T) - \gamma(t/T))$ vary smoothly over time but the coefficients $\beta(t/T)$ and $\gamma(t/T)$ restricted to be constant.

11.17 Appendix

Definition 120 *Suppose that*

$$y_t = \sigma_t g_t^{1/2} \varepsilon_t, \qquad (11.62)$$

where $\sigma_t = \sigma(t/T)$ for some smooth unknown function $\sigma(.)$, while

$$g_t = 1 - \beta - \gamma + \beta g_{t-1} + \gamma(y_{t-1}/\sigma_{t-1})^2.$$

We have $E(g_t) = 1$ and the process g_t is a **unit GARCH**. The unconditional variance of y_t is $E(y_t^2) = E(\sigma_t^2 g_t) = \sigma^2(t/T)$ as claimed. This model gives a long-run/short-run decomposition where the process g_t represents short-run deviations of volatility from its long-run level $\sigma^2(t/T)$, which itself is slowly varying over time. Spokoiny and Cizek (2009) consider the dynamic choice of window width.

11.16 Summary of Chapter

We describe the three main methods for estimating volatility: those based on derivative prices, those based on intra period observations, and those based on discrete time series models.

11.17 Appendix

Proof of (11.18). By the law of iterated expectation we have

$$\mu_4 = E\left(y_t^4\right) = E\left(\varepsilon_t^4 \sigma_t^4\right) = 3E\left(\sigma_t^4\right).$$

This must satisfy

$$\mu_4 = 3E\left(\left(\omega + \gamma y_{t-1}^2\right)^2\right) = 3E\left(\left(\omega^2 + \gamma^2 y_{t-1}^4 + 2\omega\gamma y_{t-1}^2\right)\right) = 3\left(\omega^2 + \gamma^2 \mu_4 + 2\omega\gamma\sigma^2\right).$$

Therefore,

$$\mu_4 = \begin{cases} \frac{3(\omega^2 + 2\omega\gamma\sigma^2)}{1 - 3\gamma^2} & \text{if } \gamma^2 < 1/3 \\ \infty & \text{else,} \end{cases}$$

and hence the excess kurtosis of returns satisfies

$$\kappa_4 = \frac{\mu_4}{\sigma^4} - 3$$

$$= \frac{3\omega^2 + 6\omega\gamma\sigma^2}{(1 - 3\gamma^2)} \frac{(1-\gamma)^2}{\omega^2} - 3$$

$$= \frac{6\gamma^2}{(1 - 3\gamma^2)} \geq 0.$$

□

Chapter 11 Volatility

Sketch Proof of Theorem 44. Write

$$\sigma_t^2 = \omega + \sigma_{t-1}^2 \left(\beta + \gamma \varepsilon_{t-1}^2\right) = \omega + \omega \left(\beta + \gamma \varepsilon_{t-1}^2\right) + \left(\beta + \gamma \varepsilon_{t-1}^2\right)\left(\beta + \gamma \varepsilon_{t-2}^2\right) \sigma_{t-2}^2.$$

By substituting back we have

$$\sigma_t^2 = \sigma_0^2 \prod_{i=1}^{t} \left(\gamma \varepsilon_{t-i}^2 + \beta\right) + \omega \left[1 + \sum_{j=1}^{t-1} \prod_{i=1}^{j} \left(\gamma \varepsilon_{t-i}^2 + \beta\right)\right].$$

Therefore, the behavior of the products $\prod_{i=1}^{j} \left(\gamma \varepsilon_{t-i}^2 + \beta\right)$ is crucial. We have

$$\frac{1}{j} \sum_{i=1}^{j} \log\left(\gamma \varepsilon_{t-i}^2 + \beta\right) \to E\left[\log\left(\gamma \varepsilon_{t-i}^2 + \beta\right)\right]$$

as $j \to \infty$ with probability one by the strong LLN. It follows that

$$\prod_{i=1}^{j} \left(\gamma \varepsilon_{t-i}^2 + \beta\right) = O(\exp(-\lambda j))$$

as $j \to \infty$ for some $\lambda \propto \left|E\left[\log\left(\gamma \varepsilon_{t-i}^2 + \beta\right)\right]\right|$. Therefore, the sum in $_u \sigma_t^2$ exists (and one can show that is bounded from below). Furthermore,

$$\sigma_t^2 - _u \sigma_t^2 = \sigma_0^2 \prod_{i=1}^{t} \left(\gamma \varepsilon_{t-i}^2 + \beta\right) - \omega \sum_{j=t}^{\infty} \prod_{i=1}^{j} \left(\gamma \varepsilon_{t-i}^2 + \beta\right).$$

Then

$$\log\left(\sigma_0^2 \prod_{i=1}^{t} \left(\gamma \varepsilon_{t-i}^2 + \beta\right)\right) = \log \sigma_0^2 + \sum_{i=1}^{t} \log\left(\gamma \varepsilon_{t-i}^2 + \beta\right) \to -\infty$$

with probability one. Likewise

$$\sum_{j=t}^{\infty} \prod_{i=1}^{j} \left(\gamma \varepsilon_{t-i}^2 + \beta\right) \to 0$$

with probability one. It follows that $\sigma_t^2 - _u \sigma_t^2 \to 0$ with probability one. $\square$

We derive some conditional moments for GARCH processes. Suppose that

$$\delta_j = \text{cov}(y_t^2, y_{t-j}) = E(y_t^2 y_{t-j}) = E(\varepsilon_t^2 \sigma_t^2 y_{t-j}) = E(\sigma_t^2 y_{t-j}) = \text{cov}(\sigma_t^2, y_{t-j}).$$

Then for GARCH(1,1) we have

$$\delta_j = \text{cov}(\beta \sigma_{t-1}^2, y_{t-j}) + \text{cov}(\gamma y_{t-1}^2, y_{t-j}) = (\beta + \gamma) \delta_{j-1}$$

11.17 Appendix

$$\delta_0 = \text{cov}(y_t^2, y_t) = E(\varepsilon_t^3)E\left(\sigma_t^3\right),$$

and this is zero if and only if $E(\varepsilon_t^3) = 0$. For the normal distribution, the t-distribution, and other commonly used distributions $\delta_j = 0$ for all j. When $E(\varepsilon_t^3) \neq 0$, $\delta_j = c(\beta + \gamma)^j$ for some constant c that could be positive or negative depending on the skewness of the innovation distribution.

Similarly we may show that for $j \neq k$ and $j, k > 0$, we have

$$\pi_{j,k} = E\left(y_t^2 y_{t-j} y_{t-k}\right) = E\left(\sigma_t^2 y_{t-j} y_{t-k}\right) = (\beta + \gamma)\pi_{j-1,k-1}.$$

Suppose that $j < k$, then

$$\pi_{0,k-j} = E\left(y_t^3 y_{t-(k-j)}\right) = E(\varepsilon_t^3)E\left(\sigma_t^3 y_{t-(k-j)}\right),$$

which again is zero if and only if $E(\varepsilon_t^3) = 0$.

12 Continuous Time Processes

So far we have considered discrete time series $\{X_t, t = 1, 2, \ldots, T\}$, where typically T is large, i.e., the data cover a long span. In this chapter we look at continuous time stochastic processes $\{X_t, t \in [0, T]\}$, where for each t, X_t is a random variable defined on some sample space Ω. Continuous time processes are important in mathematical finance, because they lead to simple pricing solutions, which are more complicated to implement in discrete time settings. For example, the Black and Scholes (1973) option pricing theory exploits the continuousness of time to give a simple but nonlinear formula for the price of certain options. There is now a vast literature that extends their basic setting to more general problems such as interest rate models. There is a lot of current interest in continuous time processes due to high frequency data in exchange rates, stock prices, and electricity data.

The strengths of continuous time models include that they predict behavior at all sampling frequencies: second, minute, ten minute, hourly, daily, weekly, monthly, and irregular frequency. They tie in well with economic theory. Finally, some analysis is simple as we shall see, e.g., stationarity conditions. The weaknesses of continuous time models include that: estimation and some analysis can be difficult computationally. The predictions described in the previous paragraph can be too strong and not consistent with the data, especially for very high frequency data. Finally, the statistical models are not as flexible as ones that can be achieved with discrete time models.

We start with the key building block of continuous time process, the Brownian motion or Wiener process.

12.1 Brownian Motion

Definition 121 *The standard Brownian motion process B_t has the following defining properties:*

1. *$B_t - B_s$ is independent of past information $\mathcal{F}_s = \sigma\{B_u : u \leq s\}$.*
2. *$B_t - B_s \sim N(0, t - s)$.*

Wiener (1923) proved the existence of the Brownian motion stochastic process as a well defined mathematical entity. The process B_t is nonstationary, with $\text{cov}(B_s, B_t) = \min\{s, t\}$, but has stationary increments, since the distribution of $B_t - B_s$ only depends on $t - s$ not on t. This process possesses the property of **infinite divisibility**, which is as follows. Let $B_0 = 0$. Then we can write for any sequence or partition

12.1 Brownian Motion

$0 = t_0 \leq t_1 \leq \cdots \leq t_n = t$:

$$B_t = \sum_{i=1}^{n} (B_{t_i} - B_{t_{i-1}}) = \sum_{i=1}^{n} Z_i$$

for independent random variables Z_i (in this case normals). This property is important for a number of applications.

Definition 122 *Let B_t be standard Brownian motion with $B_0 = 0$ and define the Brownian bridge to time $T > 0$*

$$\mathbb{B}_t = B_t - \frac{t}{T} B_T.$$

The process B_t is a martingale meaning $E[|B_t|] < \infty$ for all t and $E[B_t|\mathcal{F}_s] = B_s$ for all $s < t$. In fact, it is the quintessential random walk with $B_t = B_{t-1} + Z_t$, where Z_t is a sequence of i.i.d. standard normal random variables. The process $\mathbb{B}_t$ is not a martingale because $E(\mathbb{B}_T|\mathcal{F}_s) = 0 \neq \mathbb{B}_s$.

Definition 123 *The random variable B_t is just the mapping*

$$\omega \mapsto B_t(\omega), \quad t \in [0, T].$$

*The **sample path** is just the mapping*

$$t \mapsto B_t(\omega), \quad \omega \in \Omega.$$

For each realization of the stochastic process ω this is a function of time.

We next give some properties of the sample paths of Brownian motion. Brownian motion has continuous sample paths and even sample paths that are locally **Hölder continuous** up to order $\gamma < 1/2$, i.e., with probability one

$$|B_t - B_s| \leq C|t - s|^{\gamma}$$

for some finite constant C. However, the sample paths are nowhere locally Hölder continuous for any $\gamma > 1/2$ and in particular are nowhere differentiable. In fact the **modulus of continuity** is $g(\delta) = (2\delta \log(1/\delta))^{1/2}$, i.e.,

$$\Pr\left(\limsup_{\delta \downarrow 0} \frac{1}{g(\delta)} \max_{\substack{0 \leq s < t \leq 1 \\ t-s \leq \delta}} |B_s - B_t| = 1 \right) = 1.$$

Definition 124 *A function $f: [0, T] \to \mathbb{R}$ is said to be of bounded p–variation ($p > 0$) if*

$$\sup_{\text{Partitions } \{t_i: i=1,\ldots,n\} \text{ of } [0,T]} \sum_{i=1}^{n-1} |f(t_{i+1}) - f(t_i)|^p < \infty.$$

Continuously differentiable functions have bounded 1–variation (or just bounded variation). The sample paths of Brownian motion are of **unbounded variation** on any compact interval; it roams around a lot. One consequence of this is that ordinary Riemann integration of a function f with respect to B is not well defined and one has to introduce the concept of **stochastic integration**; see below.

12.1.1 Crossing Times

We next discuss the crossing times of Brownian motion. These are the times at which a process first crosses some threshold. There are many contexts where such a random variable is of interest, from option pricing to volatility measurement. See Karlin and Taylor (1981) for a review.

Univariate Case

Define for the standard Brownian motion B the **first passage** or **first crossing time**

$$\tau_a = \inf\{t : |B_t| > a\}, \tag{12.1}$$

where $B_0 = x$. This is an example of a **stopping time**, which are defined by the requirement that the decision to stop before time s, $\tau \le s$ can be determined from the values of the process at times $r \le s$. The random variable τ_a has been extensively studied.

The distribution of τ_a has been known since Bachelier (1900); it is

$$\Pr(\tau_a \le t | x, a) = 1 - \frac{2}{\pi} \sum_{j=0}^{\infty} \frac{(-1)^j}{j + \frac{1}{2}} \cos\left(\left(j + \frac{1}{2}\right) \frac{\pi x}{a}\right) \exp\left(-\left(j + \frac{1}{2}\right)^2 \pi^2 t / 2a^2\right)$$

(see Darling and Siegert (1953, p630)). The density function is calculated by differentiation, and is

$$f_0(t|x, a) = \frac{\pi}{a^2} \sum_{j=0}^{\infty} (-1)^j \left(j + \frac{1}{2}\right) \cos\left(\left(j + \frac{1}{2}\right) \frac{\pi x}{a}\right) \exp\left(-\left(j + \frac{1}{2}\right)^2 \pi^2 t / 2a^2\right). \tag{12.2}$$

Note that when $x = 0$, $E(\tau_a) = a^2/\sigma^2$.

The one-sided crossing probability is much simpler, so we give it next. Let

$$\tau_a^+ = \min\{t : B_t > a\}. \tag{12.3}$$

12.1 Brownian Motion

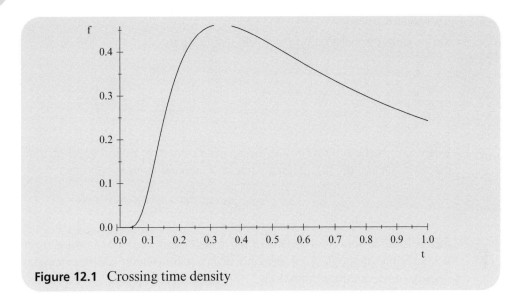

Figure 12.1 Crossing time density

Suppose that $B_0 = x = 0$. We have (Feller (1965, p171)) that:

$$\Pr\left(\tau_a^+ \leq t\right) = 2\left(1 - \Phi\left(\frac{a}{\sqrt{t}}\right)\right) \tag{12.4}$$

$$f_0(t|x, a) = \frac{a}{\sqrt{2\pi t^3}} \exp\left(-\frac{a^2}{2t}\right). \tag{12.5}$$

Note that f_0 is a bounded density on $(0, \infty)$. The shape of this density varies with a. For $a = 5$, the probability of a crossing is very low and the density rises monotonically towards the end of the period. The following shows the first passage density f_0 for $a = 1$, in which case there is an early peak followed by a slow decline.

Finally, although $E(\tau_a^+) = \infty$, we have $E(1/\tau_a^+) = \sigma^2/a^2$.

This theory has many applications in economics and finance. Andersen, Dobrev, and Schaumburg (2008) exploit this theory to define an estimator of volatility. It can also provide some framework for understand modern circuit breakers. For example, the LSE has a circuit breaker that triggers for large stocks when the price moves more than a% (was 10%) up or down from the opening price within the continuous trading segment of the day. In practice we see higher density (across stocks and days) of hits early in the day and at the end of the day. Suppose that stock prices evolve as (11.2) with parameter σ_t on day t, and that σ_t is stochastic drawn from some density h. In that case we may find days where a/σ is small (with early peak of hitting times) and other days with higher a/σ and a later peak of density. Averaging across days we could observe a U-shaped density.

Chapter 12 Continuous Time Processes

Bivariate Case

Suppose that $X_t = (X_{1t}, X_{2t})^\top$ evolves according to

$$dX_t = \Sigma^{1/2} dB_t, \quad \Sigma = \begin{pmatrix} \sigma_1^2 & \rho\sigma_1\sigma_2 \\ \rho\sigma_1\sigma_2 & \sigma_2^2 \end{pmatrix}, \quad (12.6)$$

where $\Sigma^{1/2}$ is the square root of the covariance matrix Σ, and suppose that $X_0 = x = (x_1, x_2)^\top$. Define the first passage time of X_i to zero from x_i

$$\tau_i = \inf\{t \geq 0 : X_{it} = 0\} \quad (12.7)$$

for $i = 1, 2$. The marginal density of the crossing times can be obtained from (12.4) by using the fact that $\sigma B_t > a$ if and only if $B_t > a/\sigma$. Metzler (2010) calculates the joint density of the crossing times. Let:

$$a_i = x_i/\sigma_i \ ; \quad r_0 = \sqrt{\frac{a_1^2 + a_2^2 - 2\rho a_1 a_2}{1 - \rho^2}}$$

$$\theta_0 = \begin{cases} \pi + \tan^{-1}\left(\frac{a_2\sqrt{1-\rho^2}}{a_1 - \rho a_2}\right) & a_1 < \rho a_2 \\ \frac{\pi}{2} & a_1 = \rho a_2 \ ; \\ \tan^{-1}\left(\frac{a_2\sqrt{1-\rho^2}}{a_1 - \rho a_2}\right) & a_1 > \rho a_2 \end{cases} \quad \alpha = \begin{cases} \pi + \tan^{-1}\left(-\frac{\sqrt{1-\rho^2}}{\rho}\right) & \rho > 0 \\ \frac{\pi}{2} & \rho = 0 \\ \tan^{-1}\left(-\frac{\sqrt{1-\rho^2}}{\rho}\right) & \rho < 0. \end{cases}$$

Theorem 47 *The joint density of τ_1, τ_2, denoted $f(s, t)$ is given as follows:*
For $s < t$

$$f(s,t) = \frac{\pi \sin \alpha}{2\alpha^2 \sqrt{s(t - s\cos^2\alpha)}(t-s)} \exp\left(-\frac{r_0^2}{2s} \frac{t - s\cos 2\alpha}{(t-s) + (t - s\cos 2\alpha)}\right)$$

$$\times \sum_{j=1}^{\infty} j \sin\left(\frac{j\pi(\alpha - \theta_0)}{\alpha}\right) I_{j\pi/2\alpha}\left(\frac{r_0^2}{2s} \frac{t-s}{(t-s) + (t - s\cos 2\alpha)}\right)$$

For $s > t$

$$f(s,t) = \frac{\pi \sin \alpha}{2\alpha^2 \sqrt{t(s - t\cos^2\alpha)}(s-t)} \exp\left(-\frac{r_0^2}{2t} \frac{s - t\cos 2\alpha}{(s-t) + (s - t\cos 2\alpha)}\right)$$

$$\times \sum_{j=1}^{\infty} j \sin\left(\frac{j\pi\theta_0}{\alpha}\right) I_{j\pi/2\alpha}\left(\frac{r_0^2}{2t} \frac{s-t}{(s-t) + (s - t\cos 2\alpha)}\right).$$

*Here, $I_x(.)$ is the **Bessel function** of the first kind.*

12.2 Stochastic Integrals

This says that when the Brownian motions are positively correlated, the joint density of the hitting time has a singularity on the line $s = t$, i.e., $f(s,s) = \infty$ for all $s > 0$. This prediction is made purely within a Gaussian world. This seems at odds with the feature of Gaussian distributions that extreme values are independent even for highly correlated random variables; see Chapter 14. It is consistent with the observation that circuit breaker hits on the LSE often occur simultaneously or almost simultaneously (Brugler and Linton (2014)); see the earlier discussion.

12.2 Stochastic Integrals

We next introduce the concept of stochastic integration as it allows one to define a big class of stochastic processes and to analyze their properties. Heuristically, it is easy to see how we could make sense of defining linear combinations of Brownian motion when $f(.)$ is a deterministic function, i.e., we might expect that

$$X_t = \int_a^t f(s) dB_s \sim N\left(0, \int_a^t f^2(s) ds\right). \tag{12.8}$$

The more general integral

$$X_t = \int_a^t f(B_s, s) dB_s$$

is not so obvious and needs the machinery of stochastic integration in order to define it.

Definition 125 *Stochastic integral.* Suppose that B is standard Brownian motion. Let $\{f_t, t \in [a,b]\}$ be some stochastic process adapted to the Brownian motion, i.e., f_t is a function of $\{B_s, s \leq t\}$ with $\int_a^b E(f_t^2) dt < \infty$. Let

$$I(f) = \int_a^b f_t dB_t = \lim_{n \to \infty} I_n(f), \quad I_n(f) = \sum_{i=1}^n f_{t_i}(B_{t_{i+1}} - B_{t_i}), \tag{12.9}$$

where $t_1 = a < t_2 < \cdots < t_n$ is any partition of $[a,b]$ such that $\sup_i |t_{i+1} - t_i| \to 0$ as $n \to \infty$; the limit is defined in quadratic mean, i.e.,

$$\lim_{n \to \infty} E\left((I_n(f) - I(f))^2\right) = 0.$$

The limit can alternatively be defined in probability. Stochastic integration can also be defined for more general f, and for more general stochastic processes than B: the integral $\int H dX$ is defined for a semimartingale X and locally bounded predictable process H (see Protter (2004)).

Stochastic integration like Riemann integration is a linear operator, i.e., for any functions f, g satisfying the above conditions and scalars α, β we have

$$I(\alpha f + \beta g) = \alpha I(f) + \beta I(g). \tag{12.10}$$

The stochastic integral process is itself a martingale under some conditions on the integrand process f, that is, we have with probability one

$$E(X_t|\mathcal{F}_s) = E\left(\int_a^t f_u dB_u \,\big|\, \mathcal{F}_s\right) = \int_a^s f_u dB_u = X_s \qquad (12.11)$$

for all $s < t$. The stochastic integral satisfies the **isometry property**

$$E\left(\left(\int_a^b f_t dB_t\right)^2\right) = E\left(\int_a^b E(f_t^2) dt\right), \qquad (12.12)$$

which is already present in (12.8).

12.3 Diffusion Processes

We next introduce an important general class of processes that includes Brownian motion and many processes derived from it. We first define the concept of a Markov process.

Definition 126 *A Markov process $\{X_t, t \in [0, T]\}$ satisfies*

$$\Pr(X_t \leq x|\mathcal{F}_s) = \Pr(X_t \leq x|X_s) \qquad (12.13)$$

for all x, t, s, that is, given X_s, X_t is independent of X_r, $r < s$.

For example, AR and ARCH processes in discrete time and Brownian motion in continuous time are Markov processes. However, MA and GARCH processes are not Markov in the observed returns.

Definition 127 *A diffusion process is a continuous time **strong Markov process** with continuous sample path.*

The strong Markov property is the Markov property with time replaced by a stopping time τ, that is, if, for each stopping time τ, conditioned on the event $\{\tau < \infty\}$, we have that for each $t \geq 0$, $X_{\tau+t}$ is independent of the past given X_τ. A more common way of introducing a diffusion process is through the stochastic differential equation representation.

12.3 Diffusion Processes

Definition 128 *Suppose that $X_0 = X$, and that*

$$dX_t = \mu(X_t, t)dt + \sigma(X_t, t)dB_t, \qquad (12.14)$$

where X is a given random variable and B_t is standard Brownian motion. The general diffusion process can equivalently be written as

$$X_t = X_0 + \int_0^t \mu(X_s, s)ds + \int_0^t \sigma(X_s, s)dB_s, \qquad (12.15)$$

where the second integral is a stochastic integral but the first one is an ordinary Riemann integral.

The function $\mu(.)$ is called the drift, while $\sigma^2(.)$ is the volatility function or diffusion coefficient. Note that

$$\mu(x) = \lim_{\Delta \to 0} \frac{E(X_{t+\Delta} | X_t = x) - x}{\Delta} \qquad (12.16)$$

whenever the expectation exists. Thus $\mu(x)$ is really the time derivative of the conditional expectation. The volatility function or diffusion coefficient is similarly defined as

$$\sigma^2(x) = \lim_{\Delta \to 0} \frac{E((X_{t+\Delta} - x)^2 | X_t = x)}{\Delta}, \qquad (12.17)$$

i.e., it is the rate of change of the conditional variance. Because B_t is a Gaussian process with no unknown parameters, the entire distribution of the process $\{X_t\}$ is determined solely by $\mu(.), \sigma(.)$.

The process (12.14) generalizes discrete time nonlinear stochastic difference equations $X_{t+1} - X_t = \mu(X_t, t) + \sigma(X_t, t)\varepsilon_{t+1}$, where $t = 1, 2, \ldots$ and $X_0 = X$ given. For a discrete time stochastic difference equation given an initial condition we can always define a unique solution, i.e., X_{t+1}, provided $\mu(.), \sigma(.)$ are well defined on the domain. However, without some conditions there is no guarantee that there is a unique solution to the stochastic differential equation (12.14).

Theorem 48 *The following conditions are sufficient to ensure that there is a unique solution $\{X_t, t \in [0, T]\}$ to (12.15) that is a Markov process with continuous sample paths and satisfies $\int E(X_t^2) < \infty$:*

1. $E(X^2) < \infty$.
2. *Lipschitz condition.* μ, σ *are Borel measurable functions, and there exists a finite K such that for all $x, y \in \mathbb{R}$,*

$$|\mu(x, t) - \mu(y, t)| \leq K|x - y|, \quad |\sigma(x, t) - \sigma(y, t)| \leq K|x - y|.$$

3. *Growth condition. For some K and for all $x \in \mathbb{R}$,*

$$|\mu(x, t)| \leq K(1 + x^2)^{1/2}, \quad |\sigma(x, t)| \leq K(1 + x^2)^{1/2}.$$

See Liptser and Shiryaev (2001, chapter 4). They discuss the distinction between a strong and weak solution. We note that the conditions of this theorem are needed.

Example 62 *Consider the equation*

$$X_t = 3\int_0^t X_s^{1/3} ds + 3\int_0^t X_s^{2/3} dB_s.$$

This has uncountably many solutions of the form (for any $\alpha \in [0, \infty]$),

$$X_t^\alpha = \begin{cases} 0 & 0 \leq t < \beta_\alpha \\ B_t^3 & \beta_\alpha \leq t < \infty, \end{cases}$$

where $\beta_\alpha = \inf\{s \geq \alpha, B_s = 0\}$. The Lipschitz condition on the drift is not satisfied.

Conditions 2 and 3 do not guarantee that the solution process is stationary. Many reasonable and widely used processes are non-stationary. It depends on the application whether stationarity is a desirable property for X_t. Models for stock prices X_t are typically nonstationary.

Definition 129 *The geometric Brownian motion satisfies*

$$dX_t = \mu' X_t dt + \sigma' X_t dB_t$$

for some μ', σ'.

In this case, returns, $d\log X_t$, are stationary but prices X_t are not. On the other hand, models for interest rates usually impose stationarity on X_t.

12.3.1 Stationarity

We now provide conditions that ensure that a diffusion process is strictly stationarity. We need an additional condition. Define the scale density associated with the process

$$s(x) = \exp\left[-\int_{-\infty}^x \frac{2\mu(y)}{\sigma^2(y)} dy\right].$$

Definition 130 *Suppose that the following condition is satisfied*

$$\int_{-\infty}^\infty \frac{dx}{\sigma^2(x)s(x)} < \infty. \tag{12.18}$$

Then, the process X_t is strictly stationary with density p given by

$$p(x) \propto \frac{1}{\sigma^2(x)s(x)}. \tag{12.19}$$

12.3 Diffusion Processes

Note that when $\mu(x) = 0$ for all x, the condition (12.18) becomes the simpler requirement that

$$\int_{-\infty}^{\infty} \frac{dx}{\sigma^2(x)} < \infty. \tag{12.20}$$

In this case the diffusion process is $dX_t = \sigma(X_t)dB_t$, which is a continuous time version of the heteroskedastic unit root process $X_{t+1} - X_t = \sigma(X_t)\varepsilon_{t+1}$. Clearly, if $\sigma^2(x) = \sigma^2$ for all x, the condition (12.20) can't be satisfied, but for other choices of $\sigma^2(x)$ such as $\sigma^2(x) = a + bx^2$, it can. This means that we may have a random walk that is stationary, which may appear rather confusing given the standard assumptions in the literature. This was called **volatility induced stationarity** by Conley, Hansen, Luttmer, and Scheinkman (1997).

We next rewrite the relations between p, μ, σ^2 in a more useful way. Note that

$$\log(\sigma^2 \times p)(x) = 2\int_{-\infty}^{x} \frac{\mu(y)}{\sigma^2(y)} dy - I,$$

where I is the constant of integration. Differentiating with respect to x we find

$$\frac{d}{dx}\log(\sigma^2 \times p)(x) = \frac{1}{(\sigma^2 \times p)(x)} \frac{d}{dx}(\sigma^2 \times p)(x) = 2\frac{\mu(x)}{\sigma^2(x)}.$$

It follows that $d(\sigma^2 \times p)(x)/dx = 2\mu(x)p(x)$, and therefore:

$$\mu(x) = \frac{1}{2p(x)} \frac{d}{dx}(\sigma^2 \times p)(x) \tag{12.21}$$

$$\sigma^2(x) = \frac{2}{p(x)} \int_{-\infty}^{x} \mu(y)p(y)dy. \tag{12.22}$$

These relations (12.19), (12.21), and (12.22) show that given knowledge of any two of p, μ, σ^2, we may explicitly construct the third quantity. This property was exploited by Aït-Sahalia (1996ab); see below.

12.3.2 Itô's Lemma, Rule, Formula, or Theorem

Suppose that we have some diffusion process X_t and transform it by some smooth function $f(X_t, t)$. What are the dynamics of $Y_t = f(X_t, t)$? This question was answered by Itô. His result gives a very convenient algebra for manipulating diffusion processes.

Theorem 49 *Suppose that f has continuous second order partial derivatives. Then, $Y_t = f(X_t, t)$ is a diffusion process with law of motion*

$$dY_t = \left(\frac{\partial f}{\partial X}(X_t, t)\mu(X_t, t) + \frac{\partial f}{\partial t}(X_t, t) + \frac{1}{2}\frac{\partial^2 f}{\partial X^2}(X_t, t)\sigma^2(X_t, t)\right) dt + \frac{\partial f}{\partial X}(X_t, t)\sigma(X_t, t)dB_t.$$

This follows by the heuristic argument that

$$df = \frac{\partial f}{\partial X}(X_t, t)dX_t + \frac{\partial f}{\partial t}(X_t, t)dt + \frac{1}{2}\frac{\partial^2 f}{\partial X^2}(X_t, t)(dX_t)^2$$
$$= \frac{\partial f}{\partial X}(X_t, t)\left(\mu(X_t, t)dt + \sigma(X_t, t)dB_t\right) + \frac{\partial f}{\partial t}(X_t, t)dt$$
$$+ \frac{1}{2}\frac{\partial^2 f}{\partial X^2}(X_t, t)\left(\mu(X_t, t)dt + \sigma(X_t, t)dB_t\right)^2.$$

Then, since $dB_t \sim \sqrt{dt}N(0, 1)$, we have heuristically $(dB_t)^2 \sim dt + dt \times (N(0, 1)^2 - 1)$, and the stochastic part is mean zero but with variance of order $(dt)^2$ and so is of smaller order in probability than the stochastic term dB_t. Note the difference from the usual Taylor theorem due to the $(dX_t)^2$ term. This is another manifestation of the difference between integration of ordinary deterministic functions and stochastic integration of random variables. See for example Mikosch (1998). This result applies (with some modification) to more general functions f (not necessarily twice continuously differentiable) and processes X. If f is invertible in its first argument with inverse g so that $f(g(y, t), t) = y$, then we can express Y_t as

$$dY_t = m(Y_t, t)dt + v(Y_t, t)dB_t$$

for functions m, v.

We can use Itô's lemma to calculate stochastic integrals; see the exercises for an example.

12.3.3 Examples

For stock prices, geometric Brownian motion is a common model for prices. By Itô's lemma this implies that

$$d\log X_t = \mu dt + \sigma dB_t,$$

where $\mu' = \mu + \frac{1}{2}\sigma^2$ and $\sigma' = \sigma$. That is, returns are Brownian motion with drift. This formulation ensures the positivity of prices X_t. This model is used in Black–Scholes and other derivative pricing. The specification implies that $\log X_t - \log X_0 \sim N(\mu t, \sigma^2 t)$. This model is a bit restrictive because it implies time invariant volatility amongst other things; this is no longer accepted as a reasonable empirical model for stock returns. Credible models should allow the volatility to change with time and state.

Stock Return Modelling

We might expect that stock prices are nonstationary, although returns are stationary. For stock returns, interest is often on the diffusion coefficient $\sigma(.)$, because the drift function is hard to identify.

12.3 Diffusion Processes

There is a connection between diffusion processes and discrete time GARCH and stochastic volatility processes. Nelson (1990b) constructs a sequence of GARCH processes $_hX_t$ indexed by sampling frequency h such that as $h \to 0$, $_hX_t$ converges in an appropriate sense to a diffusion process.

Theorem 50 *(Nelson (1990b). For $h > 0$ and $t = 1, 2, \ldots$ with $_hX_0 = X$ and*

$$_hX_{th} = {}_h\sigma_{th}\varepsilon_{t,h}$$

$$_h\sigma_{th}^2 = \omega_h + \beta_{hh}\sigma_{t-1,h}^2 + \gamma_{hh}X_{t-1,h}^2.$$

Suppose that $\omega_h = \omega h$, $\gamma_h = \gamma\sqrt{h/2}$, and $\beta_h = 1 - \gamma\sqrt{h/2} - \theta h$. Then, as $h \to 0$, the returns process $\{_hX_{th}\}$ converges in distribution to the continuous time process

$$dX_t = \sigma_t dB_{1t}$$

$$d\sigma_t^2 = (\omega - \theta\sigma_t^2)dt + \gamma\sigma_t^2 dB_{2t}.$$

The stochastic volatility process (11.42) has a similar limit.

Dynamic Yield Curve Modelling

The evolution of the entire yield curve is modelled as depending linearly on a small number of factors. Without loss of generality one of the factors is taken as the short term interest rate. Hence, there is a big literature on short rate evolution in continuous time. Interest rates are usually assumed to be stationary. The Federal funds rate and T-bill rates were graphed in Chapter 1. In Figure 12.2 below we give the correlogram of the daily Fed funds rate out to 1250 (about 5 years) lags along with the Bartlett 95% confidence intervals. This series is extremely persistent compared with stock returns.

Continuous time models for short term interest rates are usually included in the following class

$$dr_t = (\alpha(t) + \beta(t)r_t)\,dt + \pi(t)r_t^\gamma dB_t, \qquad (12.23)$$

where the functions α, β, and π only depend on time, but may also depend on other parameters. As far as the volatility is concerned, when $\pi(t)$ is time invariant, this is called the constant elasticity of variance class because $d\log\sigma(r)/dr = \gamma$ is constant. The drift is "homogeneous" if $\alpha = 0$ and β does not depend on time.

Some Special Cases in the Literature

Definition 131 *The Vasicek (1977) or Ornstein–Uhlenbeck process*

$$dr_t = \beta(\alpha - r_t)dt + \pi dB_t.$$

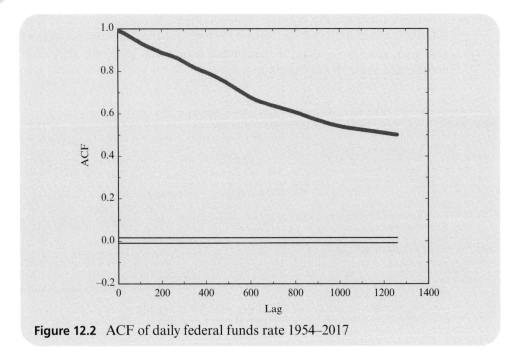

Figure 12.2 ACF of daily federal funds rate 1954–2017

The drift specification implies mean reversion, i.e., the process tends towards the equilibrium value α with speed $\beta > 0$. The solution satisfies

$$r_t = r_0 e^{-\beta t} + \alpha \left(1 - e^{-\beta t}\right) + \pi \int_0^t e^{-\beta s} dB_s.$$

Thus r_t is conditionally normal with

$$E(r_t|\mathcal{F}_s) = r_s e^{-\beta(t-s)} + \alpha \left(1 - e^{-\beta(t-s)}\right), \quad \mathrm{var}(r_t|\mathcal{F}_s) = \frac{\pi^2}{2\beta}\left(1 - e^{-2\beta(t-s)}\right).$$

As $t \to \infty$, we have

$$E(r_t|\mathcal{F}_s) \to \alpha \text{ and } \mathrm{var}(r_t|\mathcal{F}_s) \to \frac{\pi^2}{2\beta}.$$

One can find closed form solutions to derivative pricing problems for this class of processes, which is why it is popular with practitioners. The likelihood function can be easily computed.

Definition 132 *The Cox, Ingersoll, and Ross (1985) or square root process with $r_0 > 0$ and*

$$dr_t = \beta(\alpha - r_t)dt + \pi r_t^{1/2} dB_t. \qquad (12.24)$$

For this stochastic differential equation to have a unique solution it is necessary and sufficient that $2\beta\alpha > \pi^2$ (Feller (1965)).

12.3 Diffusion Processes

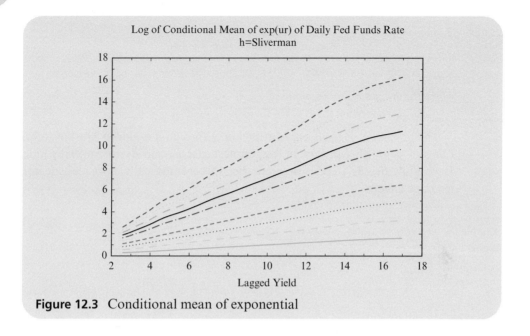

Figure 12.3 Conditional mean of exponential

Because of the square root in the volatility function, the process $\{r_t, t \in [0, T]\}$ never becomes negative, i.e., the state space is $(0, \infty)$. One can show that r_t is conditionally (on r_0) distributed as a non-central chi-squared. One can also get closed form solutions to derivative pricing problems for this class of processes. This is a member of a class of Affine models.

Definition 133 *Affine models. Suppose that for*

$$dr_t = \beta(\alpha - r_t)dt + (c + dr_t)^{1/2} dB_t.$$

This class of processes has the defining property that

$$\log E\left(\exp(iur_t) \,|r_s\right) = a(u) + b(u)r_s,$$

where $a(.)$ and $b(.)$ are functions of u and $i = \sqrt{-1}$.

There are a number of non-affine processes that authors have proposed in the literature. For example, Cox (1975), Marsh and Rosenfeld (1983), Constantinides (1992), and Longstaff (1989). These are all single factor models.

Multi-Factor Models

The class of multi-factor models have additional state variables.

Example 63 *Suppose that*

$$dX_t = \mu_t dt + \sigma_t dB_{1t}$$
$$d\sigma_t = m_t dt + v_t dB_{2t},$$

where B_{1t}, B_{2t} are Brownian motions.

The process X_t is itself not Markovian, although it is **hidden Markov**, meaning that $(X_t, \sigma_t)^\top$ is jointly Markov. In the simplest case B_1 and B_2 are mutually independent. If we allow B_{1t}, B_{2t} to be correlated, then this is empirically more realistic as it permits leverage effects.

Example 64 *The Heston (1993) model is*

$$dX_t = \mu X_t dt + \sqrt{v_t} X_t dB_t$$
$$dv_t = \kappa(\theta - v_t) dt + \xi v_t^{1/2} dW_t,$$

where B_t, W_t are standard Brownian motions with correlation $dB_t dW_t = \rho dt$. Here: μ is the rate of return of the asset, θ is the long variance, or long run average price variance; as t tends to infinity, the expected value of v_t tends to θ, κ is the rate at which v_t reverts to θ, ξ is the volatility of the volatility, or vol of vol, and determines the variance of v_t. If the parameters obey the Feller condition $2\kappa\theta > \xi^2$, then the process ν_t is well defined. Heston found a semi-analytical formula for European option prices under this specification for stock prices.

12.4 Estimation of Diffusion Models

12.4.1 Parametric, Semiparametric, and Nonparametric Models

We can broadly divide this literature into parametric, semiparametric, and nonparametric.

Definition 134 *In the parametric case we have for finite dimensional unknown quantities θ_μ and θ_σ*

$$\mu(x; \theta_\mu) \text{ and } \sigma(x; \theta_\sigma). \tag{12.25}$$

For example, Geometric Brownian motion, Cox–Ingersoll–Ross (CIR) model etc are all parametric.

Definition 135 *In the semiparametric case we may have either*

$$\mu(x; \theta_\mu) \text{ and } \sigma(x), \tag{12.26}$$

12.4 Estimation of Diffusion Models

where $\sigma(\cdot)$ is an unknown function, or

$$\mu(x) \text{ and } \sigma(x; \theta_\sigma), \qquad (12.27)$$

where $\mu(\cdot)$ is an unknown function. That is, we specify one or other functions parametrically and the other function is allowed to be of unknown functional form (Aït-Sahalia (1996a)).

Definition 136 *In the nonparametric case both drift and diffusion*

$$\mu(x) \text{ and } \sigma(x) \qquad (12.28)$$

are of unknown functional form.

Estimation methods differ according to which specification is adopted. It is also key what kind of data is used.

12.4.2 Data and Asymptotic Framework

There are several different types or levels of data. For example, in an electronic trading system, we may have a complete record of all messages sent by all traders to the matching engines of the trading system. These include the type of order, its price and quantity, as well as order cancellations and completed transactions. At a very fine level one may observe also the identity of the traders who submitted the messages. Associated with this data are the time stamp at various physical locations along the message pipeline. We are now discussing a very fine level of detail, whereas most datasets contain much less information. Most empirical work is conducted with transactions data, prices, and quantities along with the time at which the transaction was consummated. These occur at discrete intervals. The order book on the other hand is observed continuously, but one is often interested in particular time points when the best quotes change.

To put this in the context of the models we have been discussing we typically don't observe the continuous sample path. We usually get observations, say prices and quantities, at times $t_0, \ldots, t_n$. Typically the data are not equally spaced, which may be treated in several ways:

1. We may treat the observations as equally spaced. This is often called transaction time as opposed to calendar time (if the observations are transaction prices).
2. We may take account of the time spacings between observations according to the calendar time hypothesis but assume that the time spacings are random and independent of the evolution of the process itself.
3. We may instead allow that the time spacings themselves have information about the evolution of the price process. In this case, they must be jointly modelled with X.

4. The data can be aggregated or rather subsampled to be approximately equally spaced, say five minutes, and then treated according to the calendar time hypothesis.
5. The data can be subsampled to produce a sample of size $N < n$ by some other criterion. For example, one may define $\tau_0 = t_0$ and

$$\tau_{j+1} = \inf\left\{t \in \{t_1, \ldots, t_n\}, t \geq \tau_j, \; |X_t - X_{\tau_j}| \geq \delta\right\},$$

where δ is some threshold.

The analysis of estimation and inference procedures relies on large sample approximations. There are three types of asymptotics regarding the observation schemes that are used here:

1. **Infill or continuous record.** We assume that the series X is observed inside a fixed interval, say $[0, T]$, and that as $n \to \infty$ the set $\{t_1, \ldots, t_n\}$ is becoming dense in $[0, T]$, that is, $\max_j(t_j - t_{j-1}) \to 0$, typically at rate $1/n$. This is better suited to, say, intra day data than monthly data.
2. **Long span.** We assume that the largest time $t_n \to \infty$ and $\inf_j(t_j - t_{j-1}) > 0$. Usually we assume that $t_j - t_{j-1} = 1$.
3. **Mixed case.** This is where both long span and infill hold; see for example Bandi and Phillips (2003). In this case you have a triangular array of times $t_{n1}, \ldots, t_{nn}$ where $t_n \to \infty$ but $t_{nj} - t_{nj-1} \to 0$. For example, $t_{nj} - t_{nj-1} = 1/\sqrt{n}$, for $j = 1, \ldots, n$.

12.4.3 The Identification Issue

We first show that based on the infill framework (that is, data are obtained within a fixed time interval) you cannot identify the drift μ of a diffusion process. Suppose that

$$d \log X_t = \mu dt + \sigma dB_t. \tag{12.29}$$

Suppose that we observe prices at times $\{t_0, \ldots, t_n\}$, in which case $r_{t_j} = \log X_{t_j} - \log X_{t_{j-1}} \sim N(\mu(t_j - t_{j-1}), \sigma^2(t_j - t_{j-1}))$, i.e., stock returns are normally distributed. We can think of the model (12.29) as generating a regression for observed returns on the gap between price observations

$$r_j = \mu \Delta_j + \varepsilon_j \sigma \sqrt{\Delta_j},$$

for $j = 1, \ldots, n$. In the fixed span case Δ_j is small, and the ratio of signal to noise goes to zero, whereas in the long span case, Δ_j is fixed in magnitude and the signal to noise ratio stays bounded away from zero.

Therefore, the log likelihood function of the data $r_{t_1}, \ldots, r_{t_n}$ is

$$\ell(\mu, \sigma^2) = -\frac{n}{2}\log 2\pi - \frac{1}{2}\sum_{j=1}^{n} \log\left((t_j - t_{j-1})\sigma^2\right) - \frac{1}{2}\sum_{j=1}^{n} \frac{\left(r_{t_j} - \mu(t_j - t_{j-1})\right)^2}{\sigma^2(t_j - t_{j-1})}.$$

12.4 Estimation of Diffusion Models

Definition 137 *The MLE of μ, σ is:*

$$\widehat{\mu} = \frac{\sum_{j=1}^{n} r_{t_j}}{\sum_{j=1}^{n}(t_j - t_{j-1})} = \frac{\log X_{t_n} - \log X_{t_1}}{t_n - t_1} \quad (12.30)$$

$$\widehat{\sigma}^2 = \frac{1}{n}\sum_{j=1}^{n}\frac{(r_{t_j} - \widehat{\mu}(t_j - t_{j-1}))^2}{(t_j - t_{j-1})}. \quad (12.31)$$

In the infill case, $t_n \to T$ and $t_0 \to 0$, so that

$$\widehat{\mu} = \frac{\log X_{t_n} - \log X_{t_1}}{t_n - t_1} \implies \frac{\log X_T - \log X_0}{T} \sim N(\mu, \sigma^2/T),$$

which does not concentrate at a point when $T < \infty$. That is, the variance of $\widehat{\mu}$ does not go to zero as the sample size increases. The estimator is unbiased but inconsistent. In this very special case one can construct a confidence interval for μ, since the estimator is normally distributed.

In the long span case we may assume that $\sum_{j=1}^{n}(t_j - t_{j-1}) = t_n - t_1 \to \infty$, in which case $\widehat{\mu} \xrightarrow{P} \mu$ and further

$$\sqrt{\frac{t_n - t_1}{\sigma^2}}(\widehat{\mu} - \mu) \implies N(0, 1),$$

as $n \to \infty$. In this case, the MLE is consistent, and we may construct confidence intervals for μ.
We have

$$\widehat{\sigma}^2 = \frac{1}{n}\sum_{j=1}^{n}\frac{(r_{t_j} - \mu(t_j - t_{j-1}))^2}{(t_j - t_{j-1})} - (\widehat{\mu} - \mu)^2 \frac{t_n - t_1}{n},$$

and

$$E(\widehat{\sigma}^2) = \sigma^2 - E\left((\widehat{\mu} - \mu)^2\right)\frac{t_n - t_1}{n} = \sigma^2\left(1 - \frac{1}{n}\right).$$

In fact, $\widehat{\sigma}^2 \xrightarrow{P} \sigma^2$ and

$$\sqrt{n}\left(\widehat{\sigma}^2 - \sigma^2\right) \implies N\left(0, 2\sigma^4\right).$$

This is true for whatever sequence $\{t_j\}$, i.e., for long span and infill.

Note that in continuous time, non-equally spaced observations do not themselves cause any problem for deriving the properties of estimators, unlike in discrete time.

12.4.4 Maximum Likelihood Method for Parametric Diffusion Models in the Long Span Case

Linear in Parameters Models

There is literature from the 1970s about estimating linear continuous time models, i.e., where $\sigma(.)$ is constant or at least only a deterministic function of time. See Bergstrom (1988) and Bergstrom and Nowman (2007). Specifically, suppose that

$$dX_t = (aX_t + b)dt + dB_t,$$

where a, b are unknown parameters, or equivalently, given initial condition X_0

$$X_t = \int_0^t e^{a(t-r)} B_r dr + \left(X_0 + \frac{b}{a}\right) e^{at} - \frac{b}{a}.$$

Theorem 51 *Suppose that we observe the process $\{X_t, t \in [0, T]\}$ at times $t = 1, 2, \ldots$ and suppose that $X_0 = 0$. Then*

$$X_t = \alpha + \phi X_{t-1} + \varepsilon_t,$$

where $\alpha = (e^a - 1)\frac{b}{a}$, $\phi = e^a$, and

$$\varepsilon_t = \int_{t-1}^t e^{a(t-r)} B_r dr \sim N\left(0, \frac{1}{2a}\left(e^{2a} - 1\right)\right), \qquad E(\varepsilon_t \varepsilon_s) = 0 \text{ for } s \neq t.$$

The observed discrete time process is a Gaussian AR(1), which can be estimated by MLE. The parameters of the continuous time model enter the discrete time process in a nonlinear way and can be obtained from the estimated α, ϕ, by

$$a = \log \phi \quad ; \quad b = \frac{\alpha \log \phi}{\phi - 1}.$$

Nonlinear Models

When $\sigma(.)$ is non-constant and/or $\mu(.)$ is nonlinear this argument typically cannot be applied, because the derived discrete time process is much more complicated. The reason is that there is dependence of a complicated kind in the observed data. So how should one proceed?

One would like to use something like the prediction error decomposition to form the likelihood

$$\ell(\theta | X_{t_1}, \ldots, X_{t_n}) = \prod_{j=1}^T p_{X_{t_{j+1}} | X_{t_j}}(X_{t_{j+1}} | X_{t_j}; \theta) p_{X_{t_0}}(X_{t_0}; \theta) \tag{12.32}$$

based on the observed data $\{X_{t_1}, \ldots, X_{t_n}\}$. Here, we use the Markov property that only the most recent past is needed to simplify the transition densities

12.4 Estimation of Diffusion Models

$p_{X_{t_{j+1}}|X_{t_j},\ldots,X_{t_1}}(X_{t_{j+1}}|X_{t_j},\ldots,X_{t_1};\theta)$ (note that for stochastic volatility models, such as the Heston model, one needs further arguments to obtain the likelihood for the observed data since the Markov property does not hold). One may argue that for stationary processes the marginal term $p_{X_{t_0}}(X_{t_0};\theta)$ only contributes a little to the total likelihood and may be ignored. The difficult part is to obtain the transition densities $p_{X_{t_{j+1}}|X_{t_j}}(X_{t_{j+1}}|X_{t_j};\theta)$. These transition densities are not known in closed form except for very special cases. We consider several approaches to computing the likelihood or an approximation to it.

PDE approach. It is known that the transition densities satisfy two sets of **partial differential equations (PDEs)**, the Chapman–Kolmogorov so-called **forward and backwards equations**. Denote by $p(\Delta, y|x)$ the time invariant transition density from $X_t = x$ to $X_{t+\Delta} = y$, for any $\Delta \geq 0$.

Definition 138 *Forward equation:*

$$\frac{\partial p(\Delta, y\,|\,x)}{\partial \Delta} = -\frac{\partial}{\partial y}(\mu(y)p(\Delta, y\,|\,x)) + \frac{1}{2}\frac{\partial^2}{\partial y^2}\left(\sigma^2(y)p(\Delta, y|x)\right).$$

Definition 139 *Backward equation:*

$$\frac{\partial p(\Delta, y\,|\,x)}{\partial \Delta} = -\mu(x)\frac{\partial p(\Delta, y|x)}{\partial x} + \frac{1}{2}\sigma^2(x)\frac{\partial^2 p(\Delta, y\,|\,x)}{\partial x^2}.$$

These equations specify laws of motion for the transition densities in terms of $\mu(.)$ and $\sigma(.)$. These equations each require boundary conditions, which we specify next.

Definition 140 *Boundary conditions:*

$$\lim_{\Delta \to 0} p(\Delta, y\,|\,x) = \begin{cases} 1 & \text{if } x = y \\ 0 & \text{else} \end{cases}$$

$$(FWD) \lim_{y \to \partial\Omega} p(\Delta, y\,|\,x) = 0$$

$$(BWD) \lim_{x \to \partial\Omega} p(\Delta, y\,|\,x) = 0.$$

Here, Ω is the state space and $\partial\Omega$ is its boundary.

Chapter 12 Continuous Time Processes

In most cases, we consider $\Omega = \mathbb{R}$ so that $\partial \Omega = \{\pm \infty\}$. Likewise, the conditional expectations $V(x, \Delta) = E(X_{t+\Delta}|X_t = x)$ satisfy the PDE

$$\frac{1}{2}\sigma^2(x)V_{xx}(x,\Delta) + \mu(x)V_x(x,\Delta) + V_\Delta(x,\Delta) = 0 \qquad (12.33)$$

with $V(x,0) = x$.

Lo (1988) shows how to construct an approximation to the MLE for parametric models based on solving the transition density PDE equations.

Algorithm:

(i) For each θ, compute $\mu_\theta(X_{t_j})$ and $\sigma_\theta^2(X_{t_j})$.
(ii) Solve either the forward or backward equation for $p_{X_{t_{j+1}}|X_{t_j}}(X_{t_{j+1}}|X_{t_j};\theta)$.
(iii) Compute the approximate likelihood function $\ell(\theta)$.
(iv) Repeat to find maximizing value of θ.

These equations must be solved for each parameter value θ, which makes this procedure computationally demanding and potentially inaccurate since the partial differential equations must be solved by numerical methods.

Simulation methods. We next consider an alternative way of approximating the likelihood function based on simulation, sometimes called **Euler discretization**.

Definition 141 *In the interval $[t_j, t_{j+1})$ we can approximate the process by*

$$X_{t_j+(m+1)h} = X_{t_j+mh} + \mu(X_{t_j+mh};\theta)h + \sigma(X_{t_j+mh};\theta)\varepsilon_{t_j+(m+1)h}h^{1/2} \qquad (12.34)$$

for $m = 0, 1, \ldots, M-1$, where $h = (t_{j+1} - t_j)/M$, and ε_{t_j+mh} are i.i.d. standard normal random variables.

The approximation is valid as $h \to 0$ and $M \to \infty$. This says that on very small time intervals the process X can be approximated by a discrete time process, with given mean and variance and normal conditional distribution. For this discrete time process, the one-step ahead transition densities are normal

$$p_{X_{t_j+(m+1)h}|X_{t_j+mh}}(y|x;\theta) = \phi_{\mu_j(x),\sigma_j^2(x)}(y),$$

where $\mu_j(x) = x + \mu(x;\theta)h$ and $\sigma_j^2(x) = \sigma^2(x;\theta)h$. However, we need the M-step ahead densities, and to find them involves recursive integration. Thus

$$p_{X_{t_j+(m+2)h}|X_{t_j+mh}}(y|x;\theta) = \int p_{X_{t_j+(m+2)h}|X_{t_j+(m+1)h}}(y|z;\theta) p_{X_{t_j+(m+1)h}|X_{t_j+mh}}(z|x;\theta)dz$$

$$= \int \phi_{\mu_j(z),\sigma_j^2(z)}(y)\phi_{\mu_j(x),\sigma_j^2(x)}(z)dz.$$

12.4 Estimation of Diffusion Models

There is no closed form for this transition density in general.

To get the M-step ahead densities you need to compute an $M-1$-fold integral. One can do this by simulation methods as we next show. Suppose we want to compute the integral

$$I = \int_A h(x)dx,$$

where h is some given function and $A \subset \mathbb{R}^d$. Write

$$I = \int_A \frac{h(x)}{g(x)} g(x)dx,$$

where g is some density from which it is easy to draw random variables. Let $x_1, \ldots, x_S$ be S draws from g, then let

$$\widehat{I} = \frac{1}{S} \sum_{i=1}^{S} \frac{h(x_i)}{g(x_i)}. \tag{12.35}$$

Provided $E(|h(x_i)/g(x_i)|) < \infty$, the LLN says that $\widehat{I} \to I$ with probability one as $S \to \infty$. This gives a way of approximating $p_{X_{t_{j+1}}|X_{t_j}}(X_{t_{j+1}}|X_{t_j}; \theta)$, which can then be inserted in the likelihood function.

Finally, the method of **indirect inference** (Gourieroux, Monfort, and Renault (1993)) can be used. We estimate an auxiliary model whose likelihood is simple to compute. Then we simulate data from the true model for parameter θ using Euler discretization with largeish M. Then choose θ to minimize some distance between the simulated data and the auxiliary model.

12.4.5 Generalized Method of Moments Estimation for Long Span

We next consider the method of moments for estimating diffusion processes as it bypasses the technical complications around computing likelihood functions. When the process X_t is stationary we can obtain simple moment conditions that can be used to generate estimators of finite dimensional parameters. Chen, Hansen, and Scheinkman (2009) present some useful theory, which we describe next.

Definition 142 *For any measurable function ϕ, define the shift operator T_t*

$$T_t \phi(y) \equiv E(\phi(X_t) \mid X_0 = y)$$

on the space of functions $\mathcal{L}^2(P)$ that are square integrable with respect to P.

Theorem 52 *The set $\{T_t: t \geq 0\}$ is a **semigroup**, i.e., $T_{t+s} = T_t T_s = T_s T_t$ for all t, s.*

Proof. We have

$$\begin{aligned}
T_{t+s}\phi(y) &\equiv E\left(\phi(X_{t+s}) \mid X_0 = y\right) \\
&= E\left(E\left(\phi(X_{t+s}) \mid X_t, X_0\right) \mid X_0 = y\right) \\
&= E\left(E\left(\phi(X_{t+s}) \mid X_t\right) \mid X_0 = y\right) \\
&= E\left(E\left(\phi(X_s) \mid X_0 = X_t\right) \mid X_0 = y\right) \\
&= E\left(T_s\phi(X_t) \mid X_0 = y\right) \\
&= T_t\left(T_s\phi\right)(y),
\end{aligned}$$

using the law of integrated expectations, the Markov property, and time homogeneity. $\square$

Definition 143 *The **infinitesimal generator** $\mathcal{A}$ is an operator defined as*

$$\lim_{t \to 0} \frac{T_t\phi - \phi}{t} = \mathcal{A}\phi,$$

whenever the limit exists.

The operator $\mathcal{A}$ describes the local evolution of the process. Take $\phi \in \mathcal{L}^2(P)$ a smooth test function. Then, by Itô's lemma we have

$$d\phi_t = \phi'(X_t)dX_t + \frac{1}{2}\phi''(X_t)(dX_t)^2 = \left(\mu\phi' + \frac{\sigma^2}{2}\phi''\right)(X_t)dt + (\sigma\phi')(X_t)dW_t,$$

whence

$$\mathcal{A}\phi = \mu\phi' + \frac{\sigma^2}{2}\phi''.$$

Theorem 53 *For all $\phi \in D$ the domain of $\mathcal{A}$, we have*

$$E(\mathcal{A}\phi(X_t)) = 0.$$

Proof. $\{X_t\}$ is stationary implies that $E(\phi(X_t))$ is independent of t. Therefore,

$$\frac{d}{dt}E(\phi(X_t)) \equiv 0$$

for all $\phi \in \mathcal{L}^2(P)$. By the law of iterated expectation for all $\phi \in \mathcal{L}^2(P)$

$$E(\phi(X_t)) = E(E(\phi(X_t)] \mid X_0)) = E(T_t\phi)$$

12.4 Estimation of Diffusion Models

if and only if $E(\mathcal{T}_t\phi - \phi) = 0$. Now we restrict to $\phi \in D \subset \mathcal{L}^2(P)$. Then

$$E\left(\lim_{t \to 0} \frac{1}{t}(\mathcal{T}_t\phi - \phi)\right) = E(\mathcal{A}\phi)$$

exists. It can be shown that for a suitable set of ϕ,

$$E(\mathcal{A}\phi) = E\left(\lim_{t \to 0} \frac{1}{t}(\mathcal{T}_t\phi - \phi)\right) = \lim_{t \to 0} \frac{1}{t} E(\mathcal{T}_t\phi - \phi) = 0. \qquad \square$$

Xiaohong Chen is Malcolm K. Brachman Professor of Economics at Yale University. She previously taught at University of Chicago (1993–1999), London School of Economics (1999–2002), and New York University (2002–2007). She has been an elected Fellow of the Econometric Society since 2007. She was awarded the 2017 China Economics Prize, rewarding her contribution in the field of econometrics. She has done fundamental work on estimation and inference of models defined by moment conditions in a variety of complex semiparametric and nonparametric settings. She is also well known for her work on estimation of diffusion models, on the weak dependence properties of GARCH models, and on the estimation of copula based models. She was born and raised in Hubei province, China. She got her bachelor's degree in mathematics from Wuhan University and PhD in economics from University of California at San Diego.

This result can be used to deliver an estimation strategy as follows. Suppose that the drift and diffusion functions are parametric, denoted $\mu_\theta, \sigma_\theta^2$ for some unknown parameters $\theta \in \mathbb{R}^p$. Then compute a quadratic form in the vector of sample moments

$$G_{nk}(\theta) = \frac{1}{n} \sum_{j=1}^{n} \left(\phi_k'(X_{t_j}) \mu_\theta(X_{t_j}) + \frac{1}{2} \phi_k''(X_{t_j}) \sigma_\theta^2(X_{t_j}) \right) \qquad (12.36)$$

for some functions $\phi_1, \ldots, \phi_K$ chosen by the practitioner. We then define $\hat{\theta}_{GMM}$ to minimize the objective function (10.10) with W_n a $K \times K$ symmetric positive definite weighting matrix. If we take $W_{opt} = \Omega^{-1}$, then the resulting estimator is consistent and asymptotically normal, and optimal within this class of estimators. However, it is not as efficient as the MLE since it only uses the marginal distribution of the process. One can improve efficiency by using a second set of moment conditions that uses joint distribution information. For example, we know that $E(\phi(X_{t+1})\psi(X_t))$ and $E(\phi(X_t)\psi(X_{t+1}))$ do not depend on calendar time t for all $\phi, \psi \in \mathcal{L}^2(P)$. We can obtain restrictions from these second set of moment conditions, and these improve the efficiency of GMM.

12.4.6 Nonparametric and Semiparametric Approaches in Long Span

Aït-Sahalia (1996a) considered a semiparametric model where the drift was parametric $\mu(.;\theta_\mu)$ but the volatility $\sigma(.)$ was nonparametric. In particular, he considered a linear drift

$$\mu(r_t) = \beta(\alpha - r_t),$$

where α, β are unknown parameters. The volatility function, which is crucial for a lot of derivative pricing, is unspecified. The framework he considered was one with equally spaced data (daily in application) and long span asymptotics, i.e., many weeks.

Yacine Aït-Sahalia is the Otto A. Hack 1903 Professor of Finance and Economics at Princeton University where he served as the inaugural Director of the Bendheim Center for Finance from 1998 until 2014. He was previously an Assistant Professor (1993–1996), Associate Professor (1996–1998) and Professor of Finance (1998) at the University of Chicago's Graduate School of Business, where he received the Emory Williams Award for Excellence in Teaching in 1995. His research concentrates on financial econometrics, fixed income and derivative securities, and optimal portfolio selection, and has been published in leading academic journals. His research contributions in financial econometrics include various methods to estimate and test continuous-time models that are sampled at discrete time intervals, including nonparametric methods, closed-form expansions for the transition density of continuous-time models and various methods to analyze high frequency data with a particular emphasis on the presence of jumps. He recently authored *High Frequency Financial Econometrics* with Jean Jacod, served as the editor of the *Review of Financial Studies* and an associate editor for *Econometrica*, the *Journal of Finance*, and the *Annals of Statistics*. He currently serves as the co-managing editor of the *Journal of Econometrics*. Professor Aït-Sahalia is a Fellow of the Econometric Society, a Fellow of the Institute of Mathematical Statistics, a Fellow of the American Statistical Association, an Alfred P. Sloan Foundation Research Fellow, a Fellow of the Guggenheim Foundation, and a Research Associate for the National Bureau of Economic Research. He received his PhD in Economics from the Massachusetts Institute of Technology in 1993 and is a graduate of École Polytechnique in France.

He proposed the following estimation strategy. Note that the conditional expectation of $r_{t+1}|r_t$ is linear, i.e.,

$$E(r_{t+1}|r_t) = \alpha + e^{-\beta}(r_t - \alpha) = \theta_0 + \theta_1 r_t \qquad (12.37)$$

12.4 Estimation of Diffusion Models

regardless of the volatility. Therefore, the parameters $\theta = (\theta_0, \theta_1)$ can be estimated by OLS, and used to obtain estimates of the more meaningful parameters (α, β). Second, estimate the marginal density of r_t by kernel smoothing methods, i.e., let

$$\widehat{p}(r) = \frac{1}{nh} \sum_{i=1}^{n} K\left(\frac{r - r_{t_i}}{h}\right)$$

for some bandwidth h and kernel K. Under the stationarity condition and under weak dependence, this estimator is consistent as $n \to \infty$ (Bosq (1998)). Finally, one can estimate the volatility by the plug-in method

$$\widehat{\sigma}^2(r) = \frac{2}{\widehat{p}(r)} \int_0^r \mu(r; \widehat{\theta}) \widehat{p}(r) dr. \tag{12.38}$$

This estimator is also consistent and asymptotically normal under some regularity conditions. Aït-Sahalia also develops estimators for derivatives prices and the sampling theory thereof. He applies his method to short term interest rates and shows that the shape of $\sigma^2(r)$ is quite nonlinear.

Stanton (1997) constructs first-, second-, and third-order approximations for drift and diffusion term. He finds substantial nonlinearity in the drift function in daily 3 month T-bills. Fan and Zhang (2003) argue against higher order approximations. They show that higher order approximations reduce the numerical approximation errors in asymptotic biases but escalate (nearly exponentially) the asymptotic variances. Chapman and Pearson (2000) study the sample paths of simulated CIR model and conclude that the estimators of all authors display spurious nonlinearities (e.g., nonlinear estimated drift when the true drift is linear). Fan and Zhang (2003) develop a nonparametric test for nonlinearity and find that there is no strong evidence against the null hypothesis of linear drift (with weekly 1954–1999 US T-bill rate data). This is consistent with the finding of Pritsker (1998) that the test in Aït-Sahalia (1996a) rejects too often. Hong and Li (2005) develop a nonparametric specification test based on the transition density that is "robust to serial dependence and provides excellent finite sample performance." They reject all candidate models, including the one proposed by Aït-Sahalia (1996b). That is, their test has much better power than the marginal density test. Kristensen (2010) extends the estimation methodology to allow nonlinear parametric drift.

12.4.7 Nonparametric and Semiparametric Approaches: In-fill Asymptotics

Florens-Smirou (1993) considered nonparametric estimation of the volatility function under in-fill asymptotics. The idea is that

$$(dX_t)^2 = \sigma^2(X_t)dt + 2\mu(X_t)\sigma(X_t)dt dB_t + \mu^2(X_t)(dt)^2$$
$$= \sigma^2(X_t)dt + noise + smaller.$$

Interpreting $(dX_t)^2$ as squared returns, one essentially has a nonparametric regression model over small time increments.

We follow the set-up of Jiang and Knight (1997). Suppose that observations are equally spaced, infilling on $[0, T]$. Write $t_i = i\Delta_n, i = 1, \ldots, n$, where $\Delta_n = T/n$ is the spacing of the data. Their estimator is

$$\widehat{\sigma}^2(x) = \frac{1}{\Delta_n} \frac{\sum_{i=1}^{n-1} K\left(\frac{X_{t_i} - x}{h_n}\right) \left[X_{t_{i+1}} - X_{t_i}\right]^2}{\sum_{i=1}^{n-1} K\left(\frac{X_{t_i} - x}{h_n}\right)}, \tag{12.39}$$

where K is a kernel and h is a bandwidth. Florens-Smirou is the special case with uniform kernel. Jiang and Knight (1997) show that $\widehat{\sigma}^2(x)$ is consistent and has asymptotically a **mixed normal** distribution. To discuss the limiting distribution we need the concept of **local time**.

Definition 144 *The occupation measure counts the number of visitations of the Borel set B by the process X_s over the interval $[0, t]$*

$$\nu_t(B) = \int_0^t 1(X_s \in B) ds.$$

Definition 145 *The local time of the process $\{X_t\}$ at point x over the time interval $[0, t]$ is defined as the random variable*

$$L_t(x) = \lim_{\Delta \to 0} \frac{1}{2\Delta} \int_0^t 1(|X_s - x| < \Delta) ds.$$

The local time L_t can be interpreted as the Radon–Nikodym derivative of ν_t, i.e., we have $\nu_t(B) = \int_B L_t(x) dx$. If X_t is stationary, $L_t(x) = tp(x)$. Local time is continuous in both arguments and is nondecreasing in t with probability one. The inverse local time is $\tau_u(x) = \inf\{t > 0 : L_t(x) > u\}$, and is related to the crossing time defined in (12.3) for Brownian motion. A key property of local time is the following.

Theorem 54 *For semimartingale $\{X_t\}$ and for every Borel function f of (X_t), we have*

$$\int_0^T f(X_t) \, dt = \int_{-\infty}^{+\infty} f(x) L_T(x) dx.$$

We define the concept of **semimartingale** below. We next define the concept of recurrence, which is essential for any nonparametric estimation procedure.

Definition 146 *The process $\{X_t\}$ is recurrent, if at every point x on its support, $L_T(x) \to \infty$ as $T \to \infty$.*

This basically says that the process revisits the point x infinitely many times during the interval $[0, T]$ as time T goes to infinity. A stationary process is recurrent;

12.4 Estimation of Diffusion Models

indeed the rate of convergence of $L_T(x)$ is order T in this case, which means that it visits every point in its state space a positive fraction of T times. A large class of nonstationary processes are also recurrent but with lower rates of convergence. For a unit root process or Brownian motion, the rate is $T^{1/2}$, which means that the frequency of visitation divided by T goes to zero. See Phillips and Park (1999) for a discussion of local time.

Jiang and Knight (1997) show that

$$(nh_n)^{1/2}\left(\frac{\widehat{\sigma}^2(x)}{\sigma^2(x)} - 1\right) \Longrightarrow L_T^{-1/2}(x) Z \sim MN(0, L_T^{-1}(x)), \qquad (12.40)$$

where Z is a standard normal independent of the random variable $L_T(x)$. Here, MN denotes mixed normal distribution. The local time can be estimated by

$$\widehat{L}_T(x) = \frac{1}{\Delta_n n h_n} \sum_{i=1}^{n-1} K\left(\frac{X_{t_i} - x}{h_n}\right)$$

and they obtain that

$$(nh_n)^{1/2} \widehat{L}_T^{1/2}(x) \left(\frac{\widehat{\sigma}^2(x)}{\sigma^2(x)} - 1\right) \Longrightarrow Z, \qquad (12.41)$$

from which standard pointwise confidence intervals for $\sigma^2(x)$ can be produced.

Bandi and Phillips (2003) consider both infill and long span asymptotics and show how to fully exploit the advantageous features of both informational accumulations. The time span is denoted T and the number of observations n. Let $\Delta_{n,T} = T/n$, where both $n, T \to \infty$ but $\Delta_{n,T} \to 0$. That is, the time span increases but the observations are becoming dense in the interval $[0, T]$. For example, if $T(n) = \sqrt{n}$, then this condition is fulfilled. They estimate both μ, σ^2 nonparametrically:

$$\widehat{\mu}(x) = \frac{1}{\Delta_{n,T}} \frac{\sum_{i=1}^{n-1} K\left(\frac{X_{t_i} - x}{h_{n,T}}\right)(X_{t_{i+1}} - X_{t_i})}{\sum_{i=1}^{n-1} K\left(\frac{X_{t_i} - x}{h_{n,T}}\right)},$$

$$\widehat{\sigma}^2(x) = \frac{1}{\Delta_{n,T}} \frac{\sum_{i=1}^{n-1} K\left(\frac{X_{t_i} - x}{h_{n,T}}\right)(X_{t_{i+1}} - X_{t_i})^2}{\sum_{i=1}^{n-1} K\left(\frac{X_{t_i} - x}{h_{n,T}}\right)}.$$

They actually work with a slightly different estimator that updates the squared increment part by an estimator $\widetilde{\sigma}^2$

$$\widehat{\sigma}^2(x) = \frac{\sum_{i=1}^{n-1} K\left(\frac{X_{t_i} - x}{h_n}\right)\widetilde{\sigma}^2(X_{t_i})}{\sum_{i=1}^{n-1} K\left(\frac{X_{t_i} - x}{h_n}\right)}, \quad \widetilde{\sigma}^2(X_{t_i}) = \frac{1}{m_i \Delta_{n,T}} \sum_{j \in \mathcal{I}_i} \left(X_{t_{i_{j+1}}} - X_{t_{i_j}}\right)^2,$$

where I_i is the set of points for which X_{i_j} is close to X_i as measured by another bandwidth b_n, and m_i is the cardinality of I_i. They don't require stationarity but they do require a null recurrence property. They establish the mixed asymptotic normality for both $\widehat{\mu}(x)$ and $\widehat{\sigma}^2(x)$ and give explicit formulae for biases etc.

12.5 Estimation of Quadratic Variation Volatility from High Frequency Data

We continue our focus on estimation of volatility in high frequency settings, but we consider a more general setting, not necessarily a diffusion process, and we change our target of estimation to accumulated volatility over an interval of time.

Definition 147 *The quadratic variation of a square integrable process X_t is the process*

$$\langle X, X \rangle_{0:t} = \text{plim}_{\max\{t_{k+1}-t_k\} \to 0} \sum_{t_k \leq t} |X_{t_{k+1}} - X_{t_k}|^2, \qquad (12.42)$$

where $0 = t_1 < t_2 < \cdots < t_n = t$. We sometimes denote this just by QV.

This is an ex-post measure of volatility over the interval $[0, t]$. For functions of bounded variation (Definition 124), the quadratic variation exists, and is zero. Furthermore, Andersen, Bollerslev, Diebold, and Labys (2003) show that under some quite general conditions

$$E\left((X(t+h) - X(t))^2 \, |\mathcal{F}_t\right) = E\left(\langle X, X \rangle_{t:t+h} \, |\mathcal{F}_t\right),$$

that is, the conditional variance of returns is equal to the conditional expectation of the quadratic variation over the same interval. This justifies the current interest in this quantity as a parameter of interest.

Example 65 *Suppose that X_t is a diffusion process*

$$dX_t = \mu(X_t)dt + \sigma(X_t)dB_t,$$

where B_t is standard Brownian motion. Then

$$\langle X, X \rangle_{0:t} = \int_0^t \sigma^2(X_s)ds.$$

The quadratic variation is a stochastic process in general, but when $\sigma^2(X_t) = \sigma^2$ is constant, it is just $t\sigma^2$, and when $t = 1$, this is just σ^2. However, the process $\langle X, X \rangle_t$ can

12.5 Estimation of Quadratic Variation Volatility

be shown to exist for a much larger class of stochastic processes, namely, continuous square integrable semimartingales. We endeavor to define semimartingales.

Definition 148 *A local martingale M is a stochastic process that satisfies the localized version of the martingale property. That is, there exists a sequence of stopping times, τ_k with $\lim_{k \to \infty} \tau_k = \infty$, for which the stopped process $M_{\min\{t,\tau_k\}}$ is a martingale.*

Every martingale is a local martingale; every bounded local martingale is a martingale, but not every local martingale is a martingale. In particular, a driftless diffusion process is a local martingale, but not necessarily a martingale.

Definition 149 *A real valued process X is called a semimartingale if it can be decomposed as*

$$X(t) = M(t) + A(t),$$

*where M is a local martingale and A is an adapted process (depends only on the past) of locally bounded variation with sample paths that are **cadlag** (right continuous with left limits).*

The process A has zero quadratic covariation; essentially it is slower moving than M, and so its predictability does not help much. The class of semimartingales is basically the class of processes for which stochastic integration makes sense. The class of semimartingales is large; it includes all martingales as special cases, as well as other processes. This includes Brownian driven processes and processes with jumps. This class has economic meaning – there is an absence of arbitrage opportunities for this class of processes. Fundamental theorem of asset pricing states that no arbitrage means existence of an equivalent martingale measure. The **Girsanov theorem** holds for semimartingales. **Fractional Brownian motion** is not a semimartingale (Rogers (1997)) and hence allows arbitrage opportunities.

We now consider a consistent estimator of the quadratic variation, called **realized volatility**, which was already introduced in (11.8); we just use a different notation. Suppose that we have a sample of n log prices X_t observed on an equal spaced interval over the period $[0, 1]$. Let

$$RV_X^n = \sum_{i=1}^{n-1} \left(X_{\frac{i+1}{n}} - X_{\frac{i}{n}} \right)^2. \tag{12.43}$$

This consistently estimates the quadratic variation of X over the interval $[0, 1]$. Jacod and Protter (1998) establish the CLT of this quantity for Itô semimartingales. Andersen, Bollerslev, Diebold, and Labys (2001) established various useful properties. Barndorff-Nielsen and Shephard (2002) establish consistency and the limiting distribution for the following class of Brownian semimartingales.

Definition 150 *The process X_t is a Brownian semimartingale if*

$$X_t = \int_0^t \mu_u du + \int_0^t \sigma_u dB_u,$$

where the processes μ, σ are predictable (depend only on the past) and the process σ is cadlag.

Barndorff-Nielsen and Shephard (2002) work with the so-called no-leverage case, which corresponds to the process μ and σ being independent of the process B. They show that

$$n^{1/2}(RV_X^n - QV) \Longrightarrow 2^{1/2} \int_0^1 \sigma_u^2 dB_u = MN(0, 2\int_0^1 \sigma_u^4 du), \qquad (12.44)$$

i.e., the limiting distribution is a mixed normal with random variance (that is independent of the underlying normal). They also show that one can estimate the **integrated quarticity** $IQ = \int_0^1 \sigma_u^4 du$ consistently by

$$\widehat{IQ} = \frac{n}{3}\sum_{i=1}^{n-1}\left(X_{\frac{i+1}{n}} - X_{\frac{i}{n}}\right)^4,$$

and they obtain a feasible CLT

$$\frac{n^{1/2}(RV_X^n - QV)}{\sqrt{2\widehat{IQ}}^{1/2}} \Longrightarrow N(0,1). \qquad (12.45)$$

This can be used to set confidence intervals for the estimated volatility and to carry out hypothesis testing.

Neil Shephard, FBA, is a Professor of Economics and of Statistics at Harvard University. He studied economics and statistics as an undergraduate at the University of York, graduating in 1986. He did his MSc and PhD (awarded in 1990) at the LSE, where he was a faculty lecturer from 1988 to 1993 in statistics. He moved to Nuffield College, Oxford in 1991, originally as the Gatsby Research Fellow in Econometrics. He became an Official Fellow in Economics in 1993, a position he held until 2006, when he was appointed to a statutory professorship in economics at Oxford University. He was Director of the Oxford Financial Research Centre from 2006 to 2007 and with Colin Mayer (Saïd Business School, Oxford) founded Oxford University's Masters in Financial Economics (MFE). In 2007 he founded the Oxford-Man Institute, which he directed from 2007 to 2011. He moved to Harvard University in 2013. He was elected a

12.5 Estimation of Quadratic Variation Volatility

Fellow of the British Academy in 2006, a Fellow of the Econometric Society in 2004, and a Fellow of Nuffield College, Oxford in 1991. He was awarded an honorary doctorate by Aarhus University in 2009, the 2012 Richard Stone Prize in Applied Econometrics, and the 2017 Guy Medal in Silver of the Royal Statistical Society. His most well known contributions are: (i) the formalization of the econometrics of realized volatility, which nonparametrically estimates the volatility of asset prices; (ii) the introduction of the auxiliary particle filter (signal extraction); (iii) the nonparametric identification of jumps in financial economics, through multipower variation; and (iv) stochastic volatility models based on non-Gaussian Ornstein–Uhlenbeck processes, known as "Barndorff-Nielsen–Shephard" models.

Example 66 *We can derive some intuition from the special case where*

$$dX_t = \sigma_t dB_t,$$

where σ_t is a deterministic function $\sigma_t = \sigma(t)$. In this case, returns are normally distributed with

$$X_{\frac{i+1}{n}} - X_{\frac{i}{n}} \sim N\left(0, \int_{\frac{i}{n}}^{\frac{i+1}{n}} \sigma^2(t) dt\right).$$

Suppose that $\sigma(t)$ is continuously differentiable. Then by the mean value theorem we can approximate the integral by

$$\int_{\frac{i}{n}}^{\frac{i+1}{n}} \sigma^2(t) \simeq \frac{1}{n} \sigma_{i/n}^2.$$

Letting z_i be i.i.d. standard normal random variables, we have

$$n^{1/2}(RV_X^n - QV) = \frac{1}{n^{1/2}} \sum_{i=1}^{n-1} \left(\int_{\frac{i}{n}}^{\frac{i+1}{n}} \sigma^2(t) dt\right)(z_i^2 - 1) \simeq N\left(0, \frac{2}{n} \sum_{i=1}^{n-1} \sigma_{i/n}^4\right)$$

$$\simeq N\left(0, 2\int_0^1 \sigma_u^4 du\right).$$

In this case the limiting variance is non-stochastic so the distribution is normal.

12.5.1 Measurement Error Model

These theoretical results are designed to work in a perfect continuous time laboratory. In practice, as we saw in Chapter 5, quote and transaction prices take values in a discrete state space with a finite number of possible values and at discrete points in time. These market frictions mean that for very high frequency data the above approximations can be poor, as has been evidenced in the so-called **volatility signature plots**. Figure 12.4 shows the RV sampled at higher frequency plotted against frequency, which should show convergence to QV, but actually appears to show convergence to infinity.

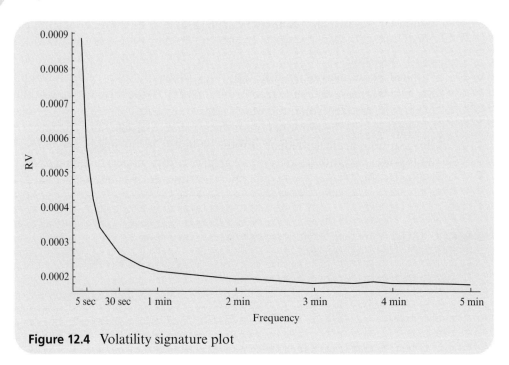

Figure 12.4 Volatility signature plot

To reflect the divergence between the predicted and actual behavior of RV, the literature has introduced a statistical model that tries to capture the effects of microstructure noise.

Definition 151 *Suppose that we observe (log) prices $\{Y_{t_j}, j=1,\ldots,n\}$, where*

$$Y_{t_j} = X_{t_j} + \varepsilon_{t_j}, \tag{12.46}$$

where ε_{t_j} are i.i.d. random variables with mean zero and variance σ_ε^2. The measurement error process ε is assumed to be independent of the process X.

The noise term represents market microstructure in a reduced form sense. Zhang, Mykland, and Aït-Sahalia (2005). In this model, the efficient price dominates the long run but in the short run the measurement error may dominate in the RV calculation based on high frequency returns.

Suppose that we calculate the realized variance based on this observed price sequence (at equally spaced time points). Then we have

$$RV_Y^n = \sum_{i=1}^{n-1} \left(Y_{\frac{i+1}{n}} - Y_{\frac{i}{n}}\right)^2$$

$$= \sum_{i=1}^{n-1} \left(X_{\frac{i+1}{n}} - X_{\frac{i}{n}}\right)^2 + \sum_{i=1}^{n-1} \left(\varepsilon_{\frac{i+1}{n}} - \varepsilon_{\frac{i}{n}}\right)^2 + 2\sum_{i=1}^{n-1} \left(\varepsilon_{\frac{i+1}{n}} - \varepsilon_{\frac{i}{n}}\right)\left(X_{\frac{i+1}{n}} - X_{\frac{i}{n}}\right).$$

12.5 Estimation of Quadratic Variation Volatility

Then note that by the LLN for i.i.d. sequences

$$\frac{1}{n}\sum_{i=1}^{n-1}\left(\varepsilon_{\frac{i+1}{n}}-\varepsilon_{\frac{i}{n}}\right)^2 \xrightarrow{P} 2\sigma_\varepsilon^2. \qquad (12.47)$$

Furthermore, by the Cauchy–Schwarz inequality

$$\left|\frac{1}{n}\sum_{i=1}^{n-1}\left(\varepsilon_{\frac{i+1}{n}}-\varepsilon_{\frac{i}{n}}\right)\left(X_{\frac{i+1}{n}}-X_{\frac{i}{n}}\right)\right| \leq \frac{1}{\sqrt{n}}\left(\frac{1}{n}\sum_{i=1}^{n-1}\left(\varepsilon_{\frac{i+1}{n}}-\varepsilon_{\frac{i}{n}}\right)^2\right)^{1/2}\left(\sum_{i=1}^{n-1}\left(X_{\frac{i+1}{n}}-X_{\frac{i}{n}}\right)^2\right)^{1/2}$$
$$\xrightarrow{P} 0.$$

It follows that as $n \to \infty$

$$RV_Y^n \xrightarrow{P} \infty,$$

i.e., the RV estimator is inconsistent.

Some argue that one can improve matters by sampling at a lower frequency. Theoretically this just reduces the effect of measurement error, it does not eliminate it, but in practice it often appears to have eliminated the substantial consequences of market frictions.

Zhang, Mykland, and Aït-Sahalia (2005) introduced the **two scales realized volatility** (TSRV) estimator. This estimates the quadratic variation using a combination of realized variances computed on two different time scales or frequencies, thereby performing an additive bias correction to eliminate the bias from microstructure. It allows consistent estimation of the underlying quadratic variation in the presence of noise, albeit with a slower convergence rate than in the case without noise. Suppose that observations are equally spaced.

Definition 152 *Write $K \times (m+1) = n$ and let the first subsample be $\{Y_0, Y_{K/n}, \ldots, Y_{mK/n}\}$ and the second be $\{Y_{1/n}, Y_{(K+1)/n}, \ldots, Y_{(mK+1)/n}\}$, and so on. In each subsample we have $m+1$ (log-)prices and hence m returns. For $j = 1, \ldots, K$, let*

$$RVsub_j = \sum_{i=1}^{m}\left(Y_{(j+iK)/n} - Y_{(j+(i-1)K)/n}\right)^2.$$

These are realized volatility computed on the lower frequency or slower time scale. This estimator would be consistent as $m \to \infty$ in the absence of microstructure noise.

In the presence of noise we have

$$\frac{1}{m} RVsub_j = \frac{1}{m} \sum_{i=1}^{m} \left(Y_{(j+iK)/n} - Y_{(j+(i-1)K)/n} \right)^2$$

$$= \frac{1}{m} \sum_{i=1}^{m} \left(X_{(j+iK)/n} - X_{(j+(i-1)K)/n} \right)^2 + \frac{1}{m} \sum_{i=1}^{m} \left(\varepsilon_{(j+iK)/n} - \varepsilon_{(j+(i-1)K)/n} \right)^2$$

$$+ \frac{2}{m} \sum_{i=1}^{m} \left(X_{(j+iK)/n} - X_{(j+(i-1)K)/n} \right) \left(\varepsilon_{(j+iK)/n} - \varepsilon_{(j+(i-1)K)/n} \right)$$

$$\xrightarrow{P} 2\sigma_\varepsilon^2$$

as $m \to \infty$. This estimator is inconsistent under (12.46), and indeed grows at the rate m. Consider the linear combination of the two scales $RVsub_j - \frac{m}{n} RV$. The leading bias term is $2m\sigma_\varepsilon^2 - \frac{m}{n} 2n\sigma_\varepsilon^2 = 0$, so that the leading bias term is knocked out by combining the full sample and the subsample estimator in this way. However, the dominant term then is

$$\sum_{i=1}^{m} \left\{ \left(\varepsilon_{(j+iK)/n} - \varepsilon_{(j+(i-1)K)/n} \right)^2 - 2\sigma_\varepsilon^2 \right\}$$

$$= \sqrt{m} \times \frac{1}{\sqrt{m}} \sum_{i=1}^{m} \left\{ \left(\varepsilon_{(j+iK)/n} - \varepsilon_{(j+(i-1)K)/n} \right)^2 - 2\sigma_\varepsilon^2 \right\},$$

which is large and contains only noise. This term is eliminated by averaging over the subsamples. Specifically, consider the term

$$T = \frac{1}{K} \sum_{j=1}^{K} \sum_{i=1}^{m} \left(\left(\varepsilon_{(j+iK)/n} - \varepsilon_{(j+(i-1)K)/n} \right)^2 - 2\sigma_\varepsilon^2 \right)$$

$$= \frac{1}{K} \sum_{j=1}^{K} \sum_{i=1}^{m} \left(\varepsilon_{(j+iK)/n}^2 - \sigma_\varepsilon^2 \right) + \frac{1}{K} \sum_{j=1}^{K} \sum_{i=1}^{m} \left(\varepsilon_{(j+(i-1)K)/n}^2 - \sigma_\varepsilon^2 \right)$$

$$- 2 \frac{1}{K} \sum_{j=1}^{K} \sum_{i=1}^{m} \varepsilon_{(j+iK)/n} \varepsilon_{(j+(i-1)K)/n}.$$

Each of these terms is mean zero. Furthermore, each term is like a sum of n independent random variables so that T is of order $\sqrt{m/K}$ in probability and satisfies a CLT after normalization. If $K/m \to \infty$, these terms are of smaller order in probability.

Definition 153 *The TSRV estimator is*

$$\widehat{\theta}_{TSRV} = \frac{1}{K} \sum_{j=1}^{K} RVsub_j - \frac{m}{n} RV_n. \tag{12.48}$$

12.5 Estimation of Quadratic Variation Volatility

The leading terms of $\widehat{\theta}_{TSRV} - \theta$, where $\theta = \langle X, X \rangle_1$, are T and

$$\sum_{i=1}^{m} \left(X_{(j+iK)/n} - X_{(j+(i-1)K)/n} \right)^2 - \theta$$

$$= \sqrt{m} \times \frac{1}{\sqrt{m}} \left(\sum_{i=1}^{m} \left(X_{(j+iK)/n} - X_{(j+(i-1)K)/n} \right)^2 - \theta \right).$$

If we choose m, K such that $1/m = m/K$ the two terms are balanced and this is the optimal configuration. In that case $m^2 = K$ so $m \simeq n^{1/3}$ and $K \simeq n^{2/3}$ so that the two leading terms are of order $n^{-1/6}$ in probability. In conclusion we have

Theorem 55 *Suppose that $K = cn^{2/3}$ for some positive finite c. Then, $\widehat{\theta}_{TSRV}$ is consistent and converges at rate $n^{1/6}$ to a mixed normal distribution*

$$n^{1/6}(\widehat{\theta}_{TSRV} - \theta) \Longrightarrow MN(0, \omega), \qquad \omega = \frac{8}{c^2 \sigma_\varepsilon^4} + c \frac{4}{3} \int_0^1 \sigma_t^4 dt.$$

The optimal choice of c can be determined from minimization of ω with respect to c. The authors also give a method for estimating ω from the data. The rate of convergence is slower than root-n because it is difficult to extract the signal from such large noise.

Zhang (2006) introduced the multi-scale realized volatility (MSRV) estimator that combines multiple ($\simeq n^{1/2}$) time scales. This is consistent and has a faster convergence rate $n^{1/4}$. This has been shown to be the optimal rate, i.e., the rate achieved by the Gaussian MLE for the special case of constant volatility. The class of estimators is as follows

$$\widehat{\theta}_{MSRV} = \sum_{\ell=1}^{L} \alpha_\ell \frac{1}{K_\ell} \sum_{j=1}^{K_\ell} RV sub_j^{K_\ell},$$

where there are restrictions on $\alpha_1, \ldots, \alpha_L$ including $\sum_{\ell=1}^{L} \alpha_\ell \simeq 1$, and growth conditions on L, K. She shows that as $n \to \infty$

$$n^{1/4}(\widehat{\theta}_{MSRV} - \theta) \Longrightarrow MN(0, \omega^*),$$

for some random ω^*. Aït-Sahalia, Mykland, and Zhang (2011) modify TSRV and MSRV estimators and achieve consistency in the presence of serially correlated microstructure noise.

There are some other popular methods, preaveraging and realised kernels.

Definition 154 *The realized kernel estimator:*

$$RK_H = \sum_{|h| < n} k\left(\frac{h}{H+1}\right) \gamma_h(Y), \qquad \gamma_h(Y) := \sum_{j=h+1}^{n} Y_{t_j} Y_{t_{j-h}}, \ h = 0, \pm 1, \ldots$$

where the kernel k satisfies $k(0) = 1$, $k(s) \to 0$ as $s \to \infty$, and the bandwidth H controls bias-variance trade-off.

Zhou (1996) was the first to consider the use of the kernel method to deal with the problem of microstructure noise in high frequency data. For the case of independent noise, Zhou proposed this with $H = 1$. Hansen and Lunde (2006) examined the properties of Zhou's estimator and showed that, although unbiased under the presence of i.i.d. microstructure noise, the estimator is not consistent. However, they advocated that, while inconsistent, Zhou's kernel method is able to uncover several properties of the microstructure noise. Barndorff-Nielsen, Hansen, Lunde, and Shephard (2008) develop some theory for this method.

Jacod, Li, Mykland, Podolskij, and Vetter (2009) propose the method of **preaveraging**, which involves averaging observed prices over a moderate number of time points to reduce the measurement error, and then applying RV to the preaveraged data. Aït-Sahalia and Jacod (2014) give a comprehensive review of volatility estimation in a continuous time framework.

We consider an example to try to relate the continuous time framework to discrete time models.

Example 67 *Suppose that volatility is constant and the measurement error is Gaussian. In this case, the observed returns satisfy for $t = 1, \ldots, n$*

$$R_t = Y_t - Y_{t-1} = \frac{1}{n^{1/2}} \sigma z_t + \varepsilon_t - \varepsilon_{t-1}.$$

This is the sum of an i.i.d. Gaussian error term with variance $1/n$ and a (unit root) MA(1) process. We have

$$\text{var}(R_t) = \frac{\sigma^2}{n} + 2\sigma_\varepsilon^2$$

$$\text{cov}(R_t, R_{t-j}) = \begin{cases} -\sigma_\varepsilon^2 & j=1 \\ 0 & j \geq 2. \end{cases}$$

This is the covariance function of an MA(1) process, that is, we may write

$$R_t = U_t - \theta_n U_{t-1},$$

where U_t is an i.i.d. mean zero shock process and the parameters θ_n and σ_U^2 satisfy two restrictions: $\sigma_U^2(1 + \theta_n) = \frac{\sigma^2}{n} + 2\sigma_\varepsilon^2$ and $\theta_n \sigma_U^2 = \sigma_\varepsilon^2$. Explicit solutions can be given for θ_n and σ_U^2 in terms of $\sigma^2, \sigma_\varepsilon^2$ and n, but note that

$$\theta_n = 1 - \frac{1}{n^{1/2}} \left(\frac{\sigma^2}{2\sigma_\varepsilon^2} \right)^{1/2} + O(1/n).$$

This means that the process R_t is an MA(1) process that has a moderate deviation from unit root (Phillips and Magdalinos (2007)). Since R_t is Gaussian there is a theory of optimal estimation here for the parameter σ^2. It can be shown that the MLE of σ^2 converges at rate $n^{1/4}$ and one can obtain the limiting variance. Therefore, it is not possible to improve on this rate of convergence.

Liu, Patton, and Sheppard (2015) compare some of these methods with 5-minute RV.

12.6 Levy Processes

We consider the class of Levy processes, which are more general than Brownian motions and are consequently very useful for modelling financial series.

Definition 155 X_t is a Levy process if

(1) $X_t - X_s$ is independent of $\mathcal{F}_s$.
(2) $X_t - X_s$ has the same distribution as $X_{t+h} - X_{s+h}$ for any h.

It follows that X_t is continuous in probability

$$\lim_{h \to 0} \Pr(|X_{t+h} - X_t| \geq \epsilon) = 0 \tag{12.49}$$

for all $\epsilon > 0$. This does not imply that the sample paths are continuous, i.e., the process can jump at certain times, although the sample path is cadlag. This is a more general class of processes than Brownian motion since it does not specify the nature of the distribution of the increments. Levy processes have the infinite divisibility property, that is, we can write for any n

$$X_t = \sum_{i=1}^{n} \varepsilon_{ni},$$

with ε_{ni} i.i.d. We have the following characterization of Levy processes.

Definition 156 *Levy-Khintchine.* The characteristic function of X_t has to be of a particular form,

$$E\left(e^{iuX_t}\right) = e^{t\psi(u)}$$

$$\psi(u) = i\gamma u - \frac{1}{2}\sigma^2 u^2 + \int \left(e^{iux} - 1 - iux 1_{|x|\leq 1}\right) \nu(dx) \tag{12.50}$$

for some Levy measure $\nu(dx)$ that satisfies $\int \min\{1, x^2\} \nu(dx) < \infty$ and parameters $\sigma^2 \geq 0$, $\gamma \in \mathbb{R}$.

A special case of this is the Brownian motion when $\nu(dx) = 0$ and $\gamma = 0$ so the class of processes includes Brownian motion. It also includes many other processes with quite different properties. A Levy process can be decomposed as

$$X_t = \gamma t + \sigma B_t + J_t + M_t,$$

where B_t is a Brownian motion, and J_t, M_t are jump processes that we consider more below.

We next consider another special case of the Levy processes, the counting process.

Definition 157 *Define the counting process*

$$N_t = \{i : T_i \leq t\} = \sum_{i=1}^{\infty} 1\,(T_i \leq t),$$

where T_i, $i = 1, \ldots$ is a strictly increasing sequence of strictly positive random variables. This process has the following properties:

(0) $N_0 = 0$ and N_t is integer valued.
(1) N_t has independent increments, i.e., $N_{t+h} - N_t$ is independent of $\mathcal{F}_t$.
(2) The increments are stationary so that $N_{t+h} - N_t$ has the same distribution as $N_{s+h} - N_s$ for any h, s, t.

It follows that N_t is continuous in probability, although its sample path is not continuous but cadlag.

Example 68 *Suppose that*

$$\Pr(N_{t+h} - N_t = k) = \frac{(\lambda h)^k}{k!} \exp(-\lambda h).$$

This is consistent with $T_i - T_{i-1}$ being exponential with mean $1/\lambda$.

For any counting process write $dN(t) = N((t+dt)-) - N(t-)$, where $N(t-)$ denotes the value of N just before time t. Define the intensity process λ

$$E(dN(t)|\mathcal{F}_{t-}) = \lambda(t)dt, \tag{12.51}$$

where $\mathcal{F}_{t-}$ contains all information available as time s for all $s < t$. In general, $\lambda(t)$ may be stochastic through their dependence on $\mathcal{F}_{t-}$. For example, in the so-called **Hawkes processes** $\lambda(t)$ has an autoregressive structure. In simple models $\lambda(t)$ is assumed to be deterministic and perhaps constant. We may further write the counting process evolution in the form of a stochastic differential equation

$$dN(t) = \lambda(t)dt + dM_t, \tag{12.52}$$

where $M_t = N_t - \int_0^t \lambda(s)ds$ is a martingale.

Definition 158 *A pure jump process (or compound Poisson process) is a marked counting process*

$$Z_t = \sum_{i=1}^{N_t} Y_i = \sum_{i=1}^{\infty} Y_i 1\,(T_i \leq t)$$

where Y_i are i.i.d. real valued random variables independent of the counting process N_t.

12.6 Levy Processes

We may combine jump processes with continuous variation processes to obtain more complex behaviors.

Definition 159 *Suppose that*

$$X_t = \int_0^t a_u du + \int_0^t \sigma_u dB_u + \sum_{s=1}^{N(t)} J_s,$$

where N is a simple counting process and J_s are the associated (positive) jumps, which happen at times $0 < \tau_1 < \tau_2 < \cdots$ then

$$QV = \int_0^1 \sigma_u^2 du + \sum_{s=1}^{N(t)} J_s^2.$$

See Cont and Tankov (2003) for a discussion of jump processes. It has been shown that under some conditions the realized volatility estimator satisfies

$$RV \xrightarrow{P} QV. \tag{12.53}$$

For some purposes one may wish to separate the two components of the quadratic variation.

Definition 160 *The bipower variation*

$$BPV = \plim_{n \to \infty} \sum_{i=1}^{n-1} \left| X_{\frac{i+1}{n}} - X_{\frac{i}{n}} \right| \left| X_{\frac{i}{n}} - X_{\frac{i-1}{n}} \right|.$$

This is useful for detecting jumps. This is because for the class of processes defined above,

$$BPV = E^2(|Z|) \int_0^1 \sigma_u^2 du = \frac{2}{\pi} \int_0^1 \sigma_u^2 du,$$

where Z is standard normal. It follows that

$$QV - \frac{\pi}{2} BPV = \sum_{s=1}^{N(t)} J_s^2.$$

One can use realized volatility and the realized BPV to determine whether there are jumps or not and to divide the variability into the continuous part and the jumpy part (Barndorff-Nielsen and Shephard (2006a,b)).

Another application of counting processes is to **time changed Brownian motion**.

Example 69 *Consider the process*

$$dX_t = \sigma dW_{N(t)},$$

where $N(t)$ is the counting process defined above. The process N represents the times at which trading occurs. We have
$$X_t|N(t) \sim N(0, \sigma^2 N(t))$$
so that the unconditional variance of X_t (or quadratic variation) is
$$\sigma^2 E(N(t)) = \sigma^2 \int_0^t \lambda(s)ds.$$

In the case where $\sigma = \sigma(t)$, we have the unconditional variance as $\int_0^t \sigma(s)\lambda(s)ds$ and the spot variance as $\sigma(t)\lambda(t)$.

12.7 Summary of Chapter

We considered the main concepts of continuous time models. These models are applied to interest rate data and stock prices. The estimation of volatility based on continuous time concepts is now well established with many useful tools.

13 Yield Curve

In this chapter we consider the measurement of interest rates and the time value of money. This plays an important role in many applications: portfolio allocation, predicting future interest rates (Campbell and Shiller (1991)), future inflation and national income (Estrella and Mishkin (1997)), asset pricing including interest rate derivatives, and testing rational expectations theories. We also discuss the discrete time modelling of the yield curve.

13.1 Discount Function, Yield Curve, and Forward Rates

We first give some definitions of the key notions of fixed income mathematics. These start with the simple case of zero bonds, which pay a fixed amount at some specified date in the future. Let n denote the horizon.

Definition 161 *The spot rate is the yield y_n on an n-period zero with payoff M and price p*

$$p = \frac{M}{(1+y_n)^n}.$$

Definition 162 *The discount factor d_n is the standardized price for unit payoff in the n-period future, and satisfies*

$$d_n = \frac{1}{(1+y_n)^n}.$$

This is a decreasing function of n.

Definition 163 *The forward rate f_n is the rate paid on a one-period investment arranged today and made at time n in the future maturing at time $n+1$*

$$1 + f_n = \frac{d_n}{d_{n+1}}.$$

Suppose a bond pays guaranteed amounts b_j (coupons plus redemption value) at times $\tau_j, j = 1, \ldots, m$ in the future. The absence of arbitrage yields the linear pricing rule

$$p = \sum_{j=1}^{m} b_j d_j = \sum_{j=1}^{m} \frac{b_j}{(1+y_j)^{\tau_j}}, \qquad (13.1)$$

where the price is equal to the present discounted value of future cash flows. The zero coupon yield curve is just the set of points $y_1, \ldots, y_n$. This can be equivalently represented by the discount factors $d_1, \ldots, d_n$ or the forward rates $f_1, \ldots, f_n$.

Definition 164 *The yield to maturity y on a coupon bond paying c in each period and 1 at the end of n periods is implied by the relation*

$$p = \frac{c}{(1+y)} + \frac{c}{(1+y)^2} + \cdots + \frac{1}{(1+y)^n}.$$

The par bond yield curve is the graph of the yield to maturity of coupon bonds that sell at par.

The choice of period matters in the above formulae. It is convenient to work with infinitesimal time periods, i.e., **continuous compounding**.

Definition 165 *The discount function at time τ denoted $d(\tau)$ is the price of one dollar received at point τ in the future. The discount function and yield curve d and y are related by*

$$d(\tau) = \exp(-\tau y(\tau)), \qquad (13.2)$$

where $y(\tau)$ is the yield, i.e., the rate of interest on a zero-coupon payment at maturity τ. The yield curve is related to the forward rate by

$$y(\tau) = \frac{1}{\tau} \int_0^\tau f(s) ds,$$

where $f(s)$ is the forward short term rate applicable at time s. The quantities d, y, and f are equivalent given the boundary conditions $d(0) = 1$ and $d(\infty) = 0$. The discount function is monotonically decreasing.

13.2 Estimation of the Yield Curve from Coupon Bonds

In practice, we do not observe a full set of zero coupon bonds, one for each date, but instead we have a sample of bonds many of which are coupon bonds. From this sample we want to extract the zero coupon yield curve. We abstract from tax, inflation risk, and other issues, as is common in the literature. In practice, we have data on prices p, payments b, and payment times τ. We want to estimate the functions d, y, and f. Provided we have enough bonds, in principle we can invert the linear system (13.1) to estimate these functions at a grid of points. In practice, one issue is that we get too many answers, i.e., there exist many different discount functions that exactly price a subset of the data. Equivalently, we can find two bonds with apparently identical

13.2 Estimation of the Yield Curve from Coupon Bonds

payment streams that have different reported prices. The reasons for this are related to liquidity effects: different prices for **on the run** (most recently issued and most liquid) versus **off the run** issues.

We suppose that the cash flow model holds up to a random error term.

Definition 166 *The statistical model is*

$$p_i = \sum_{j=1}^{m_i} b_{ij} d(\tau_{ij}) + \varepsilon_i, \qquad (13.3)$$

where: p_i is the price of bond i at the observation date, $b_{ij} = b_i(\tau_{ij})$ is the payment to be received by the holder of bond i at time τ_{ij} in the future, where $j = 1, \ldots, m_i$. Here, ε_i is a random error term that soaks up small pricing errors, liquidity effects, maturity effects, etc. We assume that it is mean zero with finite variance for simplicity. We also assume that ε_i are independent across i.

The quantity of interest is the discount function $d(\cdot)$, which is an unknown but smooth function of time. In matrix notation we may write (13.3) as

$$p = Bd + \varepsilon,$$

where p is the $n \times 1$ vector of observed bond prices, ε is the $n \times 1$ vector of error terms, d is the $m \times 1$ vector of unknown unique parameters chosen from $\{d(\tau_{ij}), j = 1, \ldots, m_i, i = 1, \ldots, n\}$. The dimension of d, m, is an integer such that $\max_{1 \leq i \leq n} m_i \leq m \leq \sum_{i=1}^{n} m_i$, and B is the $n \times m$ matrix of known (future) payments (b_{ij}). To estimate d we may consider the least squares estimator

$$\widehat{d} = (B^\top B)^{-1} B^\top p.$$

In general, m is larger than n so one can't uniquely define $\widehat{d}$ in this way. Even in the case where $m < n$, m will be quite large, and the resulting estimates will be too variable. The direct least squares estimator does not impose monotonicity on d. Another limitation of this method is that it does not directly solve the interpolation problem that one wants to compute d at points in between payment times. For these reasons, the literature has adopted regularization or smoothing methods.

13.2.1 Estimation by Series Expansion

McCulloch (1971) pioneered the nonparametric approach to estimation of d. Suppose that

$$d(t) = \sum_{\ell=1}^{\infty} \theta_\ell g_\ell(t), \qquad (13.4)$$

where g_ℓ are known basis functions and θ_ℓ are unknown coefficients. Then for finite L

$$p_i = \sum_{j=1}^{m_i} b_i(\tau_{ij}) \sum_{\ell=1}^{\infty} \theta_\ell g_\ell(\tau_{ij}) + \varepsilon_i$$

$$= \sum_{\ell=1}^{\infty} \theta_\ell \left(\sum_{j=1}^{m_i} b_i(\tau_{ij}) g_\ell(\tau_{ij}) \right) + \varepsilon_i$$

$$= \sum_{\ell=1}^{L} \theta_\ell X_{\ell i} + \varepsilon_i + \mathfrak{R}_L,$$

where $X_{\ell i} = \sum_{j=1}^{m_i} b_i(\tau_{ij}) g_\ell(\tau_{ij})$ are observable, and $\mathfrak{R}_L$ is the approximation error obtained by taking finite L. This error term is small if L is large.

J. Huston McCulloch is Professor of Economics at the Ohio State University, where he also holds an appointment with the Department of Finance. He was formerly at Boston College, has been a Faculty Research Fellow at the National Bureau of Economics Research, and served as Editor of the *Journal of Money, Credit and Banking* from 1983–1991. His primary research interests center on money, banking, and macroeconomic fluctuations, and extend to related financial and econometric issues. He is an internationally recognized authority on the term structure of interest rates, heavy-tailed stable probability distributions, and Austrian utility theory. He is the author of the book *Money and Inflation: A Monetarist Approach* and has published articles in the *American Economic Review*, the *Journal of Political Economy*, the *Quarterly Journal of Economics*, the *Journal of Monetary Economics*, the *Journal of Finance*, *Computation in Statistics*, the *Bulletin of the London Mathematical Society*, the *Zeitschrift fur National Okonomie*, and the *Tennessee Anthropologist*.

We can estimate θ_ℓ, $\ell = 1, \ldots, L$ by least squares regression assuming that $L << n$. McCulloch (1971) took g_ℓ, $\ell = 2, \ldots, L$ to be piecewise quadratic functions on intervals and $g_1(0) = 1$. McCullogh recommended taking $L = n^{1/2}$.

A number of extensions of this approach have been suggested in the literature. Schaefer (1981) suggested using Bernstein polynomials, i.e., $d(t) = 1 + \sum_{\ell=1}^{L} \theta_\ell g_\ell(t)$, where

$$g_\ell(t) = \sum_{r=1}^{L-\ell} (-1)^{r+1} \binom{L-\ell}{r} \frac{t^{\ell+r}}{\ell+r}.$$

13.2 Estimation of the Yield Curve from Coupon Bonds

Vasicek and Fong (1982) considered $g_\ell(t) = \exp(-\ell\alpha t)$, where α is an additional parameter to be estimated. Langetieg and Smoot (1989) considered

$$d(t) = \exp\left(-t \sum_{\ell=1}^{L} \theta_\ell g_\ell(t)\right).$$

Fisher, Nychka, and Zervos (1995) considered smoothing splines, which are defined by solving the penalized least squares

$$\min_{f \in \mathcal{F}} \sum_{i=1}^{n} (p_i - \widehat{p}_i(f))^2 + \lambda \int f''(s)^2 ds.$$

The US Treasury's yield curve is currently derived using a quasi-cubic Hermite spline function.

Under some regularity conditions these methods can be shown to be consistent as the sample size n increases, in the sense that

$$\sup_{\tau \in [0, \tau^{\max}]} \left|\widehat{d}(\tau) - d(\tau)\right| \xrightarrow{P} 0 \qquad (13.5)$$

as $n \to \infty$, where $\tau^{\max}$ is some finite maximum maturity. Here, the function $d(.)$ is assumed to be unknown but smooth (e.g., continuously differentiable).

13.2.2 Parametric Methods

Nelson and Siegel (1987) specify a parametric model for the forward curve

$$f(t) = \beta_0 + (\beta_1 + \beta_2(t/\tau_0))\exp(-t/\tau_0),$$

where $\theta = (\beta_0, \beta_1, \beta_2, \tau_0)^\mathsf{T}$ are unknown parameters. This implies that the discount function is

$$d_\theta(t) = \exp\left(-t\left(\beta_0 + (\beta_1 + \beta_2)(1 - \exp(-t/\tau_0)) \times \frac{\tau_0}{t} - \beta_2 \exp(-t/\tau_0)\right)\right).$$
(13.6)

This functional form is quite general and allows a combination of shapes for the long end and the short end of the yield curve. Estimation of θ is by nonlinear least squares

$$\widehat{\theta} = \arg\min_\theta \sum_{i=1}^{n} \left(p_i - \sum_{j=1}^{m_i} b_{ij} d_\theta(\tau_{ij})\right)^2,$$

which yields an estimated discount function $d_{\widehat{\theta}}(t)$. Diebold and Li (2006) estimate this model for US monthly data for 1985–2000, obtaining a yield curve each month. They then fit time series models to the time specific parameters β and use this to forecast the future yield curve.

Svensson (1994) extends this by adding two extra parameters

$$f(t) = \beta_0 + \beta_1 \exp(-t/\tau_1) + \beta_2(t/\tau_1)\exp(-t/\tau_1) + \beta_3(t/\tau_2)\exp(-t/\tau_2).$$

Again, the unknown parameters are estimated by nonlinear least squares using the sample of bonds and implied discount function model.

These methods can be shown to be consistent and asymptotically normal when the parametric model is correct, that is,

$$\sqrt{n}\left(\widehat{\theta} - \theta\right) \Longrightarrow N(0, \Omega) \tag{13.7}$$

for some covariance matrix Ω, and hence $d_{\widehat{\theta}}(\tau) \to d_\theta(\tau)$ as $n \to \infty$.

13.2.3 The Fama–Bliss Method

Fama and Bliss (1987) proposed a method to fit the yield curve that is also called, confusingly, bootstrapping. The idea is to throw out bonds whose values are too noisy and to exactly price a specific subset assuming that the forward rate curve is constant between successive bond maturities.

Definition 167 *Let the sequence of observed bonds be ordered from the shortest maturity to longest maturity, where τ^i is the time to maturity of the ith bond, and $i=1,\ldots,I$, where $I<n$. Let f^i denote the (constant) forward rate on the interval $(\tau^{i-1}, \tau^i]$ where $\tau^0 = 0$, that is $f(\tau) = f^i$ on $(\tau^{i-1}, \tau^i]$. So forward rates are piecewise constant on I intervals. The method is sequential: first compute f^1, then compute f^2, etc. To compute f^i use the ith observed bond and find that f^i that solves*

$$p_i = \sum_{j=1}^{m_i} b_i(\tau_{ij}) d(\tau_{ij}),$$

where the discount function $d(\tau_{ij})$ depends on $\{f^j\}_{j=1}^i$ and $\{f^j\}_{j=1}^i$ has been computed from previous bonds in the same fashion.

A key part of the method is the selection procedure. Only fully taxable, non-callable, and non-flower instruments are used, Treasury notes and bonds are excluded from the sample if their time to maturity is less than one year, an instrument is included if either its yield to maturity is within 0.2% absolute difference of the yield to maturities of surrounding instruments or in between them, and an instrument is included if the resulting yield curve when the instrument is included does not exhibit large yield reversals (adjacent changes that are greater that 0.2% in absolute value and in opposite directions).

The statistical properties of this method are unknown. However, my interpretation is that it is a bit like estimating linear regression using a carefully chosen subset of the observations.

Example 70 *Suppose that we have a linear regression*

$$y_i = \beta x_i + u_i,$$

where u_i are i.i.d. mean zero errors. First, estimate β by OLS $\widehat{\beta} = \sum x_i y_i / \sum x_i^2$. Then order the residuals $\widehat{u}_i = y_i - \widehat{\beta} x_i$

$$|\widehat{u}|_{(1)} < \cdots < |\widehat{u}|_{(n)}$$

according to their magnitude. Finally, let

$$\widetilde{\beta} = \frac{y_{(1)}}{x_{(1)}}.$$

Then it can be shown that $\widetilde{\beta}$ is consistent and asymptotically normal and indeed equivalent to $\widehat{\beta}$.

The interpretation is that the Fama–Bliss method of fitting yield curves has a good statistical interpretation so long as the selection criteria are as effective in throwing out noisy observations as would be a consistent initial fit.

Bliss (1997) compares a number of different yield curve estimation methods according to a variety of performance measures.

13.3 Discrete Time Models of Bond Pricing

We are concerned here with developing some discrete time models for the yield curve. How do yields behave over time? How should we price contingent claims based on the yield curve? Our approach follows Backus, Foresi, and Telmer (1998).

13.3.1 Economic Hypotheses about Interest Rates

There are several theories about how interest rates and yields are determined.

Expectations hypothesis. Today's forward rates are expectations of future one period spot rates, that is, the expected rate of return from rolling over short term bonds must be equal to the rate of return from holding the long bond to maturity.

Liquidity (risk) premium hypothesis. Today's forward rates are equal to the expectations of future short rates plus a premium to make long term bonds as liquid as short term bonds. People on average are prepared to pay a premium to hedge against future macroeconomic uncertainty, and hence prefer to pay a higher interest rate on longer term bonds. It is consistent with upward sloping yield curve.

Market segmentation hypothesis. Different maturity sectors represent distinct markets with their own demand and supply forces. For example, regulatory or other constraints may force a need for instruments with particular maturities.

Preferred habitat. Same as the market segmentation hypothesis except that investors will deviate from their desired maturity sector if offered a premium.

Modern term structure theory places restrictions on the form of the risk premium.

13.3.2 Statistical Properties of Yields

CLM examined yield data from 1952–1991 obtained from McCulloch and Kwon (1993) for fixed maturities in the range 1 month–120 months. They find that the mean yield curve is upward sloping, i.e., average interest rates increase with maturity, and the yield curve is concave. This is to be contrasted with the shapes you can find for individual yield curves, which may be: upward sloping, downward sloping, and hump shaped. They find that the (time series) standard deviations are more or less constant with respect to maturity. They find that skewness and excess kurtosis are both higher and positive for short end and decrease with maturity. They find that autocorrelation is high and increases with maturity to 0.992 for 120 months. They also considered yield spreads $y_{jt} - y_{j_1,t}$, where j_1 is the one month maturity and j is some other maturity. They found that: the mean and standard deviation of spread increases a lot with maturity. The skewness and kurtosis are larger in absolute terms for the short end but as for the raw yields they decrease with maturity. Autocorrelation is low for 3 months but increases to 0.885 for 120 months. Finally, they also examined the statistical properties of differenced yields $y_{jt} - y_{j,t-1}$. They found that: mean changes are small and increase a bit with maturity; standard deviation of changes decreases with maturity; skewness is negative and increases to a slight positive with maturity; kurtosis is high and decreases with maturity; and autocorrelation is low and positive for changes in yields.

We work with the US treasury zero coupon yields as provided on the US government web site; see the data sources page. This gives yields for 1, 3, 6, 12, 24, 36, 50, 84, 120, 240, and 360 months over the period 1990–2016, although there is some missing data. For example the one month yield is only available after 2000 and the thirty year yield was missing for 2002–2006. We plot the one month and 120 month yields for comparison with CLM's Figure 10.5. The behavior of yields in the more recent

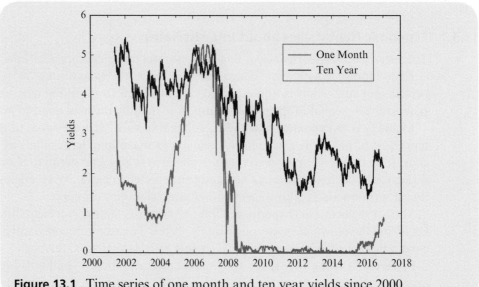

Figure 13.1 Time series of one month and ten year yields since 2000

13.4 Arbitrage and Pricing Kernels

Table 13.1 Summary statistics of daily yields

	1	3	6	12	24	36	60	120	240	360
m	1.2488	2.8712	3.0002	3.1330	3.4597	3.6945	4.1198	4.6965	4.8131	5.2964
s	1.5804	2.3632	2.3977	2.3920	2.3966	2.3266	2.1700	1.9040	1.5439	1.7906
κ_3	1.2588	0.1873	0.1671	0.1451	0.1225	0.1087	0.1007	0.1690	−0.0473	0.1408
κ_4	0.3229	−1.2892	−1.3163	−1.3089	−1.2677	−1.2108	−1.0924	−0.8705	−0.8769	−1.0555

Table 13.2 Autocorrelation of daily yields

	1	3	6	12	24	36	60	120	240	360
$\rho_\Delta(1)$	0.1907	0.1122	0.0602	0.0518	0.0215	0.0271	0.0238	0.0264	−0.0066	0.0078
$\rho_\Delta(2)$	−0.0998	−0.1109	−0.0616	−0.0221	−0.0389	−0.0324	−0.0405	−0.0273	−0.0212	−0.0132
$\rho_\Delta(3)$	−0.2034	−0.1283	−0.0613	−0.0094	−0.0052	−0.0059	−0.0052	−0.0138	−0.0093	−0.0205
$\rho_\Delta(4)$	−0.0132	0.0309	0.0440	−0.0059	−0.0203	−0.0208	−0.0211	−0.0295	−0.0251	−0.0272
$\rho_\Delta(5)$	0.0660	0.0505	0.0631	−0.0093	−0.0229	−0.0195	−0.0234	−0.0234	−0.0234	−0.0290

period is obviously very different in terms of the level and trend as compared with the period 1952–1991 covered by their graph. In particular at the short rate of one month the level was very close to zero from 2008–2016, and obviously approaching the **zero lower bound**.

We give the summary statistics of the daily yields in Table 13.1, which shows the upwards sloping average yield curve. It also shows the negative kurtosis for yields from 3 months upwards.

We also show the autocorrelations of the differenced yields out to five lags in Table 13.2, which shows that the very short horizon yields have some predictability after differencing, whereas the longer horizon yields are close to a unit root.

13.4 Arbitrage and Pricing Kernels

We now consider the application to pricing of fixed income contracts. We suppose that there is an SDF M such that (10.3) holds for all assets. In particular, let p_t^n be the dollar price at date t to a claim of one dollar at $t + n$. The one period return on an n-period bond is $R_{t+1} = p_{t+1}^{n-1}/p_t^n$. Therefore, we have

$$p_t^n = E_t(M_{t+1} p_{t+1}^{n-1}).$$

We consider various models for the state variables and SDF.

13.4.1 The Vasicek Model

Suppose that the stochastic discount factor satisfies

$$-\log M_{t+1} = \delta + z_t + \lambda \varepsilon_{t+1}, \qquad (13.8)$$

where the single state variable z satisfies

$$z_{t+1} = \varphi z_t + (1-\varphi)\theta + \sigma \varepsilon_{t+1}, \qquad (13.9)$$

where ε_{t+1} is i.i.d. standard normal. It follows that $E(z_t) = \theta$.

We can use this to price bonds. We know that:

$$p_t^0 = 1 \quad ; \quad p_t^1 = E_t(M_{t+1}) \quad ; \quad p_t^2 = E_t(M_{t+1}p_{t+1}^1),$$

etc. By taking $\delta = \lambda^2/2$ we obtain, using the properties of M_{t+1}, that $\log p_t^1 = -z_t$. It follows that the state variable is the short rate, i.e., $r_t = -\log p_t^1 = z_t$. Prices of long bonds follow by induction. We may guess that

$$-\log p_t^n = A_n + B_n z_t, \qquad P_t^n = \exp(-A_n - B_n z_t) \qquad (13.10)$$

for some coefficients A_n, B_n. One can find the coefficients recursively

$$A_{n+1} = A_n + \delta + B_n(1-\varphi)\theta - (\lambda + B_n \sigma^2)/2$$
$$B_{n+1} = 1 + B_n \varphi$$

with $A_1 = 0, B_1 = 1$. This model is restrictive in practice. One can generate an upward sloping average yield curve but with less curvature than real data.

13.4.2 Cox–Ingersoll–Ross Model

Suppose that the state variable and stochastic discount factor evolve according to the dynamic equations

$$z_{t+1} = \varphi z_t + (1-\varphi)\theta + \sigma z_t^{1/2} \varepsilon_{t+1}$$
$$-\log M_{t+1} = (1 + \lambda^2/2)z_t + \lambda z_t^{1/2} \varepsilon_{t+1}.$$

This is a discrete time version of (12.24). In this case, we obtain affine bond prices (13.10) with

$$A_{n+1} = A_n + B_n(1-\varphi)\theta \quad ; \quad B_{n+1} = 1 + \lambda^2/2 + B_n\varphi - (\lambda + B_n\sigma)^2/2.$$

In practice this model is too simple. If φ is chosen to reproduce the autocorrelation of the short rate, the mean yield curve is substantially less concave in models than it is in the data. Also the model predicts the same autocorrelation patterns across maturities, which is not the case. Normality is a bad assumption here also.

13.4.3 Affine Term Structure Models

CLM and Backus, Foresi, and Telmer (1998) look at specific cases with Gaussian innovations, e.g., single factor CIR process

$$r_{t+1} = \alpha + \beta r_t + \gamma r_t^{1/2}\varepsilon_{t+1}$$
$$M_{t+1} = a + br_t + cr_t^{1/2}\varepsilon_{t+1}$$

with ε_{t+1} being i.i.d. standard normal. There are restrictions on parameters $(a, b, c, \alpha, \beta, \gamma)$ implied by the SDF pricing property. Dai and Singleton (2000) characterize and test these overidentifying restrictions. CLM criticize this approach. They argue that affine models have the following properties, which are not realistic:

1. The covariance matrix of $K \geq n$ bonds is singular.
2. The volatility specification is limited.
3. The risk premia on long-term bonds always have the same sign.

Gourieroux, Monfort, and Polimenis (2006) define a general multifactor affine process for discrete time state variables $z_t \in \mathbb{R}^n$. This is a Markov process with

$$\log E\left(\exp(u^\top z_{t+1})|z_t\right) = a(u)^\top z_t + b(u) \tag{13.11}$$

for functions $a(.)$ and $b(.)$. In this case $z_t = (r_{t+1}, f_t)$. Positive z_{tj} implies restrictions. The stochastic discount function

$$m_{t+1} = \log(M_{t+1}) = \gamma_0 + \gamma_1^\top z_t + \gamma_2^\top z_{t+1}$$

for parameters $\gamma_0, \gamma_1, \gamma_2$. The pricing relation (10.3) implies restrictions on γ_j. This model includes many discrete time affine models as special cases: Vasicek, CIR, multifactor models. Their objective is to define a large class of discrete time models within which derivative pricing can be carried out with minimal extension of existing methods and yet which fit the data better. Gourieroux, Monfort, and Polimenis (2006) argue that criticisms of affine yield models in CLM etc apply perhaps to specific versions and not necessarily to their general class. The Gourieroux, Monfort, and Polimenis (2006) specification allows simple formulae for price of contingent claims.

Example 71 *Bond with residual maturity h. The price and yield satisfy*

$$\text{Price} \quad P_t^n = \exp(c_n^\top z_t + d_n)$$

$$\text{Yield} \quad y(t, t+n) = -\left(c_n^\top z_t + d_n\right)/n,$$

where for $n = 2, \ldots$

$$c_n = a(c_{n-1} + \gamma_2) - a(\gamma_2) - e_1$$
$$d_n = d_{n-1} - b(\gamma_2) + b(c_{n-1} + \gamma_2) - e_1$$

where $e_1 = (1,0)^\top$ and $c_1 = -e_1$ and $d_1 = 0$. This is a nonlinear difference equation. The Vasicek special case gives linear equations; the CIR special case has a quadratic term. Other prices can involve integrations.

The Gourieroux, Monfort, and Polimenis (2006) specification has the implication that all cumulants are linear functions of z, i.e.,

$$E(z_{t+1}|z_t = z) = a'(0)z + b'(0), \qquad \text{var}(z_{t+1}|z_t = z) = a''(0)z + b''(0)$$
$$\kappa_3(z_{t+1}|z_t = z) = a'''(0)z + b'''(0), \qquad \kappa_4(z_{t+1}|z_t = z) = a''''(0)z + b''''(0).$$

There is a similarity with **generalized linear models** for limited dependent variables. In the Poisson case all conditional cumulants would be equal to conditional mean. Common discrete time models of conditional variance (such as the GARCH model) have conditional variance being a quadratic function of the state and previous states, and in fact an infinitude of past states, which is ruled out in this case.

We show below in Figure 13.2 nonparametric estimates of the first four conditional cumulants of 3 month yields using local linear kernel estimators and bandwidth multiples of Silverman's (1986) rule. The conditional mean is very linear but close to unit

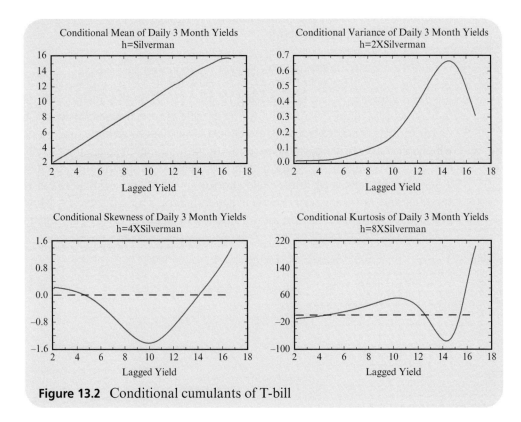

Figure 13.2 Conditional cumulants of T-bill

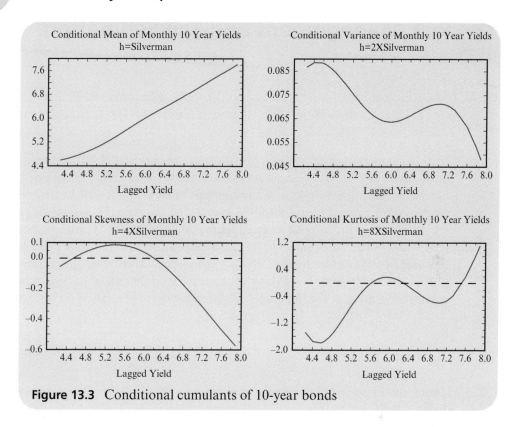

Figure 13.3 Conditional cumulants of 10-year bonds

slope. There is nonlinearity in the conditional variance and other cumulants. Similar results are shown for the 10 year yields in Figure 13.3.

13.5 Summary of Chapter

We defined some of the main concepts in fixed income. We described the leading methods for estimating the yield curve from a sample of coupon paying bonds. We described the empirical properties of the yield curve updating some of the results in CLM. We defined the leading discrete time models for the evolution of the yield curve and showed how this can deliver pricing for a range of contingent claims.

14 Risk Management and Tail Estimation

In this chapter we discuss how to measure and model risk. Traditional methods focussed on the standard deviation or variance of stock returns combined with the normal distribution to quantify the rareness of outcome levels. In that world, calculations are very easy and intuitive, everything is linear. In that world, **six sigma** events should never happen in our lifetime. Despite the early work of Mandelbrot (1963) and others, these linear Gaussian methods continue to be the main methodology used in practice (see Jorion (1997)). Recent renewed criticism, for example Turner (2009), has led to some rethinking of models for credit risk in particular, and to an acknowledgement of the limits of any mathematical model based on historical data.

14.1 Types of Risks

There are a number of different categories of risks facing financial market participants. **Credit risk**: borrowers fail to pay back money that is due. **Market risk**: adverse price changes of traded securities. **Liquidity risk**: traded securities cannot be sold or only sold at a discount. A financial institution does not obtain sufficient funding to cover current obligations. **Operational risk**: fire, natural catastrophes, IT incidents, fraud, human error, litigation. **Business risk**: margins of sold products are smaller than expected. We will consider the modelling and measurement of market risk. This field arose because of the concern of regulators and investors about the effects of large crashes, e.g., Black Monday, 1987 when there was a roughly 23% (and 23σ) drop in the stock market value, which equated to roughly $1 trillion lost in one day. The long term capital crisis of (1998) led to market wide losses of around $2 billion. The Asian crisis of 1997 and the Russian crisis of 1998 were other big events.

The first Basel Capital Accord was signed in 1988, which introduced regulatory oversight of banks. This focussed on the measure of risk called **Value-at-Risk** (VaR), which is the amount lost on a portfolio (investment) with a given small probability over a fixed time period. The regulations required the following. Banks must report their global Value-at Risk for the following 10 days to supervisory authorities every day. The VaR is to be calculated for the 99% one-sided significance level. They should use at least one year of historical data and at least quarterly updates. The implied capital requirement is the higher of the previous day VaR or the average over the last 60 trading days, times a multiplicative factor 3 plus an add-on. Banks can use their own model to calculate the VaR, but regulators can audit. There are penalties for bad models. A second and third accord have since been signed with some updates on methodology and practice. See Basel Committee on Banking Supervision (2012).

14.2 Value at Risk

Definition 168 *Define the α-quantile of a random variable X by the number q_α that satisfies*

$$q_\alpha = \inf\{q : \Pr(X \leq q) \geq \alpha\}. \tag{14.1}$$

If we think of X as being the distribution of asset values and returns over a specified period of time, then q_α is the corresponding Value-at-Risk (usually ignore the negative sign), also denoted VaR_α.

If $F(x) = \Pr(X \leq x)$ is strictly increasing (at least in the neighborhood of q_α) we can write

$$q_\alpha = F^{-1}(\alpha). \tag{14.2}$$

So if we take $\alpha = 1\%$, the number q_α is the value such that there is only a 1% chance of X being less than q_α. Note that for any $a > 0, b$ the quantile of $aX + b$ satisfies $q_\alpha(aX + b) = aq_\alpha(X) + b$. Furthermore, $q_{1-\alpha}(X) = -q_\alpha(-X)$ for continuous X.

Suppose that $X \sim N(\mu, \sigma^2)$, then

$$q_\alpha = \mu + \sigma z_\alpha,$$

where z_α is the α-quantile of a standard normal random variable Z. For the 1% case, $z_\alpha = -2.33$. We can estimate μ, σ^2 from a sample $X_1, \ldots, X_n$ and compute the estimated quantile by

$$\widehat{q}_\alpha = \overline{X} + s z_\alpha, \tag{14.3}$$

where $\overline{X}, s^2$ are the sample mean and variance respectively. The (i.i.d.) Gaussian structure implies some useful property with regard to the time horizon. If daily returns were $N(\mu_D, \sigma_D^2)$, then monthly returns (the sum of daily returns) would be $N(\mu_M, \sigma_M^2)$, where $\mu_M = T\mu_D$ and $\sigma_M^2 = T\sigma_D^2$, where T is the number of days in the month. Therefore,

$$q_\alpha(M) = \mu_M + \sigma_M z_\alpha = T\mu_D + T^{1/2}\sigma_D z_\alpha. \tag{14.4}$$

We usually assume that μ_D is small relative to σ_D, hence the square root law that monthly value at risk is approximately $\sqrt{22}$ times daily value at risk. Thus you can compute Value-at-Risk for any horizon given knowledge of the daily parameters μ_D, σ_D^2. These can be estimated from the daily data, or any horizon data in fact. This gives a lot of flexibility. One can readily provide confidence intervals for the estimated Value-at-Risk.

The problem with this approach is that we know that the distribution of daily stock returns is not well approximated by a normal distribution – the tails of the Gaussian distribution are too thin, meaning that large events are predicted to happen too rarely by this distribution. Furthermore there is quite a considerable amount of dependence over time, so that i.i.d. is not a good assumption. See Turner (2009, pp44–45) for a discussion of the implications of this. A further implication is that the square root scaling law is not a good approximation in practice.

Fama (1965) argues that a better description of daily stock returns is the stable class of distributions.

Definition 169 *The class of stable distributions S is specified by its log characteristic function*

$$\log(E(e^{iuX})) = i\mu u - |\gamma u|^{\theta}(1 + i\beta \text{sign}(u)w(u,\theta)),$$

where $i = \sqrt{-1}$ and

$$w(u,\theta) = \begin{bmatrix} \tan\frac{\pi\theta}{2} & \text{if } \theta \neq 1 \\ \frac{2}{\pi}\log|u| & \text{if } \theta = 1 \end{bmatrix}.$$

This contains four parameters $\theta, \beta, \mu, \gamma$: θ – is the characteristic exponent, $\theta = 2$ corresponds to Normal, while $\theta = 1$ corresponds to Cauchy; β – measures skewness with $-1 \leq \beta \leq 1$, and $\beta = 0$ corresponds to symmetric; μ – is a location parameter (it is the mean when $\theta > 1$ and $\beta = 0$); and γ – is a scale parameter (when $\theta = 2$, γ^2 is half the variance). The density functions, c.d.f.s and quantile functions are not given in closed form except in some special cases.

This class of distributions has the justification that all limiting sums of i.i.d. random variables are in this class, i.e., are stable. There are some aggregation results for the stable class due to its infinite divisibility. Suppose that X, Y are stable with parameters θ, γ, μ, and $\beta = 0$ (which is commonly assumed in practice), then $X + Y$ is stable with parameters θ, $(2\gamma)^{1/\theta}$, 2μ, and $\beta = 0$. It follows that for horizon T, we have

$$\text{VaR}_\alpha(T) = T^{1/\theta}\text{VaR}_\alpha(1), \tag{14.5}$$

which gives a different scaling law than the normal in general. The formula for $\text{VaR}_\alpha(1)$ is not given in closed form, however, although there are approximations for $\alpha \to 0$.

How to estimate α? One problem here is that for $\alpha < 2$, the variance is infinite, so the sample moments are not very useful. Instead estimation is based on the empirical characteristic function (see McCulloch (1986)).

The problem with this approach is that the parametric model is hard to justify, and since it is fitted from all the data, the middle of the distribution can have a big effect on the estimated parameter values and hence on the tail prediction. We now consider the nonparametric approach. Suppose that X has distribution F, where F is unknown. We may estimate the quantile of this distribution by the sample quantity, $\widehat{q}_\alpha$, defined in Chapter 4, i.e., $\widehat{q}_\alpha$ is any number such that (4.8) is satisfied. The estimator is consistent and asymptotically normal at rate square root sample size provided X has a density f and $f(q_\alpha) > 0$, i.e.,

$$T^{1/2}(\widehat{q}_\alpha - q_\alpha) \Longrightarrow N\left(0, \frac{\alpha(1-\alpha)}{f^2(q_\alpha)}\right)$$

(Koenker (2005)). This allows one to conduct inferences about q_α, i.e., to construct confidence intervals of the Value-at-Risk, by estimating the density f. In fact, a simple way of constructing approximate level τ confidence intervals for $\widehat{q}_\alpha$ is to use the intervals based on inverting the empirical distribution function intervals, which are

$$I_{\alpha,\tau} = \left[\widehat{q}_{\alpha_L(\tau)}, \widehat{q}_{\alpha_U(\tau)}\right], \tag{14.6}$$

where $\alpha_L(\tau) = \alpha - z_{\tau/2}\sqrt{\alpha(1-\alpha)/n}$ and $\alpha_L(\tau) = \alpha + z_{\tau/2}\sqrt{\alpha(1-\alpha)/n}$.

One issue with this approach is that it is focussed on moderate quantiles that are contained within the range of data that has already been observed. In some cases one is interested in **extreme quantiles** that might occur outside the range of the currently observed data, i.e., $\alpha < 1/T$ or $\alpha > 1 - 1/T$. This is because we worry about the major storm, the worst case scenario, that we have not so far experienced.

This requires some extrapolation, which effectively requires a model, at least for the extreme part of the distribution. We next discuss the theory of extreme values, which corresponds to the case $\alpha = 1/T$ or $\alpha = 1 - 1/T$. We then discuss models that allow us to extrapolate beyond this range into the deep tails.

14.3 Extreme Value Theory

What is the behavior of the sample maximum

$$M_T = \max_{1 \leq t \leq T} \{X_1, \ldots, X_T\}$$

when the sample size increases? The population maximum corresponds to the $\alpha = 1$ quantile, and the sample maximum could be interpreted as the $1 - 1/T$ quantile. Does M_T increase to infinity or does it have some finite limit? We are also interested in the sample minimum, which can be treated symmetrically, since for real valued outcomes, the minimum is the maximum of the negative of the original observations; so we just focus on the maximum. A further question arises when X_t is a vector and M_T is the coordinatewise maximum, that is, what is the relationship between the individual maxima?

The treatment of the first question goes back to pioneering work by Fisher and Tippett (1928). Gnedenko (1943) provided a definitive mathematical answer to this question. Embrechts, Klüppelberg, and Mikosch (1997) give an excellent review of this theory.

Theorem 56 *Suppose that X_t are i.i.d. Then, under some mild conditions there exist sequences of constants a_T, b_T such that*

$$a_T (M_T - b_T) \Longrightarrow \mathcal{E} \tag{14.7}$$

for some random variable $\mathcal{E}$. It is also shown that $\mathcal{E}$ can be of three main types. In particular

$$G_\gamma(x) = \Pr(\mathcal{E} \leq x) = \exp\left(-(1+\gamma x)^{-1/\gamma}\right), \quad 1 + \gamma x > 0, \tag{14.8}$$

with: the extreme value index γ with: $\gamma > 0$ being one type, $\gamma = 0$ (with $(1+\gamma x)^{-1/\gamma} = \exp(-x)$) being another, and $\gamma < 0$ being the final case. For $\gamma < 0$, the right endpoint of the distribution is $-1/\gamma$. Furthermore, the sample maximum and sample minimum are independent.

Example 72 Suppose that X_t are i.i.d. uniform on $[0,1]$, then we know that $M_T \xrightarrow{P} 1$. In fact,

$$\Pr[M_T \leq x] = \prod_{t=1}^{T} \Pr[X_t \leq x] = [F(x)]^T = [x]^T.$$

So taking $x_T = 1 - x/T$ for $x > 0$, we have

$$\Pr(T(M_T - 1) \leq -x) = \Pr(M_T \leq x_T)$$
$$= \left(1 - \frac{x}{T}\right)^T$$
$$\to \exp(-x).$$

That is, (14.7) is satisfied with $a_T = T$ and $b_T = 1$ and $\mathcal{E}$ an exponentially distributed random variable, i.e., $\gamma = -1$.

The theory for unbounded random variables is slightly more complicated because in that case we expect the maximum to diverge, and the calculation of the exact rate may be somewhat difficult.

Example 73 Suppose that $X_t \sim N(0,1)$. In this case, $M_T \xrightarrow{P} \infty$, and indeed (14.7) holds with $\gamma = 0$ and

$$a_T = \sqrt{2 \log T}$$
$$b_T = \sqrt{2 \log T} - \frac{1}{2} \frac{\log(4\pi) + \log \log T}{\sqrt{2 \log T}}$$

$\mathcal{E}$ has c.d.f. $G(x) = \exp(-\exp(-x))$

with density $g(x) = e^{-e^{-x}} e^{-x}$ defined on $x > 0$.

Example 74 Suppose that X_t is standard Cauchy with $f_X(x) = 1/\pi(1+x^2)$. In this case, $M_T \xrightarrow{P} \infty$, and indeed (14.7) holds with $\gamma = 1$ and

$$a_T = \pi/T, \quad b_T = 0$$

$\mathcal{E}$ has c.d.f. $G(x) = \exp\left(-x^{-1}\right), \quad x > 0,$

with density $g(x) = \exp(-x^{-1})/x^2$. The proof of this is given in Exercise 61.

14.3 Extreme Value Theory

We pause to give some interpretation to the extreme values of daily stock return data. The quantile function of the standard Cauchy random variable is $\tan(\pi(\alpha - 0.5))$ and the interquartile range is exactly 2. The extreme value of S&P500 daily returns on October 19, 1987 yields a *C*-score of around 12.53 (adjusting for both median and interquartile range) and a *z*-score of around 23.75 (adjusting for the mean and standard deviation). The chance of observing a value as large as or larger than this on a single day chosen at random under the Cauchy model is around 0.0125, whereas under the normal model it is preposterously small. We next see what the extreme value theory has to say about this. In fact we have $T = 17156$ observations in total between 1950 and 2018, and multiple testing considerations suggest that we should take account of the fact that we are focussing on October 19, 1987, precisely because the stock return on that day was the largest value recorded over that entire time frame. For the normal model we have

$$\Pr(M_T \geq 23.75) \simeq 1 - \exp(-\exp(-a_T(23.75 - b_T))) \simeq 1 - \exp(-\exp(-74.8)) \simeq 0.$$

Extreme value theory for Gaussian random variables says this is an extremely rare event. However, for the Cauchy model

$$\Pr(M_T \geq 12.53) \simeq 1 - \exp(-\frac{T}{12.53 \times \pi}) \simeq 1 - \exp(-(1/0.0029)) \simeq 1,$$

which is to say that the Cauchy model is not impressed by October 1987 at all if one takes account of the long sample of data for which this is the realized maximum value. The normal and Cauchy models are at polar extremes, and neither is a good match with the characteristics of stock return data.

The above theory is well developed for i.i.d. random variables, but what if the random variables are dependent, as we can expect for time series? Berman (1964) showed that if X_t is a stationary Gaussian process with marginal distribution $N(0, 1)$ and autocorrelation function $\rho(.)$ that satisfies $\rho(k)/\log k \to 0$ as $k \to \infty$, then (14.7) holds with exactly the same constants as if X_t were i.i.d. He also derives the limiting distribution of sample maxima for non-Gaussian processes, and in general the rates of convergence are not affected by the dependence, although the limiting distribution may be affected.

We next consider the multivariate case, which is important in practice since investors care about the outcomes of multiple assets and the simultaneous behavior of those outcomes. Suppose that

$$\begin{pmatrix} X_t \\ Y_t \end{pmatrix} \sim N\left(0, \begin{pmatrix} 1 & \rho \\ \rho & 1 \end{pmatrix}\right),$$

where ρ is the correlation coefficient with $|\rho| < 1$. In this case, $\max_{1 \leq t \leq T}\{X_1, \ldots, X_T\}$ and $\max_{1 \leq t \leq T}\{Y_1, \ldots, Y_T\}$ (and likewise $\min_{1 \leq t \leq T}\{X_1, \ldots, X_T\}$ and $\min_{1 \leq t \leq T}\{Y_1, \ldots, Y_T\}$) are asymptotically independent. In fact, this result continues to hold when $\{X_t, Y_t\}$ is a stationary Gaussian process, under some weak conditions on the autocorrelation function. This says that extreme values of the components of a bivariate Gaussian process are statistically independent of each other – bad news if one stock is not likely to coincide with bad news in another stock. In practice, we tend to observe large values of stock returns simultaneously, especially large negative values, which suggests the Gaussian paradigm is not well suited for understanding extreme events such as occur during crashes and bubbles. For the general case where (X_t, Y_t) have distribution F, the random variables $\max_{1 \leq t \leq T}\{X_1, \ldots, X_T\}$ and $\max_{1 \leq t \leq T}\{Y_1, \ldots, Y_T\}$ are asymptotically dependent.

In practice, we don't know what the distribution of the data is and hence we do not know what behavior to expect of the sample extremes. Without this information we are also unable to extrapolate beyond the observed data, which is a key issue we would like to address for risk management. We next consider how to estimate from the data the relevant features of the population distribution.

14.4 A Semiparametric Model of Tail Thickness

We would like to extrapolate beyond the sample data, that is, given the sample data, we can only say what the $\alpha \geq 1/T$ and $\alpha \leq 1 - 1/T$ quantiles are, but if T is quite small this doesn't allow us to consider very rare events. Using a model allows us to specify what to expect for very small α and very big α. We consider some models for the tail thickness of a distribution. The normal distribution has light tails, meaning that $\Pr(X > x)$ decays to zero very fast as $x \to \infty$ such that all moments exist. For daily stock returns we have seen that this is a little restrictive. Alternative distributions like the t-distribution allow more general tail behavior. The Pareto distribution has been widely used to model stock returns and income distributions and is quite flexible with regard to the tail behavior.

Definition 170 *The random variable X has Pareto distribution on $[L, \infty)$, if for $x \geq L$*

$$\Pr(X \leq x) = F(x) = 1 - Lx^{-\kappa},$$

where $\kappa > 0$ and $L > 0$. The density function is $f(x) = \kappa L/x^{\kappa+1}$ and the quantile function is $q_{1-\alpha} = L^{1/\kappa} \alpha^{-1/\kappa}$.

The parameter κ governs the tail thickness of the distribution. The smaller the value of κ, the heavier the tails of this distribution. We have $E(|X|^\gamma) < \infty$ if and only if $\gamma < \kappa$. One can modify this distribution to allow for positive and negative values. It does exclude thinner tails such as one finds for the normal distribution and the GED distribution, which can, however, be approximated by very large κ. This parametric distribution can be used to obtain VaR estimates using parametric estimates such as MLE. However, it models the whole distribution, whereas for many purposes we are only concerned with the tails of the distribution and even just extreme tail values.

14.4 A Semiparametric Model of Tail Thickness

We take a semiparametric approach that mimics the Pareto distribution in the tails but not everywhere. For example, stable distributions such as the Cauchy distribution have Pareto like tails, but behave differently elsewhere.

Definition 171 *We suppose that X has c.d.f. F, which is unknown, and*

$$F(x) \simeq 1 - L(x) x^{-\kappa} \qquad (14.9)$$

*as $x \to \infty$. Here, κ is called the tail index and $L(x)$ is a constant or a **slowly varying function** for which $\lim_{x\to\infty} L(ax)/L(x) = 1$ for all $a > 0$. If $F(x)$ has a density $f(x)$, then $f(x) \simeq \kappa L(x) x^{-(\kappa+1)}$ as $x \to \infty$. The quantile function satisfies $q_{1-\alpha} = L^*(1/\alpha) \alpha^{-1/\kappa}$ as $\alpha \to 0$, where $L^*(1/\alpha)$ is a constant or slowly varying function.*

One can separately model left and right tails, but usually the focus is just on the downside area. The Cauchy distribution falls into this class for both upper and lower tails with $\kappa = 1$, and more generally the t-distribution with degrees of freedom κ follows this law. We next give a more precise definition of tail thickness.

Definition 172 *Suppose that there is a $\kappa \in (0, \infty)$ such that for any $x > 0$*

$$\lim_{t\to\infty} \frac{1 - F(tx)}{1 - F(t)} = x^{-\kappa}. \qquad (14.10)$$

Then we say that X has tail thickness κ.

Example 75 *Suppose that $L(x) = c \log x$ in (14.9) (which is an example of a slowly varying function). Then from (14.9)*

$$\frac{1 - F(tx)}{1 - F(t)} \simeq \frac{c \log(t) + c \log(x)}{c \log(t)} x^{-\kappa} \simeq x^{-\kappa}$$

as $t \to \infty$, which would satisfy the formal definition (14.10).

The function $L(x)$ in (14.9), although not specified, plays a less important role than κ. Under the model assumption (14.9), $E(|X|^\gamma) = \infty$ for all $\gamma \geq \kappa$ but $E(|X|^\gamma) < \infty$ for all $\gamma < \kappa$. Thus the parameter κ is the key quantity that measures the frequency of large events and the existence of moments. This model is semiparametric, since it only concerns large values of x, and says nothing specific about how F behaves when x is close to zero, for example. The large x case behavior is what we need to understand how sample extremes would behave. The parameter κ can be related to the parameter γ of (14.8). Specifically, we have that the maximum value of a sample from the distribution F obeys (14.7) with limit (14.8) where $\kappa = 1/\gamma$. Therefore, the requirement that $\kappa > 0$ rules out the cases where $\gamma \leq 0$. That is, the model imposes heavy tails on the distribution F, distributions such as the normal with a thin tail are ruled out of this ecology. The model also rules out super heavy tailed distributions such as the **log Cauchy**, whose distribution tails decay slower than $x^{-\kappa}$ for any κ.

This Pareto tail model is used in many contexts in physical and social sciences (**Zipf's law, Gibrat's law**). It seems to describe high frequency stock returns (and trading volume) better than the Gaussian model. It is more flexible than the stable model. In fact, the strong GARCH model with Gaussian innovations implies that observed returns satisfy this model.

14.4.1 Estimation of Tail Thickness

The model (14.9) depends on the quantities κ and $L(x)$, which are generally unknown. We next consider how to estimate them. We first order the data

$$X_{(1)} > \cdots > X_{(T)}.$$

Let M be a large integer but smaller than T. For large **order statistics** $X_{(j)}$ with $j \leq M+1$, we have

$$\log \Pr(X > X_{(j)}) \simeq -\kappa \log(X_{(j)}) + \log L(X_{(j)}),$$

so that subtracting off $\log \Pr(X > X_{(M+1)})$ from both sides we obtain

$$\log(j/(M+1)) = -\kappa \log(X_{(j)}/X_{(M+1)}) + \log L(X_{(j)})/L(X_{(M+1)})$$
$$\simeq -\kappa \log(X_{(j)}/X_{(M+1)}), \qquad (14.11)$$

because $L(X_{(j)})/L(X_{(M+1)}) \to 1$ as $T \to \infty$. Furthermore, since $\int_1^t \ln(x)\,dx = t \ln t - t + 1$ we have $\log(M+1) - \sum_{j=1}^M \log(j)/M \to 1$. Therefore,

$$\frac{1}{M}\sum_{j=1}^M \log \frac{X_{(j)}}{X_{(M+1)}} = \frac{1}{\kappa}\frac{1}{M}\sum_{j=1}^M (\log(M+1) - \log j)$$
$$= \frac{1}{\kappa}\left(\log(M+1) - \frac{1}{M}\sum_{j=1}^M \log j\right)$$
$$\simeq \frac{1}{\kappa},$$

which suggests how to estimate κ.

We now define an estimator of κ and use this to define an estimator of extreme quantiles, that is, quantiles q_α with $\alpha \to 1$ (and $\alpha \to 1$ at any rate thereby allowing extrapolation outside of the sample range).

14.4 A Semiparametric Model of Tail Thickness

Definition 173 *Hill estimator. Let $M = M(T)$ be some threshold value. Compute*

$$\frac{1}{\widehat{\kappa}} = \frac{1}{M} \sum_{j=1}^{M} \log \frac{X_{(j)}}{X_{(M+1)}} \tag{14.12}$$

$$\widehat{q}_{1-\alpha} = \widehat{L}_T \alpha^{-1/\widehat{\kappa}}, \qquad \widehat{L}_T = X_{(M+1)} \left(\frac{M}{T}\right)^{1/\widehat{\kappa}}, \quad \text{for } \alpha < M/T. \tag{14.13}$$

For the quantile estimator one can extrapolate in other ways. For example, take $\widehat{L}_T = X_{(1)}/T^{1/\widehat{\kappa}}$, in which case this estimator coincides with the sample maximum when $\alpha = 1/T$.

The estimated tail thickness parameter is consistent and asymptotically normal at rate $M^{1/2}$ provided $M \to \infty$ and $M/T \to 0$.

Theorem 57 *Suppose that the von Mises condition*

$$\lim_{x \to \infty} \frac{xf(x)}{1 - F(x)} = \kappa > 0 \tag{14.14}$$

holds, where F has a density $f = F'$. Then

$$M^{1/2} \left(\widehat{\kappa} - \kappa\right) \Longrightarrow N(0, \kappa^2). \tag{14.15}$$

The quantity M has to be chosen in the Goldilocks way, not too big and not too small. There is a theory for determination of optimal M under additional conditions, but this is beyond the scope of our work here. We explore below in practice the consequence of different choices of M.

There are several alternative estimators of κ. Gabaix and Ibragimov (2011) proposed the log rank δ estimator, which is based on fitting the linear regression

$$\log(i - \delta) = a - b \log(X_{(i)}), \quad i = 1, \ldots, M \tag{14.16}$$

for $\delta \in [0, 1)$ to find the estimates $\widehat{a}_\delta, \widehat{b}_\delta$. They show that $\widehat{b}_\delta \to \kappa$ as $M \to \infty$ and indeed (14.15) holds. They recommend taking $\delta = 1/2$ as this most improves the finite sample performance of the estimator. This estimator is quite popular in practice and equally easy to apply.

Dekkers, Einmahl, and de Haan (1989) argue that the Hill estimator can only deal with the case of $\gamma > 0$ in (14.8), which rules out distributions whose extreme value behavior does not confirm to this situation. They proposed an alternative estimator of $\gamma = 1/\kappa$. Specifically, let

$$\widehat{\gamma}_{DEM} = H_T^{(1)} + 1 - \frac{1}{2}\left(1 - \frac{\left(H_T^{(1)}\right)^2}{H_T^{(2)}}\right)^{-1} \quad (14.17)$$

$$H_T^{(j)} = \frac{1}{M}\sum_{i=1}^{M}\left(\log X_{(i)} - \log X_{(M+1)}\right)^j, \quad j = 1, 2.$$

They showed that $\widehat{\gamma}_{DEM}$ consistently estimates γ in (14.8) for all $\gamma \in \mathbb{R}$, and show that $\widehat{\gamma}_{DEM}$ is asymptotically normal with the same rate of convergence. In the case where $\gamma > 0$, $\widehat{\gamma}_{DEM}$ has limiting variance $1 + \gamma^2$, i.e., it is inefficient relative to the Hill estimator in that case.

The distribution theory for these estimators has been extended to a general time series context. Under very weak conditions on the amount of dependence, the consistency result for the Hill estimator generalizes to stationary time series data (see Resnick and Stărică (1998)). Hill (2010) shows under some conditions that

$$M^{1/2}(\widehat{\kappa} - \kappa) \Longrightarrow N(0, \Phi) \quad (14.18)$$

for some variance Φ. In general $\Phi \geq \kappa^2$, and Φ depends on the dependence structure. For some special cases including strong GARCH models, $\Phi = \kappa^2$. The result holds because although the strong GARCH process has dependent extremes, a crucial stochastic array has linearly independent extremes. This property is found in many similar strong GARCH type processes. However, this property is not guaranteed to hold in, for example, a semi-strong GARCH process, and in this case Φ is not necessarily equal to κ^2. Hill (2006) proposes an estimator of Φ that is consistent under general conditions. This is

$$\widehat{\Phi} = \frac{1}{m}\sum_{s=1}^{T}\sum_{t=1}^{T}K\left(\frac{s-t}{b_T}\right)\widehat{Z}_s\widehat{Z}_t, \quad (14.19)$$

where $\widehat{Z}_t = ((\log(X_t/X_{m+1}))_+ - ((m/T)\widehat{\kappa}_T^+)$ and b_T is some bandwidth sequence, while K is a kernel.

We apply the Hill method to the S&P500 daily returns; in Figure 14.1 we show the estimated tail index against the threshold value M. The estimates depend on the threshold chosen but seem to stabilize to somewhere between three and four. The standard errors from (14.15) with $M = 100$ are of the order 0.3–0.4 and so these quantities seem to be quite well measured. This is broadly consistent with the empirical results presented in Gabaix, Gopikrishnan, Plerou, and Stanley (2005), who argue that returns follow a cubic power law ($\kappa = 3$) consistently over time and internationally. They worked with 4 years of high frequency data on around 1000 stocks from the TAQ database over 1994–1997 aggregating to 15 minutes, 1 hour, 1 day, and 1 week. They also argue that trading volume follows a half cubic law ($\kappa = 3/2$) and present a theory that explains these empirical regularities. They also argue that based on the

14.5 Dynamic Models and VAR

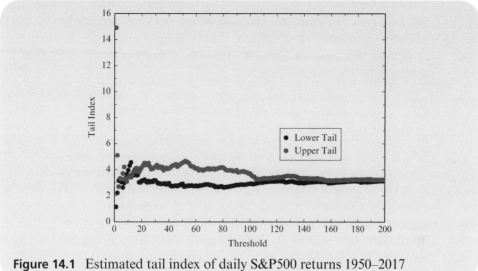

Figure 14.1 Estimated tail index of daily S&P500 returns 1950–2017

power law distribution, the 1929 and 1987 crashes are not "outliers," meaning that these values are broadly consistent with the extreme value theory associated with the cubic law. We have already seen some discussion of this for the Cauchy distribution.

14.5 Dynamic Models and VAR

We have so far considered the unconditional distribution and the Value-at-Risk and tail thickness associated with that distribution. We next consider the conditional distribution, that is conditional on currently available information. It seems reasonable to take account of the latest information in assessing the riskiness of ones current position. However, this is not without cost, since we will have to adopt a time series model and make some additional assumptions to deliver the conditional Value-at-Risk consistently.

Suppose that returns follow a GARCH(1,1) process

$$X_t = \mu + \varepsilon_t \sigma_t$$

$$\sigma_t^2 = \omega + \beta \sigma_{t-1}^2 + \gamma (X_{t-1} - \mu)^2$$

where $\varepsilon_t \sim N(0,1)$. This generates some heavy tails in the marginal distribution of X_t and allows for dependence over time as we saw in Chapter 11. The conditional α-quantile of $X_t | \mathcal{F}_{t-1}$ implied by this model is

$$q_\alpha(X_t | \mathcal{F}_{t-1}) = \mu + \sigma_t z_\alpha, \quad (14.20)$$

where now the quantile is time varying. When volatility is high, $\mu + \sigma_t z_\alpha$ will be large in magnitude. One could also allow μ to vary over time to generate additional time variation, and we just denote this by μ_t. In practice we have to replace the unknown quantities μ, σ_t by estimates obtained by the methods discussed in Chapter 11.

The assumption that $\varepsilon_t \sim N(0, 1)$ is not essential, and in practice may be too strong as we have discussed in Chapter 11. Suppose instead that ε_t is i.i.d. with unknown density f. In this case, we have

$$q_\alpha(X_t|\mathcal{F}_{t-1}) = \mu_t + \sigma_t w_\alpha, \quad (14.21)$$

where w_α is the α-quantile of ε_t. For Gaussian the 5% quantile is -1.645, whereas for a t_1 this is 6.314, and a t_6 this is -1.943. The 1% values are $-2.326, -31.821$, and -3.143 respectively. The quantity w_α has to be estimated from the data: in particular we take the sample quantile of the standardized residual $\widehat{\varepsilon}_t = (X_t - \widehat{\mu}_t)/\widehat{\sigma}_t$. This is a reasonable approach for moderately small or large quantiles. For extreme quantiles, one can use the Hill estimator (14.13) on the residuals $\widehat{\varepsilon}_t$. If $w_\alpha > z_\alpha$, then the derived conditional VaR will be higher than under the Gaussian assumption. However, the conditional quantiles vary over time in a similar fashion.

The **CaViar** approach, due to Engle and Manganelli (2004), involves dynamic modelling of the latent value at risk directly and generates more flexible time variation. Define $VaR_t(\alpha)$ as the solution to the equation

$$\Pr(X_t < -VaR_t(\alpha)|\mathcal{F}_{t-1}) = \alpha, \quad (14.22)$$

where in the sequel we drop the dependence of VaR_t on α for simplicity. This is the conditional Value-at-Risk given the previous period's information. They define various dynamic models for VaR_t itself, including the following:

$$VaR_t = VaR_{t-1} + \beta\left[1(X_{t-1} \le -VaR_{t-1}) - \alpha\right]$$

$$VaR_t = \beta_0 + \beta_1 VaR_{t-1} + \beta_2|X_{t-1} - \beta_3|$$

$$VaR_t = \beta_0 + \beta_1 VaR_{t-1} + \beta_2|X_{t-1}|.$$

These dynamic models have unknown parameters, $\beta_j, j = 0, 1, 2$, and depend on some function of lagged X as an innovation. This makes it rather hard to study from the point of view of establishing properties like stationarity and mixing from first principles. Usually there is an i.i.d. shock ε_t that then influences the dynamics, as in GARCH process or stochastic volatility process, but here there is nonsuch. This also makes it difficult to simulate from, except in some special cases. Recall that for the GARCH(1,1) case, VaR_t is a linear function of conditional standard deviation so that

$$VaR_t = \sigma_t z_\alpha = \left(\omega + \beta\sigma_{t-1}^2 + \gamma X_{t-1}^2\right)^{1/2} z_\alpha = \left(\omega + \beta(VaR_{t-1}/z_\alpha)^2 + \gamma X_{t-1}^2\right)^{1/2} z_\alpha.$$

This leads to a CaViar specification called indirect GARCH

$$VaR_t = \left(\beta_0 + \beta_1 VaR_{t-1}^2 + \beta_2 X_{t-1}^2\right)^{1/2},$$

which in this case can be obtained from a model with i.i.d. shocks.

14.6 The Multivariate Case

Estimation of CaViar models, however, is quite straightforward. One can estimate the parameters β using standard quantile regression techniques, i.e., minimize

$$Q_T(\beta) = \frac{1}{T}\sum_{t=1}^{T} \rho_\alpha(X_t - VaR_t(\beta))$$

with respect to β, where $VaR_t(\beta)$ is computed recursively from some initial value (just to recap, $VaR_t(\beta)$ depends on α too, but this dependence has been dropped from the notation). By making high level assumptions such as stationarity and mixing, one can obtain a CLT for the estimated parameters that can be used to provide confidence intervals and carry out hypothesis tests.

Chernozhukov (2005) developed a theory for extreme quantiles when there are a finite number of fixed covariates driving the quantile regression.

14.5.1 Evaluation of Value-at-Risk Models

Suppose that $VaR_t(\alpha)$ is a proposed (conditional or unconditional) VaR measure at level α based on data $\{X_{t-T},\ldots,X_{t-1}\}$. We may evaluate this method by just counting whether subsequent returns on average exceed the threshold the predicted number of times. Let $\{1,\ldots,H\}$ be an evaluation period. Then let

$$\widehat{p}_\alpha = \frac{1}{H}\sum_{t=1}^{H} \psi_\alpha(X_t - VaR_t), \tag{14.23}$$

where $\psi_\alpha(x) = \text{sign}(x) - (1 - 2\alpha)$. This quantity should be a sum of mean zero independent binary variables if $VaR_t(\alpha)$ is correct.

14.6 The Multivariate Case

So far we have considered the measurement of tail thickness and risk management from a univariate point of view, whereas in practice financial market participants care about portfolios containing possibly many risky assets. We next describe a common approach to describing multiple risks.

Suppose that X_1 and X_2 are two continuously distributed random variables. Letting $Y_1 = F_{X_1}(X_1)$ and $Y_2 = F_{X_2}(X_2)$, we know that Y_1 and Y_2 are uniformly distributed on $[0,1]$. The function

$$C(u_1, u_2) = \Pr(Y_1 \leq u_1, Y_2 \leq u_2)$$

is called the **copula** of X_1, X_2; it is a bivariate distribution function on $[0,1] \times [0,1]$. The joint distribution of X_1, X_2 is equivalently described by the copula $C(u_1, u_2)$ and the two marginal distribution functions F_{X_1} and F_{X_2}. This allows separate modelling of the dependence (by modelling C) from the marginal distributions.

Theorem 58 *(Sklar (1959)) Suppose that X_1, X_2 are continuously distributed. Then the joint distribution of X_1, X_2 can be written uniquely as*

$$\Pr(X_1 \leq x, X_2 \leq x_2) = C(F_{X_1}(x_1), F_{X_1}(x_2))$$

for some distribution function $C: [0,1]^2 \longrightarrow [0,1]$.

This approach converts marginal distributions into a standard scale, which allows modelling of the dependence through C in a common framework. If $C(u_1, u_2) = u_1 u_2$, then X_1 and X_2 are independent, but other choices of C allow for dependence.

Example 76 *The Gaussian copula is*

$$C(u_1, u_2; \rho) = \Phi_2\left(\Phi^{-1}(u_1), \Phi^{-1}(u_2); \rho\right) \quad (14.24)$$

where Φ is the standard univariate normal c.d.f., while $\Phi_2(s,t;\rho)$ is the c.d.f. of the standard bivariate normal distribution (with mean vector zero and variances equal to one) with correlation parameter ρ. This implies the model for the bivariate c.d.f. and density function of X_1, X_2

$$F(x_1, x_2) = \frac{1}{2\pi\sqrt{1-\rho^2}} \int_{-\infty}^{\Phi^{-1}(F_1(x_1))} \int_{-\infty}^{\Phi^{-1}(F_1(x_2))} \exp\left(-\frac{s^2 + t^2 - 2\rho st}{2(1-\rho^2)}\right) ds dt \quad (14.25)$$

$$f(x_1, x_2) = \frac{1}{\sqrt{1-\rho^2}} \exp\left(-\frac{\rho\Phi^{-1}(F_1(x_1))\Phi^{-1}(F_2(x_2))}{2(1-\rho^2)}\right) \times f_1(x_1) f_2(x_2). \quad (14.26)$$

This has been called **the formula that killed Wall Street**, and not because it is too complicated (needless to say, the rumours of Wall Street's death were greatly exaggerated); it is because it was widely used in credit risk modelling, because it is very flexible with regard to the marginal distributions not being Gaussian and so had the veneer of respectable generality. The weakness was that not only are extreme events likely but when they happen for one risk they tend to happen to all risks.

Andrew Patton is the Zelter Family Distinguished Professor of Economics and Professor of Finance at Duke University. Patton currently serves on the editorial boards of the *Journal of the American Statistical Association, Journal of Business and Economic Statistics, Journal of Econometrics,* and the *Review of Asset Pricing Studies*, as well as serving as a Managing Editor of the *Journal of Financial Econometrics*. He is an elected fellow of the Society for Financial Econometrics. Patton's research interests lie in financial econometrics, with an emphasis on forecasting volatility and dependence, forecast evaluation methods, and the analysis of hedge funds. He is well known for his work on

14.6 The Multivariate Case

modelling dynamic asymmetric dependence using copulas. His research has appeared in a variety of academic journals, including the *Journal of Finance*, *Journal of Econometrics*, *Journal of Financial Economics*, *Journal of the American Statistical Association*, *Review of Financial Studies*, and the *Journal of Business and Economic Statistics*. He has given hundreds of invited seminars around the world, at universities, central banks, and other institutions. Patton has previously taught at the London School of Economics, the University of Oxford, and New York University. He has been a visiting professor at the University of Sydney and served as an academic consultant to the Bank of England's Financial Stability division. He completed his undergraduate studies in finance and statistics at the University of Technology, Sydney, and his PhD in economics at the University of California, San Diego.

Patton (2006) used the symmetrized Joe-Clayton copula

$$C_{SJC}(u,v) = \frac{1}{2}\left(C_{JC}(u,v|\tau^U,\tau^L) + C_{JC}(1-u,1-v|\tau^L,\tau^U) + u + v - 1\right)$$

$$1 - C_{JC}(u,v|\tau^U,\tau^L) = \left(1 - \left((1-(1-u)^\kappa)^{-\gamma} + (1-(1-v)^\kappa)^{-\gamma} - 1\right)^{-1/\gamma}\right)^{1/\kappa}$$

where $\kappa = 1/\log_2(2-\tau^U)$ and $\gamma = -1/\log_2(\tau^L)$, where

$$\tau^L = \lim_{\varepsilon \to 0} \frac{C(\varepsilon,\varepsilon)}{\varepsilon} \quad ; \quad \tau^U = \lim_{\varepsilon \to 1} \frac{C(\varepsilon,\varepsilon)}{\varepsilon},$$

are measures of dependence known as tail dependence. Tail dependence captures the behavior of the random variables during extreme events. For a Gaussian copula $\tau^L = \tau^U = 0$, but this is not necessary in the Joe-Clayton copula. He tests for asymmetry in a model of the dependence between the Deutsche mark and the yen and finds evidence that the mark–dollar and yen–dollar exchange rates are more correlated when they are depreciating against the dollar than when they are appreciating. Chen and Fan (2006) provide a theory for estimation of copula-based models; see Patton (2012) for an up to date review.

Recently, interest has focussed on **systemic risk**, that is, how to measure the potential contribution of individual banks, for example to the riskiness of the system as a whole. One very successful measure of this was given by Adrian and Brunnermeier (2010), called **CoVaR**.

Definition 174 *We denote by $CoVaR_\alpha^{j|i}$ the VaR of institution j (or the financial system) conditional on some event $\mathbb{C}(X^i)$ of institution i. That is,*

$$\Pr\left(X^j \leq CoVaR_\alpha^{j|\mathbb{C}(X^i)}|\right) = \alpha.$$

We denote institution i's contribution to j by

$$\Delta CoVaR_\alpha^{j|i} = CoVaR_\alpha^{j|X^i = VaR_\alpha^i} - CoVaR_\alpha^{j|X^i = Median^i}.$$

14.7 Coherent Risk Measures

There is a substantial literature that questions what properties a risk measure should possess. Artzner, Delbaenm, Eber, and Heath (1999) introduced a set of axioms that they argued a coherent risk measure should satisfy. Let R be a risk measure associated with value W:

Axiom 1 *Monotonicity. If W is more risky than W^* by first order stochastic dominance, then*

$$R(W) \geq R(W^*).$$

Axiom 2 *Invariance with respect to drift. For all c, W*

$$R(W + c) = R(W) - c.$$

Axiom 3 *Homogeneity. For all $\lambda \geq 0$, W*

$$R(\lambda W) = \lambda R(W).$$

Axiom 4 *Subadditivity. For all W, W^**

$$R(W + W^*) \leq R(W) + R(W^*).$$

The axioms are rather special to banking. Why is subadditivity sensible? This reflects the idea that diversification should reduce risk. If this property would not be imposed there might be an incentive to break up the portfolio into smaller units.

Example 77 *The standard deviation satisfies homogeneity and subadditivity since*

$$\begin{aligned}(sd(X+Y))^2 &= var(X+Y) \\ &= var(X) + var(Y) + 2cov(X, Y) \\ &\leq var(X) + var(Y) + 2sd(X)sd(Y) \\ &= (sd(X) + sd(Y))^2.\end{aligned}$$

But it does not satisfy the invariance with respect to drift.

14.8 Expected Shortfall

Value at risk VaR_α satisfies some but not all of these properties. Specifically, it does not necessarily satisfy subadditivity.

Example 78 *Suppose X and Y are independent with outcomes*

$$\begin{cases} -100 & \text{with probability } 0.04 \\ 0 & \text{with probability } 0.96. \end{cases}$$

Then the 0.05 Value-at-Risk is zero for each asset. But note that

$$X + Y = \begin{cases} -200 & \text{with probability } 0.0016 \\ -100 & \text{with probability } 0.0768 \\ 0 & \text{with probability } 0.9216 \end{cases}$$

so that the Value-at-Risk for $X + Y$ is 100.

However, Ibragimov (2009) shows that for a large class of distributions including α-stable with $\alpha > 1$ the Value-at-Risk is subadditive. For example, for two independent normal distributions $X_j \sim N(\mu_j, \sigma_j^2)$, the sum has $N(\mu_1 + \mu_2, (\sigma_1^2 + \sigma_2^2)^{1/2})$. Therefore, the Value-at-Risk satisfies $\text{VaR}_\alpha(X_j) = \mu_j + \sigma_j z_\alpha$, and

$$\begin{aligned} \text{VaR}_\alpha(X_1 + X_2) &= \mu_1 + \mu_2 + (\sigma_1^2 + \sigma_2^2)^{1/2} z_\alpha \\ &\leq \text{VaR}_\alpha(X_1) + \text{VaR}_\alpha(X_2) \\ &= \mu_1 + \mu_2 + (\sigma_1 + \sigma_2) z_\alpha \end{aligned}$$

by the subadditivity for standard deviation. He shows that when the (stable) index $\alpha < 1$, Value-at-Risk is superadditive, i.e., the inequality goes the other way. Thus, for extremely heavy tails, diversification can be bad! His main results are for the independent case, but he has some results for a factor model. Danielsson, Jorgensen, Samorodnitsky, Sarma, and de Vries (2013) show a similar result based on more general conditions of a regular variation index.

14.8 Expected Shortfall

Definition 175 *The expected shortfall is defined as*

$$ES_\alpha(X) = -E(X_t | X_t \leq q_\alpha) = -\frac{E(X_t \mathbf{1}(X_t \leq q_\alpha))}{\Pr(X_t \leq q_\alpha)} = -\frac{E(X_t \mathbf{1}(X_t \leq q_\alpha))}{\alpha}. \quad (14.27)$$

For a continuous random variable $ES_\alpha(X) = \int_0^\alpha \text{VaR}_\tau(X) d\tau_\tau$.

This is an alternative measure of risk that does satisfy subadditivity, which we show below. On the other hand it requires one moment to exist for its definition (which is not needed for Value-at-Risk at a given quantile to exist).

Theorem 59 *Expected shortfall is subadditive*

Proof. Let $Z = X + Y$ where X, Y are two random variables. Then

$$\begin{aligned}
\alpha\left[ES_\alpha(X) + ES_\alpha(Y) - ES_\alpha(Z)\right] &= E\left[Z1(Z \leq q_\alpha(Z)) - X1(X \leq q_\alpha(X)) - Y1(Y \leq q_\alpha(Y))\right] \\
&= E\left[X\{1(Z \leq q_\alpha(Z)) - 1(X \leq q_\alpha(X))\}\right] \\
&\quad + E\left[Y\{1(Z \leq q_\alpha(Z)) - 1(Y \leq q_\alpha(Y))\}\right] \\
&\geq q_\alpha(X)\left[\alpha - \alpha\right] + q_\alpha(Y)\left[\alpha - \alpha\right] \\
&= 0,
\end{aligned}$$

where we use the inequalities:

$$1(Z \leq q_\alpha(Z)) - 1(X \leq q_\alpha(X)) \begin{cases} \geq 0 & \text{if } X \geq q_\alpha(X) \\ \leq 0 & \text{if } X \leq q_\alpha(X). \end{cases}$$

□

We can estimate ES_α by either:

$$\widehat{ES}_\alpha = -\frac{\sum_{t=1}^{T} X_t 1(X_t \leq \widehat{q}_\alpha)}{\sum_{t=1}^{T} 1(X_t \leq \widehat{q}_\alpha)} \tag{14.28}$$

$$\widetilde{ES}_\alpha = -\frac{1}{\alpha} \sum_{t=1}^{T} X_t 1(X_t \leq \widehat{q}_\alpha), \tag{14.29}$$

where $\widehat{q}_\alpha$ is the estimated quantile or Value-at-Risk. These estimators are both consistent and asymptotically normal as the sample size increase (Scaillet (2004, 2005)).

In Figure 14.2 we show estimates of Value-at-Risk and expected shortfall by the quantile level α for daily S&P500 returns. They tell a pretty similar story.

In a dynamic model such as the GARCH(1,1), the conditional expected shortfall is

$$ES_\alpha(X_t|\mathcal{F}_{t-1}) = \mu + \sigma_t ES_\alpha(\varepsilon),$$

where $ES_\alpha(\varepsilon)$ is the expected shortfall of the error ε_t, and so it moves around perfectly with (14.21).

14.9 Black Swan Theory

The **black swan theory** or theory of black swan events is a metaphor that describes an event that comes as a surprise, has a major effect, and is often inappropriately

14.9 Black Swan Theory

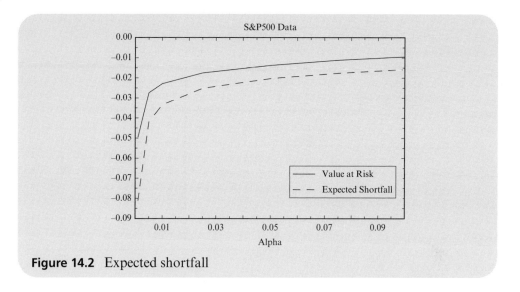

Figure 14.2 Expected shortfall

rationalized after the fact with the benefit of hindsight. The theory was developed by Nassim Nicholas Taleb to explain:

- The disproportionate role of high-profile, hard-to-predict, and rare events that are beyond the realm of normal expectations in history, science, finance, and technology.
- The non-computability of the probability of the consequential rare events using scientific methods (owing to the very nature of small probabilities).
- The psychological biases that blind people (both individually and collectively) to uncertainty and to a rare event's massive role in historical affairs.

There are also dragon kings and gray rhinos. Dragon king is a double metaphor for an event that is both extremely large in size or impact (a "king") and born of unique origins (a "dragon") relative to its peers (other events from the same system). The gray rhino is a metaphor for obvious dangers that we ignore. One explanation for the 2008 financial crisis was that there was a lurking gray rhino that transitioned into a dragon king that then ate the black swan.

In Figure 14.3 we show the occurrence of extreme events, 26 days in total when the absolute daily S&P500 stock return exceeded six sigma. The event 19550926 was a reaction to the heart attack of President Eisenhower and 19620528 was part of the **Kennedy slide** that took the stock market substantially downwards in 1962. Between 1962 and 1987 there were no such events, and perhaps one could have easily have deluded oneself in the early 1980s that such events were things of the past, although this is quite different from believing that they had never existed.

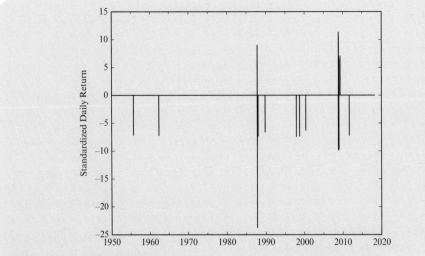

Figure 14.3 Daily S&P500 stock return events exceeding 6 sigma

14.10 Summary of Chapter

We described the concepts of Value-at-Risk and extreme shortfall and how to estimate them from a sample of data under a variety of assumptions. We provided a short introduction to extreme value theory, which can be used to describe the extreme properties of financial time series. We used the semiparametric Pareto tail model to extrapolate beyond the sample to forecast extreme values.

15 Exercises and Complements

1. First obtain daily price data on a stock index and two individual stocks from a market of your choice (some choices below). You should not choose the same index as anyone else in the class. The method for assignment is given below. The calculations can be performed in Excel and/or `Eviews`, but also in other software packages, as you prefer.

    (a) Compute the sample statistics of the stock return (computed from the daily closing price) series, i.e., the mean, standard deviation, skewness, and kurtosis. You may ignore dividends and just focus on capital gain.
    (b) Compute the first 20 autocorrelation coefficients and test whether the series is linearly predictable or not.
    (c) Does it make a difference whether you compute returns using log price differences or as actual return?

2. The (equal weighted) moving average filter of a series X_t is defined as

$$SMA_t^k = \frac{X_t + X_{t-1} + \cdots + X_{t-k}}{k},$$

where k is the number of lags to include, sometimes called a bandwidth parameter. For daily stock prices, common values include 5, 10, 20, 50, 100, and 200. Compute the SMA for the series. The exponential weighted average is defined as

$$EWMA_t = \alpha X_t + (1-\alpha)EWMA_{t-1},$$

where $EWMA_1 = X_1$ and $\alpha \in (0,1)$ can relate $\alpha = 2/(k+1)$, where k is number of time periods. These smoothed values are often used in trading strategies of the contrarian type, that is: buy when $X_t < SMA_t^k$ and sell when $X_t > SMA_t^k$, or moment type trading strategies, that is, buy when $X_t > SMA_t^k$ and sell when $X_t < SMA_t^k$. Comment on the efficacy of these trading strategies for your dataset. (Faber 2013).

3. The so-called Bollinger bands (http://en.wikipedia.org/wiki/Bollinger_Bands) are a modification of the moving average rules that allow a margin of safety by allowing for time varying volatility. They are defined as follows:

$$BB_t^U = SMA_t^k + 2\sigma_t$$
$$BB_t^L = SMA_t^k - 2\sigma_t$$
$$\sigma_t = std(X_t, X_{t-1}, \ldots, X_{t-k}).$$

Compute the Bollinger bands for your data series and compare the trading strategies: buy when $X_t < BB_t^L$ and sell when $X_t > BB_t^L$, or moment type, that is, buy when $X_t > BB_t^U$ and sell when $X_t < BB_t^L$.

4. Recently, Warren Buffett has predicted that the Dow Jones index will exceed one million in a hundred years time. Given that the current level of the index is 22400, what annual rate of return is he assuming?

5. Generate data from a random walk with normal increments and from a random walk with Cauchy increments. Graph the resulting time series and comment on its behavior.

6. We give an outline argument for Theorem 14. Suppose that $E(Y_t^4) < \infty$. With $\mu = E(Y_t)$, we write

$$\hat{\gamma}(j) = \frac{1}{T} \sum_{t=j+1}^{T} \left(Y_t - \mu - (\overline{Y} - \mu)\right)\left(Y_{t-j} - \mu - (\overline{Y} - \mu)\right)$$

$$= \frac{1}{T} \sum_{t=j+1}^{T} (Y_t - \mu)(Y_{t-j} - \mu) - (\overline{Y} - \mu)\frac{1}{T}\sum_{t=j+1}^{T}(Y_t - \mu)$$

$$- (\overline{Y} - \mu)\frac{1}{T}\sum_{t=j+1}^{T}(Y_{t-j} - \mu) + \frac{T-j}{T}(\overline{Y} - \mu)^2$$

$$\simeq \frac{1}{T} \sum_{t=j+1}^{T} (Y_t - \mu)(Y_{t-j} - \mu) - (\overline{Y} - \mu)^2$$

$$\simeq \frac{1}{T} \sum_{t=j+1}^{T} (Y_t - \mu)(Y_{t-j} - \mu),$$

because $\sqrt{T}(\overline{Y} - \mu)$ satisfies a CLT, $\sqrt{T}(\overline{Y} - \mu)^2 \xrightarrow{P} 0$. We can show that

$$\frac{1}{\sqrt{T}} \sum_{t=j+1}^{T} (Y_t - \mu)(Y_{t-j} - \mu)$$

satisfies a CLT. It follows that

$$\sqrt{T}\hat{\rho}(j) \simeq \frac{1}{\gamma(0)} \frac{1}{\sqrt{T}} \sum_{t=j+1}^{T} (Y_t - \mu)(Y_{t-j} - \mu) = \frac{1}{\sqrt{T}} \sum_{t=j+1}^{T} \widetilde{X}_t \widetilde{X}_{t-j},$$

where $\widetilde{X}_t = (Y_t - \mu)/\sqrt{\gamma(0)}$ has mean zero and variance one and is i.i.d.

Chapter 15 Exercises and Complements

7. Prove (3.20). We have

$$\tilde{\gamma}(j) = \frac{1}{T-j} \sum_{t=j+1}^{T} (Y_t - \mu)(Y_{t-j} - \mu) + \frac{1}{T-j} \sum_{t=j+1}^{T} (Y_t - \mu)(\mu - \overline{Y}_j)$$

$$+ \frac{1}{T-j} \sum_{t=j+1}^{T} (\mu - \overline{Y}_j)(Y_{t-j} - \mu) + \frac{1}{T-j} \sum_{t=j+1}^{T} (\mu - \overline{Y}_j)^2$$

$$= \frac{1}{T-j} \sum_{t=j+1}^{T} (Y_t - \mu)(Y_{t-j} - \mu) - \frac{1}{T-j} \sum_{t=j+1}^{T} (\mu - \overline{Y}_j)^2.$$

8. We explain the result of Theorem 16 here. Suppose for simplicity that Y_{it} are i.i.d. with mean zero and variance one. Then consider

$$\overline{\rho}(1) = \frac{1}{n} \sum_{i=1}^{n} \left(\frac{1}{T} \sum_{t=2}^{T} Y_{it} Y_{i,t-1} \right) = \frac{1}{nT} \sum_{i=1}^{n} \sum_{t=2}^{T} Y_{it} Y_{i,t-1}.$$

Under the null hypothesis $E(Y_{it} Y_{i,t-1}) = 0$ and $E(Y_{it} Y_{i,t-1} Y_{js} Y_{j,s-1}) = 0$, provided $t \neq s$. It follows that

$$\mathrm{var}(\overline{\rho}(1)) = \frac{1}{n^2 T^2} \sum_{i=1}^{n} \sum_{j=1}^{n} \sum_{t=2}^{T} E(Y_{it} Y_{i,t-1} Y_{jt} Y_{j,t-1}),$$

where

$$E(Y_{it} Y_{i,t-1} Y_{jt} Y_{j,t-1}) = E(Y_{it} Y_{jt} Y_{i,t-1} Y_{j,t-1}) = E(Y_{it} Y_{jt}) E(Y_{i,t-1} Y_{j,t-1}) = \omega_{ij}^2,$$

and $\omega_{ii}^2 = 1$. The result follows.

9. Consider the $VR(q)$ test with $q = 3$. Find a time series process that is not i.i.d. but for which $VR(3) = 1$. Comment on the properties of the test in this case.

10. Derive the limiting properties of the (non-overlapping) variance ratio statistic under **rw1**. First, note that provided $E(r_t^4) < \infty$,

$$\sqrt{T} \left(\hat{\sigma}_H^2 - \sigma^2 \right) \Longrightarrow N \left(0, \mathrm{var}\left((r_t - \mu)^2 \right) \right).$$

We have $\mathrm{var}((r_t - \mu)^2) = E((r_t - \mu)^4) - \sigma^4 = (\kappa_4 + 2)\sigma^4$. Likewise, in the non-overlapping case we have

$$\sqrt{T} \left(\frac{\hat{\sigma}_L^2(p)}{p} - \sigma^2 \right) \Longrightarrow N \left(0, \frac{\mathrm{var}\left((r_t(p) - p\mu)^2 \right)}{p} \right).$$

In fact,

$$\begin{aligned}
\text{var}\left((r_t(p) - p\mu)^2\right) &= E\left((r_t(p) - p\mu)^4\right) - E^2\left((r_t(p) - p\mu)^2\right) \\
&= E\left((r_t(p) - p\mu)^4\right) - p^2\sigma^4 \\
&= pE((r_t - \mu)^4) + 3p(p-1)\sigma^4 - p^2\sigma^4 \\
&= p\sigma^4(\kappa_4 + 3) + 3p(p-1)\sigma^4 - p^2\sigma^4 \\
&= p\sigma^4(\kappa_4 + 2p).
\end{aligned}$$

Note that $\text{var}(\hat{\sigma}_L^2(p)/p)/\text{var}(\hat{\sigma}_H^2) \simeq (\kappa_4 + 2p)/(\kappa_4 + 2) \geq 1$ so that $\hat{\sigma}_H^2$ is a more efficient estimator of σ^2 than $\hat{\sigma}_L^2(p)/p$; this is true whatever the distribution of returns provided the fourth moment is finite. Furthermore,

$$\begin{aligned}
\text{cov}\left(\hat{\sigma}_H^2, \hat{\sigma}_L^2(p)/p\right) &= \frac{1}{T^2} \times n \times \text{cov}\left(\sum_{j=1}^{p}(r_j - \mu)^2, \left(\sum_{j=1}^{p}(r_j - \mu)\right)^2\right) \\
&= \frac{1}{T} \times \frac{1}{p} \times \text{var}\left(\sum_{j=1}^{p}(r_j - \mu)^2\right) \\
&= \frac{1}{T} \times \text{var}\left((r_j - \mu)^2\right).
\end{aligned}$$

In fact, $\hat{\sigma}_H^2(p), \hat{\sigma}_L^2(p)$ are jointly asymptotically normal. By Taylor expansion

$$\sqrt{T}\left(\widehat{VR}(p) - 1\right) = \frac{\sqrt{T}\left(\hat{\sigma}_L^2(p)/p - \sigma^2\right)}{\sigma^2} - \frac{\sqrt{T}\left(\hat{\sigma}_H^2 - \sigma^2\right)}{\sigma^2} + \mathfrak{R}_T,$$

where $\mathfrak{R}_T \xrightarrow{P} 0$. It follows that

$$\begin{aligned}
\text{var}\left(\sqrt{T}\left(\widehat{VR}(p) - 1\right)\right) &\simeq \frac{\text{var}\left(\sqrt{T}\left(\hat{\sigma}_L^2(p)/p - \sigma^2\right)\right)}{\sigma^4} + \frac{\text{var}\left(\sqrt{T}\left(\hat{\sigma}_H^2 - \sigma^2\right)\right)}{\sigma^4} \\
&\quad -2 \frac{\text{cov}\left(\sqrt{T}\left(\hat{\sigma}_H^2 - \sigma^2\right), \sqrt{T}\left(\hat{\sigma}_L^2(p)/p - \sigma^2\right)\right)}{\sigma^4} \\
&= \frac{\text{var}\left(\sqrt{T}\left(\hat{\sigma}_L^2 - \sigma^2\right)\right)}{\sigma^4} - \frac{\text{var}\left((r_t - \mu)^2\right)}{\sigma^4} \\
&= (\kappa_4 + 2p) - (\kappa_4 + 2) = 2p - 2.
\end{aligned}$$

11. It has been said that: "Economics is just astrology for dudes." Discuss this statement with reference to the efficient markets hypothesis.

Chapter 15 Exercises and Complements

12. Suppose that you have a cross section of stock returns observed at the daily frequency. Consider the test statistic

$$\tau(p) = \sum_{i=1}^{p} \widehat{\rho}_i^2(k),$$

where $\widehat{\rho}_i(k)$ is the sample autocorrelation for the ith stock computed from a sample of size T. Suppose that stock returns R_{it} are i.i.d. with mean zero and variance σ_i^2 across time and are mutually independent. Show that $\tau(p)$ is approximately χ_p^2 distributed when p is fixed. Suppose now that returns are correlated contemporaneously with

$$\text{cov}(R_{it}, R_{js}) = \begin{cases} \sigma_{ij} & \text{if } t = s \\ 0 & \text{if } t \neq s. \end{cases}$$

What is the large sample distribution of $\tau(p)$?

13. Suppose that

$$R_t = \mu_t(\theta) + \varepsilon_t,$$

where ε_t are i.i.d. with mean zero and finite variance, while $\mu_t(\theta)$ is some nonlinear time varying mean depending on parameters $\theta \in \mathbb{R}^p$. Suppose that you have an estimator $\widehat{\theta}$ that satisfies

$$\widehat{\theta} - \theta = \frac{1}{T} \sum_{t=1}^{T} \psi_t + \mathfrak{R}_T,$$

where ψ_t are i.i.d. with mean zero and finite variance, and $\sqrt{T} \mathfrak{R}_T \xrightarrow{P} 0$. What is the limiting distribution of the sample autocorrelaiton coefficient in this case?

14. The **semivariogram**, which is widely used in spatial statistics, can also be used to test the EMH. This is defined as follows for each j

$$\widetilde{sv}(j) = \frac{1}{2(T-j)} \sum_{t=j+1}^{T} (Y_t - Y_{t-j})^2.$$

Show that this statistic is an unbiased estimator of

$$sv(j) = E\left[(Y_t - Y_{t-j})^2\right] = \gamma(0) + \gamma(j)$$

for all j. Under the EMH, $sv(j) = \gamma(0)$ for all j. For example, we may test the hypothesis that $sv(1) - sv(2) = 0$ by looking at $\widetilde{sv}(1) - \widetilde{sv}(2)$. What is the limiting distribution of this test statistic under **rw1**?

15. Suppose that true returns $r_1, \ldots, r_T$ are recorded as

$$0, \ldots, 0, r_1 + \cdots + r_k, 0, \ldots, r_{k+1} + \cdots + r_{2k}, \ldots, 0, \ldots, 0, r_{T+1-k} + \cdots + r_T,$$

where $T = j \times k$. Let $\tilde{r}_t$ denote the typical member of this sequence. Compare

$$\bar{r} = \frac{1}{T}\sum_{t=1}^{T} r_t, \quad s_r^2 = \frac{1}{T-1}\sum_{t=1}^{T}(r_t - \bar{r})^2, \quad \gamma_r(s) = \frac{1}{T-s}\sum_{t=s+1}^{T}(r_t - \bar{r})(r_{t-s} - \bar{r})$$

with

$$\frac{1}{T}\sum_{t=1}^{T}\tilde{r}_t, \quad \frac{1}{T-1}\sum_{t=1}^{T}(\tilde{r}_t - \bar{\tilde{r}})^2, \quad \frac{1}{T-s}\sum_{t=s+1}^{T}(\tilde{r}_t - \bar{\tilde{r}})(\tilde{r}_{t-s} - \bar{\tilde{r}}).$$

What can you say in general? What happens when $k = T$ and $j = 1$? How does this relate to the non trading model considered in CLM?

16. Show that in the Roll model with $\mu = 0$

$$\text{cov}((\Delta P_t)^2, (\Delta P_{t-1})^2) = 0.$$

How would you go about testing this implication of the Roll model? For the data you obtained in the first exercise, check whether this implication seems reasonable when prices or log prices are used.

17. Extend the Roll model to allow the spread s to vary over time so that

$$P_t = P_t^* + \frac{1}{2}Q_t s_t,$$

where Q_t is as before. Suppose that $s_1, \ldots, s_T$ are i.i.d. independent of $Q_1, \ldots, Q_T$ with mean μ_s and variance σ_s^2. Calculate

$$\text{cov}(\Delta P_t, \Delta P_{t-1}) \quad ; \quad \text{cov}((\Delta P_t)^2, (\Delta P_{t-1})^2).$$

Now suppose that $s_1, \ldots, s_T$ are deterministic and don't vary. Show that

$$\text{cov}(\Delta P_t, \Delta P_{t-1}) = -\frac{1}{4}s_{t-1}^2.$$

Typically, we expect spreads to widen at the open and the close of a market; what should this say about the predictability of returns during the day?

18. Suppose that fundamental prices satisfy

$$P_t^* = \mu + P_{t-1}^* + \varepsilon_t,$$

where ε_t is i.i.d. with mean zero and variance σ_ε^2. Observed prices satisfy

$$P_t^o = \begin{cases} P_{t-1}^o & \text{if there is no trade at } t \text{ (i.e., } \delta_t = 1) \\ P_t^* + \frac{1}{2}Q_t s & \text{if there is a trade at } t \text{ (i.e., } \delta_t = 0), \end{cases}$$

where Q_t is the ± 1 trade indicator, and

$$\delta_t = \begin{cases} 1 \text{ (no quote update)} & \text{with probability } \pi \\ 0 \text{ (quote update)} & \text{with probability } 1 - \pi. \end{cases}$$

Derive the properties of $P_t^O - P_{t-1}^O$.

19. Suppose you want to apply event study methodology to detect insider trading. Explain some of the issues that may be involved. Specifically, what type of data would you need? What event window would you choose? What econometric methods would you use? You may focus on the country that you chose in the first exercise sheet. For comparison, read Wong (2002).

20. For the data you selected in Problem 1:

    (a) Using the full sample regress the excess returns of the individual stocks on the index return and perform tests that the intercept is zero. Report the point estimates, t-statistics, and whether or not you reject the CAPM.
    (b) For each stock perform the same test over each of two subsamples of equal size, and report the point estimates, t-statistics, and whether or not you reject the CAPM in each subperiod.
    (c) Perform joint tests of the CAPM using both stocks using the F-test statistic for the whole period and each subperiod.

21. Suppose that X and Y are mean μ random variables with $\text{var} X = \sigma_X^2$ and $\text{var} Y = \sigma_Y^2$ and suppose that $\text{cov}(X, Y) = \sigma_{XY} = \sigma_X \sigma_Y \rho_{XY}$ with $|\rho_{XY}| \leq 1$. Invest a fraction w of your wealth in X and $1 - w$ in Y, called portfolio $P(w)$. Show that:

    (a) For all $w \in [0, 1]$
    $$\text{var}(P(w)) \leq \max\{\sigma_X^2, \sigma_Y^2\}.$$

    (b) The optimal w satisfies
    $$w_{opt} = \frac{\sigma_Y^2 - \sigma_X \sigma_Y \rho_{XY}}{\text{var}(X - Y)}$$
    (provided $\text{var}(X - Y) > 0$), and for this value
    $$\text{var}(P(w_{opt})) \leq \min\{\sigma_X^2, \sigma_Y^2\}.$$

    (c) Under what conditions would w be negative?
    Solution. We have
    $$X_w = w X_1 + (1 - w) X_2$$
    $$\mu_w = E(X_w) = w \mu_1 + (1 - w) \mu_2$$
    $$\sigma_w^2 = w^2 \sigma_1^2 + (1 - w)^2 \sigma_2^2 + 2w(1 - w) \sigma_{12}.$$

Two main features: risk return trade-off and diversification. Consider the special case where $\mu_1 = \mu_2$. Then minimize variance

$$\sigma_w^2 = w^2\sigma_1^2 + (1-w)^2\sigma_2^2 + 2w(1-w)\sigma_{12}$$

with respect to w gives the first order condition

$$2w\sigma_1^2 - 2(1-w)\sigma_2^2 + 2(1-2w)\sigma_{12} = 0$$

which can be solved to give

$$w = \frac{\sigma_2^2 - \sigma_{12}}{\sigma_1^2 + \sigma_2^2 - 2\sigma_{12}}.$$

The second order derivative is $2\sigma_1^2 + 2\sigma_2^2 - 4\sigma_{12} = 2\text{var}(X_1 - X_2)$, which is always non-negative and should be greater than zero. We need $\sigma_{12} \leq \sigma_2^2$ and $\sigma_{12} \leq \sigma_1^2$ for an interior solution so

$$\sigma_{12} \leq \min\{\sigma_1^2, \sigma_2^2\}.$$

Consider the special case where $\mu_1 = \mu_2$ and $\sigma_1^2 = \sigma_2^2 = 1$. Then minimize variance

$$\sigma_w^2 = w^2 + (1-w)^2 + 2w(1-w)\rho$$

with respect to w. The first order condition is

$$2w - 2(1-w) + 2(1-2w)\rho = 0.$$

Solution is

$$w_{GMV} = \frac{1}{2} \quad ; \quad \sigma_{GMV}^2 = \frac{1}{2}(1+\rho) \leq 1.$$

Suppose that $\mu_1 = \mu_2$ but $\sigma_1^2 \neq \sigma_2^2$ but $\rho = 0$

$$\sigma_w^2 = w^2\sigma_1^2 + (1-w)^2\sigma_2^2.$$

In this case the first order condition is

$$2w\sigma_1^2 = 2(1-w)\sigma_2^2$$

$$w_{GMV} = \frac{\sigma_2^2}{\sigma_1^2 + \sigma_2^2} \quad ; \quad \sigma_{GMV}^2 = \frac{\sigma_1^2\sigma_2^2}{\sigma_1^2 + \sigma_2^2} \leq \min\{\sigma_1^2, \sigma_2^2\}.$$

Consider the case where $\rho = 1$ with $\sigma_1^2 \neq \sigma_2^2$

$$\sigma_w^2 = w^2\sigma_1^2 + (1-w)^2\sigma_2^2 + 2w(1-w)\sigma_1\sigma_2$$
$$= (w\sigma_1 + (1-w)\sigma_2)^2.$$

The first order condition is

$$2(w\sigma_1 + (1-w)\sigma_2)(\sigma_1 - \sigma_2) = 0$$

with second order condition $2\sigma_1^2 + 2\sigma_2^2 - 4\sigma_1\sigma_2 = 2(\sigma_1 - \sigma_2)^2 > 0$. The solution is

$$w = \frac{\sigma_2}{\sigma_2 - \sigma_1}.$$

22. Suppose that the $n \times n$ covariance matrix satisfies

$$\Omega = B\Sigma B^\mathsf{T}$$

where the $n \times K$ matrix B is of rank $K \leq n$ and $\Sigma = I_K$. Show that there exists an $n \times K$ orthonormal matrix B^* with $B^* B^{*\mathsf{T}} = I_n$ and a diagonal matrix Σ^* with

$$\Omega = B^* \Sigma^* B^{*\mathsf{T}}.$$

Show the converse.

23. Explain what you think of the following statements regarding the efficient markets hypothesis (EMH).

(a) Although the EMH claims investors cannot outperform the market, analysts such as Warren Buffet have done exactly that. Hence the EMH must be incorrect. *This interpretation is incorrect because the EMH implies that investors cannot consistently outperform the market; there will be times when, out of luck, an asset will outperform the market. Additionally, it is possible for one to consistently outperform the market by chance. Suppose a fund manager has a 50% chance of beating the market next year. His probability of beating the market two years in a row is 25%; of beating the market eight years in a row, 4%. Consequently, out of 1000 fund managers, 4 will consistently outperform the market eight years in a row. Hence not only is it possible to outperform the market on occasion, it is possible to consistently outperform the market by luck.*

(b) According to the weak form of the EMH, technical analysis is useless in predicting future stock returns. Yet financial analysts are not driven out of the market, so their services must be useful. Hence, the EMH must be incorrect. *This statement is incorrect because a financial analyst can put together a portfolio that matches the risk-tolerance of each client, whereas a random collection of stocks will most likely not cater to particular risk preferences. Secondly, financial analysis is essential to the efficiency of markets because it allows investors to take advantage of new information to identify mispriced stocks. When there is competition among many investors, arbitrage opportunities vanish, i.e., stock prices adjust immediately to incorporate any new information, leading to market efficiency.*

(c) The EMH must be incorrect because stock prices are constantly fluctuating randomly. *The fact that stock prices are constantly changing is evidence in support of the EMH, because new information appears almost continuously in the form of opinions, news stories, announcements, expectations, and even* lack of news. The constant arrival of new information causes the continuous adjustment of prices, as the EMH claims.

(d) If the EMH holds, then all investors must be able to collect, analyze, and interpret new information to correctly adjust stock prices. However, most investors are not trained financial experts. Therefore, the EMH must be false. The EMH does not require that all traders be informed. A relatively small core of experts is necessary to analyze new information and adjust stock prices correctly, after which other investors can trade on the new prices without having followed the underlying causes for the price change.

24. The efficient markets hypothesis is untestable! Describe some common tests for market efficiency, give their properties, and describe the common findings.
25. What determines the bid–ask spread for stock returns?
26. What can explain the finding of negative individual stock autocorrelation and positive portfolio or index autocorrelation?
27. Non-trading model with stochastic volatility. Consider the non-trading model of Chapter 2. Suppose that we have a stationary stochastic volatility process σ_t independent of everything and

$$r_t = \sigma_t \varepsilon_t,$$

where ε_t is i.i.d. standard normal. Show that $E((r_t^O)^2) = \sigma^2$ and derive $\text{cov}((r_t^O)^2, (r_{t+1}^O)^2)$.

Solution. As we saw $E(r_t^O) = 0$. We also have

$$E\left((r_t^O)^2\right) = (1-\pi) E\left(E\left(\left(\sum_{k=0}^{d_t} r_{t-k}\right)^2 \mid d_t, \sigma(.)\right)\right)$$

$$= (1-\pi) E\left(\text{var}\left(\sum_{k=0}^{d_t} r_{t-k} \mid d_t, \sigma(.)\right)\right)$$

$$= (1-\pi) \sum_{j=0}^{\infty} p_j \sum_{k=0}^{j} E(\sigma_{t-k}^2)$$

$$= \sigma^2,$$

where $\sigma^2 = E(\sigma_t^2)$ and $p_j = (1-\pi)\pi^j$. Furthermore,

$$\text{cov}((r_t^O)^2, (r_{t+1}^O)^2) = E\left((r_t^O)^2 (r_{t+1}^O)^2\right) - E^2\left((r_t^O)^2\right)$$

$$= E\left[(r_t^O)^2 (r_{t+1}^O)^2\right] - \left(\text{var}(r_t^O) + E^2(r_t^O)\right)^2$$

$$= (1-\pi)^2 E\left(r_{t+1}^2 \left(\sum_{i=0}^{d_t} r_{t-i}\right)^2 \mid d_t, \sigma(.)\right) - \sigma^4$$

Chapter 15 Exercises and Complements

$$= (1-\pi)^2 \sum_{j=0}^{\infty} p_j \sum_{k=0}^{j} E(\sigma_{t+1}^2 \sigma_{t-k}^2) - \sigma^4$$

$$= (1-\pi)^2 \sigma^4 \sum_{j=0}^{\infty} p_j \sum_{k=0}^{j} + (1-\pi)^2 \sum_{j=0}^{\infty} p_j \sum_{k=0}^{j} \gamma_{\sigma^2}(k+1) - \sigma^4$$

$$= -\pi\sigma^4 + (1-\pi)^2 \sum_{j=0}^{\infty} p_j \sum_{k=0}^{j} \gamma_{\sigma^2}(k+1),$$

where $\gamma_{\sigma^2}(k) = E(\sigma_t^2 \sigma_{t-k}^2) - \sigma^4$. With an explicit assumption on σ_t^2 such as it is AR(1), one could obtain a formula for this covariance and compare it with the covariance of squared true returns

$$\text{cov}(r_t^2, r_{t+1}^2) = \gamma_{\sigma^2}(1).$$

Specifically, suppose that $\gamma_{\sigma^2}(k) = \alpha \beta^k$ for some $\alpha, \beta > 0$ with $\beta < 1$. Then

$$(1-\pi)^2 \sum_{j=0}^{\infty} p_j \sum_{k=0}^{j} \gamma_{\sigma^2}(k+1) = (1-\pi)^2 \alpha \sum_{j=0}^{\infty} p_j \sum_{k=0}^{j} \beta^{k+1}$$

$$= (1-\pi)^3 \alpha \beta \sum_{j=0}^{\infty} \pi^j \left(\frac{1-\beta^{j+1}}{1-\beta} \right)$$

$$= \frac{(1-\pi)^2 \alpha \beta}{1-\beta} - \frac{(1-\pi)^3}{1-\beta} \alpha \beta^2 \frac{1}{1-\pi\beta}$$

$$= \alpha\beta \frac{(1-\pi)^2}{1-\pi\beta}$$

$$< \alpha\beta,$$

and so

$$\text{cov}((r_t^o)^2, (r_{t+1}^o)^2) < \text{cov}(r_t^2, r_{t+1}^2).$$

This shows that the LM non-trading model can reduce the persistence in observed volatility. As $\pi \to 1$, the bias becomes larger.

28. Consider the market model

$$R_{it} = \alpha_i + \beta_i R_{mt} + \varepsilon_{it}.$$

Show that the covariance matrix of returns satisfies

$$\Omega = \sigma_m^2 \beta \beta^\top + D$$

for some matrix D.

Chapter 15 Exercises and Complements

29. Suppose that the fundamental price P^* satisfies

$$P_{it}^* = P_{it-1}^* + \varepsilon_{it},$$

where ε_{it} are i.i.d. with mean zero across both i and t. Buy and sell orders arrive randomly. The full spread is s_i and the half-spread is $s_i/2$. We have for each firm

$$P_{it} = P_{it}^* + Q_{it}\frac{s_i}{2},$$

where Q_{it} is a trade direction indicator, $+1$ for buy and -1 if customer is selling. Assume that Q_{it} is i.i.d. with equal probability of $+1$ and -1 and unrelated to P_{it}^*. Suppose that one considers a portfolio with weights $\{w_i\}_{i=1}^n$ and let $P_t^w = \sum w_i P_{it}$ denote the value of the portfolio at time t. Obtain an expression for the autocorrelation of P_t^w in two cases:

(a) The portfolio is well diversified, i.e., $w_i = 1/n$ with n very large.
(b) The portfolio consists of a large position in the first asset and a diversified position across the remaining large set of assets, i.e., $w_1 = 0.5$ and $w_j = 1/2n$ for $j = 2, 3, \ldots$.

30. The Roll model assumes that trade directions are uncorrelated with changes in the efficient price, i.e., $\text{cov}(Q_t, \varepsilon_t) = 0$. Suppose that $\text{cov}(Q_t, \varepsilon_t) = \rho$ where $\rho \in (0, 1)$. This reflects the notion that a buy order is associated with an increase in the security value. Calculate $\text{cov}(P_t, P_{t-1})$ and $\text{var}(\Delta P_t)$. Show that the usual Roll model estimate of s is upward biased in this case.

31. Roll model with $E(Q_t) = \vartheta \neq 0$. In this case

$$\Delta P_t = \varepsilon_t + (Q_t - Q_{t-1})\frac{s}{2}.$$

We have

$$E(\Delta P_t) = 0$$

$$E\left[(\Delta P_t)^2\right] = E(\varepsilon_t^2) + E\left[(Q_t - Q_{t-1})^2\right]\frac{s^2}{4} = E(\varepsilon_t^2) + \left[1 - \vartheta^2\right]\frac{s^2}{2}$$

$$E[\Delta P_t \Delta P_{t-1}] = E[(Q_t - Q_{t-1})(Q_{t-1} - Q_{t-2})]\frac{s^2}{4} = \frac{s^2}{4}\left[\vartheta^2 - 1\right].$$

Whence

$$\text{corr}(\Delta P_t, \Delta P_{t-1}) = \frac{-\tau\frac{s^2}{4}}{\sigma_\varepsilon^2 + \tau\frac{s^2}{2}}$$

$$\tau = 1 - \vartheta^2.$$

Derive a formula for s^2.

32. Suppose that the world is simpler than you thought. There are only 2 securities whose returns R_{1t} and R_{2t} are observed over time periods $t = 1, \ldots, T$. The market portfolio is formed from the equal mixture of 1 and 2, i.e., $R_{mt} = 0.5R_{1t} + 0.5R_{2t}$.

Chapter 15 Exercises and Complements

The risk free rate is exactly zero. Describe how you would test the CAPM in this world. The market model satisfies

$$R_{1t} = \alpha_1 + \beta_1 R_{mt} + \varepsilon_{1t}$$
$$R_{2t} = \alpha_2 + \beta_2 R_{mt} + \varepsilon_{2t},$$

where $\varepsilon_{1t}, \varepsilon_{2t}$ are i.i.d. with mean zero and mutually uncorrelated with variances s_{jj}. The CAPM is that $\alpha_j = 0$. The only issue here is that there is no independent variation across the two assets. Specifically,

$$R_{mt} = \frac{1}{2}[\alpha_1 + \beta_1 R_{mt} + \varepsilon_{1t} + \alpha_2 + \beta_2 R_{mt} + \varepsilon_{2t}]$$
$$= \frac{\alpha_1 + \alpha_2}{2} + \frac{\beta_1 + \beta_2}{2} R_{mt} + \frac{\varepsilon_{1t} + \varepsilon_{2t}}{2}$$

so that $\alpha_1 + \alpha_2 = 0$, $\beta_1 + \beta_2 = 2$, and $\varepsilon_{1t} + \varepsilon_{2t} = 0$ with probability one. Therefore, one can only use information from one equation, say the first one, and one would do a t-test of the hypothesis of $\alpha_1 = 0$.

33. What role does the assumption of normality play in testing the CAPM? What is the evidence regarding normality in stock returns? If stock returns are not normal and indeed have heavy tailed distributions with some extreme outliers, what are the properties of the standard normal-based tests of this hypothesis?

34. Suppose that (logarithmic) returns r_t are observed at the daily frequency. Consider the (forward) aggregated returns, for $K = 1, 2, \ldots$ and for $t = 1, \ldots$

$$r_{t,K} = r_t + r_{t+1} + \cdots + r_{t+K-1}.$$

Suppose that daily returns are i.i.d. normally distributed with $E(r_t) = \mu$ and $\text{var}(r_t) = \sigma^2$. What are the properties of $r_{t,K}$? In particular, calculate the mean, the variance, and the autocovariance function. Suppose I compute the backward aggregation

$$r_{t,-K} = r_t + r_{t-1} + \cdots + r_{t-K+1}.$$

What is the cross covariation $\text{cov}(r_{t,K}, r_{s,-K})$? Suppose that

$$r_{t+1} = \alpha + \beta x_t + \varepsilon_t$$

where x_t, ε_t are i.i.d. and mutually independent. Then consider the regression

$$r_{t+1,K} = a + b x_{t,K} + e_t.$$

What are the values of a, b and what are the properties of e_t? Then consider the regression

$$r_{t+1,K} = a + b x_t + e_t$$
$$r_{t+1} = a + b x_{t,-K} + e_t.$$

35. Consider the model

$$p_t = p_{t-1} - \alpha\left(p_{t-1} - p^*_{t-1}\right) + \varepsilon_t$$

$$p^*_t = p^*_{t-1} + \eta_t.$$

Calculate the autocorrelation function of observed returns. Relate this to the Roll model.

36. In the hedge fund industry, reported returns are often highly serially correlated. Suppose that true returns r_t are i.i.d. and normally distributed with $E(r_t) = \mu$ and $\text{var}(r_t) = \sigma^2$. Suppose that reported returns r^o_t satisfy

$$r^o_t = \alpha r_t + (1-\alpha) r_{t-1}$$

for some $\alpha \in (1/2, 1)$, that is, firms only report smoothed returns rather than actual returns. Calculate the properties of r^o_t including the mean, the variance, and $\text{cov}(r^o_t, r^o_{t-s})$. How could you estimate the true returns from data on observed returns? That is, given a sample $\{r^o_1, \ldots, r^o_T\}$ how would you estimate $\{r_1, \ldots, r_T\}$? We have

$$\text{var}(r^o_t) = \left(\alpha^2 + (1-\alpha)^2\right) \sigma^2_r < \sigma^2_r$$

$$\text{cov}(r^o_t, r^o_{t-1}) = \text{cov}(\alpha r_t + (1-\alpha) r_{t-1}, \alpha r_{t-1} + (1-\alpha) r_{t-2}) = \alpha(1-\alpha) \sigma^2_r$$

$$\text{corr}(r^o_t, r^o_{t-1}) = \frac{\alpha(1-\alpha)}{\alpha^2 + (1-\alpha)^2} = \frac{\tau}{1 - 2\tau}.$$

Therefore,

$$\tau = \frac{\text{corr}(r^o_t, r^o_{t-1})}{1 + 2\text{corr}(r^o_t, r^o_{t-1})}.$$

Then in the region that α lives the equation $\tau = \alpha(1-\alpha)$ has the unique solution

$$\alpha = \frac{1}{2} + \frac{1}{2}\sqrt{1 - 4\tau}.$$

Suppose we have α. Then let us assume that r_{t-1} is known, then let

$$r_t = \frac{1}{\alpha}\left(r^o_t - (1-\alpha) r_{t-1}\right).$$

We can take $r_1 = \overline{r^o_t}$, which is a consistent estimator of μ. Or we can take $r_1 = r^o_1$. Another way of seeing this uses the lag polynomial. We have

$$r^o_t = \alpha\left(1 + \beta L\right) r_t,$$

where $\beta = (1-\alpha)/\alpha < 1$. Then

$$r_t = \frac{1}{\alpha} \frac{r^o_t}{1 + \beta L} = \frac{1}{\alpha}\left[r^o_t - \beta r^o_{t-1} + \beta^2 r^o_{t-2} - \cdots\right].$$

Chapter 15 Exercises and Complements

37. Suppose that true returns are i.i.d. normally distributed with $E(r_t)=\mu$ and $\text{var}(r_t)=\sigma^2$. Suppose that reported returns r_t^o satisfy

$$r_t^o = \begin{cases} r_t & \text{if } |r_t| > \alpha \\ 0 & \text{else,} \end{cases}$$

where α is some constant. What are the properties of r_t^o, i.e., what is $E(r_t^o)$, $\text{var}(r_t^o)$, and $\text{cov}(r_t^o, r_{t-s}^o)$? You may use the fact that for a standard normal density ϕ

$$\int_0^a x\phi(x)dx = \frac{1}{\sqrt{2\pi}}\left(1 - e^{-\frac{1}{2}a^2}\right).$$

If you cant give explicit expressions, say whether the quantity is smaller or larger than the corresponding quantity for the observed series.

38. Glosten–Harris model with uniform value. Suppose that Value V is chosen from the distribution

$$V \sim U[0, T].$$

The type of investor is chosen from

$$\begin{cases} I & \text{with probability } \mu \\ U & \text{with probability } 1 - \mu. \end{cases}$$

Strategies: if informed (I), buy if value V is high (relative to ask price) and sell if value is low relative to bid price; if uninformed (U), buy or sell with probability 1/2; dealer sets B, A to make zero profits (this is forced on her). Derive B and A.
Solution. We may show that

$$\Pr(\text{Buy}|V) = \mu 1(V > A) + \frac{1-\mu}{2}$$

$$\Pr(\text{Buy}) = \mu \Pr(V > A) + \frac{1-\mu}{2} = \mu \frac{T-A}{T} + \frac{1-\mu}{2}$$

$$f(V|\text{Buy}) = \frac{\Pr(\text{Buy}|V)f(V)}{\Pr(\text{Buy})} = \frac{\mu 1(V > A) + \frac{1-\mu}{2}}{T\left[\mu \frac{T-A}{T} + \frac{1-\mu}{2}\right]} 1(0 < V < T)$$

$$E(V|\text{Buy}) = \frac{\mu E[V1(V > A)] + \frac{T}{2}\frac{1-\mu}{2}}{\mu \frac{T-A}{T} + \frac{1-\mu}{2}} = \frac{\mu \frac{T^2 - A^2}{2T} + \frac{T}{2}\frac{1-\mu}{2}}{\mu \frac{T-A}{T} + \frac{1-\mu}{2}}$$

$$= \frac{T^2(1+\mu) - 2\mu A^2}{2[T(1+\mu) - 2\mu A]}$$

Chapter 15 Exercises and Complements

$$\Pr(Sell|V) = \mu 1(V<B) + \frac{1-\mu}{2}$$

$$\Pr(Sell) = \mu \Pr(V<B) + \frac{1-\mu}{2} = \mu\frac{B}{T} + \frac{1-\mu}{2}$$

$$f(V|Sell) = \frac{\Pr(Sell|V)f(V)}{\Pr(Sell)} = \frac{\mu 1(V<B) + \frac{1-\mu}{2}}{T\left[\mu\frac{B}{T} + \frac{1-\mu}{2}\right]} 1(0<V<T)$$

$$E(V|Sell) = \frac{\mu E[V1(V<B)] + \frac{T}{2}\frac{1-\mu}{2}}{\mu\frac{B}{T} + \frac{1-\mu}{2}} = \frac{\mu\frac{B^2}{2T} + \frac{T}{2}\frac{1-\mu}{2}}{\mu\frac{B}{T} + \frac{1-\mu}{2}}$$

$$= \frac{T^2(1-\mu) + 2\mu B^2}{2[T(1-\mu) + 2\mu B]}.$$

Under the zero profit condition, this means that A must solve

$$\frac{T^2\mu - 2A^2\mu + T^2}{2T - 4A\mu + 2T\mu} = A,$$

which is equivalent to the quadratic equation

$$2\mu A^2 - (2T + 2T\mu)A + (T^2\mu + T^2) = 0,$$

which has solution

$$A = \begin{cases} T\left(\frac{1+\mu-\sqrt{1-\mu^2}}{2\mu}\right) & \text{if } \mu \neq 0 \\ \frac{T}{2} & \text{if } \mu = 0. \end{cases}$$

Likewise we have

$$\frac{2B^2\mu - T^2\mu + T^2}{2T + 4B\mu - 2T\mu} = B$$

$$B = T\frac{\left(\mu - 1 + \sqrt{1-\mu^2}\right)}{2\mu}.$$

Then

$$A - B = \frac{\left(1 - \sqrt{1-\mu^2}\right)}{\mu} T.$$

Note that $A + B = T$. The bid–ask spread increases with μ.

In this case there is also the possibility of no order, which arises when $A > V$ or $B < V$

$$\Pr(No|V) = \mu 1(B<V<A)$$

$$\Pr(No) = \mu \Pr(B<V<A) = \mu\frac{A-B}{T}$$

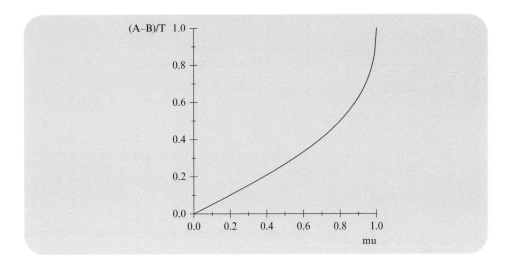

$$f(V|No) = \frac{1(B < V < A)}{A - B}$$

$$E(V|No) = \frac{\mu E[V1(B < V < A)]}{\mu \frac{A-B}{T}} = \frac{\mu \frac{A^2-B^2}{2T}}{\mu \frac{A-B}{T}} = \frac{1}{2}A + \frac{1}{2}B = \frac{T}{2}.$$

39. Suppose that
$$r_t = \rho r_{t-1} + \varepsilon_t.$$

We can write
$$(1 - \rho L)r_t = \varepsilon_t.$$

Consider the process r_t^2. We have
$$r_t^2 = \rho^2 r_{t-1}^2 + \varepsilon_t^2 + 2\rho r_{t-1}\varepsilon_t$$

so that
$$(1 - \rho^2 L)r_t^2 = \sigma_\varepsilon^2 + u_t,$$

where $u_t = \varepsilon_t^2 - E(\varepsilon_t^2) + 2\rho r_{t-1}\varepsilon_t$ is a martingale difference sequence. It follows that
$$r_t^2 = \frac{\sigma_\varepsilon^2}{1 - \rho^2} + \sum_{j=0}^{\infty} \vartheta^j u_t,$$

where $\vartheta = \rho^2$. It follows that
$$\text{corr}(r_t^2, r_{t-j}^2) = \vartheta^j = \rho^{2j}.$$

It follows that for this process the autocorrelation of the squares dies out faster than the autocorrelation of the original series.

Chapter 15 Exercises and Complements

40. Suppose that returns follow a moving average process

$$r_t = \varepsilon_t + \theta\varepsilon_{t-1},$$

where $\varepsilon_t \sim N(0, \sigma_\varepsilon^2)$. Recall that

$$\operatorname{corr}(r_t, r_{t-j}) = \begin{cases} \frac{\theta}{(1+\theta^2)} & \text{if } j=1 \\ 0 & \text{else.} \end{cases}$$

We have

$$r_t^2 = (1+\theta^2)\sigma_\varepsilon^2 + \varepsilon_t^2 - E(\varepsilon_t^2) + \theta^2(\varepsilon_{t-1}^2 - E(\varepsilon_{t-1}^2)) + 2\theta\varepsilon_t\varepsilon_{t-1}.$$

It follows that

$$\operatorname{var}(r_t^2) = \left(1+\theta^4\right)\operatorname{var}(\varepsilon_t^2) + 4\theta^2\sigma_\varepsilon^4$$

$$\operatorname{cov}(r_t^2, r_{t-j}^2) = \begin{cases} \theta^2 \operatorname{var}(\varepsilon_t^2) & \text{if } j=1 \\ 0 & \text{else.} \end{cases}$$

For a normal distribution $\operatorname{var}(\varepsilon_t^2) = 2\sigma_\varepsilon^4$. Therefore,

$$\operatorname{corr}(r_t^2, r_{t-j}^2) = \begin{cases} \frac{\theta^2}{(1+\theta^2)^2} & \text{if } j=1 \\ 0 & \text{else.} \end{cases}$$

This yields a positive autocorrelation, but it dies out after one lag. It follows that for this process the autocorrelation of the squares dies out faster than the autocorrelation of the original series.

41. What is a martingale difference sequence, and what is its significance in empirical finance?
42. Describe the testable restrictions of the Black version of the CAPM.
43. Suppose that

$$Z_{it} = \beta_i Z_{mt} + \varepsilon_{it},$$

where the usual assumptions apply. Now suppose that we estimate

$$\widehat{\beta}_i = \frac{\sum_{t=1}^T Z_{mt} Z_{it}}{\sum_{t=1}^T Z_{mt}^2}$$

for $i = 1, \ldots, N$. Let $\widehat{\beta}_{\max} = \widehat{\beta}_{i_{\max}}$ and $\widehat{\beta}_{\min} = \widehat{\beta}_{i_{\min}}$, where

$$i_{\max} = \arg\max_{1 \leq i \leq N} \widehat{\beta}_i \quad ; \quad i_{\min} = \arg\min_{1 \leq i \leq N} \widehat{\beta}_i.$$

Chapter 15 Exercises and Complements

Define the estimated risk premium from the cross-sectional regression of excess returns

$$\widehat{\gamma} = \frac{\overline{Z}_{i_{\max}}\widehat{\beta}_{\max} + \overline{Z}_{i_{\min}}\widehat{\beta}_{\min}}{\widehat{\beta}_{\max}^2 + \widehat{\beta}_{\min}^2}.$$

Investigate the performance of $\widehat{\gamma}$ for simulated data. Specifically, suppose that $\beta_i \sim U[0,2]$, while ε_{it}, Z_{mt} are all standard normal and independent of each other. Consider different size T, N.

44. Slowly varying expected returns. Suppose that

$$r_{t+1} = \mu + x_t + \varepsilon_{t+1},$$

where

$$x_{t+1} = \phi x_t + \xi_{t+1}, \quad -1 < \phi < 1$$

and ε_t, ξ_s are mutually independent for all t, s and are individually i.i.d. with mean zero and variances σ_ε^2 and σ_ξ^2. Calculate $E(r_t), \text{var}(r_t), E_t(r_{t+1})$, and $\text{var}_t(r_{t+1})$. Compute the unconditional autocorrelation function

$$\rho(k) = \frac{\text{cov}(r_t, r_{t-k})}{\text{var}(r_t)}.$$

Is this consistent with the empirical evidence regarding autocorrelation of return series? What about the evidence for the autocorrelation of squared returns?

45. The **Durbin–Wu–Hausman** test is based on comparing two estimators. Consider the market model and define the vector of unrestricted and restricted estimators of β, $\widehat{\beta}$, and $\widetilde{\beta}$ respectively. Construct a test of the CAPM, that $\alpha = 0$ using this approach. Does this test have power against all alternatives?

$$DWH = T\left(\widehat{\beta} - \widetilde{\beta}\right)^\top \left(\frac{1}{\widehat{\sigma}_m^2} - \frac{1}{\widehat{\mu}_{m2}}\right)^{-1} \widehat{\Sigma}^{-1}\left(\widehat{\beta} - \widetilde{\beta}\right)$$

$$= T\left(\frac{\widehat{\sigma}_m^2 \widehat{\mu}_{m2}}{\widehat{\mu}_{m2} - \widehat{\sigma}_m^2}\right)\left(\widehat{\beta} - \widetilde{\beta}\right)^\top \widehat{\Sigma}^{-1}\left(\widehat{\beta} - \widetilde{\beta}\right)$$

$$= T\left(\frac{\widehat{\mu}_{m2}^2}{\overline{Z}_m^2} - \widehat{\mu}_{m2}\right)\left(\widehat{\beta} - \widetilde{\beta}\right)^\top \widehat{\Sigma}^{-1}\left(\widehat{\beta} - \widetilde{\beta}\right).$$

46. Suppose that asset returns satisfy

$$R_{it} - E(R_{it}) = f_t b_i + \varepsilon_{it}$$

where ε_{it} are i.i.d. with mean zero and variance σ_ε^2. The time series of scalar factors $\{f_t\}$ and the cross section of scalar loadings $\{b_i\}$ are unobserved. Calculate the two matrices Ω, Σ with typical elements

$$\Omega_{ij} = \frac{1}{T} \sum_{t=1}^{T} (R_{it} - E(R_{it})) (R_{jt} - E(R_{jt}))$$

$$\Sigma_{ts} = \frac{1}{n} \sum_{i=1}^{n} (R_{it} - E(R_{it})) (R_{is} - E(R_{is})).$$

Suppose that f_t are i.i.d. $N(0,1)$ and b_i are i.i.d. $N(0,1)$ and both processes are independent of all of ε. Obtain the probability limit of Ω as T gets big and the probability limit of Σ as n gets big.

$$\Sigma_{tt} = f_t^2 + \sigma^2$$
$$\Sigma_{ts} = f_t f_s, \quad \text{for } t \neq s.$$

Suppose that $n = 2$. Then

$$\begin{bmatrix} f_1^2 + \sigma^2 \\ f_2^2 + \sigma^2 \\ f_1 f_2 \end{bmatrix} = \begin{bmatrix} \sigma_{11} \\ \sigma_{22} \\ \sigma_{12} \end{bmatrix}.$$

Let $f_1 = \sigma_{12}/f_2$ and $\sigma^2 = \sigma_{22} - f_2^2$. Then we have the quadratic equation

$$f_2^4 + (\sigma_{11} - \sigma_{22}) f_2^2 - \sigma_{12}^2 = 0$$

which has solution

$$f_2^2 = \frac{-\frac{\sigma_{11} - \sigma_{22}}{2} \pm \sqrt{(\sigma_{11} - \sigma_{22})^2 + 4\sigma_{12}^2}}{2}$$

provided $(\sigma_{11} - \sigma_{22})^2 + 4\sigma_{12}^2 > 0$. The solution for f_2^2 does not determine the sign of f_2. Likewise the sign of f_1 is not determined.

47. Suppose that

$$R_{it} - R_{ft} = \alpha_i + \beta_i (R_{mt} - R_{ft}) + \varepsilon_{it},$$

where we don't observe R_{mt} but we observe a proxy f_t that obeys

$$f_t = \pi_0 + \pi_1 (R_{mt} - R_{ft}) + \eta_t,$$

where η_t is mean zero given all the right hand side variables. How can one test the CAPM ($\alpha_i = 0$) in this case? Suppose that the risk free rate is not observed and

$$R_{it} = \alpha_i + \beta_i R_{mt} + \varepsilon_{it}.$$

Chapter 15 Exercises and Complements

How can one test the Black version of the CAPM (that $\alpha_i = (1-\beta_i)\gamma$ for some γ) in this case?

Hint. In the first case we have

$$E_L(R_{it} - R_{ft}|z_t) = a_i + b_i f_t$$

$$b_i = \frac{\text{cov}(R_{it} - R_{ft}, f_t)}{\text{var}(f_t)} = \frac{\pi_1 \sigma_{ff}}{\pi_1^2 \sigma_{ff} + \sigma_{\eta\eta}} \beta_i = c\beta_i,$$

which is a linear function of β_i with fixed coefficients. We also have

$$\begin{aligned}
a_i &= E(R_{it} - R_{ft}) - b_i E(f_t) \\
&= \alpha_i + \beta_i E(R_{mt} - R_{ft}) - b_i(\pi_0 + \pi_1 E(R_{mt} - R_{ft})) \\
&= \alpha_i + \beta_i((1 - c\pi_1) E_m - c\pi_0) \\
&= \alpha_i + c^* \beta_i.
\end{aligned}$$

Suppose that $\alpha_i = 0$. Then $a_i = C^* \beta_i$. This implies that

$$a_i = \delta b_i$$

for some unknown $\delta = c^*/c$. Provided $\delta \neq 0, \infty$, one can test this restriction with a finite set of assets.

Provide the details.

In the second case, suppose that $\alpha_i = (1-\beta_i)\gamma$ and

$$R_{it} = \alpha_i + \beta_i R_{mt} + \varepsilon_{it},$$

so we do not observe the risk free rate returns, i.e., Black version of CAPM holds. Then $a_i = \gamma_0 + \gamma_1 \beta_i$. In that case, for any pair of assets i, j

$$a_i - a_j = \gamma_1(\beta_i - \beta_j)$$

$$b_i - b_j = c(\beta_i - \beta_j)$$

and so

$$a_i - a_j = \frac{\gamma_1}{c}(b_i - b_j),$$

which can also be tested. Provide the details.

48. Blanchard and Watson (1982) model. Suppose that

$$B_{t+1} = \begin{cases} \frac{1+R}{\pi} B_t + \eta_{t+1} & \text{with probability } \pi \\ \eta_{t+1} & \text{with probability } 1 - \pi \end{cases}$$

where η_t is i.i.d. with mean zero and variance one. What are the properties of the bubble process? What is $E_t(B_{t+1})$? What is $\text{var}_t(B_{t+1})$? What is the chance that the bubble lasts for more than 5 periods? Suppose that we observed prices satisfying

$$P_t = P_t^* + B_t$$

$$P_t^* = P_{t-1}^* + u_t$$

where u_t is normally distributed with mean zero and variance one. Suppose also that you have a method for identifying observations for which the bubble is in operation (i.e., the first regime of B is in operation). How would you test for the presence of the bubble?

49. Compare the following two models for stock prices using the S&P500 daily data

$$\log(P_t) = \alpha + \beta t + \varepsilon_t$$

$$\log(P_t) = \alpha + \log(P_{t-1}) + \varepsilon_t.$$

50. What is the risk return tradeoff?

51. Suppose that you have a time series of daily returns on a stock i that is traded in a different time zone from stock j. Specifically, the trading day for i is the first third of the day, and the trading day for j is the second third of the day. The final third of the day contains no trading. We observe the closing prices for each asset on their respective "trading days," which we denote by $P_{i1}, P_{i4}, \ldots$, and $P_{j2}, P_{j5}, \ldots$. We want to calculate the contemporaneous return covariance. We assume that each stock has i.i.d. return and that the contemporaneous covariance between return on stock i and stock j is γ, that is,

$$\text{cov}(p_{it} - p_{i,t-1}, p_{it+s} - p_{i,t+s-1}) = 0$$

$$\text{cov}(p_{it} - p_{i,t-1}, p_{jt} - p_{j,t-1}) = \gamma.$$

Then show that

$$\text{cov}(P_{i4} - P_{i1}, P_{j5} - P_{j2}) = \text{cov}(P_{i4} - P_{i2}, P_{j4} - P_{j2}) = 2\text{cov}(P_{i2} - P_{i1}, P_{j2} - P_{j1}).$$

So we use stale price adjustment

$$\frac{2}{3}\text{cov}(r_{it}, r_{jt}) + \frac{1}{3}\text{cov}(r_{it}, r_{j,t-1}).$$

52. The Sharpe–Lintner CAPM predicts that

$$\text{var}(R_{mt})E(R_{it} - R_{ft}) - \text{cov}(R_{it} - R_{ft}, R_{mt} - R_{ft})E(R_{mt} - R_{ft}) = 0$$

Chapter 15 Exercises and Complements

for each asset i. Provide a test of this restriction using the estimated quantities

$$\widehat{\text{var}}(R_{mt}) = \frac{1}{T-1}\sum_{t=1}^{T}(R_{mt}-\overline{R}_m)^2$$

$$\widehat{E}(R_{it}-R_{ft}) = \frac{1}{T}\sum_{t=1}^{T}(R_{it}-R_{ft})$$

$$\widehat{\text{cov}}(R_{it}-R_{ft},R_{mt}-R_{ft}) = \frac{1}{T-1}\sum_{t=1}^{T}(R_{it}-\overline{R}_i)(R_{mt}-\overline{R}_m)$$

$$\widehat{E}(R_{mt}-R_{ft}) = \frac{1}{T}\sum_{t=1}^{T}(R_{mt}-R_{ft}),$$

where $\overline{R}_m = \sum_{t=1}^{T} R_{mt}/T$ and $\overline{R}_i = \sum_{t=1}^{T} R_{it}/T$.

53. Suppose that the return to holding painting i in period t is

$$r_{it} = \mu_t + \varepsilon_{it},$$

where μ_t is the common component and ε_{it} is an error term that is i.i.d. across time with mean zero and variance σ_i^2. Suppose that painting i is bought at time t_{bi} and sold at time t_{si} with $t_{bi} < t_{si}$ and the prices are only observed at these times. The holding return on painting i (assuming it was bought and sold exactly once) is

$$r_i = \sum_{t=t_{bi}}^{t=t_{si}} \mu_t + \sum_{t=t_{bi}}^{t=t_{si}} \varepsilon_{it}, \quad i=1,\ldots,n.$$

Therefore write

$$r = A\mu + u,$$

where r is the $n \times 1$ vector containing $r_1, \ldots, r_n$ and A is a known $n \times T$ matrix of zeros and one, while u is an $n \times 1$ vector of error terms. Thereby, show how to estimate μ_t, $t=1,\ldots,T$ when $n > T$.

54. Suppose that

$$y_{it} = \mu + u_{it}, \quad i=1,\ldots,N, \ t=1,\ldots,T$$

where $E(u_{it}) = 0$. Suppose that

$$u_{it} = \theta_i^\mathsf{T} f_t + \varepsilon_{it},$$

where ε_{it} is i.i.d. and θ_i and f_t are also i.i.d. random variables with mean zero. Calculate the covariance matrix Σ of the $NT \times 1$ vector $u = (u_{11},\ldots,u_{NT})^\mathsf{T}$. What is the covariance matrix of the $N \times 1$ and $T \times 1$ vectors

$$\overline{u}_{time} = \left(\sum_{t=1}^{T} u_{1t},\ldots,\sum_{t=1}^{T} u_{Nt}\right)^\mathsf{T} \quad ; \quad \overline{u}_{cross} = \left(\sum_{i=1}^{N} u_{i1},\ldots,\sum_{i=1}^{N} u_{iT}\right)^\mathsf{T} \ ?$$

55. Suppose that y follows a GARCH(1,1) process

$$y_t = \sigma_t \varepsilon_t$$
$$\sigma_t^2 = \omega + \beta\sigma_{t-1}^2 + \gamma y_{t-1}^2,$$

where ε_t is i.i.d. with some distribution for which $E(\varepsilon_t^4) = \mu_4(\varepsilon)$. Then show that

$$E(y_t^4) < \infty \iff \gamma^2 < \frac{1}{\mu_4(\varepsilon)}$$
$$\kappa_4(y) = \frac{\mu_4(\varepsilon) - 3 + 2\mu_4(\varepsilon)\gamma^2}{(1 - \mu_4(\varepsilon)\gamma^2)}.$$

We have $\mu_4(\varepsilon) \geq 1$ by Cauchy–Schwarz inequality so $\kappa_4(y) \geq -2$. In principle there is no restriction on γ so long as $\mu_4(\varepsilon)$ is close to one.

56. Consider the IGARCH(1,1) process,

$$\sigma_t^2 = \omega + \beta\sigma_{t-1}^2 + (1 - \beta)y_{t-1}^2.$$

This process is not weakly stationary. The differenced process

$$\sigma_t^2 - \sigma_{t-1}^2 = \omega + (1 - \beta)(\varepsilon_{t-1}^2 - 1)\sigma_{t-1}^2$$

has mean ω for all t (given starting values) as does

$$y_t^2 - y_{t-1}^2 = \sigma_t^2 - \sigma_{t-1}^2 + (\varepsilon_t^2 - 1)\sigma_t^2 - (\varepsilon_{t-1}^2 - 1)\sigma_{t-1}^2.$$

Recall that linear nonstationary processes, like unit root processes, can be made stationary by differencing. Is this process y_t^2 difference stationary under some conditions? It depends. If the innovation is i.i.d., then the answer is no! But if the innovation has a particular type of time varying distribution then it can be possible to induce weak stationarity through differencing. Defining the MDS $\eta_t = \sigma_t^2(\varepsilon_t^2 - 1)$, we write the squared returns as $y_t^2 = \sigma_t^2 + \eta_t$ and

$$y_t^2 - y_{t-1}^2 = \omega + \eta_t - \beta\eta_{t-1}.$$

Consider a semi-strong IGARCH model with

$$\varepsilon_t = \text{sign}(z_t)\left\{1 + \frac{v_t^2 - 1}{1 + \sigma_t^2}\right\}^{1/2},$$

where z_t, v_t are mutually independent random variables with v_t mean zero and variance one, hence $E(\varepsilon_t^2|\mathcal{F}_{t-1}) = 1$. If z_t symmetric about zero, it follows that $E(\varepsilon_t|\mathcal{F}_{t-1}) = 0$. Furthermore,

$$\eta_t = (v_t^2 - 1)\frac{\sigma_t^2}{1 + \sigma_t^2}$$

Chapter 15 Exercises and Complements

satisfies $E(|\eta_t|) < \infty$, and so we have $E(|\Delta y_t^2|) < \infty$. Provided v_t has finite fourth moment then $E((\Delta y_t^2)^2) < \infty$. However, $E(\sigma_t^2) = E(y_t^2) = \infty$.

57. Consider the GARCH (1,1) model:

$$r_t = h_t^{1/2} \eta_t$$

$$h_t = \omega + \beta h_{t-1} + \gamma r_{t-1}^2$$

where h_t is the conditional variance of time t returns and η_t is a mean zero series. a) Explain the restrictions on the parameters of the GARCH(1,1) model required to ensure that the long-run unconditional variance exists. b) Describe the unconditional variance in terms of these parameters. c) Discuss how the values of the parameters affect the persistence of the response of dynamic volatility to a return shock.

58. Suppose that

$$y_t = \varepsilon_t \sigma_t$$

$$\sigma_t^2 = \omega + \beta \sigma_{t-1}^2 + \gamma y_{t-1}^2.$$

What is $E(y_t^2 | y_{t-j})$? Intuitively expect that it is a quadratic function. If $\beta = 0$ (ARCH(1) case) we have

$$E(y_t^2 | y_{t-1}) = \omega + \gamma y_{t-1}^2$$

$$E(y_t^2 | y_{t-2}) = \omega + \gamma E(y_{t-1}^2 | y_{t-2}) = \omega + \gamma(\omega + \gamma y_{t-2}^2)$$

$$E(y_t^2 | y_{t-j}) = \omega \sum_{l=0}^{j-1} \gamma^l + \gamma^j y_{t-j}^2.$$

In this case the univariate regression functions are all quadratic with different coefficients. But this is not necessarily true for GARCH or even ARCH(p). See Tong (1990). He considers an MA process $y_t = \varepsilon_t - \theta \varepsilon_{t-1}$ and gives an example where $E(y_t | y_{t-1})$ is nonlinear. (Obviously if ε_t is normal then $E(y_t | y_{t-1})$ is linear.) Recall that in GARCH, y_t^2 is an ARMA(1,1) with non normal innovations. In ARCH(p) case, the corresponding property is that $E(y_t^2 | y_{t-j}, \ldots, y_{t-j-p})$ is a quadratic function for any j.

59. Show that

$$\int_0^t B_s dB_s = \frac{1}{2} B_t^2 - \frac{1}{2} t \sim \frac{t}{2}(\chi_1^2 - 1).$$

Proof. Let $Y_t = B_t^2/2$. Then

$$dY_t = d\left(\frac{1}{2} B_t^2\right) = B_t dB_t + \frac{1}{2}(dB_t)^2 = B_t dB_t + \frac{1}{2} dt$$

$$\frac{1}{2} B_t^2 = \int_0^t B_s dB_s + \frac{1}{2} t.$$

60. Consider the common discrete time model

$$z_{t+1} = \mu(z_t) + \sigma(z_t)\varepsilon_{t+1}$$

with ε_t i.i.d. mean zero and variance. Show that this process cannot be in the Affine class unless $\mu(z_t), \sigma(z_t)$ are both linear and ε_{t+1} is Gaussian.

61. Suppose that X is Cauchy with density and c.d.f.

$$f(x) = \frac{1}{\pi} \frac{1}{1+x^2}$$

$$F(x) = \frac{1}{2} + \frac{1}{\pi} \arctan(x).$$

Let $M_n = \max_{1 \leq i \leq n} X_i$. Then derive the limiting distribution of M_n.

Solution: We have

$$G_n(x) = \Pr\left(a_n^{-1} M_n \leq x\right) = \Pr\left(M_n \leq a_n x\right) = F(a_n x)^n = \left(\frac{1}{2} + \frac{1}{\pi} \arctan(a_n x)\right)^n,$$

where $a_n = O(n) \to \infty$. We recall that

$$\arctan\left(\frac{1}{x}\right) = \begin{cases} \frac{\pi}{2} - \arctan(x) & \text{if } x > 0 \\ -\frac{\pi}{2} - \arctan(x) & \text{if } x < 0 \end{cases}.$$

Therefore,

$$G_n(x) = \begin{cases} \left(1 - \frac{1}{\pi} \arctan\left(\frac{1}{a_n x}\right)\right)^n & \text{if } x > 0 \\ \left(-\frac{1}{\pi} \arctan\left(\frac{1}{a_n x}\right)\right)^n & \text{if } x < 0 \end{cases}.$$

We then use the asymptotic expansion for small t, $\arctan(t) = t - \frac{t^3}{3} + \cdots$, and obtain

$$G_n(x) \simeq \begin{cases} \left(1 - \frac{1}{a_n \pi x}\right)^n & \text{if } x > 0 \\ \left(-\frac{1}{a_n \pi x}\right)^n & \text{if } x < 0 \end{cases}.$$

We suppose that $a_n = n/\pi$ in which case

$$G_n(x) \to \begin{cases} \exp\left(-\frac{1}{x}\right) & \text{if } x > 0 \\ 0 & \text{if } x < 0 \end{cases}.$$

Therefore,

$$\Pr\left(\frac{\pi}{n} M_n \leq x\right) \to G(x) = \begin{cases} \exp\left(-\frac{1}{x}\right) & \text{if } x > 0 \\ 0 & \text{if } x < 0 \end{cases}.$$

The density of $G(x)$ is $g(x) = \frac{1}{x^2} \exp\left(-\frac{1}{x}\right)$, which is shown below.

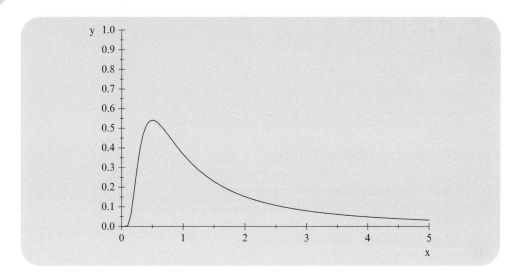

62. Pensions regulators emphasize the solvency probability for some portfolio with random return $w^T X$, $\Pr(w^T X > s)$, where s is some solvency threshold. Suppose that one must achieve at least 67% solvency probability. Suppose that $X \sim N(\mu, \Sigma)$. What is the optimal (in terms of mean return) unit cost portfolio weighting vector w subject to the restriction that the solvency probability must be greater than 0.67?

63. The satirist **Rabelais** wrote that a swan's neck was the best toilet paper he had encountered, although he had never seen or heard of a black swan. Did the discovery of the black swan have a major effect on the paper industry? On financial markets? On anything? What is the connection between black swan theory and Knightian uncertainty? Can general to specific modelling be used in the presence of a black swan?

16 Appendix

16.1 Common Abbreviations

ABS Asset-backed securities
ACD Autoregressive conditional duration
ACF Autocorrelation function
ADR American depository receipts
AMEX American exchange
APC Asymptotic principal components
APT Arbitrage pricing theory
ARCH Autoregressive conditional heteroskedasticity
ARMA Autoregressive moving average
BoE Bank of England
CAPM Capital asset pricing model
CAR Cumulative abnormal return
CARA Constant absolute risk aversion
CBOE Chicago Board Options Exchange
CBPS Corporate Bond Purchase Scheme
CCC Constant conditional correlation
CD Certificate of deposit
c.d.f. Cumulative distribution function
CDO Collaterized debt obligation
CFTC Commodities Futures Trading Commission
CIR Cox–Ingersoll–Ross
CLM Campbell, Lo, and MacKinlay (1997)
CLT Central limit theorem
CME Chicago Mercantile Exchange
CML Capital market line
CPI Consumer price index
CRRA Constant relative risk aversion
CRSP Center for Research in Security Prices
DARA Decreasing absolute risk aversion
DCC Dynamic conditional correlation
DID Differences in differences
DJIA Dow Jones Industrial Average
DRRA Decreasing relative risk aversion
EGARCH Exponential GARCH
EIS Elasticity of intertemporal substitution

16.1 Common Abbreviations

EMH Efficient markets hypothesis
ETF Exchange-traded Fund
EUT Expected utility theory
EWMA Exponentially weighted moving average
FARIMA Fractional autoregressive integrated moving average
FESE World Federation of Exchanges (yes, it is in French)
FF Fama and French
FIGARCH Fractional integrated GARCH process
FM Fama and Macbeth (1973)
GARCH Generalized autoregressive conditional heteroskedasticity
GDR Global depository receipt
GED Generalized error distribution
GJR Glosten, Jagannathan, and Runkle (1993)
GLS Generalized least squares
GMM Generalized method of moments
GMV Global minimum variance
HAR Heterogeneous autoregressive model
HARA Hyperbolic absolute risk aversion
HFT High frequency trading
IARA Increasing absolute risk aversion
I/B/E/S Institutional Brokers' Estimate System
i.i.d. Independent and identically distributed
IOSCO International Organisation of Securities Commissions
IPO Initial public offering
IRRA Increasing relative risk aversion
LAD Least absolute deviation
LASSO Least absolute selection and shrinkage operator
LIBOR London Interbank Offered Rate
LLN Law of large numbers
LM Lagrange multiplier
LMnt Lo and MacKinlay (1990b) non trading model
LR Likelihood ratio
LSE London Stock Exchange
MCMC Markov chain Monte Carlo
MDS Martingale difference sequence
MiFID Markets in Financial Instruments Directive
MLE Maximum likelihood estimator
MSRV Multi-scales realized volatility
MV Mean-variance
NBBO National Best Bid and Offer
NBER National Bureau of Economic Research
NMS National Market System
NYSE New York Stock Exchange
OLS Ordinary least squares
OPEC Oil producing and exporting countries

OTC Over the counter
PCE Personal Consumption Expenditure
PDE Partial differential equation
P/E Price to earnings ratio
QMLE Quasi-maximum likelihood estimator
QV Quadratic variation
reg NMS Regulation National Market System
RP Repurchase agreements
RV Realized volatility
rw Random walk
S&P500 Standard and Poors's 500
SDF Stochastic discount factor
SEC Security and Exchange Commission
SEO Seasoned equity offering
SMEs Small and medium sized enterprises
SML Security market line
SPV Special purpose vehicle
SV Stochastic volatility
SWnt Scholes and Williams (1977) non trading model
T-bills Treasury bills
TSRV Two scales realized volatility
VAR Vector autoregression
VARMA Vector autoregression moving average
VaR Value-at-Risk
VMA Vector moving average
VNM Von Neumann–Morgenstern
VR Variance ratio
VWAP Volume weighted average price

16.2 Two Inequalities

Theorem 60 *Cauchy–Schwarz inequality.* Suppose that X, Y are random variables. Then

$$|E(XY)| \leq \left(E(|X|^2)\right)^{1/2} \left(E(|Y|^2)\right)^{1/2}.$$

Proof. Let $h(t) = E\left((tX - Y)^2\right)$. Then for all $t \in \mathbb{R}$

$$0 \leq h(t) = t^2 E(X^2) + E(Y^2) - 2t E(XY).$$

Where $h(t)$ is a quadratic function in t, which increases as $t \to \pm\infty$. It has a minimum at t_{min} where

$$h'(t_{min}) = 2t_{min} E(X^2) - 2E(XY) = 0,$$

whereby
$$t_{\min} = \frac{E(XY)}{E(X^2)}.$$

Substituting back into $h(t)$ we have
$$0 \leq \frac{E^2(XY)}{E(X^2)} + E(Y^2) - \frac{2E^2(XY)}{E(X^2)},$$

which implies
$$0 \leq E(Y^2) - \frac{E^2(XY)}{E(X^2)}, \text{ i.e., } E^2(XY) \leq E(X^2)E(Y^2). \qquad \square$$

Theorem 61 *Jensen's inequality says that function f is convex if and only if*
$$E(f(X)) \geq f(E(X)).$$

Similarly, the function g is concave if and only if
$$E(g(X)) \leq g(E(X)).$$

Furthermore, if the convexity/concavity is strict and X is not degenerate, i.e. $\Pr(X \neq E(X)) > 0$, then the inequalities are strict.

16.3 Signal Extraction

Theorem 62 *Suppose that $X \sim N(\mu_x, \sigma_x^2)$ and $Y|X = x \sim N(x, \sigma_u^2)$. Then*
$$X|Y = y \sim N(m, v)$$
$$m = \frac{\sigma_u^2}{\sigma_u^2 + \sigma_x^2}\mu_x + \frac{\sigma_x^2}{\sigma_u^2 + \sigma_x^2}y, \quad v = \frac{\sigma_x^2 \sigma_u^2}{\sigma_u^2 + \sigma_x^2}.$$

Proof. We have $Y \sim N(\mu_x, \sigma_x^2 + \sigma_u^2)$, so that:
$$f_X(x) = \frac{1}{\sqrt{2\pi\sigma_x^2}} \exp\left(-\frac{1}{2}(x - \mu_x)^2/\sigma_x^2\right)$$

$$f_{Y|X}(y|x) = \frac{1}{\sqrt{2\pi\sigma_u^2}} \exp\left(-\frac{1}{2}(y - x)^2/\sigma_u^2\right)$$

$$f_Y(y) = \frac{1}{\sqrt{2\pi\sigma_x^2 + \sigma_u^2}} \exp\left(-\frac{1}{2}(y - \mu_x)^2/(\sigma_x^2 + \sigma_u^2)\right).$$

The Bayes theorem says that

$$f_{X|Y}(x|y) = \frac{f_{Y|X}(y|x)f_X(x)}{f_Y(y)}.$$

We have by elimination

$$\frac{(y-\mu_x)^2}{\sigma_u^2+\sigma_x^2} - \frac{(y-x)^2}{\sigma_u^2} - \frac{(x-\mu_x)^2}{\sigma_x^2} = -\frac{1}{\sigma_u^2\sigma_x^2(\sigma_u^2+\sigma_x^2)}\left(x\left(\sigma_u^2+\sigma_x^2\right) - y\sigma_x^2 - \sigma_u^2\mu_x\right)^2$$

$$= -\frac{(\sigma_u^2+\sigma_x^2)}{\sigma_u^2\sigma_x^2}\left(x - \frac{\sigma_x^2}{(\sigma_u^2+\sigma_x^2)}y - \frac{\sigma_u^2}{(\sigma_u^2+\sigma_x^2)}\mu_x\right)^2.$$

The result follows. □

16.4 Lognormal Random Variables

Suppose that X is log normally distributed with

$$\log X \sim N(\mu, \sigma^2),$$

that is, $\log X = \mu + \sigma Z$, where Z is standard normal. Then

$$E(X) = E(\exp(\log X)) = E(\exp(\mu + \sigma Z)) = \exp\left(\mu + \frac{1}{2}\sigma^2\right)$$

using the moment generating function of the normal. Therefore

$$\log E(X) = \mu + \frac{1}{2}\sigma^2 \neq E(\log X) = \mu. \qquad (16.1)$$

The discrepancy between $\log EX$ and $E \log X$ is often called the **Jensen's inequality** term.

Suppose that X and Y are jointly normally distributed, i.e.,

$$\begin{pmatrix} \log X \\ \log Y \end{pmatrix} \sim N\left(\begin{pmatrix} \mu_X \\ \mu_Y \end{pmatrix}, \begin{pmatrix} \sigma_X^2 & \sigma_{XY} \\ \sigma_{XY} & \sigma_Y^2 \end{pmatrix}\right).$$

Then XY is lognormally distributed since

$$\log(XY) = \log(X) + \log(Y) \sim N\left(\mu_X + \mu_Y, \sigma_X^2 + \sigma_Y^2 + 2\sigma_{XY}\right).$$

Hence:

$$E(XY) = E(\exp(\log XY)) = \exp\left(\mu_X + \mu_Y + \frac{1}{2}\left(\sigma_X^2 + \sigma_Y^2 + 2\sigma_{XY}\right)\right)$$

$$\log(E(XY)) = \mu_X + \mu_Y + \frac{1}{2}\left(\sigma_X^2 + \sigma_Y^2 + 2\sigma_{XY}\right) \neq E(\log(XY)) = \mu_X + \mu_Y.$$

16.5 Data Sources

There are many entities that provide some data for free download:

1. European Central Bank's Statistical Data Warehouse. http://sdw.ecb.europa.eu/
2. St. Louis Fed. Lots of freely downloadable macro and financial data. https://fred.stlouisfed.org/
3. Federal Reserve Bank of New York. Regional economic data; markets operation data; national economic data. Note also electronic data bases attached to published papers. www.newyorkfed.org/data-and-statistics
4. US Treasury Yield curves. www.treasury.gov/resource-center/data-chart-center/interest-rates/Pages/TextView.aspx?data=yield
5. World Bank Open Data. Lots of data that can be browsed by country or indicator. http://data.worldbank.org
6. Bank of England Statistical Interactive Database. Economic, financial, interest rate, and exchange rate data by: country, financial instrument, business category, and economic and industrial sector. Note also electronic databases attached to published papers. www.bankofengland.co.uk/boeapps/iadb/NewInterMed.asp?Travel=NIxIRx
7. IMF. The IMF publishes a range of time series data on IMF lending, exchange rates, and other economic and financial indicators. www.imf.org/en/data
8. BLOOMBERG and DATASTREAM. Bloomberg provides current and historical financial quotes, business newswires, and descriptive information, research, and statistics on over 52,000 companies worldwide. You can only access the database from one of their terminals.
9. WRDS (Wharton Research Data Services). WRDS provides the user with one location to access over 250 terabytes of data across multiple disciplines including Accounting, Banking, Economics, Finance, ESG, and Statistics. WRDS provides access to S&P Capital IQ, CRSP, NYSE, Thomson Reuters, Bureau van Dijk, Global Insight, OptionMetrics, and other important business research databases. Using a standard query structure, for example company names, dates of interest, financial parameters, students can create customized reports. These reports can be stored, revisited, and exported. https://wrds-web.wharton.upenn.edu/wrds/index.cfm?
10. FAMA–FRENCH Data Library. This site has a large amount of freely downloadable historical data compiled by Eugene Fama and Kenneth French. The data is updated regularly, and the Fama–French 3-factor data is

especially useful for analyzing fund and portfolio performance. http://mba.tuck.dartmouth.edu/pages/faculty/ken.french/data_library.html
11. Robert Shiller Data Library. This site has Robert Shiller's PE10 data which was used in the book *Irrational Exuberance*. PE10 data is updated regularly. www.econ.yale.edu/~shiller/data.htm
12. William Schwert. Monthly stock returns from 1802–1925 and daily stock returns from 1885–1962. http://schwert.ssb.rochester.edu/data.htm
13. YAHOO FINANCE and GOOGLE FINANCE. Provide historical price information at the daily frequency for thousands of assets, easy to download in csv or excel format, no login etc. https://uk.finance.yahoo.com/ and www.google.co.uk/finance?
14. ASSET MACRO. Historical macroeconomic data and market data (stocks indices, bonds, commodities, FX). www.assetmacro.com/market-data/
15. OANDA. A great resource for historical FX rates that you can easily download. www.oanda.com/solutions-for-business/historical-rates/main.html
16. THINKNUM. US macro data at the zip code level. Requires signup. www.thinknum.com/
17. GITHUB. This is an open source community for programmers. You can find lots of useful software and tools there, and also public datasets. Need to create an account, but free. https://github.com/caesar0301/awesome-public-datasets#finance and https://vincentarelbundock.github.io/Rdatasets/datasets.html
18. LOBSTER. Trade and Quote data from NASDAQ.
19. Caltech Quantitative Finance Group. Gives some tips about data. http://quants.caltech.edu/research.html
20. Quandl. Includes some free data but need to signup. www.quandl.com/
21. CME. High frequency trade and quote data from their platforms (modest fees apply). www.cmegroup.com/market-data.html
22. Things that R can do. https://cran.r-project.org/web/views/Finance.html
23. Amit Goyal's website. www.hec.unil.ch/agoyal/
24. Andrew Patton's website. http://public.econ.duke.edu/~ap172/

16.6 A Short Introduction to Eviews

You can access most of the `Eviews` functionality via menus. Just browse through the menus, and find the appropriate command. You will then be guided through several windows that prompt you for the information required to perform the command. The most important and difficult step (in any software) is to read in data. To do this you need to create a **workfile**. We show here how to do this for dated daily data. We assume here that you have downloaded a csv file with 7 columns with a header describing each (as one would have obtained from Yahoo finance).

How to create an S&P500 daily workfile in EVIEWS 9:

1. Go to https://finance.yahoo.com
2. In the Quote lookup dialog box enter ^GSPC.

16.6 A Short Introduction to Eviews

3. Click Historical Data.
4. Change time period to Max and then enter Done and Apply in the right hand box.
5. Click download data. You should now have an excel file ^GSPC.csv. It should have 7 columns with a header describing each with the first date being 03/01/1950 (January 3rd, 1950) or 20150103.
6. Open Eviews 9 and create a new workfile click `File > New > Workfile`.
7. Select the `dated > regular > frequency` option and further specify `daily frequency` (five day week) and the start date and end date using American dating convention (you may have downloaded in European dating convention).
8. The workfile is created with two preexisting variables, c and resid. They stand for "coefficients" and "residuals." Every time you estimate something, the coefficients are stored in c and the residuals in resid.
9. Import the data: `Select File > Import from File` and click the file on your drive, and you should see a box with title `Text Read - Step 1 of 4`. Click `Finish` box (if you like, you may click `Next` and see the options available, which I don't think you need at this point). You should see a box with `Link imported series to external source`. Click Yes, why not!
10. You now have a workfile (UNTITLED – you can save it later with a name) with 8 named series: adj_close, c, close, high, low, open, resid, volume. Do a cross-check with your raw data file by clicking. Compare the number of observations; check if each series got the correct name. Eviews should have transposed the dates into the right order. You should see a number of observations with NA (which means not available), these are mostly just holidays when the stock price was not returned. We have to eliminate these next.
11. From the workfile window, `Proc > Copy Extract from Current Page` (by value to new page or workfile). You then see a box called `Workfile copy by Value`; you enter the following into the open box headed `Sample - observations to copy`

$$\text{@all if close<>NA}$$

Then click `Page Destination` and `New workfile`. You now get to enter the name of the file you want to keep. Lets call it sp500daily. Enter. To be on the safe side use `File > Save > sp500daily`. The dates with missing observations have been eliminated.
12. You next want to create a new variable for return. Click the `Genr` button on the workfile window and then insert the defining equation of your new variable. For example `Ret=(adj_close-adj_close(-1))/adj_close(-1)`. Note that when you write x(-1), Eviews understands that you want x lagged one period. For logarithmic return define `r=log(close)-log(close(-1))`.
13. Generating the day of the week dummy. Use the `Genr` command and enter `D1=@weekday=1`, which gives the Monday dummy, likewise for the other days of the week
14. You can now get by with the menu driven commands.

15. Exporting Eviews graphs. Eviews generates graphs, but they may not look the way you want them to. In order to use an Eviews graph in another program you can save it as a Windows metafile ("*.wmf"). After generating the graph, click Object > View Options and then you may copy to clipboard or save to disk. You may copy the graph to clipboard and then insert in your word processing document. Alternatively, you may choose to save to disk, in which case you will be offered several formats.

Bibliography

Adrian, T. and M. Brunnermeier (2010), "CoVaR," *American Economic Review* 106.7, 1705–41.

Aït-Sahalia, Y. (1996a), "Nonparametric Pricing of Interest Rate Derivative Securities," *Econometrica* 64, 527–60.

Aït-Sahalia, Y. (1996b), "Testing Continuous-Time Models of the Spot Interest Rate," *Review of Financial Studies* 9, 385–426.

Aït-Sahalia, Y. and J. Jacod (2014), *High-Frequency Financial Econometrics* (Princeton University Press).

Aït-Sahalia, Y., P. A. Mykland, and L. Zhang (2011), "Ultra High Frequency Volatility Estimation with Dependent Microstructure Noise," *Journal of Econometrics* 160, 160–175.

Aït-Sahalia, Y. and M. Saglam (2013), "High Frequency Traders: Taking Advantage of Speed," NBER Working Paper No. 19531.

Allais, M. (1953), "Le comportement de l'homme rationnel devant le risque: critique des postulats et axiomes de l'école Américaine," *Econometrica* 21.4, 503–46.

Amihud, Y. (2002), "Illiquidity and Stock Returns: Cross-Section and Time Series Effects," *The Journal of Financial Markets* 5, 31–56.

Amihud, Y. and H. Mendelson (1986), "Asset Pricing and the Bid-Ask Spread," *Journal of Financial Economics* 17, 223–49.

Andersen, P. K., O. Borgan, R. Gill, and N. Keiding (1993), *Statistical Models Based on Counting Processes* (Springer Verlag).

Andersen, T. G. and T. Bollerslev (1998), "Answering the Skeptics: Yes, Standard Volatility Models do Provide Accurate Forecasts," *International Economic Review*, 39.4: *Symposium on Forecasting and Empirical Methods in Macroeconomics and Finance*, 885–905.

Andersen, T. G., T. Bollerslev, F. X. Diebold, and P. Labys (2001), "The Distribution of Exchange Rate Realised Volatility," *Journal of the American Statistical Association* 96, 42–55.

Andersen, T. G., T. Bollerslev, F. X. Diebold, and P. Labys (2003), "Modeling and Forecasting Realized Volatility," *Econometrica* 71, 579–626.

Andersen, T. G., D. Dobrev, and E. Schaumburg (2008), "Duration-Based Volatility Estimation," Working Paper, Northwestern University.

Anderson, N., F. Breedon, M. Deacon, A. Derry, and G. Murphy (1996), *Estimating and Interpreting the Yield Curve* (John Wiley Series in Financial Economics and Quantitative Analysis).

Anderson, R. L. (1942), "Distribution of the Serial Correlation Coefficient," *Annals of Mathematical Statistics* 13, 1–12.

Anderson, T. W. (1984), *An Introduction to Multivariate Statistical Analysis* (Wiley & Sons).

Andrews, D. W. (1991), "Heteroskedasticity and Autocorrelation Consistent Covariance Matrix Estimation," *Econometrica* 59.3, 817–58.

Ang, A. and D. Kristensen (2012), "Testing Conditional Factor Models," *Journal of Financial Economics* 106.1, 132–56.

Bibliography

Ang, A., J. Liu, and K. Schwarz, "Using Stocks or Portfolios in Tests of Factor Models," AFA 2009 San Francisco Meetings Paper. Available at SSRN: https://ssrn.com/abstract=1106463 or http://dx.doi.org/10.2139/ssrn.1106463

Angrist, J. D. and J. S. Pischke (2009), *Mostly Harmless Econometrics: An Empiricist's Companion* (Princeton University Press).

Arrow, K. (1965), *Aspects of the Theory of Risk Bearing* (Yrjo Jahnssonin Saatio).

Artzner, P. F., J. Delbaenm, M. Eber, and D. Heath (1999), "Coherent Measures of Risk," *Mathematical Finance* 9, 203–28.

Baba, Y., R. F. Engle, D. F. Kraft, and K. F. Kroner (1990), "Multivariate Simultaneous Generalized ARCH," Department of Economics, University of California, San Diego.

Bachelier, L. (1900), *Theorie de la speculation* (Gauthier-Vllars).

Backus, D., S. Foresi, and C. Telmer (1998), "Discrete-Time Models of Bond Pricing," NBER Working Paper No. 6736.

Bai, J. (2003), "Inferential Theory for Factor Models of Large Dimension," *Econometrica* 71, 135–71.

Bai, J. (2004), "Estimating Cross-Section Common Stochastic Trends in Nonstationary Panel Data," *Econometrica* 122.1, 137–83.

Bai, J. and Ng, S. (2002), "Determining the Number of Factors in Approximate Factor Models," *Econometrica* 70, 191–221.

Baillie, R.T., T. Bollerslev, and H. O. Mikkelsen (1996), "Fractionally Integrated Generalized Autoregressive Conditional Heteroskedasticity," *Journal of Econometrics* 74, 3–30.

Bampinas, G., K. Ladopoulos, and T. Panagiotidis (2018), "A Note on the Estimated GARCH Coefficients from the S&P1500 Universe," *Applied Economics* 50, 3647–3653.

Bandi, F. M., L. Lian, and J. R. Russell (2012), "How Effective are Realized Effective Spreads?," available at https://pdfs.semanticscholar.org/2238/2c534f244d25941e7d04990281d9abe662d2.pdf

Bandi, F. M. and Phillips, P. C. B. (2003), "Fully Nonparametric Estimation of Scalar Diffusion Models," *Econometrica* 71, 241–83.

Bansal, R. and A. Yaron (2004,) "Risks for the Long Run: A Potential Resolution of Asset Pricing Puzzles," *Journal of Finance* 59, 1481–509.

Banz, R.W. (1981), "The Relationship Between Return and Market Value of Common Stocks," *Journal of Financial Economics* 9, 3–18.

Barber, B. M. and J. D. Lyon (1997) "Detecting Long-Run Abnormal Stock Returns: The Empirical Power and Specification of Test Statistics," *Journal of Financial Economics* 43, 341–72.

Barndorff-Nielsen, O., P. Hansen, A. Lunde, and N. Shephard (2006), "Designing Realized Kernels to Measure the Ex-Post Variation of Equity Prices in the Presence of Noise," Unpublished Paper: Nuffield College, Oxford.

Barndorff-Nielsen, O., P. Hansen, A. Lunde, and N. Shephard (2008), "Designing Realised Kernels to Measure the Ex-Post Variation of Equity Prices in the Presence of Noise," *Econometrica* 76, 1481–536.

Barndorff-Nielsen, O. and N. Shephard (2002), "Econometric Analysis of Realised Volatility and its Use in Estimating Stochastic Volatility Models," *Journal of the Royal Statistical Society: Series B* 64, 253–80.

Barndorff-Nielsen, O. and N. Shephard (2006a), "Econometrics of Testing for Jumps in Financial Economics using Bipower Variation," *Journal of Financial Econometrics* 4, 1–30.

Barndorff-Nielsen, O. and N. Shephard (2006b), "Impact of Jumps on Returns and Realised Volatility: Econometric Analysis of Time-Deformed Levy Processes," *Journal of Econometrics* 131, 217–52.

Bartlett, M. S. (1950), "Periodogram Analysis and Continuous Spectra," *Biometrika* 37, 1–16.

Bibliography

Bartram, S. M., G. W. Brown, and R. M. Stulz (2012), "Why Are U.S. Stocks More Volatile?," *Journal of Finance* 67.4, 1329–70.

Basel Committee on Banking Supervision (2012), "Fundamental Review of the Trading Book: Consultative Document," available at http://www.bis.org/publ/bcbs219.pdf

Bauwens, L., S. Laurent, and J. V. K. Rombouts (2004), "Multivariate GARCH Models: A Survey," *Journal of Applied Econometrics* 21, 79–109.

Bergstrom, A. R. (1988), "The History of Continuous-Time Econometric Models," *Econometric Theory* 4, 365–83.

Bergstrom, A. R. and B. Nowman (2007), *A Continuous Time Econometric Model of the United Kingdom with Stochastic Trends* (Cambridge University Press).

Berk, J. (1997), "Necessary Conditions for the CAPM," *Journal of Economic Theory* 73, 245–57.

Berman, S. M. (1964), "Limit Theorems for the Maximum Term in Stationary Sequences," *Annals of Mathematical Statistics* 35, 502–16.

Berndt, E. R. and N. E. Savin (1977), "Conflict among Criteria for Testing Hypotheses in the Multivariate Linear Regression Model," *Econometrica* 45.5, 1263–77.

Bertrand, M., E. Duflo, and S. Mullainathan (2004), "How Much Should We Trust Differences-in-Differences Estimates?," *The Quarterly Journal of Economics* 119.1, 249–75.

Bhattacharya, U. and H. Daouk (2002), "The World Price of Insider Trading," *The Journal of Finance* 57, 75–108.

Black, F. (1972), "Capital Market Equilibrium with Restricted Borrowing," *The Journal of Business* 45.3, 444–55.

Black, F. and M. Scholes (1973), "The Pricing of Options and Corporate Liabilities," *Journal of Political Economy* 81.3, 63–54.

Blanchard, O. J. and M. W. Watson (1982), "Bubbles, Rational Expectations and Financial Markets," in Paul Wachtel (ed.), *Crises in the Economic and Financial Structure* (D. C. Heath and Company), 295–316.

Bliss, R. (1997), "Testing Term Structure Estimation Methods," *Advances in Futures and Options Research* 9, 197–231.

Bollerslev, T. (1986) "Generalized Autoregressive Conditional Heteroskedasticity," *Journal of Econometrics* 31.3, 307–27.

Bollerslev, T. (1987), "A Conditional Heteroskedastic Time Series Model for Speculative Prices and Rates of Returns," *Review of Economics and Statistics* 69, 542–47.

Bollerslev, T. (1990), "Modelling the Coherence in Short-Run Nominal Exchange Rates: A Multivariate Generalized ARCH Model," *Review of Economics and Statistics* 72, 498–505.

Bollerslev, T., R. F. Engle, and D. Nelson (1994), "ARCH Models," in D. F. McFadden and R. F. Engle (eds.), *The Handbook of Econometrics, Volume 4* (North Holland).

Bollerslev, T., R. F. Engle, and J. M. Wooldridge (1988), "A Capital Asset Pricing Model with Time-Varying Covariances," *The Journal of Political Economy* 96, 116–31.

Bollerslev, T. and J. M. Wooldridge (1992), "Quasi-Maximum Likelihood Estimation and Inference in Dynamic Models with Time-Varying Covariances," *Econometric Reviews* 11, 143–72.

Bosq, D. (1998), *Nonparametric Statistics for Stochastic Processes* (Springer Verlag).

Boudoukh, J., M. P. Richardson, and R. E. Whitelaw (1994), "A Tale of Three Schools: Insights on Autocorrelations of Short-Horizon Stock Returns," *Review of Financial Studies* 7.3, 539–73.

Boudoukh, J., M. P. Richardson, and R. E. Whitelaw (2008), "The Myth of Long-Horizon Predictability," *The Review of Financial Studies* 21, 1577–605.

Bibliography

Breidt, F. J., N. Carto, and P. de Lima (1998), "The Detection and Estimation of Long Memory in Stochastic Volatility Models," *Journal of Econometrics* 83, 325–48.

Brennan, M.J., T. Chordia, and A. Subrahmanyam (1998), "Alternative Factor Specifications, Security Characteristics and the Cross Section of Expected Returns," *Journal of Financial Economics* 49, 345–73.

Brockwell, P. J. and R.A. Davis (2006), *Time Series: Theory and Methods* (2nd edition; Springer Verlag).

Brogaard, J., T. Hendershott, and R. Riordan (2014), "High Frequency Trading and Price Discovery," *Review of Financial Studies* 27, 2267–306.

Brooks, C., S. P. Burke, and G. Persand (2001), "Benchmarks and the Accuracy of GARCH Model Estimation," *International Journal of Forecasting* 17, 45–56.

Brown, S. J. and J. B. Warner (1980), "Using Daily Stock Returns: The Case of Event Studies," *Journal of Financial Economics* 14.1, 3–31.

Brugler, J. and O. Linton (2014), The Cross-Sectional Spillovers of Single Stock Circuit Breakers," Cemmap working paper CWP07/14.

Brunnermeier, Markus K. (2008), "Bubbles," in Lawrence Blume and Steven Durlauf (eds.), *New Palgrave Dictionary of Economics* (Palgrave).

Bryzgalova, S. (forthcoming), "Spurious Factors in Linear Asset Pricing Models," *The Review of Financial Studies*.

Budish, E., P. Cramton, and J. Shim (2015), "The High-Frequency Trading Arms Race: Frequent Batch Auctions as a Market Design Response," *The Quarterly Journal of Economics* 130.4, 1547–621.

Campbell, J. (1993), "Intertemporal Asset Pricing without Consumption Data," *American Economic Review* 83, 487–512.

Campbell, J. Y. and J. Cochrane (1999), "By Force of Habit: A Consumption-Based Explanation of Aggregate Stock Market Behavior," *Journal of Political Economy* 107.2, 205–51.

Campbell, J. Y., M. Lettau, B. Malkiel, and Y. Xu (2001), "Have Individual Stocks Become More Volatile?," *Journal of Finance* 56, 1–43.

Campbell, J. Y., A. W. Lo, and A. C. MacKinlay (1997), *The Econometrics of Financial Markets* (Princeton University Press).

Campbell, J. Y. and R. Shiller (1988a), "The Dividend–Price Ratio and Expectations of Future Dividends and Discount Factors," *Review of Financial Studies* 1, 195–228.

Campbell, J. Y. and R. Shiller (1988b), "Stock Prices, Earnings and Expected Dividends," *Journal of Finance* 43, 661–76.

Campbell, J. Y. and R. Shiller (1991), "Yield Spreads and Interest Rate Movements: A Bird's Eye View," *Review of Economic Studies* 58, 495–514.

Campbell, J. Y. and S. B. Thompson (2008), "Predicting Excess Stock Returns Out of Sample: Can Anything Beat the Historical Average?," *The Review of Financial Studies* 21.4, 1509–31.

Carhart, M. (1997), "On Persistence in Mutual Fund Performance," *Journal of Finance* 52, 57–82.

Carr, P. and L. Wu (2006), "A Tale of Two Indices," *The Journal of Derivatives* Spring, 13–29.

Carrasco, M. and X. Chen (2002), "Mixing and Moment Properties of Various GARCH and Stochastic Volatility Models," *Econometric Theory* 18, 17–39.

Castura, J., R. Litzenberger, and R. Gorelick (2012), "Market Efficiency and Microstructure Evolution in US Equity Markets: A High Frequency Perspective," available at https://www.sec.gov/comments/s7-02-10/s70210-364.pdf

CBOE (2010), "White Paper: Cboe Volatility Index", available at http://www.cboe.com/micro/vix/vixwhite.pdf

Chamberlain, G. (1983), "A Characterization of the Distributions that Imply Mean-Variance Utility Functions," *Journal of Economic Theory* 29, 185–201.

Chamberlain, G. (1984), "Panel Data," in Z. Griliches and M. D. Intriligator (eds.), *Handbook of Econometrics, Volume* 2 (North Holland), 1247–318.

Chamberlain, G. and M. Rothschild (1983), "Arbitrage, Factor Structure, and Mean-Variance Analysis on Large Asset Markets," *Econometrica* 51.5, 1281–304.

Chan, K. C., Nai-fu Chen, and D. A. Hsieh (1985), "An Exploratory Investigation of the Firm Size Effect," *Journal of Financial Economics* 14.3, 451–71.

Chapman, D. and N. Pearson (2000), "Is the Short Rate Drift Actually Nonlinear?," *Journal of Finance* 55, 355–88.

Chen, N.-F., R. Roll, and S. A. Ross (1986), "Economic Forces and the Stock Market," *Journal of Business* 59.3, 383–403.

Chen, X. (2007), "Large Sample Sieve Estimation of Semi-Nonparametric Models," in J. J. Heckman and E. E. Leamer (eds.), *The Handbook of Econometrics, vol. 6B* (North-Holland).

Chen, X. and Y. Fan (2006), "Estimation of Copula-Based Semiparametric Time Series Models," *Journal of Econometrics* 130, 307–35.

Chen, X., L. P. Hansen, and J. Scheinkman (2009), "Nonlinear Principal Components and Long-Run Implications of Multivariate Diffusions," *Annals of Statistics* 37 (2009), 4279–312.

Chen, X. and E. Ghysels (2010), "News – Good or Bad – and its Impact on Volatility Predictions over Multiple Horizons," *Review of Financial Studies* 24, 46–81.

Chernozhukov, V. (2005), "Extremal Quantile Regression," *The Annals of Statistics* 33.2, 806–39.

Chordia, T., R. Roll, and A. Subrahmanyam (2011), "Recent Trends in Trading Activity and Market Quality," *Journal of Financial Economics*, 101.2, 243–63.

Chou, R. Y., H. C. Chou, and N. Liu (2009), "Range Volatility Models and their Applications in Finance," in Cheng-Few Lee and John Lee (eds.), *The Handbook of Quantitative Finance and Risk Management* (Springer), Part V, 1273–81.

Cochrane, J. H. (1999), "New Facts in Finance," NBER Working Paper No. 7169.

Cochrane, J. H. (2001), *Asset Pricing* (Princeton University Press).

Conley, T. G., L. P. Hansen, E. G. J. Luttmer, J. A. Scheinkman (1997), "Short-Term Interest Rates as Subordinated diffusions," *Review of Financial Studies* 10.3, 525–77.

Connor, G. (1984), "A Unified Beta Pricing Theory," *Journal of Economic Theory* 34, 13–31.

Connor, G., L. R. Goldberg, and R. A. Korajczyk (2010), *Portfolio Risk Analysis* (Princeton University Press).

Connor, G., M. Hagmann, and O. Linton (2012), "Efficient Semiparametric Estimation of the Fama–French Model and Extensions," *Econometrica* 80.2, 713–54.

Connor, G. and R. A. Korajczyk (1988), "Risk and Return in an Equilibrium APT: Application of a New Test Methodology," *Journal of Financial Economics* 21.2, 255–89.

Connor, G. and R. A. Korajczyk (1993), "A Test for the Number of Factors in an Approximate Factor Model," *Journal of Financial Economics* 48, 1263–91.

Connor, G., R. A. Korajczyk, and O. Linton (2006), "The Common and Specific Components of Dynamic Volatility," *Journal of Econometris* 132, 231–55.

Constantinides, G. M. (1990), "Habit Formation: A Resolution of the Equity Premium Puzzle," *Journal of Political Economy* 98.3, 519–43.

Constantinides, G. M. (1992), "A Theory of the Nominal Term Structure of Interest Rates," *Review of Financial Studies* 5, 531–52.

Constantinides, G. M. and A. Ghosh (2011), "Asset Pricing Tests with Long-Run Risks in Consumption Growth," *Review of Asset Pricing Studies* 1.1, 96–136.

Cont, R. and P. Tankov (2003), *Financial Modelling with Jump Processes* (Chapman & Hall and CRC Financial Mathematics Series).

Copeland, T. and D. Galai (1983), "Information Effects on the Bid-Ask Spread," *The Journal of Finance* 38.5, 1457–69.

Cowan, A. R. (1992), "Nonparametric Event Study Tests," *Review of Quantitative Finance and Accounting* 2.4, 343–58.

Cowles, A. and H. Jones (1937), "Some A Posteriori Probabilities in Stock Market Action," *Econometrica* 5, 280–94.

Cox, J. C. (1975), "Notes on Option Pricing I: Constant Elasticity of Variance Diffusions," Unpublished Note.

Cox, D. R. and D. V. Hinkley (1979), *Theoretical Statistics* (CRC Press).

Cox, J. C., J. E. Ingersoll, and S. A. Ross (1985), "A Theory of the Term Structure of Interest Rates," *Econometrica* 53, 385–407.

Cutler, D. M., J. M. Poterba, and L. H. Summers (1989), "What Moves Stock Prices?," NBER Working Paper No. 2538 (Also Reprint No. r1232), *The Journal of Portfolio Management* 15.3, 4–12.

Dahlhaus, R. (1997), "Fitting Time Series Models to Nonstationary Processes," *Annals of Statistics* 25.1, 1–37.

Dahlhaus, R. and S. Subba Rao (2006), "Statistical Inference for Time-Varying ARCH Processes," *Annals of Statistics* 34.3, 1075–114.

Dai, Q. and K. J. Singleton (2000), "Specification Analysis of Affine Term Structure Models," *The Journal of Finance* 55.5, 1943–78.

Danielsson, J. (1994), "Stochastic Volatility in Asset Prices: Estimation with Simulated Maximum Likelihood," *Journal of Econometrics* 64: 375–400.

Danielsson, J. (2011), *Financial Risk Forecasting* (Wiley).

Danielsson, J. and C. G. de Vries (1997), "Tail Index and Quantile Estimation with Very High Frequency Data," *Journal of Empirical Finance* 4, 241–57.

Danielsson, J., B. N. Jorgensen, G. Samorodnitsky, M. Sarma, and C. G. de Vries (2013), "Fat Tails, VaR and Subadditivity," *Journal of Econometrics* 172.2, 283–91.

Darling, D. A. and A. J. F. Siegert (1953), "The First Passage Problem for a Continuous Markov Process," *Annals of Mathematical Statistics* 24.4, 624–39.

Davis, J. L. (1994), "The Cross-Section of Realized Stock Returns: The Pre-COMPUSTAT Evidence," *The Journal of Finance* 49, 1579–93.

De Bondt, W. F. M. and R. Thaler (1985), "Does the Stock Market Overreact?," *Journal of Finance* 40, 793–805.

De Bondt, W. F. M. and R. Thaler (1987), "Further Evidence of Investor Overreaction and Stock Market Seasonality," *Journal of Finance* 42, 557–81.

Dekkers, A. L. M., J. H. J. Einmahl, and L. de Haan (1989), "A Moment Estimator for the Index of an Extreme-Value Distribution," *Annals of Statistics* 17, 1833–55.

Diebold, F. X. and C. Li (2006), "Forecasting the Term Structure of Government Bond Yields," *Journal of Econometrics* 130, 337–64.

Diebold, F.X. and R. S. Mariano (1995), "Comparing Predictive Accuracy," *Journal of Business and Economic Statistics* 13.3, 253–63.

Diebold, F. X. and M. Nerlove (1989), "The Dynamics of Exchange Rate Volatility: A Multivariate Latent Factor ARCH Model," *Journal of Applied Econometrics* 4, 1–21.

Dimson, E., P. Marsh, and M. Staunton (2008), "The Worldwide Equity Premium: A Smaller Puzzle," in R. Mehra (ed.), *Handbook of the Equity Risk Premium* (Elsevier).

Dolley, J. C. (1933), "Characteristics and Procedure of Common Stock Split-Ups," *Harvard Business Review* 11, 316–26.

Doukhan, P. (1994), *Mixing: Properties and Examples* (Springer Verlag).

Bibliography

Drost, F. C., C. A. J. Klaassen, and B. J. M. Werker (1997), "Adaptive Estimation in Time-Series Models," *Annals of Statistics* 25.2, 786–817.

Drost, F. C. and T. Nijman (1993), "Temporal Aggregation of GARCH Processes," *Econometrica* 61.4, 909–27.

Dubow, B. and N. Monteiro (2006), "Measuring Market Cleanliness," FSA Occasional Paper no 23.

Dufour, J. M., M. Hallin, and I. Mizera (1998), "Generalized Runs Tests for Heteroscedastic Time Series," *Nonparametric Statistics* 9, 39–86.

Easley, D., N. M. Kiefer, M. O'Hara, and J. B. Paperman (1996), "Liquidity, Information and Infrequently Traded Stocks," *Journal of Finance* 51.4, 1405–36.

Easley, D. and M. O'Hara (1987), "Price, Trade Size, and Information in Securities Markets," *Journal of Financial Economics* 19.1, 69–90.

Easley, D. and M. O'Hara (1992), "Time and the Process of Security Price Adjustment," *Journal of Finance* 47, 576–605.

Easley, D. and M. O'Hara (2004), "Information and the Cost of Capital," *Journal of Finance* 59(4), 1553–83.

Efron, B. (1979), "Bootstrap Methods: Another Look at the Jackknife," *Annals of Statistics* 7.1, 1–26.

Eicker, F. (1967), "Limit Theorems for Regressions with Unequal and Dependent Errors," in L. M. Le Can and J. Neyman (eds.), *Proceedings of the Fifth Berkeley Symposium on Mathematical Statistics and Probability, Volume 1* (University of California Press), pp. 59–82.

Embrechts, P., C. Klüppelberg, and T. Mikosch (1997), *Modelling Extremal Events* (Springer Verlag).

Engle, R. F. (1982), "Autoregressive Conditional Heteroscedasticity with Estimates of the Variance of United Kingdom Inflation," *Econometrica* 50, 987–1007.

Engle, R. F. (1984), "Wald, Likelihood Ratio, and Lagrange Multiplier Tests in Econometrics," in Z. Griliches and M. D. Intriligator (eds.), *Handbook of Econometrics, Volume 2* (North Holland), 775–826.

Engle, R. F. and G. Gonzalez-Rivera (1991), "Semiparametric ARCH Model," *Journal of Business and Economic Statistics* 9, 345–60.

Engle, R. F., D. M. Lilien, and R. P. Robins (1987), "Estimating Time Varying Risk Premia in the Term Structure: The ARCH-M Model," *Econometrica* 55, 391–407.

Engle, R. F. and S. Manganelli (2004), "CAViaR: Conditional Autoregressive Value at Risk by Regression Quantiles," *Journal of Business and Economic Statistics* 22, 367–81.

Engle, R. F. and V. K. Ng (1993), "Measuring and Testing the Impact of News on Volatility," *The Journal of Finance* 48, 1749–78.

Engle R. F., V. K. Ng, and M. Rothschild M (1990), "Asset Pricing with a Factor-ARCH Covariance Structure: Empirical Estimates for Treasury Bills," *Journal of Econometrics* 45, 213–38.

Engle, R. F. and J. G. Rangel (2008), "The Spline-GARCH Model for Low-Frequency Volatility and its Global Macroeconomic Causes," *Review of Financial Studies*, 21.3, 1187–222.

Engle, R. F. and Jeffrey R. Russell (1998), "Autoregressive Conditional Duration: A New Model for Irregularly Spaced Transaction Data," *Econometrica* 66.5, 1127–62.

Engle, R. F. and K. Sheppard (2001), *Theoretical and Empirical Properties of Dynamic Conditional Correlation Multivariate GARCH* (Mimeo, UCSD).

Epps, T. W. (1979), "Comovements in Stock Prices in the Very Short Run," *Journal of the American Statistical Association* 74, 291–96.

Epstein, L. and S. E. Zin (1989), "Substitution, Risk Aversion, and the Temporal Behavior of Consumption and Asset Returns: A Theoretical Framework," *Econometrica* 57, 937–69.

Estrella, A. and F. S. Mishkin (1997), "The Predictive Power of the Term Structure of Interest Rates in Europe and the United States: Implications for the European Central Bank," *European Economic Review* 41, 1375–401.

Faber, M. (2010), "Relative Strength Strategies for Investing," available at SSRN: https://ssrn.com/abstract=1585517 or http://dx.doi.org/10.2139/ssrn.1585517

Faber, M. (2013), "Quantitative Approach to Tactical Asset Allocation," available at SSRN: https://ssrn.com/abstract=962461

Fair, Ray C. (2002), "Events that Shook the Market," *Journal of Business* 75.4, 713–31. Available at SSRN: https://ssrn.com/abstract=334004

Fama, E. (1963), "Mandelbrot and the Stable Paretian Hypothesis," *Journal of Business* 36, 420–29.

Fama, E. (1965), "The Behaviour of Stock Market Prices," *Journal of Business* 38, 34–105.

Fama, E. (1970), "Efficient Capital Markets: A Review of Theory and Empirical Work," *Journal of Finance* 25.2, 383–417.

Fama, E. and R. Bliss (1987), "The Information in Long Maturity Forward Rates," *The American Economic Review* 77, 680–92.

Fama, E., L. Fisher, M. Jensen, and R. Roll (1969), "The Adjustment of Stock Prices to New Information," *International Economic Review* 10.1, 1–21.

Fama, E. and K. R. French (1992), "The Cross-Section of Expected Stock Returns," *Journal of Finance* 47, 427–65.

Fama, E. and K. R. French (1993), "Common Risk Factors in the Returns on Bonds and Stocks," *Journal of Financial Economics* 33, 3–56.

Fama, E. and K. R. French (1996), "Multifactor Explanations of Asset Pricing Anomalies," *Journal of Finance* 51, 55–84.

Fama, E. and K. R. French (1997), "Industry Costs of Equity," *Journal of Financial Economics* 43, 153–93.

Fama, E. and K. R. French (1998), "Value Versus Growth: The International Evidence," *Journal of Finance* 53, 1975–99.

Fama, E. and K. R. French (2004), "The Capital Asset Pricing Model: Theory and Evidence," *Journal of Economic Perspectives* 18, 25–46.

Fama, E. and K. R. French (2006), "The Value Premium and the CAPM," *Journal of Finance* 61, 2163–85.

Fama, E. and K. R. French (2015), "A Five-Factor Asset Pricing Model," *Journal of Financial Economics* 116.1, 1–22.

Fama, E. and M. R. Gibbons (1982), "Inflation, Real Returns, and Capital Investment," *Journal of Monetary Economics* 9, 297–323.

Fama, E. and J. MacBeth (1973), "Risk, Return, and Equilibrium: Empirical Tests," *Journal of Political Economy* 71, 607–36.

Fan, J., Y. Liao, and M. Mincheva (2013), "Large Covariance Estimation by Thresholding Principal Orthogonal Complements," *Journal of the Royal Statistical Society Series B* 75, 603–80.

Fan, J. and Q. Yao (1998), "Efficient Estimation of Conditional Variance Functions in Stochastic Regression," *Biometrika* 85, 645–66.

Fan, J. and C. Zhang (2003), "A Reexamination of Diffusion Estimations with Applications to Financial Model Validation," *Journal of the American Statistical Association* 13, 965–92.

Faust, J. (1992), "When are Variance Ratio Tests for Serial Dependence Optimal?," *Econometrica* 60.5, 1215–26.

Feller, W. (1965), *An Introduction to Probability Theory and its Applications, Volume* 2 (Wiley).

Fernandez, P. (2015), "CAPM: An Absurd Model," available at SSRN: https://ssrn.com/abstract=2505597 or http://dx.doi.org/10.2139/ssrn.2505597

Fernandez, P., A. Ortiz Pizarro, and I. Fernández Acín (2015), "Discount Rate (Risk-Free Rate and Market Risk Premium) Used for 41 Countries in 2015: A Survey," available at SSRN: https://ssrn.com/abstract=2598104 or http://dx.doi.org/10.2139/ssrn.2598104

Fisher, M., D. Nychka, and D. Zervos (1995), "Fitting the Term Structure of Interest Rates with Smoothing Splines," Board of Governors of the Federal Reserve System, Finance and Economics Discussion Series 95–1.

Fisher, R. A. and L. H. C. Tippett (1928), "Limiting Forms of the Frequency Distribution of the Largest or Smallest Member of a Sample," *Proceedings of the Cambridge Philosophical Society* 24, 180–290.

Flood, R. P. and R. J. Hodrick (1986), "Asset Price Volatility, Bubbles, and Process Switching," NBER Working Paper No. w1867. Available at SSRN: https://ssrn.com/abstract=307107

Flood, R. P. and R. J. Hodrick (1990), "On Testing for Speculative Bubbles," *Journal of Economic Perspectives* 4.2, 85–101.

Florens-Zmirou, D. (1993), "On Estimating the Diffusion Coefficient from Discrete Observations," *Journal of Applied Probability* 30, 790–804.

Foresight (2012), "The Future of Computer Trading in Financial Markets: UK Government Office for Science Final Project Report," available at https://www.gov.uk/government/publications/future-of-computer-trading-in-financial-markets-an-international-perspective

Foster, D. P. and D. B. Nelson (1996), "Continuous Record Asymptotics For Rolling Sample Variance Estimators," *Econometrica* 64.1, 139–74.

Foucault, T., M. Pagano, and A. Röell (2013), *Market Liquidity: Theory, Evidence, and Policy* (Oxford University Press).

Frahm, G. (2013), "Absorbability of Financial Markets," available at https://arxiv.org/abs/1304.3824

French, K. R. and R. Roll (1986), "Stock Return Variances: The Arrival of Information and the Reaction of Rraders," *Journal of Financial Economics* 17.1, 5–26.

Friedman, M. and L. J. Savage (1948), 'Utility Analysis of Choices Involving Risk," *Journal of Political Economy* 56.4, 279–304.

Froot, K. A. and K. Rogoff (1995), "Perspectives on and Long-Run Real Exchange Rates," in G. Grossman and K. Rogoff (eds.), *Handbook of International Economics, Volume 3* (Elsevier), 1647–88.

Froot, K. A. and J. C. Stein (1998), "Risk Management, Capital Budgeting, and Capital Structure Policy for Financial Institutions: An Integrated Approach," *The Journal of Financial Economics* 47, 55–82.

Gabaix, X. (2008), "Power Laws," in in Lawrence Blume and Steven Durlauf (eds.), *New Palgrave Dictionary of Economics* (Palgrave).

Gabaix, X., P. Gopikrishnan, V. Plerou, and H. Eugene Stanley (2005), "Institutional Investors and Stock Market Volatility," MIT Department of EconomicsWorking Paper No. 03-30. Available at SSRN: https://ssrn.com/abstract=442940

Gabaix, X. and R. Ibragimov (2011), "Rank-1/2: A Simple Way to Improve the OLS Estimation of Tail Exponents," *Journal of Business and Economic Statistics* 29, 24–39.

Gallant, A. R. and G. E. Tauchen (1996), "Which Moments to Match?," *Econometric Theory* 12, 657–81.

Garber, P. M. (1990), "Famous First Bubbles," *The Journal of Economic Perspectives* 4.2, 35–54.

Garman, M. B. and M. J. Klass (1980), "On the Estimation of Security Price Volatilities from Historical Data," *Journal of Business* 53.1, 67–78.

Gatarek, L. and S. Johansen (2014), "Optimal Hedging with the Cointegrated Vector Autoregressive Model," Discussion Papers 14–22, Department of Economics, University of Copenhagen.

Gatev, E., W. N. Goetzmann, and K. G. Rouwenhorst (2006), "Pairs Trading: Performance of a Relative Value Arbitrage Rule," *The Review of Financial Studies* 19.3, 797–827.

Ghysels, E., A. Harvey, and E. Renault (1996), "Stochastic Volatility," in G. S. Maddala and C. R. Rao (eds.), *Handbook of Statistics 14: Statistical Methods in Finance* (Elsevier Science), 119–91.

Gibbons, M. R. (1982), "Multivariate Tests of Financial Models: A New Approach," *Journal of Financial Economics* 10, 3–27.

Gibbons, M. R., S. Ross, and J. Shanken (1989), "A Test of the Efficiency of a Given Portfolio," *Econometrica* 57, 1121–52.

Giraitis, L. and P. M. Robinson (2001), "Whittle Estimation of ARCH Models," *Econometric Theory* 17, 608–31.

Glosten, L. R. (1987), "Components of the Bid-Ask Spread and the Statistical Properties of Transaction Prices," *The Journal of Finance*, 42, 1293–307.

Glosten, L. R. (1994), "Is the Electronic Open Limit Order Book Inevitable?," *Journal of Finance* 49, 1127–61.

Glosten, L. R. and L. E. Harris (1988), "Estimating the Components of the Bid-Ask Spread," *Journal of Financial Economics* 21, 123–42.

Glosten, L. R., R. Jagannathan, and D. E. Runkle (1993), "On the Relation between Expected Value and the Volatility of the Nominal Excess Return on Stocks," *Journal of Finance* 48, 1779–801.

Glosten, L. R. and P. R. Milgrom (1985), "Bid, Ask, and Transaction Prices in a Specialist Market with Heterogeneously Informed Traders," *Journal of Financial Economics* 14, 71–100.

Gnedenko, B. (1943), "Sur la distribution limite du terme maximum d'une série aléatoire," *Annals of Mathematics* 44, 423–53.

Gourieroux, C., A. Monfort, and V. Polimenis (2006), "Affine Models for Credit Risk Analysis," *Journal of Financial Econometrics* 4.3, 494–530.

Gourieroux, C., A. Monfort, and E. Renault (1993), "Indirect Inference," *Journal of Applied Econometrics* 8 (Supplement), S85–S118.

Goyal, A. and P. Santa-Clara (2003), "Idiosyncratic Risk Matters!," *The Journal of Finance* 58, 975–1007.

Goyal, A. and I. Welch (2003), "Predicting the Equity Premium with Dividend Ratios," *Management Science* 49.5, 639–54.

Goyal, A. and I. Welch (2008), "A Comprehensive Look at the Empirical Performance of Equity Premium Prediction," *The Review of Financial Studies* 21.4, 1455–508.

Goyenko, R. Y., C. W. Holden, and C. A. Trzcinka (2009), "Do Liquidity Measures Measure Liquidity?," *Journal of Financial Economics* 92.2, 153–81.

Graham, B. and D. Dodd (1928) *Security Analysis* (6th edition with a foreword by Warren Buffett, published by Security Analysis Prior Editions in 2008).

Granger, C. W. J. (1969), "Investigating Causal Relations by Econometric Models and Cross-Spectral Methods," *Econometrica* 37.3, 424–38.

Grinold, R. C. and R. N. Kahn (1999), *Active Portfolio Management : A Quantative Approach for Producing Superior Returns and Selecting Superior Money Managers* (Wiley).

Grossman, S. J. and J. E. Stiglitz (1980), "On the Impossibility of Informationally Efficient Markets," *The American Economic Review* 70.3, 393–408.

Hafner, C. M. and J. V. K. Rombouts (2007), "Semiparametric Multivariate Volatility Models," *Econometric Theory* 23, 251–80.

Hall, P. and C. C. Heyde (1980), *Martingale Limit Theory and its Applications* (Academic Press).

Hall, P. and P. Yao (2003), "Inference in ARCH and GARCH Models with Heavy-Tailed Errors," *Econometrica* 71, 285–317.

Han, H., O. B. Linton, T. Oka, and Y. J. Whang (2014), "The Cross-Quantilogram: Measuring Quantile Dependence and Testing Directional Predictability between Time Series," Working Paper, available at SSRN: http://ssrn.com/abstract=2338468 or http://dx.doi.org/10.2139/ssrn.2338468

Hansen, L. P. (1982), "Large Sample Properties of Generalized Method of Moments Estimators," *Econometrica* 50.4, 1029–54.

Hansen, L. P. and R. J. Hodrick (1980), "Forward Exchange Rates as Optimal Predictors of Future Spot Rates: An Econometric Analysis," *The Journal of Political Economy* 88.5, 829–53.

Hansen, L. P. and K. J. Singleton (1982), "Generalized Instrumental Variables Estimation of Nonlinear Rational Expectations Models," *Econometrica* 50, 1269–86.

Hansen, L. P. and K. J. Singleton (1983), "Stochastic Consumption, Risk Aversion and the Temporal Behavior of Asset Returns," *Journal of Political Economy* 91, 249–65.

Hansen, P. R. and A. Lunde (2005), "A Forecast Comparison of Volatility Models: Does Anything Beat a GARCH(1,1)?," *Journal of Applied Econometrics* 20, 873–89.

Hansen, P. R. and A. Lunde (2006), "Realized Variance and Market Microstructure Noise," *Journal of Business and Economic Statistics* 24.2, 127–61.

Härdle, W. and O. B. Linton (1994), "Applied Nonparametric Methods," in D. F. McFadden and R. F. Engle (eds.), *The Handbook of Econometrics, Volume* 4 (North Holland), 2295–339.

Härdle, W. and A. B. Tsybakov (1997), "Locally Polynomial Estimators of the Volatility Function," *Journal of Econometrics* 81, 223–42.

Harris, L. (2003), *Trading and Exchanges: Market Microstructure for Practitioners* (Oxford University Press).

Harvey, A. C. (1989), *Forecasting, Structural Time Series Models and the Kalman Filter* (Cambridge University Press).

Harvey A. C., E. Ruiz, and N. Shephard (1994), "Multivariate Stochastic Variance Models," *Review of Economic Studies* 61, 247–64.

Hasbrouck, J. (2007), *Empirical Market Microstructure* (Oxford University Press).

Hasbrouck, J. (2009), "Trading Costs and Returns for US Equities: Estimating Effective Costs from Daily Data," *Journal of Finance* 64, 1445–77.

Hendershott, T., J. Brogaard, and R. Riordan (2014), "High Frequency Trading and Price Discovery," *The Review of Financial Studies* 27.8, 2267–306.

Herrndorf, N. (1984), "A Functional Central Limit Theorem for Weakly Dependent Sequences of Random Variables," *Annals of Probability* 12 (1984), 141–53.

Hertzberg, D. (2010), "The Great Stock Myth: Why the Market's Rate of Return – and Your Nest Egg – May Never Recover,' *The Atlantic*. Working paper available at http://www.theatlantic.com/magazine/archive/2010/09/the-great-stock-myth/308178/

Heston, S. (1993), "A Closed-Form Solution for Options with Stochastic Volatility with Applications to Bond and Currency Options," *Review of Financial Studies* 6.2, 327–43.

Hill, B. M. (1975), "A Simple General Approach to Inference about the Tail of a Distribution," *Annals of Statistics* 3, 1163–74.

Hill, J. B. (2010), "On Tail Index Estimation for Dependent Heterogeneous Data," *Econometric Theory* 26, 1398–436.

Ho, T. and H. Stoll (1981), "Optimal Dealer Pricing under Transactions and Return Uncertainty," *Journal of Financial Economics* 9, 47–73.

Ho, T. and H. Stoll (1983), "The Dynamics of Dealer Markets under Competition," *Journal of Finance* 38, 1053–74.

Hodrick, R. (1992), "Dividend Yields and Expected Stock Returns: Alternative Procedures for Inference and Measurement," *Review of Financial Studies* 5, 357–86.

Hodrick, R., D. Ng, and P. Sengmueller (1999), "An International Dynamic Asset Pricing Model," *International Taxation and Public Finance* 6, 597–620.

Högh, N., O. Linton, and J. P. Nielsen (2006), "The Froot–Stein Model Revisited," *Annals of Actuarial Science* 1.1, 37–47.

Holt, C. C. (1957), "Forecasting Seasonals and Trends by Exponentially Weighted Moving Averages,"Office of Naval Research Memorandum, 52. Reprinted as "Forecasting Seasonals and Trends by Exponentially Weighted Moving Averages," *International Journal of Forecasting* 20.1 (2004), 5–10.

Hong, Y. and H. Li (2005), "Nonparametric Specification Testing for Continuous-Time Models with Applications to Term Structure of Interest Rates," *The Review of Financial Studies* 18.1, 37–84.

Hou, K., C. Xue, and L. Zhang (2015), "Digesting Anomalies: An Investment Approach," *The Review of Financial Studies* 28.3, 650–705.

Ibragimov, R. (2009), "Portfolio Diversification and Value at Risk under Thick-Tailedness," *Quantitative Finance* 9, 565–80.

Ikenberry, D. L., G. Rankine, and E. K. Stice (1996), "What Do Stock Splits Really Signal?," *The Journal of Financial and Quantitative Analysis* 31.3, 357–75.

Imbens, G. W. and J. M. Wooldridge (2009), "Recent Developments in the Econometrics of Program Evaluation," *Journal of Economic Literature* 47.1, 5–86.

Ingersoll, J. (1987), *Theory of Financial Decision Making* (Rowan & Littlefield).

Jacod, J., Y. Li, P. A. Mykland, M. Podolskij, and M. Vetter (2009), "Microstructure Noise in the Continuous Case: The Pre-Averaging Approach," *Stochastic Processes and their Applications* 119, 2249–76.

Jacod, J. and P. Protter (1998), "Asymptotic Error Distribution for the Euler Method for Stochastic Differential Equations," *Annals of Probability* 26, 267–307.

Jacquier, E., N. G. Polson, and P. E. Rossi (1994), "Bayesian Analysis of Stochastic Volatility Models (with Discussion)," *Journal of Business and Economic Statistics* 12, 371–417.

Jagannathan, R., G. Skoulakis, and Z. Wang (2010), "The Analysis of the Cross Section of Security Returns," in Y. Aït-Sahalia and L. P. Hansen (eds.), *Handbook of Financial Econometrics, Volume* 2 (North Holland), 73–134.

Jarque, C. M. and A. K. Bera (1980), "Efficient Tests for Normality, Heteroskedasticity, and Serial Independence of Regression Residuals," *Economics Letters* 6, 255–59.

Jegadeesh, N. and S. Titman (1993), "Returns to Buying Winners and Selling Losers: Implications for Stock Market Efficiency," *The Journal of Finance* 48.1, 65–91.

Jensen, S. T. and A. Rahbek (2004), "Asymptotic Normality of the QMLE Estimator of ARCH in the Nonstationary Case," *Econometrica* 72.2, 641–46.

Jiang, G. and J. Knight (1997), "A Nonparametric Approach to the Estimation of Diffusion Processes with an Application to a Short-Term Interest Rate Model," *Econometric Theory* 13, 615–45.

Jones, C. S. (2001), "Extracting Factors from Heteroskedastic Asset Returns," *Journal of Financial Economics* 62.2, 293–325.

Jorion, P. (1997), *Value at Risk* (McGraw Hill).

Jovanovic, B. and A. Menkveld (2012), "Middlemen in Limit-Order Markets," Social Science Research Network. Available at SSRN: http://papers.ssrn.com/sol3/papers.cfm?abstract_id=1624329

Kahneman, D. and A. Tversky (1979), "Prospect Theory: An Analysis of Decision under Risk," *Econometrica* 47.2, 263–91.

Karlin, S. and H. M. Taylor (1981), *A Second Course in Stochastic Processes* (Academic Press).

Keim, D. (1983), "Size Related Anomalies and Stock Return Seasonality: Further Empirical Evidence," *Journal of Financial Economics* 12, 13–32.

Bibliography

Khandani, A. and A. Lo (2011), "What Happened to the Quants in August 2007? Evidence from Factors and Transactions Data," *Journal of Financial Markets* 14.1, 1–46.

Kim, J. and J. Park (2014), "Mean Reversion and Unit Root Properties of Diffusion Models," Working Paper, Indiana University.

Kim, S., N. Shephard, and S. Chib (1998), "Stochastic Volatility: Likelihood Inference and Comparison with ARCH Models," *Review of Economic Studies* 65, 361–93.

Koenker, R. (2005), *Quantile Regression* (Cambridge University Press).

Kirilenko, A., M. Samadi, A. S. Kyle, and T. Tuzun (2017), "The Flash Crash: The Impact of High Frequency Trading on an Electronic Market," *Journal of Finance* 72.3, 967–98.

Kristensen, D. (2010), "Pseudo-Maximum Likelihood Estimation in Two Classes of Semiparametric Diffusion Models," *Journal of Econometrics* 156, 239–59.

Kristensen, D. and O. Linton (2006), "A Closed-Form Estimator for the GARCH(1,1)-Model," *Econometric Theory* 22, 323–27.

Kyle, A. S. (1985), "Continuous Auctions and Insider Trading," *Econometrica*, 53.6, 1315–35.

Kyle, A. S. and A. A. Obizhaeva (2016a), "Market Microstructure Invariance: Empirical Hypotheses," *Econometrica* 84.4, 1345–404.

Kyle, A. S. and A. A. Obizhaeva (2016b), "Large Bets and Stock Market Crashes," available at SSRN: https://ssrn.com/abstract=2023776 or http://dx.doi.org/10.2139/ssrn.2023776

Laibson, D. (1997), "Golden Eggs and Hyperbolic Discounting," *Quarterly Journal of Economics* 112.2, 443–77.

Langetieg, T. C. and J. S. Smoot (1989), "Estimation of the Term Structure of Interest Rates," *Journal of Financial Services Research* 1, 181–222.

Ledoit, O. and M. Wolf (2003), "Improved Estimation of the Covariance Matrix of Stock Returns with an Application to Portfolio Selection," *Journal of Empirical Finance* 10.5, 603–21.

Lee S. W. and B. E. Hansen (1994), "Asymptotic Properties of the Maximum Likelihood Estimator and Test of the Stability of Parameters of the GARCH and IGARCH Models," *Econometric Theory* 10, 29–52.

Lettau, M. and S. C. Ludvigson (2001), "Resurrecting the C(CAPM): A Cross-Sectional Test when Risk Premia are Time-Varying," *Journal of Political Economy* 109, 1238–87.

Levine, R. (2002), "Bank-Based or Market-Based Financial Systems: Which is Better?," NBER Working Paper 9138. Available at http://www.nber.org/papers/w9138

Levy, H. (2007), "Risk and Rerun: An Experimental Analysis," *International Economic Review* 38, 119–49.

Levy, H. (2006), *Stochastic Dominance Investment Decision Making under Uncertainty* (Springer Verlag).

Levy, H. (2010), "The CAPM is Alive and Well: A Review and Synthesis," *European Financial Management* 16.1, 43–71.

Levy, M. and R. Roll (2010), "The Market Portfolio may be Mean/Variance Efficient After All: The Market Portfolio," *The Review of Financial Studies* 23.6, 2464–91.

Lintner, J. V., Jr. (1956), "Distribution of Incomes of Corporations Among Dividends, Retained Earnings, and Taxes," *American Economic Review Proceedings* 46, 97–113.

Linton, O. (1993), "Adaptive Estimation of ARCH Models," *Econometric Theory* 9, 539–69.

Linton, O. (2016), *Probability, Statistics and Econometrics* (Academic Press).

Linton, O., M. O'Hara, and J.-P. Zigrand (2013), "The Regulatory Challenge of High-Frequency Markets," in David Easley, Marcos López de Prado, and Maureen O'Hara (eds.), *High-Frequency Trading – New Realities for Traders, Markets and Regulators* (Risk Books), 207–30.

Linton, O. and Y. J. Whang (2007), "The Quantilogram: With an Application to Evaluating Directional Predictability," *Journal of Econometrics* 141.1, 250–82.

Linton, O. and J. Wu (2016), "A Coupled Component GARCH Model for Intraday and Overnight Volatility," available at SSRN: https://ssrn.com/abstract=2874631

Liptser, R. and A. Shiryaev (2001), *Statistics of Random Processes I. General Theory* (Springer Verlag).

Liu, L., A. Patton, and K. Sheppard (2015), "Does Anything Beat 5-Minute RV? A Comparison of Realized Measures Across Multiple Asset Classes," *Journal of Econometrics* 187.1, 293–311.

Lo, A. W. (1988), "Maximum Likelihood Estimation of Feneralized It6 Processes with Discretely-Sampled Data," *Econometric Theory* 4, 231–47.

Lo, A. W. (1991), "Long Term Memory in Stock Market Prices," *Econometrica* 59, 1279–313.

Lo, A. W. (2004), "The Adaptive Markets Hypothesis," *Journal of Portfolio Management* 30.5, 15–29.

Lo, A. W. (2005), "Reconciling Efficient Markets with Behavioral Finance: The Adaptive Markets Hypothesis," *Journal of Investment Consulting* 7.2, 21–44.

Lo, A. W. and J. Hasanhodzic (2010), *The Evolution of Technical Analysis: Financial Prediction from Babylonian Tablets to Bloomberg Terminals* (Wiley).

Lo, A. W. and A. C. MacKinlay (1988), "Stock Market Prices do not Follow Random Walks: Evidence from a Simple Specification Test," *Review of Financial Studies* 1.1, 41–66.

Lo, A. W. and A. C. MacKinlay (1990a), "When are Contrarian Profits due to Stock Market Overreaction?," *Review of Financial Studies* 3.2, 175–205.

Lo, A. W. and A. C. MacKinlay (1990b), "An Econometric Analysis of Nonsynchronous Trading," *Journal of Econometrics* 45, 181–212.

Lo, A. W. and A. C. MacKinlay (1999), *A Non-Random Walk Down Wall Street* (Princeton University Press).

Lobato, I. (2001), "Testing that a Dependent Process is Uncorrelated," *Journal of the American Statistical Association* 96, 1066–76.

Longstaff, F. A. (1989), "A Nonlinear General Equilibrium Model of the Term Structure of Interest Rates," *Journal of Financial Economics* 23, 195–224.

Lumsdaine, R. L. (1996), "Consistency and Asymptotic Normality of the Quasi-Maximum Likelihood Estimator in IGARCH(1,1) and Covariance Stationary GARCH(1,1) Models," *Econometrica* 64.3, 575–96.

Lumsdaine, R. L. and S. Ng (1999), "Testing for ARCH in the Presence of a Possibly Misspecified Conditional Mean," *Journal of Econometrics* 93.2, 257–79.

McCulloch, J. H. (1971), "Measuring the Term Structure of Interest Rates," *The Journal of Business* 44, 19–31.

McCulloch, J. H. (1986), "Simple Consistent Estimators of Stable Distribution Parameters," *Communications in Statistics – Simulation and Computation* 15 1109–36.

McCulloch, J. H. and H. C. Kwon (1993), "US Term Structure Data, 1947–1991," Working Paper 93-6, Ohio State University.

MacKinlay, A. C. (1987), "On Multivariate Tests of the CAPM," *Journal of Financial Economics* 18, 342–72.

MacKinlay, A. C. (1995), Multifactor Models do not Explain Deviations from the CAPM," *Journal of Financial Economics* 38, 3–28.

Magnus, J. R. and H. Neudecker (1988), *Matrix Differential Calculus with Applications in Statistics* (Wiley).

Malkiel, B. G. (2015), *A Random Walk Down Wall Street: The Time-Tested Strategy for Successful Investing* (W. W. Norton).

Malthus, T. (1798), *An Essay on the Principle of Population, as it Affects the Future Improvement of Society with Remarks on the Speculations of Mr. Godwin, M. Condorcet, and Other Writers* (Printed for J. Johnson, in St. Paul's Church-Yard).

Mandelbrot, B. (1963), "The Variation of Certain Speculative Prices," *Journal of Business* 36, 394–419.

Markowitz, H. M. (1952a), "The Utility of Wealth," *Journal of Political Economy* 60.2, 151–58.

Markowitz, H. M. (1952b), "Portfolio Selection," *Journal of Finance* 7.1, 77–91.

Markowitz, H. M. (1959), *Portfolio Selection: Efficient Diversification of Investments* (John Wiely & Sons).

Marsh, T. A. and R. C. Merton (1984), "Earnings Variability and Variance Bounds Tests for the Rationality of Stockmarket Prices," Working Paper No. 1559-84.

Marsh, T. A. and R. C. Merton (1986), "Dividend Variability and Variance Bounds Tests for the Rationality of Stock Market Prices," *American Economic Review* 76, 483–98.

Marsh, T. A. and E. R. Rosenfeld (1983), "Stochastic Processes for Interest Rates and Equilibrium Bond Prices," *The Journal of Finance* 38, 635–46.

Martin, I. (2017), "What is the Expected Return on the Market?," *Quarterly Journal of Economics* 132.1, 367–433.

Mehra, R. and E. C. Prescott (1985), "The Equity Premium: A Puzzle," *Journal of Monetary Economics* 15, 145–61.

Mehra, R. and E. C. Prescott (2008), "Non-Risk-Based Explanations of the Equity Premium," in Rajnish Mehra (ed.), *Handbook of the Equity Risk Premium* (Elsevier), 101–16.

Meitz, M. and P. Saikkonen (2008), "Stability of nonlinear AR–GARCH Models," SSE/EFI Working Paper Series in Economics and Finance No. 632Meitz, *Journal of Time Series Analysis* 29.3, 453–75.

Menkveld, A. J. (2013), "High Frequency Trading and the New-Market Makers," *Journal of Financial Markets* 16, 571–603.

Merton, R. C. (1972), "An Analytic Derivation of the Efficient Portfolio Frontier," *Journal of Financial and Quantitative Analysis* 7, 1851–72.

Merton, R. C. (1973), "An Intertemporal Capital Asset Pricing Model," *Econometrica* 41.5, 867–87.

Merton, R. C. (1992), *Continuous-Time Finance* (Wiley).

Metzler, A. (2010), "On the First Passage Problem for Correlated Brownian Motion," *Statistics & Probability Letters* 80, 277–84.

Mikosch, T. (1998), *Elementary Stochastic Calculus with Finance in View* (World Scientific).

Mikosch, T., and C. Stărică (2000), "Limit Theory for the Sample Autocorrelations and Extremes of a GARCH(1,1) Process," *Annals of Statistics* 28, 1427–51.

Mikosch, T., and C. Stărică (2004), "Nonstationarities in Financial Time Series, the Long-Range Dependence, and the IGARCH Effects," *Review of Economics and Statistics* 86, 378–90.

Muth, J. F. (1960), "Optimal Properties of Exponentially Weighted Forecasts," *Journal of the American Statistical Association* 55.290, 299–306.

Muthen, B. (1990), "Moments of the Censored and Truncated Bivariate Normal Distribution," *British Journal of Mathematical and Statistical Psychology* 43, 131–43.

Nadaraya, E. A. (1964), "On Estimating Regression," *Theory of Probability and its Applications* 10, 186–90.

Nelson, C. R. and A. F. Siegel (1987), "Parsimonious Modeling of Yield Curve," *Journal of Business* 60, 473–89.

Nelson, D. B. (1990a), "ARCH Models as Diffusion Approximations," *Journal of Econometrics* 45, 7–38.

Nelson, D. B. (1990b), "Stationarity and Persistence in the GARCH(1, 1) Model," *Econometric Theory* 6, 318–34.

Nelson, D. B. (1991), "Conditional Heteroskedasticity in Asset Returns: A New Approach," *Econometrica* 59, 347–70.

Nelson, D. B. and C. Q. Cao (1992), "Inequality Constraints in the Univariate GARCH Model," *Journal of Business and Economic Statistics* 10, 229–35.

Newey, W. K. and D. G. Steigerwald (1997), "Asymptotic Bias for Quasi-Maximum-Likelihood Estimators in Conditional Heteroskedasticity Models," *Econometrica* 65.3, 587–99.

Newey, W. K. and K. D. West (1987), "A Simple, Positive Semi-Definite, Heteroskedasticity and Autocorrelation Consistent Covariance Matrix," *Econometrica* 55.3, 703–08.

Ng, R. (2014), *An Empirical Study of Risk and Return in Chinese A- and H-share Markets* (Mimeo).

O'Hara, M. (1995), *Market Microstructure Theory* (Blackwell).

O'Hara, M. and M. Ye (2009), "Is Fragmentation Harming Market Quality?," *Journal of Financial Economics* 100.3, 459–74.

Ohlson, J. and S. H. Penman (1985), "Volatility Increases Subsequent to Stock Splits: An Empirical Aberration," *Journal of Financial Economics* 14.2, 251–66.

Onatskiy, A. (2012), "Asymptotics of the Principal Components Estimator of Large Factor Models with Weakly Influential Factors," *Journal of Econometrics* 168, 244–58.

Owen, J. and R. Rabinovitch (1983), "On the Class of Elliptical Distributions and their Applications to the Theory of Portfolio Choice," *Journal of Finance* 38, 745–52.

Øksendahl, B. (1985), *Stochastic Differential Equations: An Introduction with Applications* (Springer Verlag).

Pagan, A. R. (1984), "Econometric Issues in the Analysis of Regressions with Generated Regressors," *International Economic Review* 25.1, 221–47.

Pagan, A. R. (1996), "The Econometrics of Financial Markets," *Journal of Empirical Finance* 3, 15–102.

Pagan, A. R. and Y. S. Hong (1991), "Nonparametric Estimation and the Risk Premium," in W. Barnett, J. Powell, and G. E. Tauchen (eds.), *Nonparametric and Semiparametric Methods in Econometrics and Statistics* (Cambridge University Press), 51–75.

Pagan, A. R. and G. W. Schwert (1990), "Alternative Models for Conditional Stock Volatility," *Journal of Econometrics* 45, 267–90.

Parkinson, M. (1980), "The Extreme Value Method for Estimating the Variance of the Rate of Return," *Journal of Business* 53.1, 61–65.

Parzen, E. (1957), "On Consistency of the Spectrum of a Stationary Time Series," *Annals of Mathematical Statistics* 28.2, 329–34.

Pastor, L. and R. F. Stambaugh (2003), "Liquidity Risk and Expected Stock Returns," *Journal of Political Economy* 111.3, 642–85.

Pastor, L. and P. Veronesi (2006), "Was there a Nasdaq Bubble in the Late 1990s?," *Journal of Financial Economics* 81, 61–100.

Patton, A. (2006), "Modelling Asymmetric Exchange Rate Dependence," *International Economic Review* 47.2, 527–56.

Patton, A. (2009), "Are 'Market Neutral' Hedge Funds Really Market Neutral?," *Review of Financial Studies* 22, 2495–530.

Patton, A. (2012), "A Review of Copula Models for Economic Time Series," *Journal of Multivariate Analysis* 110, 4–18.

Peng, L. and Q. Yao (2003), "Least Absolute Deviations Estimation for ARCH and GARCH Models," *Biometrika* 90, 967–75.

Pesaran, M. H. (2006), "Estimation and Inference in Large Heterogeneous Panels with a Multifactor Error Structure," *Econometrica* 74, 967–1012.

Pesaran, M. H. and A. Timmermann (2007), "Selection of Estimation Window in the Presence of Breaks," *Journal of Econometrics* 137.1, 134–61.

Pesaran, M. H. and T. Yamagata (2012), "Testing CAPM with a Large Number of Assets (Updated 28th March 2012)," Cambridge Working Papers in Economics 1210, Faculty of Economics, University of Cambridge.

Pesaran, M. H. and T. Yamagata (2017), "Testing for Alpha in Linear Factor Pricing Models with a Large Number of Securities," CESifo Working Paper Series No. 6432. Available at SSRN: https://ssrn.com/abstract=2973079

Peterson, R. L., C. K. Ma, and R. J. Ritchey (1992), "Dependence in Commodity Prices," *Journal of Futures Markets* 12.4, 429–46.

Phillips, P. C. B. and S. Jin (2014), "Testing the Martingale Hypothesis," *Journal of Business and Economic Statistics* 32.4, 537–54.

Phillips, P. C. B. and T. Magdalinos (2007), "Limit Theory for Moderate Deviations from a Unit Root," *Journal of Econometrics* 136.1, 115–30.

Phillips, P. C. B. and J. Y. Park (1999), "Asymptotics for Nonlinear Transformations of Integrated Time Series," *Econometric Theory* 15, 269–98.

Phillips, P. C. B., S. P. Shi, and J. Yu (2012), "Testing for Multiple Bubbles," Cowles Foundation Discussion Paper no. 1843.

Phillips, P. C. B. and V. Solo (1992), "Asymptotics for Linear Processes," *Annals of Statistics* 20.2, 97–1001.

Phillips, P. C. B. and J. Yu (2010), "Dating the Timeline of Financial Bubbles during the Subprime Crisis," *Quantitative Economics* 2.3, 455–91.

Poterba, J. M. and L. H. Summers (1988), "Mean Reversion in Stock Prices: Evidence and Implications," *Journal of Financial Economics* 22.1, 27–59.

Pratt, J. W. (1964), "Risk Aversion in the Small and in the Large," *Econometrica* 32, 122–36.

Press, S. J. (1967), "A Compound Events Model for Security Prices," *The Journal of Business* 40, 317–35.

Pritsker, M. (1998), "Nonparametric Density Estimation and Tests of Continuous Time Interest Rate Models," *Review of Financial Studies* 11, 449–87.

Protter, P. E. (2004), *Stochastic Integration and Differential Equations* (Springer Verlag).

Psarakis, S. and J. Panaretos (2001), "On Some Bivariate Extensions of the Folded Normal and the Folded-T Distributions," *Journal of Applied Statistical Science* 10.2, 119–36. Available at SSRN: https://ssrn.com/abstract=947441

Rashes, M. S. (2001), "Massively Confused Investors Making Conspicuously Ignorant Choices (MCI-MCIC)," *The Journal of Finance* 56.5, 1911–27.

Resnick, S. and C. Stărică (1998), "Tail Index Estimation for Dependent Data," *Annals of Applied Probability* 8.4, 1156–83.

Robinson, P. M. (1983), "Nonparametric Estimators for Time Series," *Journal of Time Series Analysis* 4, 185–207.

Robinson, P. M. (1994), "Semiparametric Analysis of Long-Memory Time Series," *Annals of Statistics* 22.1, 515–39.

Rogers, L. C. G. (1997), "Arbitrage with Fractional Brownian Motion," *Mathematical Finance* 7.1, 95–105.

Rogers, L. C. G. and S. E. Satchell (1991), "Estimating Variances from High, Low, and Closing Prices," *Annals of Applied Probability* 1.4, 504–12.

Roll, R. (1977), "A Critique of the Asset Pricing Theory's Tests; Part i: On Past and Potential Testability of the Theory," *Journal of Financial Economics* 4, 129–76.

Roll, R. (1983), "On Computing Mean Returns and the Small Firm Premium," *Journal of Financial Economics* 12, 371–86.

Roll, R. (1984a), "A Simple Implicit Measure of the Effective Bid-Ask Spread in an Efficient Market," *The Journal of Finance* 39.4, 1127–39.

Roll, R. (1984b), "Orange Juice and Weather," *American Economic Review* 74.5, 861–80.

Rosenberg, B. (1972), "The Behavior of Random Variable with Nonstationary Variance and the Distribution of Stock Prices," Research Program in Finance Working Papers 11, University of California at Berkeley.

Rosenberg, B. (1974), "Extra-Market Components of Covariance in Security Returns," *The Journal of Financial and Quantitative Analysis* 9.2, 263–74.

Ross, S. A. (1976), "The Arbitrage Theory of Capital Asset Pricing," *Journal of Economic Thoery* 13, 341–60.

Ross, S. A. (2005), *Neoclassical Finance* (Princeton University Press).

Rothschild, M. and J. E. Stiglitz (1970), "Increasing Risk I: A Definition," *Journal of Economic Theory* 2.3, 225–43.

Rothschild, M. and J. E. Stiglitz (1971), "Increasing Risk II: Its Economic Consequences," *Journal of Economic Theory* 3.1, 66–84.

Rydberg, T. V. and N. Shephard (2003), "Dynamics of Trade-by-Trade Price Movements: Decomposition and Models," *Journal of Financial Econometrics* 1.1, 2–25.

Scaillet, O. (2004), "Nonparametric Estimation and Sensitivity Analysis of Expected Shortfall," *Mathematical Finance* 14, 115–29.

Scaillet, O. (2005), "Nonparametric Estimation of Conditional Expected Shortfall," *Insurance and Risk Management Journal* 74, 382–406.

Schaefer, S. M. (1981), "Measuring a Tax Specific Term Structure of Interest Rates in the Market for British Government Securities," *Economic Journal* 91, 415–38.

Scheinkman, J. A. (2014), *Speculation, Trading, and Bubbles* (Columbia University Press).

Scholes, M. and J. Williams (1977), "Estimating Betas from Nonsynchronous Data," *Journal of Financial Economics* 5.3, 309–27.

Schwert, G. W. (1989), "Why Does Stock Market Volatility Change Over Time?," *Journal of Finance* 44, 1115–53.

Schwert, G. W. (2011), "Stock Volatility during the Recent Financial Crisis," *European Financial Management* 17, 789–805.

Sentana, E. (1998), "The Relation between Conditionally Heteroskedastic Factor Models and Factor GARCH Models," *Econometrics Journal* 1, 1–9.

Shanken, J. (1985), "Multivariate Tests of the Zero-Beta CAPM," *Journal of Financial Economics* 14, 327–48.

Shanken, J. (1987), "Multivariate Proxies and Asset Pricing Relations: Living with the Roll Critique," *Journal of Financial Economics* 18, 91–110.

Shanken, J. (1992), "On the Estimation of Beta-Pricing Models," *Review of Financial Studies* 5, 1–33.

Sharpe, W. (1964), "Capital Asset Prices: A Theory of Market Equilibrium under Conditions of Risk," *Journal of Finance* 19, 425–42.

Shiller, R. J. (1981), "Do Stock Prices Move Too Much to be Justified by Subsequent Changes in Dividends?," *The American Economic Review* 71.3, 421–36.

Shiller, R. J. (2000), *Irrational Exuberance* (Princeton University Press).

Shin, H. S. (2010), *Risk and Liquidity. Clarendon Lectures in Finance* (Oxford University Press).

Shleifer, A. and R. W. Vishny (1997), "The Limits of Arbitrage," *The Journal of Finance*, 52.1, 35–55.

Silvain, S. (2013), "Fama–MacBeth 1973: Replication and Extension," available at http://home.uchicago.edu/~serginio/research/FamaPaper_fm73replication_extension.pdf

Silverman, B. W. (1986), *Density Estimation for Statistics and Data Analysis* (Chapman and Hall).

Sklar, A. (1959), "Fonctions de répartition à n dimensions et leurs marges," *Publications de l'Institut Statistique de l'Université de Paris* 8, 229–31.

Snow, J. (1855), *On the Mode of Communication of Cholera* (2nd edition; Churchill). Excerpted in B. MacMahon and T. F. Pugh (1970), *Epidemiology* (Little Brown).

Snowdon, C. (2009), "The Myth of the Smoking Ban Miracle," *Spiked*, September 24, 2009.

Bibliography

Solnik, B. H. (1974), "Why Not Diversify Internationally Rather Than Domestically?," *Financial Analysts Journal* 30.4, 48–54.

Sornette, D. (2003), *Why Stock Markets Crash* (Princeton University Press).

Spokoiny, V. and P. Cizek (2009), "Varying Coefficient GARCH Models, in T. G. Andersen, R. A. Davis, J.-P. Kreiss, and T. Mikosch (eds.), *Handbook of Financial Time Series* (Springer Verlag, 2009), 169–85.

Stambaugh, R. F. (1982), "On the Exclusion of Assets from Tests of the Two-Parameter Model: A Sensitivity Analysis," *Journal of Financial Economics* 10, 237–68.

Stambaugh, R. F. (1999), "Predictive Regressions," *Journal of Financial Economics* 54.3, 375–421.

Stambaugh, R. F. and Y. Yuan (2017), "Mispricing Factors," *Review of Financial Studies* 30, 1270–315. Available at SSRN: https://ssrn.com/abstract=2626701 or http://dx.doi.org/10.2139/ssrn.2626701

Stanton, R. (1997), "A Nonparametric Model of Term Structure Dynamics and the Market Price of Interest Rate Risk," *Journal of Finance* 52, 1973–2002.

Starica, C. (2003), "Is Garch(1,1) as Good a Model as the Accolades of the Nobel Prize Would Imply?," available at SSRN: https://ssrn.com/abstract=637322 or http://dx.doi.org/10.2139/ssrn.637322

Stone, C. J. (1980), "Optimal Rates of Convergence for Nonparametric Estimators," *Annals of Statistics* 8, 1348–60.

Straumann, D. (2004), *Estimation in Conditionally Heteroscedastic Time Series Models* (Springer).

Svensson, L. E. O. (1994), "Estimating and Interpreting Forward Rates: Sweden 1992–4," National Bureau of Economic Research Working Paper #4871.

Sylvain, S. (2013), "Fama–MacBeth 1973: Replication and Extension," available at http://home.uchicago.edu/~serginio/research/FamaPaper_fm73replication_extension.pdf

Taleb, N. N. (2010), *The Black Swan* (2nd edition; Penguin).

Taylor, S. J. (1982), Financial Returns Modelled by the Product of Two Stochastic Processes, a Study of Daily Sugar Prices 1961–79," in O. D. Anderson (ed.), *Time Series Analysis: Theory and Practice, Volume 1* (North-Holland), 203–26.

Taylor, S. J. (1986), *Modelling Financial Time Series* (Wiley).

Thaler, R. H. (2016), "Behavioral Economics: Past, Present, and Future," *American Economic Review* 106.7, 1577–600.

Thompson, S. B. (2011), "Simple Formulas for Standard Errors that Cluster by Both Firm and Time," *Journal of Financial Economics* 99, 1–10.

Tibshirani, R. J. (1996), "Regression Shrinkage and Selection via the Lasso," *Journal of the Royal Statistical Society, Series B* 58.1, 267–88.

Timmerman, A. (2008), "Elusive Return Predictability," *International Journal of Forecasting* 24.1, 1–18.

Tong, H. (1990), *Non-Linear Time Series. A Dynamical System Approach* (Oxford University Press).

Turner, A. (2009), "The Turner Review: A Regulatory Response to the Global Banking Crisis," Financial Services Authority. Available at http://www.fsa.gov.uk/pubs/other/turner_review.pdf

Van Praag, B. and A. Wesselman (1987), "Elliptical Regression Operationalized," *Economics Letters* 23, 269–74.

Vasicek, O. A. (1977), "An Equilibrium Characterization of the Term Structure," *Journal of Financial Economics* 5, 177–88.

Vasicek, O. A. and H. G. Fong (1982), "Term Structure Modeling Using Exponential Splines," *Journal of Finance* 37, 177–88.

Vorkink, K. (2003), "Return Distributions and Improved Tests of Asset Pricing Models," *The Review of Financial Studies* 16, 845–74.

Watson, G. S. (1964), "Smooth Regression Analysis," *Sankhya Series A* 26, 359–72.

Weld, W. C., R. Michaely, R. H. Thaler, and Shlomo Benartzi (2009), "The Nominal Share Price Puzzle," *Journal of Economic Perspectives* 23.2, 121–42.

Whistler, D. E. N. (1990), "Semiparametric Models of Daily and Intradaily Exchange Rate Volatility (PhD thesis, University of London).

White, H. (1980), "A Heteroskedasticity-Consistent Covariance Matrix Estimator and a Direct Test for Heteroskedasticity," *Econometrica* 48, 817–38.

White, H. (2000), "A Reality Check for Data Snooping," *Econometrica* 68, 1097–126.

White, H. and I. Domowitz (1984), "Nonlinear Regression with Dependent Observations," *Econometrica* 52.1, 143–61.

Whittaker, E. T. (1923), "On a New Method of Graduation," *Proceedings of the Edinburgh Mathematical Society* 41, 63–75.

Whittaker, E. T. and G. Robinson (1967), "Normal Frequency Distribution," in their *The Calculus of Observations: A Treatise on Numerical Mathematics* (4th edition; Dover), 164–208.

Wiener, N. (1923), "Differential-Space," *Journal of Mathematics and Physics* 2, 131–74.

Winters, P. R. (1960), "Forecasting Sales by Exponentially Weighted Moving Averages," *Management Science* 6.3, 324–42.

Wong, E. (2002) "Investigation of Market Efficiency: An Event Study of Insider Trading in the Stock Exchange of Hong Kong," Unpublished, Stanford University. Available at https://pdfs.semanticscholar.org/a800/29a2a911f068eb35d32e8c7b4e6b6697ca4c.pdf

Zellner, A. (1962), "An Efficient Method of Estimating Seemingly Unrelated Regression Equations and Tests for Aggregation Bias," *Journal of the American Statistical Association* 57, 348–68.

Zhang, L. (2006), "Efficient Estimation of Stochastic Volatility Using Noisy Observations: A Multi-Scale Spproach," *Bernoulli* 12, 1019–43.

Zhang, L., P. Mykland, and Y. Aït-Sahalia (2005), "A Tale of Two Timescales: Determining Integrated Volatility with Noisy High-Frequency Data," *Journal of the American Statistical Association* 100, 1394–411.

Zhou, G. (1993), "Asset Pricing Tests under Alternative Distributions," *Journal of Finance* 48, 1927–42.

Zhou, B. (1996), "High Frequency Data and Volatility in Foreign-Exchange Rates," *Journal of Business and Economic Statistics* 14, 45–52.

Index

Aït-Sahalia, Yacine, 174, 195, 431, 437, 446, 447, 454, 457
Adverse selection, 174, 179, 182, 194, 226
Algorithmic trading, 9
Andersen, Torben, 72, 365, 425, 450, 451
Arbitrage, 49, 279, 321, 471
Ask price, 10
Asset-backed securities, 5
Autocorrelation function (ACF), 62, 95, 98, 127, 134
Autoregressive conditional heteroskedasticity (ARCH), 377

Behavioral finance, 151, 199
Best linear predictor, 56
Bid price, 10, 150, 168, 173, 187, 189
Big data, 80, 130, 266
Black and Scholes Formula, 164, 194, 432
Black Swan theory, 494
Bollerslev, Tim, 72, 367, 380, 396, 405, 407, 410, 450, 451
Bootstrap, 59
  Multiplier, 59
Box–Pierce statistic, 88, 91, 95, 97, 98, 108, 111, 114, 133
Brownian motion, 164, 359, 366, 422, 424, 427, 428, 436, 448, 450, 459
Bubbles, 151, 192, 286, 314, 316, 321, 482
  Rational, 319, 320, 335, 339

Calendar time, 206, 213, 437, 445
  Hypothesis, 14, 62, 84, 107, 115, 130, 182, 372
Capital asset pricing model, 45, 238, 241, 339
Cauchy distribution, 83, 139, 385, 478, 480, 483, 487, 498, 522
Cauchy–Schwarz inequality, 218, 265, 284, 313, 455, 520
Chaos theory, 329
Chen, Xiaohong, 148, 386, 414, 443, 445, 491
Cointegration, 132, 321
Connor, Gregory, 282, 286, 300, 304, 307, 311, 371
Constant relative risk aversion (CRRA), 32, 34, 35, 341, 350

Contrarian, 79, 100, 124, 126
Copula, 445, 489, 491
Coupon bonds, 4, 464
Covariance matrix, 39, 54, 56, 106, 154, 209, 213, 214, 217, 220, 227, 238, 243, 257, 258, 275, 276, 279, 280, 282, 290, 291, 301–303, 305, 306, 308, 345, 397, 426, 468, 473
  Conditional, 407, 408, 410
Curate's egg
  Portfolio grouping and, 268

Depth of market, 174, 185, 190
Diebold, Frank, 71, 410, 450, 451, 467
Discount function, 464, 467
  Stochastic, 338, 473
Dividends, 6, 12, 13, 16, 130, 231, 244, 271, 316, 319, 325, 327, 330, 331, 353, 355, 356

E-mini futures, 7
Efficient markets, 75, 77–81, 86, 87, 91, 95, 102, 106, 114, 118, 129, 130, 132, 134, 148, 151, 179, 199, 202, 203, 210, 501, 505
Endogenous risk, 314
Engle, Robert F., 164, 377, 393, 398, 407, 409, 410, 412, 418, 488
Epps effect, 102, 163
Epstein–Zin–Weil, 351, 354
Exchange-traded funds, 6
Expected shortfall, 493

Factor models, 49, 51, 52, 203, 204, 238, 241, 266, 272, 279, 286, 288–291, 293, 294, 296, 297, 300, 302, 304, 308, 312
Fan, Jianqing, 266, 282, 414, 447
Federal Funds rate, 4

Generalized autoregressive conditional heteroskedasticity (GARCH), 62, 110, 270, 380, 382, 386, 388, 393, 395, 398, 404
Generalized method of moments (GMM), 258, 341, 342, 344, 351, 403, 445
Goodhart's Law, 150

Härdle, Wolfgang, 147, 148, 412, 414
Hawkes processes, 460

Index

Head and Shoulders, 78
High frequency trading (HFT), 9, 189, 190, 226, 373

Information, 45, 79, 80, 126
  Asymmetric, 2, 174, 179, 196
  Private, 372
  Public, 372
  Set, 68, 70, 75–78, 81, 90, 129, 130, 132, 147, 167, 210, 323, 329, 337, 340, 341, 363, 377, 387, 407, 422, 460, 487
Initial public offering, 7
Insider trading, 2, 150
Itô's rule, 431, 444, 451

Jagger, Joseph, 144
January effect, 150
Jensen's inequality, 20, 30, 33, 384, 385, 527

Kernel estimation, 147, 412, 414, 417, 418, 447, 448, 457, 486
Knightian uncertainty, 20
Kronecker product, 250
Kurtosis, 22, 112, 161, 244, 274, 376, 379, 387, 414, 417, 470
Kyle, Albert "Pete", 183, 185

Law of iterated expectation, 77, 78, 153, 154, 158, 178, 317, 323, 342, 419, 444
Leverage, 11, 110, 372
  Hypothesis, 389, 392, 402, 436, 452
Levy processes, 459
LIBOR, 4
Linton, Oliver, xxii, 139, 141, 266, 300, 320, 371, 373, 400, 412, 414, 427
Liquidity, 1, 8–10, 165, 183, 185
  Measurement of, 185
Local time, 448, 449
Lognormal distribution, 22, 155, 344, 345, 351, 359, 403, 528
Long memory, 331, 404
Long run variance, 65, 66, 71, 89, 114, 132, 159
LSE, 9, 10, 16, 164, 187, 193, 425, 427

Manski's law of decreasing credibility, 84
Market in Financial Instruments Directive (MiFID), 9, 10
martingale, 66, 83, 84, 110, 125, 178, 179, 258, 274, 317, 322, 338, 388, 397–399, 423, 451, 460
  semimartingale, 427, 448, 451
Maximum likelihood, 101, 249, 302, 304, 344
Maximum likelihood (MLE), 60
McCulloch, J. Huston, 465, 466, 470, 478

Mean reversion, 67, 79, 151, 434
Mid quote, 171, 179
Mixed normal distribution, 448, 449, 452, 457
Mixing process, 66, 103, 120, 311, 386, 408, 488
  Beta, 62
  Covariance, 62
  Strong, 62, 108, 282
Momentum, 79, 100, 124, 126, 271, 296
Monday effect, 130, 150
Moore's law, 193
Moral hazard, 3
Multiple testing, 87, 250, 481

NASDAQ, 8
National Best Bid and Offer (NBBO), 189
News impact curve, 388, 390, 403
Nonlinear
  Dynamic system, 326, 329
  Models, 376, 384, 415
  Objective function, 342
Nonparametric tests
  Cowles and Jones, 135
  Quantilogram, 140
  Runs tests, 144
  Wilcoxon signed rank test, 229
NYSE, 9

O'Hara, Maureen, 182, 229, 373
Over the counter, 6, 8, 153, 164

Patton, Andrew, 273, 458, 490, 491, 530
Pesaran, Mohammed Hashem, 220, 253, 266
Phillips, Peter Charles Bonest, 321, 322, 438, 449, 458
Poisson process, 167, 376, 460, 474
Portfolio choice, 39, 43, 238, 279, 299, 304, 366, 407
Portfolio grouping, 238, 253, 263, 265–268, 273, 277
Power law, 486, 487

Quantile, 59, 122, 140, 141, 358, 477–479, 481, 487, 493
  Extreme, 482, 484
  Regression, 222

Random walk, 67, 81–83, 90, 106, 122, 126, 132, 169, 171, 179, 181, 297, 318, 372, 384, 405, 423, 431
Realized range, 366
Recurrent process, 156, 384, 448
Risk aversion, 27, 31, 32, 34, 80, 126, 174, 341, 346, 349, 351–354, 356, 393

Index

Rolling window, 74, 121, 268, 270, 286, 322, 333, 347, 348

Sample selection, 231, 348
Sharpe ratio, 21, 46, 252
Shephard, Neil, 167, 411, 451, 458, 461
Sieve methods, 148
Size effect, 126
Sparsity, 216, 217, 231, 282
Spread, 194, 202
   Bid–ask, 9, 10, 150, 152, 165, 168, 171, 174, 176, 179, 182, 184, 187, 188
   Effective, 189
   Realized, 189, 190
St Petersburg Paradox, 28, 29
Stationary process, 85
   Strong, 61, 385, 430
   Weak, 61, 378
Stochastic dominance, 37, 38, 492
Systemic risk, 378, 491

Tail thickness, 477, 482–484, 487
Tick size, 10, 16, 164, 165, 167, 168, 236
Trading time
   Hypothesis, 14, 84, 130, 372
Treasury Bills, 3

Unit root, 67, 321, 322, 331, 383, 405, 431, 449, 458, 471, 520

Value-at-Risk, 358, 477, 479, 487, 489, 493, 494
Vector autoregression, 72, 73, 344, 345, 352, 407, 408
VIX, 360–362
Volatility, 11, 31, 110, 149, 153, 169, 171, 174, 177, 194, 224, 296, 314, 328, 355, 358
Von Neumann and Morgenstern, 28, 29, 350

White's standard errors, 58, 103